Advances in Spatial Science

Springer-Verlag Berlin Heidelberg GmbH

Titles in the Series

Cristoforo S. Bertuglia, Manfred M. Fischer and *Giorgio Preto* (Eds.)
Technological Change, Economic Development and Space
XVI, 354 pages. 1995. ISBN 3-540-59288-1

Harry Coccossis and *Peter Nijkamp* (Eds.)
Overcoming Isolation
VIII, 272 pages. 1995. ISBN 3-540-59423-X

Luc Anselin and *Raymond J.G.M Florax* (Eds.)
New Directions in Spatial Econometrics
XIX, 420 pages. 1995. ISBN 3-540-60020-5

Heikki Eskelinen and *Folke Snickars* (Eds.)
Competitive European Peripheries
VIII, 271 pages. 1995. ISBN 3-540-60211-9

Jeroen C.J.M. van den Bergh, Peter Nijkamp and *Piet Rietveld* (Eds.)
Recent Advances in Spatial Equilibrium Modelling
VIII, 392 pages. 1996. ISBN 3-540-60708-0

Peter Nijkamp, Gerard Pepping and *David Banister*
Telematics and Transport Behaviour
XII, 227 pages. 1996. ISBN 3-540-60919-9

David F. Batten and *Charlie Karlsson* (Eds.)
Infrastructure and the Complexity of Economic Development
VIII, 298 pages. 1996. ISBN 3-540-61333-1

Tönu Puu
Mathematical Location and Land Use Theory
IX, 294 pages. 1997. ISBN 3-540-61819-8

Yee Leung
Intelligent Spatial Decision Support Systems
XV, 470 pages. 1997. ISBN 3-540-62518-6

Cristoforo S. Bertuglia, Silvana Lombardo and *Peter Nijkamp* (Eds.)
Innovative Behaviour in Space and Time
X, 437 pages. 1997. ISBN 3-540-62542-9

Anna Nagurney and *Stavros Siokos*
Financial Networks
XVI, 492 pages. 1997. ISBN 3-540-63116-X

Manfred M. Fischer · Arthur Getis (Eds.)

Recent Developments in Spatial Analysis

Spatial Statistics, Behavioural Modelling, and Computational Intelligence

With 80 Figures
and 49 Tables

Springer

Professor Dr. Manfred M. Fischer
Vienna University of Economics and Business Administration
Department of Economic and Social Geography
Augasse 2-6, A-1090 Vienna, Austria
and
Director, Institute for Urban and Regional Research
Austrian Academy of Sciences
Postgasse 7/4/2, A-1010 Vienna, Austria

Professor Dr. Arthur Getis
Stephen and Mary Birch Foundation
Chair of Geographical Studies
and
San Diego State University
Department of Geography
San Diego, CA 92182-4493, USA

ISBN 978-3-642-08321-1

Cataloging-in-Data applied for
Die Deutsche Bibliothek - CIP-Einheitsaufnahme
Recent developments in spatial analysis : spatial statistics, behavioural modelling, and computational intelligence; with 49 tables / Manfred M. Fischer; Arthur Getis (ed.).
(Advances in spatial science)
ISBN 978-3-642-08321-1 ISBN 978-3-662-03499-6 (eBook)
DOI 10.1007/978-3-662-03499-6

Originally published by Springer-Verlag Berlin Heidelberg New York in 1997
Softcover reprint of the hardcover 1st edition 1997

Hardcover design: Erich Kirchner, Heidelberg
SPIN 10632833 42/2202-5 4 3 2 1 0 - Printed on acid-free paper

Preface

In recent years, spatial analysis has become an increasingly well established field, as evidenced by the establishment of educational and research programs at many universities. Its popularity can be attributed to two important societal and academic trends: (1) an increasing recognition of the opportunities created by new technologies (geographic information systems and remote sensing technologies) and the availability of digital spatial data; and (2) the growing emphasis on the special role of geographic information within the broader context of the information society, and the corresponding research initiatives toward the development of a spatial data infrastructure in North America and Europe.

This volume illustrates some recent developments in the field of spatial analysis and aims to disseminate recent research into graduate classrooms. It originated from special invited sessions held at the 28th International Geographical Congress held in The Hague, The Netherlands, August 5-10, 1996, under the auspices of the International Geographical Union's (IGU) Commission on Mathematical Models. This book includes a selection of contributions taken from these sessions, complemented with a few solicited additional papers.

In producing the book, we have profited by our friendship with each other and with our colleagues whose work is represented here. These scholars offered advice and assistance in the form of referee reports. The soundness of their comments and ideas have contributed immensely to the quality of this volume. In addition, we would like to acknowledge the timely manner in which most of the contributing authors responded to all of our requests, and their willingness to follow stringent editorial guidelines. Moreover, we acknowledge the support provided by the Department of Geography at San Diego State University, the Department of Economic and Social Geography at the Vienna University of Economics and Business Administration, and the IGU. Finally, we have benefitted greatly from the editorial assistance provided by Laura Martin Makey, a student enrolled in the joint doctoral program at San Diego State University and University of California, Santa Barbara. Her expertise in handling several word processing systems, formatting, and indexing, together with her care and attention to detail helped immeasurably.

May 1997

MANFRED M. FISCHER, Vienna
ARTHUR GETIS, San Diego

Contents

List of Figures

1 Advances in Spatial Analysis

Manfred M. Fischer[1] and Arthur Getis[2]

[1]Department of Economic & Social Geography, Vienna University of Economics and Business Administration, A-1090 Vienna, Augasse 2-6, Austria and Institute for Urban and Regional Research, Austrian Academy of Sciences. A-1010 Vienna, Postgasse 4, Austria
[2]Department of Geography, San Diego State University, San Diego, CA, 92182-4493, USA

1.1 Prologue

The origins of modern spatial analysis lie in the development of quantitative geography and regional science in the late 1950s. The use of quantitative procedures and techniques to analyse patterns of points, lines, areas and surfaces depicted on analogue maps or defined by co-ordinates in two- or three-dimensional space characterise the initial stage. Later on, more emphasis was placed on the indigenous features of geographical space, on spatial choices and processes, and their implications for the spatio-temporal evolution of complex spatial systems. Spatial analysis, as it has become over the past four decades, is more than spatial statistics and data analysis, and goes far beyond data sampling, data manipulation, exploratory and confirmatory spatial data analysis, into areas of spatial modelling encompassing a large and diverse set of models in both the environmental and the social sciences. In the environmental sciences, they range from physical dispersion models (e.g. for suspended particulates), chemical reaction models (e.g. photochemical smog), and biological systems models (e.g. for ecosystems in water), through deterministic process and stochastic process models of regional environmental quality management, and integrated models for environment-energy-economic assessment. For many models, especially the physical dispersion models, chemical reaction models, and biological systems models, the geographical location of the site is not considered to be of overriding importance. The variation in model results over space can be achieved by obtaining different inputs to the area in question. Process models include deterministic versions which attempt to describe a particular sequence in terms of known physical laws, and stochastic model versions which aim to describe a particular process such as erosion, groundwater movement and absorption of pollutants in terms of probability theory. For such processes, there is an interaction between the spatial process and a substrate. This substrate provides a one-, two- or three-dimensional framework within which the process model can operate. The space in which the process is modelled is disaggregated into a set of finite elements which are usually assumed to be internally homogeneous. Two-dimensional models generally use as their data small pixels (see Fischer et al., 1996).

In the social sciences in general, and in quantitative geography as well as in regional economics in particular, a wide range of spatial models have been developed in the past decades for the purposes of describing, analyzing, forecasting, and policy appraising economic developments within a set of localities or regions. Such models deal not only with internal structures or regions and relationships within one region, but also with interregional interrelationships. The tradition of spatial analysis has shown a strong orientation towards location-allocation problems in space (e.g., siting of retail centres, employment location), witness the wide variety of spatial interaction models dealing with flows of people, commodities and resources between regions. Most of these models are static in nature and regard space as a set of discrete points (areas, grids) rather than as a continuum. The early studies in spatial choices and processes during the sixties and early seventies were dominated by spatial interaction models of the gravity type justified by probability and entropy maximizing formulations. This lack of a behavioural context was criticised in the 1970s, and gave rise to the study of individual choice behaviour in various contexts, such as journey-to-work, journey-to-shop, and migration. This, in conjunction with the parallel development of discrete choice models made it possible to propose new disaggregate choice based alternatives. Unlike spatial interaction models, these alternatives link individual decisions at the micro-level with population flows and other observables at the macro-level (see Fischer et al., 1990).

In the last few years, we have witnessed a major change in the key technologies that are relevant to spatial analysis and which promise new styles of spatial analysis models and methods in order to cope with spatial data rich environments, such geographical information systems (GIS) and remote sensing systems (RS). Computational intelligence (CI) technologies provide a new paradigm for spatial analysis, one that is likely to slowly evolve rather than revolutionize the spatial science fields over a short time frame. Computational intelligence denotes the lowest-level forms of intelligence which exhibit some form of computational adaptivity and fault tolerance, without explicitly representing knowledge in the artificial intelligence sense. CI-based spatial analysis provides a basis for improving the spatial data analysis techniques and models to meet the large-scale data processing needs of the emerging new era of data-driven exploratory searches for patterns and relationships in the context of an analysis process increasingly driven by the availability of very large quantities of spatial data (e.g., RS data, census products, GIS data, for more information see Fischer, 1996).

Currently, spatial analysis appears to be a rich field with many linkages to urban and regional problems, and marketing, transportation, and natural resource problems. GIS- and RS-technologies are greatly increasing the need for spatial analysis. In light of the foregoing remarks, the book is divided into three major parts representing the areas of

- spatial statistics and data analysis,
- behavioural modelling, and
- CI-based spatial analysis,

all of which will be briefly discussed in this introductory chapter.

1.2 Spatial Statistics and Data Analysis

Over the past forty years, an important series of methodological developments have appeared in statistics, based on the need to deal with the special nature of spatial data sets. This has led to a large variety of specialized analytical techniques and models and to the evolution of the field of spatial statistics and data analysis. Traditional statistical theory bases its models on assumed independent observations. Although common sense tells us that in most real world situations independence among observations on a single variable is more the exception than the rule, independence is still a suitable benchmark from which to identify statistically significant non-independent phenomena. The field of spatial statistics is based on the non-independence of observations; that is, the research is based on the assumption that nearby units are in some way associated. Sometimes this association is due to a spatial spillover effect, such as the obvious economic relationship between city and suburb. Sometimes the association is a distance decline effect; that is, as distance increases from a particular observation, the degree of association between observations lessens.

The types of statistical methods popular today are a function of both the nature of the problems being studied, the nature of the data, and the availability of computers. The Cliff-Ord (1973) monograph explored the concept of spatial autocorrelation, enabling researchers to assess statistically the degree of spatial dependence in their data, and, in so doing, to search for additional or more appropriate variables, and avoid many of the pitfalls that arise from correlated data. Much of today's interest in spatial analysis derives directly from both the Cliff-Ord monograph and their more complete discussion (1981). They shed light on the problem of model misspecification owing to autocorrelation and demonstrated statistically how one can test residuals of a regression analysis for spatial randomness. They explicated the nature of the spatial weight matrix and provided step-by-step procedures for applying statistical tests on Moran's I and Geary's c, the two major autocorrelation statistics.

Finding the degree of spatial association (autocorrelation) among data representing related locations is fundamental to the statistical analysis of dependence and heterogeneity in spatial patterns. Like Pearson's product moment correlation coefficient, Moran's statistic is based on the covariance among designated associated locations, while Geary's takes into account numerical differences between associated locations. The tests are particularly useful on the mapped residuals of an ordinary least squares regression analysis. Statistically significant spatial autocorrelation implies that the regression model is not properly specified and that one or more new variables or parameters should be entered into the regression model.

Many of the areas of current research can be traced to an interest in problems that the study of spatial autocorrelation was designed to address. The six main contributions to spatial statistics and data analysis in Part A of the present volume are representative of this pattern. They are:

- Bayesian regression and the expansion method (Chapter 2),
- Local versus global patterns of spatial association in regression analysis (Chapter

3),

- Measuring spatial variations in relationship with geographically weighted regression (Chapter 4),
- Regionalisation tools for the exploratory spatial statistical analysis of health data (Chapter 5),
- A structural time series approach to forecasting the space-time incidence of infectious diseases (Chapter 6),
- Estimating the impact of preventative action on the space-time incidence of HIV/ AIDS (Chapter 7).

These six contributions will now be discussed briefly.

Analysts often prioritize the independent variables in their models in the sense that they entertain presuppositions as to which variables are more or less important explanators of the dependent variables. The models constructed by expansions, i.e. by redefining the parameters of an initial model into functions of expansion variables, often reflect a prioritization. There are at least two different rationales for constructing models by expansions. The first centers on the extension of models to encompass their contexts, and the second on the addition of complexities to initial simplified formulations. Both the model-context and the complexity-building applications of the expansion methodology involve in most cases an ordering of variables by importance. Then it is useful to bring this prioritization to bear upon the models' estimation. How to do so and why, is the central theme of Chapter 2, authored by EMILIO CASETTI. When models generated by expansions are estimated, multicollinearity is often a problem. This contribution presents an approach to coping with multicollinearity based on a variant of Theil-Goldberger's mixed estimation. The variant exploits the analyst's presuppositions as to which independent variables are more or less important explanators of the dependent variable(s). The approach is demonstrated upon a spatial model of the AIDS epidemic.

In the third chapter, LUC ANSELIN and SHUMING BAO discuss computer techniques of exploratory spatial data analysis in order to: describe spatial distributions, visualize the pattern of spatial autocorrelation, and assess the presence of global and local spatial association. The integration of these is based on loose coupling, an approach in which data or results are passed between software modules using auxiliary files to minimize the amount of user intervention in the process. The technical aspects of the chapter explain an operational implementation of a linked, loose coupled GIS and spatial data analysis framework using the popular GIS software, ArcView, and the well-regarded spatial analysis package, SpaceStat. Conceptual and practical issues related to the implementation of the linked framework are discussed and illustrated with an exploratory analysis of the spatial pattern of housing values in West Virginia counties.

The fourth chapter is written by STEWART FOTHERINGHAM, MARTIN CHARLTON and CHRIS BRUNSDON, and suggests a statistical technique, termed Geographical Weighted Regression. This technique utilizes a distance-decay-based spatial weighting structure and can be used both to account for and to examine the presence of spatial non-stationarity in regression relationships between one or more independent variables and a single dependent variable. It is important to note that the points for which the local regression results are computed need not to be those at which data have been collected. The technique is illustrated by an application using health data from the 1991 UK Census of Population. The design of zoning systems is a fundamental step in area-based spatial analysis. Standard census units are not always suitable, since they may be too small (such as, e.g., the British enumeration districts which normally contain about 150 households) or too large or too inhomogeneous. The availability of digital boundaries and Geographical Information Systems software capable of manipulating them opens up the possibility of making regionalisation an intrinsic part of the analysis of spatial data, rather than a one-off, time-consuming exercise.

Chapter 5, written by STEVE WISE, ROBERT HAINING and JINGSHENG MA, describes work to develop a suite of classification and regionalisation tools suitable for the interactive spatial analysis of health data. The research forms part of a larger project which is developing a software system for the analysis of health data, with the emphasis on rapid interactive visualisation of data supported by a suite of cartographical, graphical and analytical exploratory tools. The system uses a GIS for the storage, manipulation and cartographical display of the spatial data and is called SAGE - Spatial Analysis in a GIS Environment. Models typically used to forecast space-time incidence of infectious epidemic diseases tend either to overestimate the total number of cases occurring in each outbreak, or to handle inadequately the spacing between epidemics.

In Chapter 6, produced by J.D. LOGAN and ANDREW D. CLIFF, a structural time series approach is developed to tackle these two problems. In contrast to ARMA approaches structural time series models are composed of unobservable components, such as trend and seasonality, and observable components, the explanatory and intervention variables. This type of model illustrated by applications to quarterly measles data for Iceland (1965-85) provides a basis for making predictions of future observations, but also provides a description of the relative importance of significant features of a time-series. An important characteristic of this model type is that it can be transformed into state-space form which facilitates estimation using the Kalman filter.

The final chapter in Part A has been written by RICHARD THOMAS and FIONA SMYTH, and covers the issue of estimating the impact of preventative action on the space-time incidence of HIV/AIDS. Preventative actions against the transmission of HIV/AIDS are construed as being either direct or passive. The former include medical measures linked to the serological test and social interventions, like community education, intended to modify high risk behaviour. Passive protection is conferred on a population through their collective behaviours, which might prevent sustained transmission, or through their distance from social networks where infection is present. This contribution draws together results obtained from epidemic modelling systems which allow both these forms of prevention to be evaluated in the sequel. The discussion sets these findings within the context of the more general debate about the appropriateness of the various intervention strategies against HIV/AIDS.

1.3 Behavioural Modelling

Spatial decisions and processes are fundamental to the understanding of spatial structure. During earlier stages of spatial analysis, when proposed explanations were characteristically at an aggregate, phenomenological level, the study of spatial decisions and processes was relatively neglected. However, as the field evolved and matured, the need for a deeper understanding of spatial structure gave rise to behavioural geography which begun in the sixties, grew in the seventies, and continues to foster and to rally around it the strong interest of many researchers in spatial decisions and processes. Since spatial decisions and processes belong to the more general realm of decision theory in the behavioural sciences, it was inevitable that their study would involve work extending over several fields such as economics, geography, transportation science, marketing, psychology and regional science (Fischer et al., 1990).

The behavioural modelling of spatial decisions and processes affects, and is affected by, the larger trends which characterize the modes of explanation in the field. These trends which are well represented in the contributions of the book include transitions from the aggregate to the disaggregate, from the deterministic to the probabilistic, from the static to the dynamic, and from the analytical to the computational. Part B of the book is subdivided into seven major constituent chapters, viz:

- Longitudinal approaches to analysing migration behaviour in the context of personal histories (Chapter 8),
- Computational process modelling of disaggregate travel behaviour (Chapter 9),
- Modelling non-work destination choices with choice sets defined by travel-time constraints (Chapter 10),
- Space-time consumer modelling, store wars and retail policy in Australia (Chapter 11),
- Integrated models of spatial search and impact analysis for developing retail networks (Chapter 12),
- Recent developments in the modelling of strategy formulation (Chapter 13),
- Some implications of behaviour in agricultural markets (Chapter 14).

Most studies of individual migration behaviour have been based on cross-sectional data that provide very little information about the locational histories of individuals. Cross-sectional data are limited to information on the whereabouts of individuals on only two dates, and reveal whether an individual migrated at least one time during the interval between the dates. Many of the analyses that have used cross-sectional data have been carried out by applying the discrete choice modelling approach which has a solid fundament in utility theory. Chapter 8, written by JOHN ODLAND, shows how migration and mobility behaviour can be analysed in an explicitly longitudinal context by treating incidents of migration and relocation as transitions in the personal locational histories of individuals. These locational histories unfold simultaneously with histories of other aspects of individual lives, such as work histories and marital histories, and analyses of migration can be carried out within a general framework that centers on questions about interdependencies between locational histories and the development of other histories for the same individuals. This framework lends itself to formal specification in terms of discrete-state processes in which histories are treated as sequences of distinct episodes; with each episode beginning and ending with transition events. The hypotheses that have dominated research on migration behaviour can be reformulated within this framework, where they make up only a portion of the possible array of hypotheses about interdependencies between related histories. Other categories of hypotheses, which depend on this explicitly temporal framework, are developed, especially hypotheses that center on the timing of events in two or more histories. Methods for investigating interdependencies in the timing of events are presented, along with empirical results from migration and employment histories for the United States. Spatial decisions and processes are psychologically complex phenomena, involving interactions among the processing capacity of an individual, his/her motivation, attention and perception, his/her information acquisition and evaluation, the quality of his/her memory, the decision processes (s)he employs, and his/her capacity to learn. The typically analytical nature of current spatial choice models either constrains their scope to a small part of the above scheme, or forces drastic simplifications to be employed, or both. Of course, simplifications are an integral part of modelling activity, and judging the utility of a model by the realism of its assumptions alone is a naive practice (Papageorgiou, 1982). Nevertheless, the possibility of computational process models on a different order of detail is very attractive.

Chapter 9, by MEI-PO KWAN and REGINALD G. GOLLEDGE, provides a comprehensive review of this type of model in the context of travel behaviour. Computational process models represent a linked set of computer programs together with appropriate databases which are designed to capture the essence of human decision making in different spatial contexts. Used primarily for way finding and to simulate and predict travel behaviour, this model type allows greater emphasis on the cognitive components of decision making including cognitive maps, preferences, and departure from utility maximizing and linearity in the considerations of alternative paths and destinations. The computational process model illustrated in this contribution focuses on multi person households and models travel behaviour before and after telecommuting takes place in the household. Sets of feasible alternative destinations for travel purpose are derived using GIS procedures such as buffering and path selection. Shortcomings and possible future developments of such models are then discussed.

In Chapter 10, JEAN-CLAUDE THILL and JOEL L. HOROWITZ present a new destination choice model based on a two-stage decision process. In the general theory of choice behaviour, alternative choices are reduced in the first stage to a smaller choice set, and in the second stage, the smaller choice set is analysed and a final selection is made. The new model, Approximate Nested Choice-Set Destination Choice, assumes that the choice set is contained within a travel perimeter defined by the individual's time budget. The nesting of choices and the travel time constraint make the estimation of the model computationally tractable. An empirical example using shopping destination choice information from Minneapolis-St. Paul reveals the superior fit of the new model over the conventional multinomial logit model of destination choice. The authors are able to demonstrate the importance of travel-time constraints in destination choice modelling and the ability of the new model to disentangle the effect of travel impedance on choice formation.

Chapter 11, written by ROBERT G.V. BAKER, addresses important substantive and policy relevant research issues in the field of consumer behaviour, such as the implications of deregulation of trading hours policy, in addition to the concerns with space-time and socio-economic characteristics of consumer behaviour in a deregulated trading environment, using retail space-time trip and shopping aggregate space-time visitation models applied to the Sydney shopping environment.

Chapter 12 is written by THEO A. ARENTZE, ALOYS W. BORGERS and HARRY J.P. TIMMERMANS and presents an heuristic model that is designed to support locational decision making in retail planning. The core of the model is a set of heuristic rules that define marginal changes of the current state of a retail system. A spatial interaction/choice model is used to predict the likely flows of retail expenditures in changed situations. The equilibrium state that results from iteratively applying the heuristics is presented as a possible plan scenario. This heuristic approach is based on the argument that it is not possible to formulate a mathematical objective function that can capture the ill-structured nature of the retail planning problem. A case study demonstrates the application of the model. The results suggest that the face validity and the numerical properties of the model are satisfactory. The authors argue that the model can generate significant added value to spatial decision support systems.

Chapter 13 by JAMES O. HUFF and ANNE S. HUFF deals with recent developments in the modelling of strategy reformulation. The authors inspired by the research tradition of behavioural residential mobility modelling based on the notions of stress (dissatisfaction) and cumulative inertia (resistance to moving) present a model of strategy reformulation as an ordered sequence of decisions relating to the assessment of current strategy, the timing of strategic change, the direction of change, and the strategic repositioning of the firm with respect to other firms within the industry. The model serves to highlight some recent concerns in behavioural modelling. Decisions are embedded in an ongoing decision process in which the outcomes are historically contingent and constitute the conditions of future actions. Decisions are also context dependent in that the configuration of the competitive environment and the positioning of the individual firm within that environment influence the magnitude and direction of strategic change.

The final chapter in Part B is written by GORDON F. MULLIGAN, and discusses some implications of behaviour in agricultural markets. The author derives these implications from a closed von Thünen model for a two-activity, one-dimensional economy. Equilibrium solutions are generated through matrix algebra for prices, land areas, and outputs. Both the cases of a dimensionless and a space-filling market town are addressed. Attention is also given to the two competing strategies of maximum average return and maximum guaranteed return under two states of environmental uncertainty.

1.4 CI-Based Spatial Analysis

The 1990s are witnessing the beginnings of a new quantitative revolution based upon computational intelligence (CI) technologies very relevant to spatial analysis. The stimulus is four fold:

- first, the GIS data revolution which greatly increases the relevance for spatial analysis in applied and empirical contexts,
- second, the urgent need for a new generation of data-driven geographical exploration and modelling tools which can cope rather than ignore the multidimensional complexity of data rich environments,
- third, the advent of high performance computers, especially the emerging era of parallel supercomputing, and
- fourth, the availability of practical and applicable CI tools.

The *raison d'être* of CI-based spatial analysis is to exploit the tolerance for imprecision and uncertainty in large-scale spatial analysis problems in an analysis process which is increasingly driven by the availability of huge quantities of spatial data, and to achieve tractability, robustness, computational adaptivity, real-time analysis and low cost. There are two principal areas of CI that are particularly relevant to spatial analysis: evolutionary computation which includes genetic algorithms and artificial life; and neural networks which are also known as neurocomputing. Biologically inspired evolutionary computation (i.e. genetic algorithms, genetic programming, non-Darwinian evolutionary algorithms etc.) provides the basis for developing new solutions to complex spatial optimisation problems as well as building blocks for new kinds of spatial analysis techniques and models; for example, the artificial life based pattern hunting creatures or the automated modelling system developed by Openshaw and associates (see Openshaw, 1988, 1994).

Much of the recent interest in neural network modelling in spatial analysis stems from the growing realization of the limitations of conventional tools and models as vehicles for exploring patterns and relationships in GIS and RS (remote sensing) environments and from the consequent hope that these limitations may be overcome by judicious use of neural net approaches. The attractiveness of these approaches extends far beyond the high computation rates provided by massive parallelism and essentially stems from the following features:

- the greater representational flexibility and freedom from linear model design constraints,
- the built-in capability to incorporate rather than to ignore the special nature of spatial data,
- the greater degree of robustness or fault tolerance to deal with noisy data, missing and fuzzy information, and
- the ability to deal efficiently with very large data sets and thus to provide the prospect to obtain better results by being able to process finer resolution data or

real-time analysis (see Fischer, 1996 for more details).

Part C includes the following four contributions which reflect the evolution of CI-based spatial analysis:

- Neurocomputing for earth observation: Recent developments and future challenges (Chapter 15),
- Fuzzy ARTMAP - a neural classifier for multispectral image classification (Chapter 16),
- Feedforward neural network models for spatial pattern analysis (Chapter 17),
- Building fuzzy spatial interaction models (Chapter 18).

Each of these chapters will be reviewed briefly.

Neural networks which may be viewed as non-linear extensions of conventional spatial statistical models are applicable to two major domains: first, as universal function approximators to areas such as spatial regression, spatial interaction, spatial choice and space-time series analysis; and second, as pattern recognizers and classifiers to intelligently allow the user to sift through the data, reduce dimensionality, and find patterns of interest in data-rich environments. There is no doubt that neural pattern classifiers have an important role to play in high dimensional problems of pattern recognition and classification in remote sensing environments. Chapter 15, written by GRAEME G. WILKINSON, provides a general review of applications of neurocomputing techniques in remote sensing. There has been a rapid growth in the use of neural networks in satellite remote sensing over the last five years. Most of the applications for image classification in thematic mapping were based on error-based learning systems such as the multilayer perceptron and neglected to control model complexity which is crucial for good generalization performance and, thus, for the success of neural classification. Evidently, more principled rather than ad hoc applications are needed in future. Apart from classification problems, neural network techniques have potential as general data transformation systems which can be exploited in a number of ways in remote sensing, e.g., in signal inversion and geometrical operations for image geocoding or stereo matching. The general reluctance to adopt the neural network approach more widely in Earth observation is attributed to the lack of user-oriented systems and the lack of fast training procedures for many neural network architectures. Recently developed neural network hardware systems are just beginning to be applied in remote sensing and may hold considerable potential especially if combined with appropriately engineered software systems for geographical users.

It is the major objective of Chapter 16, written by SUCHARITA GOPAL and MANFRED M. FISCHER, to analyse the capabilities and applicability of the neural pattern recognition system, called fuzzy ARTMAP, to generate high quality classifications of urban land cover using remotely sensed images. Fuzzy ARTMAP synthesizes fuzzy logic and adaptive resonance theory (ART) by exploiting the formal similarity between the computations of fuzzy subsethood and the dynamics of category choice, search and learning. The contribution describes design features, system dynamics and simulation algorithms of this learning system, which is trained and tested for classification (eight a priori given classes) of a multispectral image of a Landsat-5 Thematic Mapper scene (270 x 360 pixels) from the city of Vienna on a pixel-by-pixel basis. Fuzzy ARTMAP performance is compared with that of an error-based learning system based upon the multi-layer perceptron, and the Gaussian maximum likelihood classifier as conventional statistical benchmark on the same database. Both neural classifiers outperform the conventional classifier in terms of classification accuracy. Fuzzy ARTMAP leads to out-of-sample classification accuracies, very close to maximum performance, while the multi-layer perceptron - like the conventional classifier - shows difficulties to distinguish between some land use categories.

Chapter 17, written by YEE LEUNG, provides a most useful review of a basket of relevant neural networks such as linear associative memories, multi-layer perceptron and radial basis function networks along with various parameter estimation procedures such as back propagation and unsupervised competitive learning. The author also demonstrates how evolutionary computation, specifically genetic algorithms, can be employed to choose an optimal model out of the family of three-layer feedforward networks (i.e. both the number and size of the hidden layers).

In Chapter 18, STAN OPENSHAW describes the building process for the construction of fuzzy interaction models. He provides a most useful service by explaining in a straightforward way concepts of fuzzy sets and fuzzy logic and how they can be used as model building tools. He then assesses the performance of a number of fuzzy spatial interaction models and compares them with the results provided by more conventional models.

1.5 Conclusion

From the above discussion of recent developments in spatial statistics, behavioural modelling and neurocomputing, we show that spatial analysis has been able to base its research on solid methodological grounds and that at the same time it has managed to develop refreshing and novel analytical contributions. Thus, we may conclude from the above samples of recent advances in spatial analysis that this field has a valuable and rich heritage with a promising future. Like most good research, the contributions in this volume raise as many questions as they answer. The interaction between a sound theoretical foundation and careful data analysis seems to us to be the key to continued progress in spatial analysis in order to meet the new challenges of the expanding geocyberspace.

References

Cliff A.D. and Ord J.K. 1973. *Spatial Autocorrelation*. Pion, London.

Cliff A.D. and Ord J.K. 1981. *Spatial Processes: Models and Applications*. Pion, London.

Fischer M.M. 1993. Travel demand. in Polak J and Heertie J. (eds.) *European Transport Economics*, Blackwell, Oxford (UK), 6-32.

Fischer M.M. 1997. Spatial analysis: Retrospect and prospect, in Longley P., Goodchild M., Maguire D., and Rhind D., eds., *Geographical Information Systems: Principles, Techniques, Management and Applications*. GeoInformation International, Cambridge.

Fischer M.M., Nijkamp P., and Papageorgiou Y.Y. 1990. Current trends in behavioural modelling. in Fischer M.M., Nijkamp P., Papageorgiou Y.Y., eds. *Spatial Choices and Processes*, North-Holland, Amsterdam, 1-14.

Fischer M.M., Scholten H.J. and Unwin D. 1997. Geographic information systems, spatial data analysis and spatial modelling: Problems and possibilities,.in Fischer M.M., Scholten H.J. and Unwin D., eds. *Spatial Analytical Perspectives on GIS*, Taylor & Francis, London, (in press).

Openshaw S. 1994. Two exploratory space-time attribute pattern analysers relevant to GIS. in Fotheringham S. and Rogerson P., eds., *Spatial Analysis and GIS*, Taylor & Francis, London, 83-104.

Openshaw S. 1988. Building an automated modelling system to explore a universe of spatial interaction models. *Geographical Analysis*, **20**:31-46.

Papageorgiou Y.Y. 1982. Some thoughts about theory in the social sciences. *Geographical Analysis*, **14**:340-6.

Part A

Spatial Data Analysis

2 Mixed Estimation and the Expansion Method: An Application to the Spatial Modelling of the AIDS Epidemic

Emilio Casetti

Department of Economics, Odense University, Odense M, Denmark

2.1 Introduction

Analysts often 'prioritize' the independent variable(s) in their models in the sense that they entertain presuppositions as towhich variables are more/less important explanators of the dependent variables. The models constructed by expansions, are arrived at by redefining the parameters of an initial model into functions of expansion variables. Often these expanded models reflect a prioritization. This is the case if the initial model includes higher-priority explanators of the dependent variable(s), while the expansion variables and the terms in which they appear are lower priority refinements. Also, the initial model often articulates the primary theory that the expansions complement or widen.

Upon estimation, a terminal model generated by expansions that implement a prioritization should be congruent both with the primary theory and with its complements and expansions. Suppose that the terminal model is the true model. In the presence of strong multicollinearity we can have that the biased estimators of the unexpanded initial model are congruent with theory, while the unbiased estimators of the terminal model are inconsistent with it.

Circumstances such as these call for estimators capable of insuring that the lower priority variables will add as much as they possibly can to the explanation provided by the primary variables without obliterating it. The central theme of this paper is an implementation of the Thiel-Goldberger Mixed Estimation (ME) suited to accomplish this objective by bringing the analyst's prioritizations inside the estimation process.

The application discussed in the final portion of this paper involves adding by expansions a spatial dimension to a model of the dynamic of the AIDS epidemic. The analyses presented show that while a straightforward OLS estimation of the model yields estimated coefficients that are inconsistent with pertinent theory, satisfactory estimates are arrived at using the ME approach described in this paper.

The plan of the paper is as follows. First the interfaces among expansions prioritizations and collinearity are discussed. Mixed Estimation is then reviewed, and its application to bringing the prioritizations inside the estimation process is placed into focus. Next, an application to estimating a spatial temporal model of the dynamics of the AIDS epidemic is presented. A section of conclusions caps the paper.

2.2 Expansions and Prioritizations

2.2.1 The Expansion Method

The expansions approach to model construction (Casetti 1972, 1991, 1996) involves converting the parameters of an 'initial model' into deterministic or stochastic functions of relevant expansion variables. This process generates a 'terminal' model that can in turn become the initial model in a second round of expansions, and so on, iteratively.

Anything mathematical that has parameters can be expanded. Any models can be improved by expansions irrespective of whether they are intended for estimation optimization or solving, and also irrespective of whether they are stochastic deterministic or mixed. This paper is concerned with a narrow subset of these possibilities, and more specifically, with the expansion of econometrics models.

Many, if not most econometrics models are made of a deterministic nucleus that formalizes substantive theory or 'interesting' relationships from economics or elsewhere, complemented by RVs. These RVs represent error and/or temporally and/or spatially autoregressive terms (Anselin 1988, 1992) that are designed to mediate between deterministic abstractions and real world data. The parameters of both the deterministic and the stochastic portions of an econometrics model can be expanded. This paper focuses upon the expansion of the deterministic portion of econometrics models.

There are a number of diverse rationales for constructing models by expansions (Casetti 1996). Two of these rationales are of interest here. The first centers on the extension of models to encompass their 'contexts', the second on the addition of complexities to initial simplified formulations. Let us address them in sequence.

Much could be said about the formalization by expansions of the model context nexus, and on its potential for improving the scope and performance of the current social science models (Casetti 1993, 1995). The realities of mathematical modelling in the contemporary social sciences is dominated by constructs small in scope and tightly linked to disciplines and subfields, such as for example demand or production functions in economics, spatial interaction models in geography, and population dynamics models in demography/sociology.

The usefulness of an established model can be often increased by expanding its parameters with respect to contextual variables referring to dimensions environments or phenomena 'outside' it, and perhaps external to the discipline in which the model originated. The terminal models generated by these expansion encompass both the initial model and its variation across the context(s) considered. Searching for contextual variation is likely to result into richer theory and a better grasp of complex realities.

In geography and in the spatial peripheries of other social sciences, the expansion of models born a-spatial with respect to spatial variables and to indices of social and physical environments are especially promising for the mathematical of geographical phenomena.

The other rationale of the expansions reflects the most basic and pervasive approach by which humans grapple with complexity. To grasp complex realities the human mind constructs simplified depictions that include only perceived essentials. Other dimensions are subsequently reintroduced when needed. The expansions mimic in the construction of mathematical models this approach to grasping complexities. The initial models results from a drastic simplification. The expansions brings in dimensions that had been abstracted out during the process of simplification from which the initial models resulted. Iterated expansions can add in stages further dimensions of reality of increasingly lower 'priority'.

Both the model-context and the complexity-building applications of the expansion methodology involve in most cases the sequential introduction of aspects of reality in decreasing order of importance. When the creation of models by expansions is based on an ordering of variables by importance it is useful to bring this prioritization to bear upon the models' estimation. How to do so, and why, is the central theme of this paper.

Prioritization. Let us define 'prioritization' as the ordering or ranking of the explanatory variables appearing in a regressions in terms of their perceived importance. In the common practice prioritization is the rule, not the exception. The analyst confronted by the task of constructing or estimating a model is likely to have some notion as to which independent variables are more important. There are differences among disciplines, though.

Perhaps it is in research situations often occurring for instance in sociology or political science that prioritization has the strongest role. In these disciplines the analyst starts from a large number of candidate variables which reflect overlapping and interconnected theories and rationales, and the determination of which of the many models possible is eventually used results from specification searches guided by loose prioritizations that are not necessarily fully explicit in the analyst's mind.

The provision for forcing in a group of variables while leaving others free to enter into a regression equation, common in data analysis packages such SPSS, BMDP, SYSTAT, and others, is designed to accommodate a prioritization. However, even when these provisions are not used, a prioritization may be implemented by experimentations that leave a core of higher priority variables in all the equations tried out, while the other lower priority variables are rotated in and out.

But even in disciplinary environments in which formalized theory dominates, such as economics, a substantial component of empirical research may reflect the notion that multiple theories are to different degrees relevant to explaining a given empirical reality, or that variables that are not suggested by economic theory should be used also. In both cases a prioritization of the variables available to the analyst emerges.

In econometrics, sensitivity analysis, extreme bounds analysis (EMB), and the search for robust as opposed to fragile relationships (Leamer 1983, 1985; Leamer and Leonard 1983) are based on prioritization notions. These techniques and their variants center on the investigation of how much the coefficients of the essential variables change when alternative bundles of conditioning variables are entered in the equation. An interesting and thorough application of these concepts can be found in Levine and Renelt (1992); an expansion method application of the EMB is discussed in Bowen (1994, 1995).

Also, in the econometrics practice, implicit or explicit prioritizations can shape an empirical analysis by suggesting a minimal model to which lower priority variables are added in order to obtain well behaved residuals, or in response to tests aimed at deciding whether non-core variables in a maximal model should be retained or removed. It is useful to note that any scheme that involves the distinction between core variables and auxiliary variables can be easily conceptualized as the intercept expansion of an initial model that includes only core variables.

In any prioritization scheme, if all the core and non-core independent variables are orthogonal, it is a simple matter to decide whether the non-essential non-core variables should be entered or removed. This is not the case, however, if degrading multicollinearity occurs.

Multicollinearity. The construction of models by expansions adds variables and interaction terms to the variables appearing in the initial model and it can produce 'degrading' multicollinearity (Kochanowski 1990; Kristensen 1994). This is not to say that multicollinearity is necessarily a problem with models generated by expansions. Parsimonious expansions of simple models and in general expansions of a small number of parameters do not necessarily result into a substantial multicollinearity. However, we can expect to be confronted by strong multicollinearity when we attempt to estimate models constructed from many expansion variables and/or involving iterated expansions. Generally, the more variables are added by expansions, the stronger the collinearity.

When multicollinearity occurs, many very different parameter estimates are almost equivalent in the R-squares they produce, and minor changes in the sample consisting in the addition or removal of a variable or of a few observations can result into wildly different parameter estimates. Multicollinearity inflates the variance of the regression coefficients which results into low 't' values even when the variance explained is high. This makes it impossible to determine the individual effects of the independent variables, and in models generated by expansions, whether drift occurs, and if so, of what kind.

Under conditions of severe multicollinearity, the actual parameter estimates obtained may be substantively meaningless and also may be unrelated to the theory encapsulated into the initial model. Namely, multicollinearity can produce parameter estimates that are theoretically unacceptable because of their values an/or signs.

The prioritization that led to placing the more important variables in the initial model intrinsically implies that the expansion variables should improve the explanatory power of the initial model, not obliterate it. And yet, under conditions of strong multicollinearity this obliteration is precisely what can happen. Clearly, there is a need to bring the prioritization inside the estimation process in order to insure that the more important and the less important variables are not placed on the same footing. Rather, the coefficients of the non-core variables should reflect what these variables can add to a minimal model in which only the core variables are included.

Bringing the prioritization of the independent variables inside the estimation itself is the key point of this paper. In a capsule, this is accomplished through a variant of the Thiel-Goldberger Mixed Estimation, by complementing a model comprised of both core and non-core independent variables with a set of stochastic constraints reflecting the analyst's presupposition that the coefficients of the non-core variables are close to zero. The variances of the error terms of these constraints index the strength of the presuppositions, namely the degree of the analyst's belief in them. These variances operationalize the elusive notion of 'how close' is 'close to zero.'

In the next few sections the Mixed Estimation is placed in context and reviewed, and then its application to implement prioritizations is elaborated upon.

2.3 Mixed Estimation

A literature concerned with parameter estimation based on sample and non sample information within the framework of classical (non-Bayesian) econometrics has evolved in the past forty odd years. In this literature we can distinguish two strands, centered respectively on combining sample information with previous parameter estimates (Durbin 1953), and on combining sample information with 'priors' suggested by theory or by presuppositions. The Mixed Estimation (Theil and Goldberger 1961, Theil 1963, Theil 1971) generalized the technique pioneered by Durbin to combining sample data and priors. In the classical ME a linear model assumed to be 'true' is complemented by a set of linear stochastic constraints, also assumed to be true, that contain non sample information concerning the values of the model's parameters.

Over the years the ME has been enveloped into a sizable number of contributions that extended many of the assumptions and aspects of the original formulation. These contributions addressed exact constraints, incorrect exact constraints, biased stochastic constraints, inequality constraints, and biased linear models, among others. A great deal of attention has been given to determining under which conditions biased estimators are better than BLUE estimators in terms of risk functions and MSE criteria. Extensive discussions and reviews of these literatures have appeared (Koutsoyiannis 1977; Vinod and Ullah 1981; Toutenburg 1982; Fomby et al 1984; Judge et al 1985; Rao and Toutenburg 1995). In this paper we shall focus on the use of the classical ME as a tool to bring inside the estimation process the prioritization of the independent variables appearing in an expanded econometrics model.

The Basic ME Formalism. A brief presentation of the ME is the necessary starting point of the discussion that follows. Let Y and X be sample data, where the first column of X is vector of 1's, Y is n by 1, and X is n by p+1. The number of variables in X is denoted by p. Assume that the model

$$Y = X\beta + \epsilon \tag{2.1}$$

$$E(\epsilon) = 0 \tag{2.2}$$

$$E(\epsilon\epsilon') = \sigma^2 I \tag{2.3}$$

is 'true' and that we have information regarding β that is incorporated into the stochastic constraints

$$r = R\beta + \eta \tag{2.4}$$

$$E(\eta) = 0 \tag{2.5}$$

$$E(\eta\eta') = \gamma^2 I \tag{2.6}$$

where R is j by p+1, and r and η are j by 1.

For simplicity we assume in (2.3) and (2.6) that the observations errors and the constraints errors are independent. No major difficulties, conceptual or otherwise are involved in extending the results that follow to dependent ϵs and ηs. However we need to assume that

$$E(\epsilon\eta') = 0; \text{ and} \tag{2.7}$$

$$E(\eta\epsilon') = 0 \tag{2.8}$$

where the zeros above stand for appropriately dimensioned zero matrices.

The Mixed GLS Estimator b(τ). Combine the sample and prior information as follows:

$$\begin{bmatrix} Y \\ r \end{bmatrix} = \begin{bmatrix} X \\ R \end{bmatrix} \beta + \begin{bmatrix} \varepsilon \\ \eta \end{bmatrix} \tag{2.9}$$

$$E \begin{bmatrix} \varepsilon \\ \eta \end{bmatrix} = 0 \tag{2.10}$$

$$E \begin{bmatrix} \varepsilon \\ \eta \end{bmatrix} \begin{bmatrix} \epsilon \\ \eta \end{bmatrix}' = \begin{bmatrix} \sigma^2 I & 0 \\ 0 & \gamma^2 I \end{bmatrix} \tag{2.11}$$

Define τ as the ratio of σ^2 to γ^2:

$$\tau = \sigma^2/\gamma^2 \tag{2.12}$$

If we assume that σ and γ are known, it can be shown that the GLS 'mixed estimator' of β and its variance covariance matrix are:

$$b(\tau) = [X'X+\tau R'R]^{-1}[X'Y+\tau R'r] \tag{2.13}$$

$$var(b(\tau)) = \sigma^2[X'X+\tau R'R]^{-1} \tag{2.14}$$

It is well known that $b(\tau)$ is BLUE (Judge et al. 1985 p.59; Toutenburg 1982 p.52).

Let us place in perspective what the classical derivation of the mixed estimator due to Theil and Goldberger is and means. The $b(\tau)$ estimator is obtained by GLS from an extended data set that includes sample observations and quasi observations incorporating prior estimates or presuppositions. The key point here is that prior estimates or presuppositions are translated into data points. The 'quasi' in quasi-observations refers to the fact that the origin of these observations is something other than the measurements that produced the sample data. In the case of the analyst's presuppositions these observations are produced by introspection. The analyst translates the perceptions that result from his/hers previous work and/or from collective research into a data point. Thus, the non-sample information that in a Bayesian frame of reference is formalized into a prior probability distribution of the model's parameters, in the ME is instead translated into quasi-observations that are added to the sample data. The ME carries within itself a restructuring of the notion of 'data'.

Let us consider how and why the prioritization of the explanatory variables in an econometrics model can be operationalized as quasi observations. If the parameter of an independent variable has a zero value, the variable in question is irrelevant to explaining the dependent variable. However, if we set up the presupposition that the parameter β_i associated with the independent variable X_i equals zero plus an error using the stochastic constraint $r_i=\beta_i+\eta_i=0$, we still have not prioritized X_i. To do so we need to specify the variance γ_i^2 of η_i. A large γ_i implies a small confidence in the presupposition that $\beta_i=0$, and in the limit a very large γ_i is tantamount to removing the presupposition. Conversely, $\gamma_i=0$ implies certainty in the presupposition, and corresponds to a deterministic rather than stochastic constraint. In general, a small/large γ_i^2 denotes an low/high confidence in X_i's explanatory power.

Of two independent variables for which we presuppose that $r_i=\beta_i+\eta_i=0$ the one with a smaller γ_i has lower priority, namely, is presupposed to matter less for explaining the dependent variable. Also, the absence of a presupposition that β equals zero is equivalent to entertain this presupposition while assuming an infinitely large γ, which implies that all the variables for which we specify $r_i=\beta_i+\eta_i=0$ and a finite γ_i are prioritized lower than the other independent variables. This shows that ME can be used to bring the presuppositions of the analyst as to which independent variables are more important for explaining the dependent variable inside the estimation process.

$b(\tau)$ as a Matrix Combination of b_U and b_R. Two properties of the mixed estimator $b(\tau)$ that will prove useful in the discussions that follow are: (1) $b(\tau)$ is a weighted matrix combination of the restricted and unrestricted estimators b_R and b_U, and (2) $b(\tau)$ can be extracted from a constrained minimum formulation. Let us consider these properties in sequence.

Denote by b_U the unrestricted OLS estimator based only on the sample data:

$$b_U = (X'X)^{-1} X'Y \tag{2.15}$$

Denote by b_R the 'restricted' estimator (Fomby et al. 1984 p.83; Rao and Toutenburg 1995 p.113) based on the sample data and on the deterministic constraint $R\beta=r$:

$$b_R = b_U - (X'X)^{-1} R'(R(X'X)^{-1} R')^{-1} (Rb_U-r) \tag{2.16}$$

If we premultiply equation (2.15) by X'X and equation (2.16) by R we obtain

$$X'Y = (X'X)b_U \tag{2.17}$$

and

$$r = Rb_R \tag{2.18}$$

Substitution of the right hand sides of (2.17) and (2.18) into (2.13) yields

$$b(\tau) = [X'X+\tau R'R]^{-1} [X'Xb_U+\tau R'Rb_R] \tag{2.19}$$

which shows that the mixed estimator $b(\tau)$ is a matrix combination of the restricted and unrestricted estimators b_R and b_U. It ought to be noted that under assumptions (2.1) through (2.8) and for a known τ, $b(\tau)$ is BLUE, while b_R is biased and b_U is unbiased but inefficient. Hence we have that the BLUE estimator is a combination of two estimators, one biased and the other inefficient. It can be shown that as τ approaches zero $b(\tau)$ approaches b_U; and as τ approaches infinity $b(\tau)$ approaches b_R. Summing up, $b(\tau)$ is a combination of b_U and b_R and approaches these estimators if τ approaches zero or infinity.

The recognition that the mixed estimators are a combination of constrained and unconstrained estimators, and that the comparative contributions of these within a mixed estimator depends upon the analyst's degree of belief in the presuppositions formalized in the constraints, can change substantially our perception of the estimation choices available. Let us discuss why so with respect to the prioritization issue. Conventionally, our choice of regressors involves including them or excluding them. Regressors deemed to be important are included, while the unimportant ones are excluded. With mixed estimation, including or excluding a regressor are not the only alternatives, rather, they are the end points of a continuum encompassing infinitely many alternatives.

Implied in the notion of prioritizing the independent variables in a model is the possibility of having different levels of priority, and this possibility can be brought into the estimation process using mixed estimation. The mixed estimators permit to realize the intuitive notion that the less important variables should not be excluded but should play a less important role in explaining the dependent variable. Since the less important variables do not appear in the constrained estimator and the mixed estimator is a combination of the constrained and unconstrained estimator, the mixed estimation provides a mechanism for having the less important variables in the estimated equation but with a lesser role.

$b(\tau)$ as a Constrained Minimum. In order to show that $b(\tau)$ will arise from a constrained optimization consider the following optimum problem:

$$\text{Minimize } D(B) = [r-RB]'[r-RB] \tag{2.20}$$

$$\text{s.t. } K(B) = [B-b_U]'X'X[B-b_U] = \phi$$

This problem involves obtaining the vector B that minimizes D(B) subject to the condition that K(B) is equal to a constant ϕ. Since [r-RB] are constraints' residuals produced by B, here the objective function to be minimized is the sum of the squared constraints residuals: the similarity with respect to OLS is obvious, as in both cases the minimum of a sum of squared residuals is being sought.

As regards K(B) it can be shown (see Goldberger 1968 p.65 and p.73 ff) that

$$K(B) = [B-b_U]'X'X[B-b_U] = e_B'e_B - e_U'e_U \tag{2.21}$$

where

$$e_B = Y-XB \tag{2.22}$$

$$e_U = Y-Xb_U \tag{2.23}$$

Consequently, K(B) denotes the increment in the sum of squared residuals produced by the vector of coefficients B with respect to the minimum OLS sum of squared residuals. Thus, the optimum problem considered involves seeking the estimator B that minimizes the sum of squared constraints' residuals while limiting the increment of the sum of squared sample residuals to a prespecified value ϕ.

The solution of the optimum problem is straightforward. We define the Lagrangian function L

$$L = D(B) + \lambda(K(B)-\phi) \tag{2.24}$$

set to zero the partial derivatives of L with respect to B and to λ, and solve obtaining

$$B = [X'X+\lambda^{-1}R'R]^{-1}[X'Y+\lambda^{-1}R'r] \tag{2.25}$$

plus the constraint $K(B)=\phi$ as it appears in (2.20). Clearly, if we set $\tau=1/\lambda$ then $B=b(\tau)$. Which shows that the mixed estimator $b(\tau)$ corresponds to the solution of a constrained extremum problem.

The constrained minimum derivation of $b(\tau)$ is helpful to place in perspective the meaning and potential uses of ME. Let us show why so, in the situations addressed in this paper, namely, when ME is used to bring the analyst's prioritizations inside the estimation process.

In the setting considered here the low priority variables, that are presumed to be less important explanators, have zero coefficients in b_R, and coefficients that are not necessarily close to zero in b_U. The constrained minimization that yields $b(\tau)$ pulls the coefficients of the low priority variables toward zero subject to the condition that the resulting increase in the sum of squared sample residuals be limited to a specified value. Alternatively, we can say that of the infinitely many coefficient vectors that produce a given increase in the sum of squared sample residuals over the b_U levels, this constrained minimization selects the coefficient vector in which the low priority variables are as irrelevant as explanators as possible. Hence, the minimization of the objective function in the constrained minimization implements the analyst's priorities by shrinking b_U toward b_R. Also we could say that this constrained minimization insures that the low priority variables are given an explanatory role in the form of non-zero coefficients only to the extent that they add to the explanation given by the high priority variables.

Types of priors. Three classes of priors are consistent with the assumptions in (2.4) and (2.5) to the effect that $E(r)=R\beta$. These are (A) the priors involving unbiased estimates of some or all the β parameters from other data sets possibly of a different type; (B) the priors inspired by theoretical assumptions; and (C) the priors reflecting an analyst prioritization of the explanatory variables.

The (A) priors encompass any unbiased estimates of some or all the elements in β based on empirical data other than the sample's. Typical examples of situations falling into this category include using the estimate of a model obtained from one data set as priors in the mixed estimation of the same model from a second data set. Also in this category belongs the technique whereby parameters estimated, say, from time series, are used as priors in mixed estimations employing crossectional data.

The (B) priors reflect presuppositions of the analyst originating from theory. The typical example is the presupposition that the capital and labor elasticities in a Cobb Douglas production function add up to one, which corresponds to constant returns to scale.

The (C) priors are those appearing for instance in the ridge and generalized ridge regressions, in the Bayesian Vector Autoregression (BVAR), or in implementations of Shiller's Smoothness Priors. All of these translate into stochastic constraints with $r=0$ and with a variety of stipulations concerning the variances of the constraints' errors. Essentially, the analyst is interested in determining to what extent these zero presuppositions are overruled by the sample data, and are effective against multicollinearity.

To clarify this point, consider the BVAR: in it, the variable(s) to be forecasted are regressed against lagged variables, with many consecutive lags entering into the model used. For all variables with lags greater one, the conventional prior is that their parameters have zero values, with a confidence in the prior assumed to be stronger for longer lags. This implies that the variances of the constraints' errors are smaller for longer lags. The rationale of these priors is to break the strong collinearity intrinsic to this type of models without removing any variables and allowing the data to override 'selectively' the presupposition of zero valued coefficients. Clearly, the priors implementing the prioritizations discussed in this paper are (C) priors.

2.4 Estimation of σ,γ and τ

In the preceding sections σ, γ, and τ were assumed known. Let us now consider the issues that arise when they have to be estimated. No special problems are involved in the estimation of σ^2 from the sample data. Consequently here we need to focus on γ^2 and τ.

In the applications of ME involving the combining of information from different samples γ^2 is estimated from the alternative sample. In the applications in which the prior information originates from theory, for instance economic theory, γ^2 can be extracted directly from the assumptions made. For instance, in the estimation of a Cobb Douglas production function, if we assume constant returns to scale, γ^2 can be evaluated from statements such as that the sum of the capital and labor elasticities falls within the 0.8-1.2 interval with a .95 probability. This statement, in combination with the assumption that η is $N(0,\gamma^2)$ yields readily an estimate of γ^2.

Given an estimated σ^2 and an estimated γ^2, their ratio is an estimated tau. Estimated σ^2, γ^2, and τ are here denoted by a hat superscript. If τ in (2.13) is replaced by $\hat{\tau}$ the $b(\hat{\tau})$ obtained is a feasible GLS estimator. However $b(\hat{\tau})$ is asymptotically unbiased under non-restrictive conditions, and is 'essentially' the same as the GLS estimator for large samples (Judge et al. 1985 p.59;Fomby et al. 1984 p.92 and p.150 ff; Theil 1963; Nagar and Kakwani 1964; Yancey et al. 1974).

The approach to the implementation of ME that use estimates of γ^2 are sensitive to the size of the sample. Such sensitivity may be appropriate or tolerable in some circumstances but not in others. It is not desirable when ME is used to bring a prioritization inside the estimation process, since the presuppositions of the analyst concerning which independent variables are more/less important should not produce results dependent upon sample size. The approach discussed and applied in this paper is designed to overcome the sensitivity to sample size of the conventional implementation of ME.

Consider the following question. How can the analyst's presuppositions be operationalized so as to insure that they will 'matter' as much as the sample? Suppose that the sample consists of n observations, and that the priors are incorporated into a single quasi-observation. For the presuppositions to matter as much as the sample the quasi-observation will have to be more precise, namely, it should have an error term with a smaller variance. Specifically, we need to have $\gamma^2=\sigma^2/n$. Since $\tau=\sigma^2/\gamma^2$ we need $\tau=n$ for the prior to matter as much as the sample. Thus, if the quasi-observation is weighted n as compared to the sample's observations that are weighted one the analyst's presuppositions will matter as much as the sample in the mixed estimator $b(\tau)$.

If there are j quasi-observations and n observations, by following through the same line of reasoning we will have to conclude that the condition for the priors to matter as much as the sample is $\gamma^2/j=\sigma^2/n$, which yields a $\tau=n/j$. If we still have n observations and j stochastic constraints, but we wish that the analyst's presuppositions matter h times as much as the sample's information we need to stipulate that $\gamma^2/j=\sigma^2/nh$ which yields a $\tau=hn/j$. With this definition of τ if h=1 the priors and data weigh equally, if h>1 the priors will weigh more than the data by a factor of h, and if h<1 the data will weigh more than the priors by a factor of 1/h. Namely, the 'parameter' h determines the comparative influence of priors and data and is defined over the interval zero to infinity.

The implementation of ME on the basis of the assumption

$$\tau = hn/j \tag{2.26}$$

has several conspicuous advantages. It links the analyst's degree of belief in the priors to a decision as to how much the priors should matter visa a vis the sample. It is insensitive to the number of constraints and to the sample size and consequently it allows to use the same yardstick in different analyses much in the same way as we use same standard levels of significance across analyses. It does not require to 'estimate' γ^2, and therefore does not confront us with an estimated τ and with a feasible GLS $b(\hat{\tau})$. Instead, since τ=hn/j is assumed and not estimated we can always obtain a GLS mixed estimator $b(\tau)$.

The selection of the value of h may be based on at least two alternative rationales. The simplest one involves choosing a meaningful default value that is well suited in a broad range of situations. One convenient such value could be h=1, that renders sample and priors equally influential. Alternatively, a range of sequential values of h could be used to produce parameter 'traces' reminiscent of the ones used in ridge regression. These traces portray the effects of a range of degrees of belief in the ME priors, and can be used (a) to identify the h values yielding 'satisfactory' results, and/or (b) to rate the comparative performance of the low priority regressors. Such ratings can help to respecify the model, if needed. The criteria in terms of which the evaluations of the ME results and the comparative performance rating of the regressors are carried out reflect and formalize the researcher's judgement and may include substantive considerations such as the congruence with pertinent substantive theory. The usefulness of the ME traces based on the h parameter and the reasoning behind their use are demonstrated in the application that follows.

2.5 Application to AIDS Modelling

In the sections that follow I will describe a simple spatial dynamic model of the AIDS epidemic, and outline the shortcomings of its estimation by OLS. Let us start with the construction of the model. The approach used involves taking as 'initial model' a simple theoretically grounded relationship of the dynamic of an epidemic, and then expanding it in terms of a spatial variable to produce a 'terminal model' better suited to depict the spread of the AIDS epidemic in the US at the state level of resolution.

The initial model selected is

$$y = \beta_0 + \beta_1 x + \epsilon \tag{2.27}$$

where y denote the natural logarithm of the increment in AIDS counts for a state and over a year, and x stands for the natural logarithm of the cumulated AIDS counts up to beginning of the one year interval for which y is defined. This model is suited for forecasting, since it will 'predict' the increment in AIDS cases for an areal unit as a function of the aggregate of AIDS cases recorded in it. However its forecasting performance will not be addressed here.

This relationship represents a simplified and condensed version of the classical epidemiological differential equations in terms of susceptible and infective (Bailey 1957, 1975; Anderson and May 1991; Brown and Rothery 1993). It ought to be noted, though, that because of the AIDS' lengthy average incubation period, the classical epidemiological equations cannot be used as such, as the dynamics of the AIDS counts reflects only imperfectly and with a time lag the actual spread of the epidemic, namely, the diffusion of the HIV virus (Jewell et al. 1992; Brookmeyer and Gail 1994; Isham and Medley 1996).

Let us consider briefly why log transformations of the increments of AIDS counts and of cumulated AIDS counts were used. The transformation reduces the skewness in the distribution of the two variables. Also it renders the analysis less influenced by the observations with larger AIDS counts and larger increments in AIDS counts. Hence, the transformations makes it possible to better estimate the changes in AIDS prevalence in states in which the AIDS epidemic has been felt to a lesser degree.

The parameter β_1 in equation (2.27) is an elasticity, and represents the percentage increase in the increment in AIDS counts associated with one per cent increase in cumulated AIDS counts. Consider the parameter β_0 next. For any given value of x, a larger/smaller value of β_0 is associated with a larger/smaller y. Which suggests that in the spread of the AIDS epidemic the 'external' effects would manifest themselves primarily through β_0, and the 'internal' effects primarily through β_1.

Model (2.27) is obviously inadequate, since it implies that the y(x) relationship holds with the same parameters throughout the United States, which is untenable. There are several reasons suggesting that (2.27) should vary across geographical space. Different states are at different stages of evolution of the epidemic. In states with a large number of AIDS cases social and psychological mechanisms set into motion by the fear of the disease will tend to lower the increment in AIDS cases. Also, in these states, the increment in AIDS cases is overwhelmingly due to internal processes rather than to outside influences. The opposite is true for states in which the impact of the epidemic has been felt to a lesser degree. In these states the influences from the outside can be presumed to have a greater role, and these influences tend to correspond to a comparatively higher intercept.

In terms of its spatial distribution in the US, the AIDS epidemic is characterized by a minimum in the northern interior region centered approximately on the Dakotas. Instead, large numbers of AIDS cases can be found along the Atlantic and Pacific coasts, and in the South. Relative maxima are in California, Texas, Florida and in several locations on the East Coast.

The variable selected for its ability to position a state within this AIDS landscape is the mean distance to AIDS cases in other states, s. A value of s was computed for every state and for every time period. To clarify, consider how s would be calculated for, say, the state of Virginia in 1985. To this effect, the distance between Virginia's centroid and the centroids of the other 47 conterminous states in the continental US is computed, and multiplied by the 1985 AIDS counts for these states. These products are then added up and divided by the total AIDS count for the US minus the AIDS counts for Virginia. The value obtained is the mean distance to the out of state AIDS counts for Virginia in 1985.

This s variable has larger values for states remote from where the AIDS epidemic has been felt more strongly, and vice versa. Consequently, s is larger for the states in northern central portion of the US where the AIDS epidemic has began more recently and is less prevalent, and tends instead to be smaller on the Atlantic, Gulf of Mexico, and Pacific coast states. In other words s is a good proxy for the relative location of each state within the AIDS epidemic landscape, and is very well suited to represent a spatial dimension along which equation (2.27) can be expected to vary.

Model (2.27) was expanded in terms of the variable z=ln(s) on the basis of the following expansion equations:

$$\beta_0 = \beta_{00} + \beta_{01}z \qquad (2.28)$$
$$\beta_1 = \beta_{10} + \beta_{11}z$$

to yield the terminal model

$$y = \beta_{00} + \beta_{10}x + \beta_{01}z + \beta_{11}xz + \epsilon \qquad (2.29)$$

The initial and terminal models above were estimated by OLS using data by state by year for the period 1985 to 1991. The log of the mean distance to AIDS measure, z, was calculated for each year and each state using coordinates of the states' centroids. The results of these estimates are shown in columns 1 and 2 of Table 2.1.

The measure of multicollinearity reported in Table 2.1 is the Condition Number (CN) of the unit scaled design matrix. Let W be the design matrix X transformed to 'unit scale'. Let Lmax and Lmin be the respectively the largest and the smallest eigenvalues of W'W. Then the condition number CN of X is the square root of Lmax/Lmin. Degrading multicollinearity is said to occur when the CN of a design matrix is greater than 30 (Belsley et al. 1980).

The strong multicollinearity that can occur when we estimate models generated by expansions is apparent if we compare the CNs for the initial model which is 6, to that for the model produced by expansions which is 486. Clearly, the expanded model is plagued by a huge multicollinearity.

Multicollinearity inflates the variance of the regression coefficients, and renders these coefficients unstable, so that they vary widely across samples or with changes in the variables included in the regression. The latter effect of multicollinearity is very conspicuous here. The coefficients of the logarithm of cumulated AIDS counts, x, is .771 in the unexpanded initial model and -1.01 in the expanded model. More extensive analyses involving more equations than are reported here indicated that the instability of the estimated coefficients is a definite problem, and that multicollinearity causes it.

The estimate of the unexpanded model (regression 1 in Table 2.1) constitutes a good initial result consistent with theoretical presuppositions. Both the estimates of β_0 and β_1 are positive and significant, the condition number is low, and the R-square is very high at .895.

Instead, the estimates of the expanded model are very poor. All the coefficients are significant or very close to significance, but their signs and values are substantively meaningless. The coefficient of the distance to AIDS variable is significant and negative, implying that the external effects on the increment of AIDS cases are smaller where the epidemic is in an early stages than where AIDS is well established, which is absurd. Also, the coefficients of x and xz are not consistent with the presuppositions that size of the epidemic has a positive effect on the increment in AIDS cases which is born out by the estimated initial model, and that geographical location proxied by z should bring only somewhat minor corrections to this effect. The very high condition number for this regression suggests degrading multicollinearity as the cause of these results.

The conventional approach to remedying this type of situations involves specification searches whereby many variants of the expanded model or of subsets of it are tried out and evaluated in terms of congruence with theory and statistical significance. If a satisfactory regression is obtained the analyst presents it as the result of the investigation, often leaving undescribed and unjustified the steps of the specification search that produced it. The ME approach employed here, and in general its Bayesian and quasi-Bayesian counterparts are designed to bring the criteria that drive the specification searches inside the estimation process, and to document them.

In the approach followed here the x, z, and xz variables are prioritized by stochastic constraints. Specifically, the independent variable appearing in the initial model, x, is assumed to be more important for explaining y, than the expansion variables z and xz. This prioritization is operationalized using the assumptions (2.4) through (2.8). Here r and R are respectively a 2 by 1 vector and a 2 by 4 matrix. Denote by r_1 and r_2 the elements of r, and by R_1 and R_2 the rows of R. The prioritization is accomplished by setting $r_1=r_2=0$, $R_1=[0,0,1,0]$, and $R_2=[0,0,0,1]$. Using (2.12) and (2.26) a range of γ^2 are specified implicitly by assuming suitable values of h. In this example n=304, and j=2.

Analyses were carried out for a sequence of values of h, namely for h=.01,.1,1,10,100. Let us restate for the sake of clarity that an h=.01 means that we wish the sample information to matter 100 times as much as the priors, while h=100 means the opposite, that the priors should weigh 100 times more than the sample information. The results of these analyses are shown in Table 2.2.

In order to place these results in context we have to think the sequence of h values from which they are derived as points on an interval bounded by 0 and infinity. For each point in the interval we have an h value and a vector of parameter estimates. When an h and implicitly the γ^2 that goes with it are 'assumed' the associated parameter estimates are BLUE. The prioritation operationalized here is a statement to the effect that we believe that the parameters associated with z and xz are close to zero, with h and γ^2 specifying 'how close'. As we specify/assume 'how close' we also define a set of BLUE coefficients. Since we can choose infinitely many levels of closeness we also have not one but infinitely many results, which is a strength of the approach discussed here. The availability of infinitely many BLUE results is what opens the door to bringing inside the estimation process the criteria that usually drive the specification searches.

Table 2.1. OLS Regression Results

Vars	REG#1	REG#2	REG#3
Const	0.84624	11.0760	0.87721
	(9.14)**	(2.99)**	(9.40)**
x	0.77115	-1.01180	0.42823
	(50.8)**	(-1.84)	(2.52)*
z	——	-1.35190	——
	——	(-2.75)**	——
xz	——	0.23548	0.04498
	——	(3.25)**	(2.02)*
R-sq	0.895	0.899	0.896
CN[1]	6.29	486.04	86.92

t values are shown in parentheses under their respective coefficients. Two/one asterisks denote significance at the one/five percent levels.
[1]Condition Number

Table 2.2. ME Regression Results

Vars	REG#1 h = 0.01	REG#2 h = 0.1	REG#3 h = 1	REG#4 h = 10	REG#5 h = 100
Const	4.8727	1.4565	0.90018	0.84890	0.84639
	(2.09)*	(1.58)	(2.89)**	(6.42)**	(8.83)**
x	-1.13356	0.35875	0.50456	0.68607	0.76017
	(-0.36)	(1.73)	(3.34)**	(8.08)**	(22.7)**
z	-0.52848	-0.07682	-0.00401	0.00067	0.00011
	(-1.72)	(-0.63)	(-0.10)	(0.05)	(0.03)
xz	0.11930	0.05418	0.03498	0.01116	0.00143
	(2.46)*	(1.99)*	(1.77)	(1.02)	(0.37)
R-sq	0.898	0.897	0.896	0.896	0.895
CN[1]	309.54	139.87	90.21	49.81	17.73

t values are shown in parentheses under their respective coefficients. Two/one asterisks denote significance at the one/five percent levels.
[1]Condition Number

Table 2.3. ME Regression Results

Vars	REG#1 h = 0.01	REG#2 h = 0.1	REG#3 h = 1	REG#4 h = 10	REG#5 h = 100
Const	0.87704 (9.40)**	0.87545 (9.39)**	0.86555 (9.32)**	0.85064 (9.22)**	0.84675 (9.19)**
x	0.43030 (2.54)*	0.44775 (2.71)**	0.55738 (4.14)**	0.72248 (11.0)**	0.76557 (29.1)**
z	——— ———	——— ———	——— ———	——— ———	——— ———
xz	0.04471 (2.02)*	0.04242 (1.97)*	0.02803 (1.60)	0.00638 (0.76)	0.00073 (0.25)
R-sq	0.896	0.896	0.896	0.895	0.895
CN[1]	86.66	84.41	68.67	32.98	12.08

t values are shown in parentheses under their respective coefficients. Two/one asterisks denote significance at the one/five percent levels.
[1]Condition Number

Consider a specification search aimed at obtaining regression results congruent with theory and in which the estimated parameters are statistically significant. The counterpart of the implementation of these criteria within the mixed estimation approach considered here consists in selecting from the range of h values evaluated those that yield mixed estimators congruent with theory and statistically significant. Since there can be infinitely many values of h for which these conditions are met, an additional principle is needed to extract from them a single 'optimum' h.

One such principle is suggested by the aversion that most econometricians and data analysts have to using priors. This aversion provides a solid basis for an optimum h, since it suggest selecting the 'least intrusive' prior. Larger h's produce regression results reflecting the sample information to a lesser degree. Consequently, the least intrusive priors are the ones associated with the smallest h. Thus, we can select the smallest h yielding 'acceptable' mixed estimators. Here estimators are regarded as acceptable if they are congruent with theory and statistically significant.

In the application discussed in this paper let us stipulate that congruence with theory requires that the coefficients of x and z be non-negative. None of the regressions reported in Table 2.2 satisfies this requirement and at the same time has significant coefficients. However, if we compare the coefficients of the expansion variables in terms of their performance with respect to the significance and congruence with theory criteria, it becomes readily apparent that the coefficient of z is the worst performer. Which suggests to respecify the model using only x and the expansion variable xz. Analyses in which the higher priority variable is x and the lower priority variable is xz for the same values of h used previously are shown in Table 2.3. The OLS estimation with x and xz as independent variables is given in regression 3 of Table 2.1. Table 2.3 suggest that there are a range of values of h ranging from h=0 to circa h=.1 that are acceptable because have coefficient that are significant and congruent with theory. Of these, the one that has the smallest h is the OLS regression 3 in Table 2.1, that therefore constitutes the final result of the analysis.

2.6 Conclusion

One application of the expansion methodology is to construct more complex models from simpler ones by a process that implements the analyst's prioritizations. If the initial model is based on more important explanatory variables and more important theory, the lower priority dimensions are introduced later, by expansions. Upon estimation, the terminal model thus constructed may be confronted by collinearity, the more so, the more variables are added by expansions. In turn, collinearity may bring about meaningless results, inconsistent with the theories that inspired the initial model and the expansions. Under these circumstances it is desirable to bring the analyst's prioritizations inside the estimation to provide the additional information that can break the multicollinearity.

This paper shows that this can be accomplished by a variant of the Mixed Estimation. To this effect, the notion that some variables are less important is translated into stochastic constraints formalizing the presupposition that their parameters are close to zero. In the implementation of ME discussed here, how close is close is specified by a parameter h, the value of which is assumed. When a value of h is selected, the mixed estimator corresponding to it is BLUE, and it can be obtained directly by GLS rather than by the two step procedure leading to feasible GLS required by other ME implementations. Since h can be assumed to have any values between zero and infinity, this approach yields infinitely many BLUE mixed estimators. Of these the best can be selected on the basis of criteria external to the estimation process.

In the AIDS application discussed in this paper it is stipulated that the 'best' mixed estimator should be consistent with pertinent substantive theory, should produce statistically significant regression coefficients, and should be based on priors as unintrusive as possible. The estimators based on the least intrusive priors are the ones that result as much as possible from sample information rather than from the analyst's presuppositions.

In closing, let us note that the implementation of ME discussed in this paper can be used for combining any non-sample information with sample information: namely is not confined to the prioritizations that constitute the primary objective of this paper.

Acknowledgements
Financial support by the NSF under grant no. SBR-9223985 is gratefully acknowledged.

References

Anderson R.M. and May R.M. 1992., *Infectious Diseases of Humans: Dynamics and Control*, Oxford University Press.

Anselin L. 1988. *Spatial Econometrics: Methods and Models*, Dordrecht: Kluwer Academic.

Anselin L.1992 Spatial Dependence and Spatial Heterogeneity: Model Specification Issues in the Spatial Expansion Paradigm, pp.334- 354 in *Applications of the Expansion Method*, J.P. Jones III and E. Casetti (eds), Routledge, London and New York

Bailey N.T.J.1957. *The Mathematical Theory of Epidemics*, Griffin: London

Bailey N.T.J.1975. *The Mathematical Theory of Infectious Diseases*, Griffin: London

Belsley D.A., Kuh E., and Welsch R.E. 1980. *Regression Diagnostics: Identifying Influential Data and Sources of Collinearity*, Wiley

Bowen J.L. 1994. *Investigating the Relationship Between Foreign Aid and Economic Growth in Recipient Countries*, Ph.D. Dissertation, UMI: Ann Arbor, Michigan

Bowen J.L. 1995. Foreign Aid and Economic Growth: An Empirical Analysis, *Geographical Analysis*, **27**:249-261.

Brookmeyer R. and Gail M.H. 1994. *AIDS Epidemiology: A Quantitative Approach*, Oxford University Press: New York Oxford

Brown D. and Rothery P.1993. *Models in Biology: Mathematics, Statistics and Computing*, Wiley

Casetti E. 1972. Generating Models by the Expansion Method: Applications to Geographic Research, *Geographical Analysis*, **4**:81-91.

Casetti E. 1991. The Investigation of Parameter Drift by Expanded Regressions: Generalities and a 'Family Planning' Example, *Environment and Planning A*, **23**:1045-1061.

Casetti E. 1993. Spatial Analysis: Perspectives and Prospects, *Urban Geography*, **14**:526-537.

Casetti E. 1995. Spatial Mathematical Modeling and Regional Science, *Papers In Regional Science*, **74**:3-11.

Casetti E. 1996. *The Expansion Method, Mathematical Modeling and Spatial Econometrics*, Manuscript

Durbin J.1953. A Note on Regression when there is Extraneous Information about One of the Coefficients, *Journal of the American Statistical Association*, **48**:799-808.

Fomby T.B., Hill R.C. and Johnson S.R.1984. *Advanced Econometrics Methods*, Springer-Verlag

Goldberger A.S. 1968. *Topics in Regression Analysis*, Macmillan: London

Isham V. and Medley G. eds. 1996. *Models for Infectious Human Deseases: Their Structure and Relation to Data*, Cambridge University Press: Cambridge

Jewell N.P., Dietz K. and Farewell V.T. eds. 1992. *AIDS Epidemiology: Methodological Issues*, Birkhauser: Boston Basel Berlin

Judge G.G., Griffith W.E., Hill R.C., Lutkepohl H., and Lee T.C. 1985. *The Theory and Practice of Econometrics*, New York: Wiley

Kochanowski P. 1990. The Expansion Method as a Tool of Regional Analysis, *Regional Science Perspectives*, **20**:52-65.

Kristensen G. and Tkocz Z. 1994. The Determinants of Distance to Shopping Centers in an Urban Model Context. *Journal of Regional Science*, **34**:425-443.

Koutsoyiannis A. 1977. *Theory of Econometrics*, 2nd ed. Macmillan

Leamer E.E. 1983. Let's Take the Con Out of Econometrics, *American Economic Review*, **73**:31-43.

Leamer E.E., 1985. Sensitivity Analyses would Help, *American Economic Review*, **75**:308-313.

Leamer E.E. and Leonard H.1983. Reporting the Fragility of Regression Estimates, *Review of Economic and Statistics*, **65**:306-317.

Levine R. and Renelt D. 1992. A Sensitivity Analysis of Cross-Country Growth Regressions. The *American Economic Review*. **82**:942-963.

Nagar A.L. and Kakwani N.C. 1964. The Bias and Moment Matrix of a Mixed Regression Estimator. *Econometrica*, **32**:389-402.

Rao C.R. and Toutenburg H. 1995. *Linear Models: Least Squares and Alternatives*, Springer Verlag: New York

Theil H. and Goldberger A.S. 1961. On Pure and Mixed Statistical Estimation in Economics, *International Economic Review*, **2**:65-78.

Theil H.1963. On the Use of Incomplete Prior Information in Regression Analysis, *Journal of the American Statistical Association*, **58**:401-414.

Theil H. 1971. *Principles of Econometrics*, Wiley

Toutenburg H. 1982. *Prior Information in Linear Models*, Wiley

Vinod H.D. and Ullah A.1981. *Recent Advances in Regression Methods*, Dekker: New York

Yancey T.A. Judge G.G. and Bock M.E. 1974. A Mean Square Error Test When Stochastic Restrictions Are Used in Regression. *Communications in Statistics*, **3**:755-768.

3 Exploratory Spatial Data Analysis Linking SpaceStat and ArcView

Luc Anselin[1] and Shuming Bao[2]

[1] Regional Research Institute and Department of Economics, West Virginia University, Morgantown, WV 26506-6825, USA
[2] Data Analysis Products Division, MathSoft, Inc., Seattle, WA 98109, USA

3.1 Introduction

The extension of the functional capacity of geographic information systems with tools for spatial analysis has been an increasingly active area of research in recent years. Following Goodchild's (1987) call to action [see also Goodchild (1992)], a growing number of conferences and workshops has been devoted to the topic in the academic GIS community, resulting in many articles and several edited volumes [among others, Fischer and Nijkamp (1993), Fotheringham and Rogerson (1994), Painho (1994), Fischer, Scholten and Unwin (1996)]. Among the types of analyses suitable for inclusion within the functionality of a GIS, spatial *data* analysis in particular has received considerable attention. Several conceptual frameworks have been suggested, proposing different degrees of linkages between GIS and spatial analysis, and outlining the types of spatial statistical techniques that would be most suitable for inclusion [examples are Openshaw (1991), Anselin and Getis (1992), Goodchild et al. (1992), Bailey (1994), Haining (1994), Openshaw and Fischer (1995)]. In broad terms, the proposed linkages can be described as either tight vs. loose [Goodchild et al. (1992)] or as encompassing vs. modular [Anselin and Getis (1992)] depending on the degree to which a module with "traditional" GIS functionality and a data analysis module are integrated into a single software environment.

While the early discussions stressed conceptual frameworks and taxonomies of techniques, lately several operational implementations of these ideas have been carried out as well. Early such efforts followed what Anselin et al. (1993) call a *one-directional* form of integration, in which data from a GIS are efficiently transferred to a statistical system for analysis, or results from a statistical package are moved to the GIS for mapping and visualization. Such attempts typically follow the loose-coupling or modular paradigm. Early illustrative examples of this approach, among many others, are the joint use of the Grass GIS with the S statistical package to carry out exploratory data analysis in Farley et al. (1990), and Williams et al. (1990), and the use of Poisson regression results from Glim as a basis for areal interpolation in the Arc/Info GIS in Flowerdew and Green (1991). More recently, a close coupling approach was outlined in Symanzik et al. (1994), in which information from the Arc/Info GIS is efficiently passed to the XGobi software environment for exploring multivariate data by means of dynamic graphics, brushing, linking and the grand tour [for a recent review of the XGobi environment, see Buja et al. (1996)]. However, while sophisticated in its implementation, the link remained one-directional, in the sense that all analysis and further data visualization is carried out in XGobi and the results are not passed back to the GIS.

A *dynamic* integration necessitates a software environment in which spatial data and the results of spatial data analyses efficiently move back and forth between the GIS and a statistical module to allow a truly interactive data analysis [Bailey and Gatrell (1995), Anselin (1997a)]. A prototype for such a dynamic integration was suggested in Anselin et al. (1993), consisting of Arc/Info as the GIS module and SpaceStat [Anselin (1992, 1995a)] as the spatial data analysis module. This implementation was based on a loose coupling or modular approach, in which data (or results) are passed between the two modules using auxiliary files with standardized file names, to minimize the amount of user intervention in the process. A similar framework is outlined in Zhang et al. (1994), where four software modules are linked in a loose fashion across both workstation and PC platforms, to allow GIS processing (in Arc/Info), cluster analysis (Mclust-Plus), traditional exploratory data analysis (XGobi) as well as spatial data analysis (SpaceStat).

Most recent implementations of dynamic integration have not been based on loose coupling, however, but have focused instead on extending the functionality of existing commercial GIS with spatial data analysis routines, by taking advantage of the system's macro or scripting language facilities (e.g., AML for Arc/Info) as well as by calling pre-compiled functions. This approach is fully integrated within the GIS user interface and typically hides the linked nature of the analysis routines from the user. Recent examples are the extension of Arc/Info with functions for the implementation of non-spatial EDA tools, such as scatterplots [e.g., Batty and Xie (1994)], as well as routines for the computation of global and local indicators of spatial autocorrelation [Ding and Fotheringham (1992), Bao et al. (1995), Can (1996)]. A well-recognized drawback of this integrated approach is a performance penalty, since the macro or scripting languages are not optimized to handle the computations necessary for spatial statistical analysis, resulting in a loss of speed and/or limitations on the size of the problems that can be handled. An alternative to this that retains the close-coupling paradigm is the linkage between two software packages that allow remote procedure calls (in unix) or dynamic data exchange (in a Windows environment). This approach is taken in the only commercial implementation that exists to date of an integrated data analysis and GIS environment, the S+Gislink between the S-Plus statistical software and the Arc/Info GIS. In the S+Gislink add-on to the S-Plus software for unix workstations, a bi-directional link is established with the Arc/Info GIS that allows data to be imported and exported between the two packages in the native data formats (i.e., the internal data frame format in S-Plus and the Info data table in Arc/Info). In addition, the linkage allows users to call S-Plus functions from within Arc/Info, although the reverse is not possible [MathSoft (1996a)]. A similar perspective is taken in the recent efforts at the Statistics Laboratory of Iowa State University to implement a linked environment between ArcView and XGobi on unix workstations, focused on the use of XGobi for exploratory data analysis, such as the visualization and brushing of scatterplots and cumulative distribution functions [e.g., Majure et al. (1996), Symanzik et al. (1996), Cook et al. (1996), Majure and Cressie (1997)].

In this chapter, we extend and further operationalize the framework of Anselin et al. (1993), using the ArcView GIS [ESRI (1995a)] in a Windows environment as the visualization engine, and Spacestat as the spatial data analysis engine. Our approach differs from the efforts described above in three respects: (a) it has an explicit focus on the so-called lattice perspective towards spatial data analysis, in contrast to the geostatistical viewpoint that currently dominates in the literature; (b) it uses the GIS (ArcView) for the visualization of the statistical results, not the statistical software (compared to the use of respectively S-Plus and XGobi for visualization in the S+Gislink and Iowa State implementations); and (c) it is primarily targeted at a PC platform, although in the right network environment, it could be implemented in unix as well (ArcView exists for both unix and Windows platforms)

In terms of methodology, our emphasis is on tools for exploratory *spatial* data analysis (ESDA) in which the spatial dependence of the data is taken into account explicitly. This contrasts with most other implementations of linked EDA and GIS, where traditional EDA tools such as box plots, histograms and scatterplots are the focus of attention and the complicating effects of spatial autocorrelation are typically ignored [e.g., MathSoft (1996b, Chapter 3)].

In the remainder of the chapter, we first define our vision of ESDA and outline the types of techniques that are central to it. This is followed by a more detailed conceptual overview of the linkage between spatial data analysis and GIS and its implementation for ArcView and SpaceStat. We illustrate this framework with a few examples in which the spatial pattern of housing values in West Virginia counties is examined. We close with some remarks on future directions.

3.2 Exploratory Spatial Data Analysis

Exploratory Data Analysis, or EDA, has become increasingly popular as a methodology to generate insight into patterns and associations in data (especially large data sets), without strong prior assumptions and taking into account the potentially misleading influence generated by "extreme" or "atypical" observations. Since the pioneering work of Tukey (1977), EDA has gained considerable influence as a paradigm in applied statistics and it now forms the basis for many of the visualization and graphical features of modern statistical software [Good (1983), Cleveland (1993), Venables and Ripley (1994)]. However, none of the traditional tools of EDA are especially geared to dealing with spatial data, in the sense that the effects of location, spatial dependence and spatial heterogeneity are ignored. Moreover, many EDA techniques for the initial exploration of bivariate and multivariate relationships, such as smoothed scatterplots, may yield indications that are invalid in the presence of spatial autocorrelation [Anselin (1990), Anselin and Getis (1992)]. In contrast, Exploratory *Spatial* Data Analysis (ESDA), focuses explicitly on these spatial effects and consists of techniques to describe spatial distributions, identify atypical locations (spatial outliers), discover patterns of spatial association (spatial clustering) and suggest different spatial regimes or other forms of spatial instability (spatial non-stationarity) [Anselin (1994)]. We follow Anselin (1994) in drawing the distinction between true ESDA and so-called *spatialized* EDA, in which standard EDA features are displayed at particular locations on a map, or shown as a function of a given distance metric. Familiar examples of the latter are the mapping of Chernoff faces in two-dimensional space [e.g., Haining (1990, p. 226), Fotheringham and Charlton (1994)], or the display of box plots by distance bands [Haining (1990, p. 212, 224)].

Central to ESDA is the concept of spatial association or spatial autocorrelation. In this respect, it is important to draw a distinction between methods appropriate for the geostatistical or distance-based perspective on the one hand and those geared to the lattice or neighborhood view of spatial data on the other hand [see Cressie (1993), Anselin (1994)]. Our focus will be on the latter, in which data are observed for a given discrete set of fixed locations. In contrast to the geostatistical perspective, these locations are not considered to form a sample of an underlying continuous distribution, but the data are conceptualized as a single realization of a spatial stochastic process, similar to the approach taken in time series analysis. In this view, spatial interaction is conceptualized as a step function, where a location interacts with a given set of neighbors. The overall interaction (or covariance) in the observed data is then obtained by imposing (assuming) a particular form for the spatial stochastic process. This approach requires the formal expression of a neighborhood structure for each observation (i.e., the topology or spatial arrangement of the data) in the form of a spatial weights matrix **W**. In this matrix, nonzero elements indicate the presence of a neighbor relationship, which may be expressed as a simple binary variable (e.g., $w_{ij} = 1$) or may take on a more general form, as a prior for the strength of interaction between observation i and its neighbor j. By convention, the diagonal elements of the weights matrix (w_{ii}) are set to zero [for a more extensive discussion, see, e.g., Cliff and Ord (1981), Upton and Fingleton (1985), Anselin (1988)]. For realistically sized problems, the construction of such a weights matrix cannot be carried out by visual inspection of a map and must rely on the data structures present in a GIS [see, e.g., Anselin (1995a)].

In the traditional approach to spatial autocorrelation, the overall pattern of dependence in the data is summarized into a single indicator, such as the familiar Moran's I, Geary's c or Gamma indicators of spatial association [for details, see, e.g., Cliff and Ord (1981), Upton and Fingleton (1985), Haining (1990)]. We refer to this as *global* spatial autocorrelation, in contrast to *local* indicators of spatial association (LISA) which we consider below. The various global measures of spatial association can be used to assess the range of spatial interaction in the data and can be easily visualized by means of a spatial correlogram (a series of spatial autocorrelation measures for different orders of contiguity) [see, e.g., Oden (1984), and for a recent application, Lam et al. (1996)]. A major drawback of global indicators of spatial association is that they are based on an assumption of spatial stationarity, which among other requirements necessitates a constant mean (no spatial drift) and constant variance (no outliers) across space. While this may have been useful in the analysis of small data sets, such as in the classic example of 26 Irish counties in Cliff and Ord (1981), it is not very meaningful or may even be highly misleading in analyses of spatial association for hundreds or thousands of spatial units that characterize current

GIS applications. In addition, most of these global measures of spatial association were developed in an era of scarce computing power, small data sets and minimal computer graphics, and their implementation takes only limited advantage (if at all) of the data storage, retrieval and visualization capabilities embodied in a modern GIS. Hence, rather than stressing these global statistics as useful additions to the analytical capabilities of a GIS [as in Griffith (1993)], we see the main contribution of ESDA techniques in measuring and displaying *local* patterns of spatial association, indicating local non-stationarity and discovering islands of spatial heterogeneity [Anselin (1994)].

In the subsections below, we elaborate on three classes of ESDA techniques that are implemented in the linked SpaceStat-ArcView framework outlined in section 3: the description and visualization of spatial distributions; the visualization of global spatial association and detection of spatial non-stationarity; and local indicators of spatial association. The illustration of these techniques is deferred to section 4.

3.2.1 Describing Spatial Distributions

The description of spatial distributions has become increasingly integrated within the interactive and dynamic visualization techniques of EDA, such as scatterplot brushing [Becker and Cleveland (1987)] and plot windows [Stuetzle (1987)], in which multiple *views* of the data (such as tables, charts and plots) are presented simultaneously. The views are shown in different windows on a computer screen, and are dynamically linked in the sense that when a location in any one of the windows (e.g., a bar on a bar chart, or a set of points in a plot) is selected by means of a pointing device (so-called brushing), the corresponding locations in the other windows are highlighted. This allows a highly interactive approach to data analysis, which is particularly effective in detecting unexpected patterns in high-dimensional data [Buja et al. (1991, 1996)].

While geographic locations have always played an important role in dynamic graphics [e.g., the various examples in Cleveland and McGill (1988)], it is only recently that the map was introduced explicitly as an additional view on the data, e.g., in Monmonier (1989), Haslett et al. (1990) and MacDougall (1991). Particularly in the Spider-Regard software tools of Haslett, Unwin and associates, the distribution of data in a spatial subset of observations can be effectively visualized by means of a linked map, histogram and box plot [for a recent overview, see Unwin (1994)]. In other words, for any subset of locations highlighted on the map, the corresponding distribution of the data is highlighted in a histogram and/or box plot and can thereby be contrasted to the overall distribution. Also, for any subset of data highlighted in a non-spatial view, such as the histogram, the corresponding locations are highlighted on the map.

A slightly different approach towards visualizing the distribution of spatial data is by means of a spatial cumulative distribution function (SCDF). In the implementation of this idea in Majure et al. (1996) and Cook et al. (1996), a continuous density function is estimated for all observations in a given region. By linking a map in ArcView and the SCDF plot in XGobi, it becomes possible to highlight subregions of the data and to find the corresponding portion of the SCDF plot, and vice versa.

When spatial data pertain to aggregate areal units, it is straightforward to visualize the distribution of a variable by means of a *quantile map*, i.e., a choropleth map in which each color corresponds to observations within a given quantile of the spatial distribution. For a quartile map (four quantiles), this matches the grouping of observations in a box plot, except that the latter also indicates outliers in the distribution as locations outside the "fences" in the plot [see Cleveland (1993, pp. 25–27)]. It is straightforward to construct an equivalent device in the form of a *box map*, in which the quartile map is augmented with highlighted outliers (e.g., in different colors for the lower and upper outliers). The comparison of box maps for different variables provides an initial look at potential multivariate associations, in the sense that maps with matching quartiles (and outliers) are likely to correspond to correlated variables [e.g., Talen (1997)].

In terms of the integration of GIS with spatial data analysis to describe spatial distributions, it becomes clear that the capacity to dynamically link different views of the data is crucial. This is further explored in section 3 and illustrated in section 4.

3.2.2 Visualizing Patterns of Spatial Association

The typical approach towards visualizing spatial dependence taken in the literature is based on the geostatistical perspective and uses the variogram as a parameter of spatial association. The basic tools are outlined in Cressie (1993) and include a *variogram cloud* (a scatterplot of squared differences between pairs of observations, sorted by distance band), a *variogram box plot* (a box plot of the variogram cloud for each distance band) and a *spatial lag scatterplot* (a scatterplot for a given distance band where the horizontal axis corresponds to the value at each location and the vertical axis shows the corresponding value at the spatially lagged locations). The variogram cloud is integrated with the dynamically linked windows in the Spider-Regard software system to investigate patterns of spatial association in subsets of the data (e.g., by brushing points in the variogram cloud, the corresponding locations on the map are highlighted), with a special focus on detecting local "pockets" of spatial non-stationarity [see, e.g., Haslett et al. (1991), Haslett (1992), Bradley and Haslett (1992), Haslett and Power (1995)]. A recent extension of the use of the spatial lag scatterplot to a multivariate setting is given in Majure and Cressie (1997).

When data consist of aggregate areal units, i.e., when the lattice or neighborhood view is taken, the variogram is less meaningful as a device to model spatial dependence, since it relies on the assumption of an underlying continuous spatial process. Instead, a crucial role in the visualization of spatial association is played by the concept of a *spatial lag*. In Anselin (1988), this is defined as a weighted average of the values observed for the neighbors of a given location, where the weights are taken from a spatial weights matrix. More formally, when w_{ij} are row-standardized spatial weights (i.e., such that $\Sigma_j w_{ij} = 1$), the spatial lag for y_i would be $\Sigma_j w_{ij}.y_j$. The structure of the weights matrix ensures that only those values of y_j are taken into account for which the locations *j* are neighbors to *i* (since $w_{ij} = 0$ for locations that are not neighbors). In matrix notation, the vector of observations on a spatially lagged variable consists of the product of the spatial weights matrix with the vector of observations, **W.y**.

The matching of a value observed at each location with its spatial lag provides useful insight into the local as well as the global pattern of spatial association in the data. Specifically, when a high degree of positive local spatial autocorrelation is present, the observed value at a location and its spatial lag will tend to be similar. In the extreme case, the value at a location would be predicted exactly by the observed values in the neighboring locations (implying a spatial autoregressive coefficient of 1). In a global sense, spatial clusters would be indicated by subsets of locations with great similarity between the values observed and their spatial lags. In the opposite situation, when a high degree of negative spatial autocorrelation is evident, low values at a location would tend to be surrounded by higher values for the neighbors (i.e., a higher weighted average for the neighbors compared to the value at the location), or high values would tend to be surrounded by lower values for the neighbors. If the magnitude of the difference is sufficient, both instances could be classified as spatial "outliers." The presence of many spatial outliers intermixed with an overall pattern of positive spatial association may provide evidence of local non-stationarity.

The association between a variable and its spatial lag is easily visualized by means of so-called *spatial lag pies* and *spatial lag bar charts* [Anselin et al. (1993), Anselin (1994)]. The former are only appropriate when the observed values are all strictly positive, since they rely on the division of a circle into two pies, each of which is proportional to the relative share of the variable and its spatial lag in their sum. More precisely, the share corresponding to each pie is $y_i/[y_i + (Wy)_i]$ and $(Wy)_i/[y_i + (Wy)_i]$. Hence, for $y_i = (Wy)_i$, each pie would equal half the circle. Negative spatial autocorrelation for high values would be indicated by the dominance of the pie share corresponding to y_i and vice versa. The circles themselves may be graduated, to indicate the overall magnitude of the value observed at each location [for an illustration, see Anselin et al. (1993)]. In a spatial bar chart, the graph simply consists of two bars at each location, one of which corresponds to y_i, the other to $(Wy)_i$. Other visualization schemes are possible as well, for example, based on the difference, absolute difference, squared difference or ratio between y_i and $(Wy)_i$ at each location. The spatial lag bar chart is illustrated in section 4.

A more quantified approach towards visualizing local and global spatial association is based on the concept of a *Moran scatterplot* [Anselin (1995b, 1996)]. It follows from the interpretation of the familiar Moran's *I* statistic for spatial autocorrelation as a regression coefficient in a bivariate spatial lag scatterplot. More precisely, for a row-standardized weights matrix, the normalizing constants in Moran's *I* cancel out and the statistic reduces to $I = z'Wz / z'z$, with z as the vector of observations in deviations from the mean. This is the slope coefficient in a regression of Wz on z. The interpretation of Moran's *I* as a regression coefficient has three interesting implications. First, the statistic can easily be visualized as the slope of a straight line in a scatterplot, which is especially insightful when the pattern of spatial association is evaluated for different variables, or for the same variable over time [for an example, see O'Loughlin and Anselin (1996)]. Secondly, since the statistic is obtained from least squares estimation, the usual battery of diagnostics may be applied to identify outliers and observations with high leverage or influence on the slope. This often provides an intuitive check for border effects and other potential consequences of a poorly specified spatial weights matrix [for technical details, see Anselin (1996)]. Thirdly, the extent to which more general nonlinear scatterplot smoothers, such as a loess regression, provide a different fit than the linear regression may suggest spatial regimes and other forms of non-stationarity.

A second use of the Moran scatterplot is as a device to decompose the global spatial autocorrelation statistic into four types of association. These four types correspond to the four quadrants in the scatterplot when the variable z is normalized, such that its mean is zero and its standard deviation equals one. The upper right and lower left quadrants correspond to instances of positive spatial association [in the traditional sense of Cliff and Ord (1981)], i.e., the presence of similar values in neighboring locations. For the upper right quadrant, this association is between values of z_i above the mean whose spatial lag is also above the mean, and in the lower left quadrant, these are values of z_i below the mean whose spatial lag is also below the mean. The other two quadrants correspond to negative spatial association, values of z_i below the mean whose spatial lag is above the mean (upper left quadrant), and values of z_i above the mean whose spatial lag is below the mean (lower right quadrant). Both of these could indicate potential spatial outliers, provided that the values involved are "extreme" enough. The latter is easy to assess, since the scatterplot is constructed for standardized variates. Consequently, two units on the scatterplot correspond with two standard deviations from the mean, which can be used to identify outliers as those points outside a circle with radius equal to two that is centered on the origin. The decomposition of the global spatial association into the four quadrants of the Moran scatterplot can also be mapped in a straightforward fashion, with each quadrant corresponding to a different color or shading on the map (possibly only for those points identified as outliers). The resulting *Moran scatterplot map* is illustrated in section 3.4.

3.2.3 Local Indicators of Spatial Association

More recently, in part to address the need to develop techniques to analyze the large spatial data bases that are becoming increasingly available, attention has focused on *local* indicators of spatial association, or *LISA* [Getis and Ord (1992), Anselin (1995b), Ord and Getis (1995), Bao and Henry (1996), Unwin (1996)]. Following the definition of Anselin (1995b), a LISA is an indicator that achieves two objectives: (a) it allows for the detection of significant patterns of local spatial association (i.e., association around an individual location, such as hot spots and spatial outliers); and (b) it can be used as a diagnostic for stability of a global diagnostic (i.e., to assess the extent to which the global pattern of association is reflected uniformly throughout the data set). Not all local statistics suggested in the literature fit the two requirements. For example, the G_i and G_i* statistics of Getis and Ord (1992) primarily satisfy the first objective, while the Moran scatterplot discussed in the previous section is geared to the detection of local "pockets" of non-stationarity in the computation of Moran's *I*.

The first indicators of local spatial association that gained wide acceptance were the distance-based G_i and G_i^* statistics of Getis and Ord (1992). These indicators can be computed for each location in the data set as the ratio of the sum of values in neighboring locations (defined to be within a given distance band or order of contiguity) to the sum over all the values. The two statistics differ with respect to the inclusion of the value observed at *i* in the calculation (not included for G_i). Formally, a G_i statistic is thus $\Sigma_j w_{ij}(d)y_j / \Sigma_j y_j$, with $w_{ij}(d)$ as a distance-based binary spatial weights matrix. Locations with a statistically significant G_i or G_i^* statistic can easily be mapped and used in an exploratory analysis to detect hot spots or spatial clusters [e.g., Ding and Fotheringham (1992), Anselin et al. (1993), Bao et al. (1995), O'Loughlin and Anselin (1996), Unwin (1996)]. The interpretation of the statistics is slightly different from standard practice in spatial autocorrelation analysis, in the sense that a negative statistic points to association between similar *small* values (and not to association between dissimilar values). Consequently, the G_i and G_i^* statistics measure association in the upper right and lower left quadrants of the Moran scatterplot, but not in the others. A global Moran's *I* statistic can be expressed as a weighted average of G_i^* statistics [Ord and Getis (1995)].

The local Moran statistic outlined in Anselin (1995b) satisfies both criteria for a LISA and can thus effectively be used to identify both local spatial clusters as well as to assess the stability of the global Moran's *I*. The former use is more appropriate when no global association is found in the data, while the latter is more appropriate in the opposite case. When local Moran statistics are computed in the presence of global spatial autocorrelation, their significance must be interpreted with caution [see Anselin (1995b), Ord and Getis (1995), for technical details] although they remain useful as an exploratory technique. Formally, the local Moran statistic for observation *i* is $z_i[\Sigma_j w_{ij}.z_j]$, where w_{ij} are the elements of a spatial weights matrix and the z_i are standardized variates. Significance of the local Moran can be derived analytically under the null hypothesis of no spatial association, or by means of a conditional permutation approach [for technical details, see Anselin (1995b), Bao and Henry (1996)]. The local Moran is particularly useful as an ESDA tool when used in conjunction with a Moran scatterplot. Specifically, the quadrant in a Moran scatterplot indicates what type of spatial relationship is found for locations with significant local Moran statistics. The mapping of significant LISA statistics in a GIS, together with a Moran scattermap and/or overlays of maps for other variables of interest provides the basis for a substantive interpretation of spatial clusters or spatial outliers [for examples, see Barkley et al. (1995), Talen and Anselin (1997)]. This is further illustrated in section 4.

3.3 Linking Exploratory Spatial Data Analysis and GIS

The linkage between exploratory spatial data analysis and GIS considered here builds upon the general framework outlined in Anselin and Getis (1992). Following the

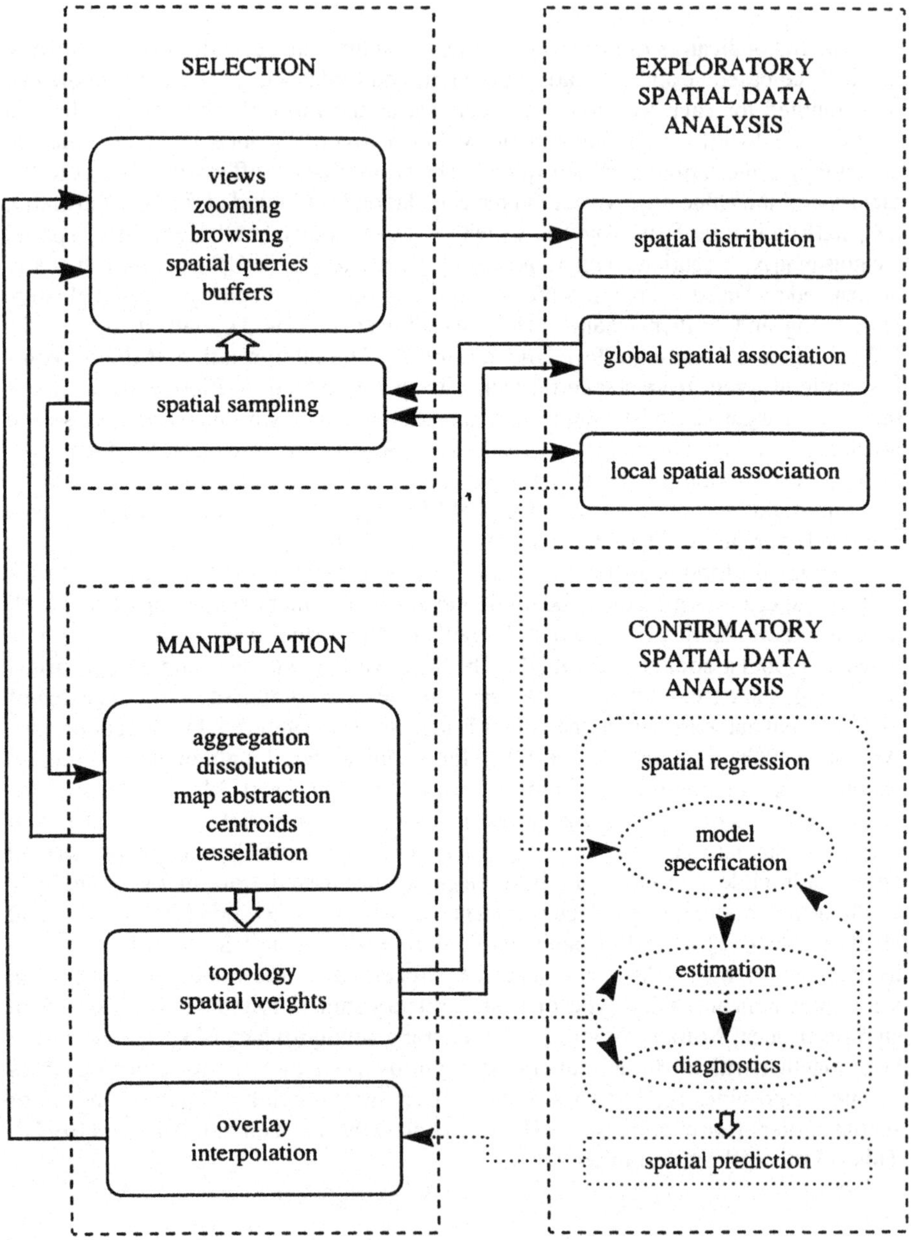

Fig. 3.1. Integration of GIS Module with Spatial Data Analysis Module

usual classification of GIS functionality into four broad groups — input, storage, analysis and output — they further subdivided the analysis function into *selection, manipulation, exploration* and *confirmation*. In Anselin et al. (1993), the first two of these functions were considered to form a "GIS module" while the latter two formed a "Data Analysis module" to emphasize the practical division of labor between typical commercial GIS software and the specialized (add-on) software needed to carry out spatial data analysis. This distinction has become increasingly tenuous, since many statistical software packages now have some form of mapping (or even GIS) functionality, and a growing number of statistical functions are included in GIS software. More important than classifying these functions as belonging to one or the other module is to stress their interaction and the types of information that must be interchanged between them.

A schematic overview of the linkages between the four functions is given in Figure 3.1, with an emphasis on the links between the selection and manipulation functions and exploratory spatial data analysis [for a more extensive discussion of the various linkages, see also Anselin (1997b)]. While many taxonomies are possible, the main point of the classification in Figure 3.1 is that selection and manipulation (shown on the left) are present in virtually all systems and have become known as "spatial analysis" in the commercial world [e.g., ESRI (1995b, Lesson 8)]. In contrast, the spatial *data* analysis functions (shown on the right) are much less prevalent in commercial systems.

The selection functions include operators necessary to obtain the values (attributes) of a set of variables for particular locations in a spatial data base. This ranges from simple zooming and browsing functions and traditional relational data base queries to spatial queries, buffering and spatial selection. This is the starting point for any ESDA and is particularly relevant for the computation and display of the distribution of values over a given spatial subset of locations (spatial density functions, box map). Related to this is a spatial sampling function, in which a subset of locations is selected to represent a spatial "population" in further statistical analysis. Due to the prevalence of spatial autocorrelation, standard random sampling techniques may not be appropriate for spatial data sets. In order to carry out proper spatial sampling, often an initial data analysis is needed to assess the range and significance of spatial autocorrelation. This would be the result of an exploratory spatial data analysis, as illustrated by the link between global spatial association and spatial sampling in Figure 3.1.

The manipulation function includes all operations to "create spatial data." The virtual limitless ability of GIS to produce maps of data at any scale and for any level of areal aggregation is often seen as its most powerful "analysis" feature. However, though typically hidden from the user, such operations are themselves based on specific functions, algorithms and models and often involve a prior statistical sampling and/or analysis of the data. The data manipulation operations can be broadly classified into three groups. The first contains those pertaining only to attribute values, i.e., traditional data summaries and transformations (aggregation, averaging, etc.). A second group consists of those operations that pertain only to spatial information, i.e., a manipulation of the coordinates of the points, lines and polygons in a spatial data base to perform spatial transformations, map abstraction, spatial aggregation and dissolution, the computation of topology (determination of neighbors), centroids, area, perimeter, etc. The most important aspect of this second group for data analysis is the construction of topology or spatial arrangement for a set of areal units. This information is crucial for the computation of any statistic for spatial autocorrelation, for which a spatial weights matrix or spatial lag operator is essential. As pointed out earlier, the spatial lag is a central element in the visualization and exploration of both global and local spatial association (in the ESDA module). Finally, a third group of functions combines both spatial and non-spatial information and is commonly referred to as "data integration." This capability allows for the construction of "data" for a particular unit of analysis by combining information on different variables and at different levels of spatial aggregation by means of polygon overlay and spatial interpolation operations. The flexibility to move between different levels of spatial aggregation, to relate multiple variables in a spatial data base and to interactively select subsets of observations provides a powerful platform to carry out ESDA. The particular implementation of these linkages between the ArcView GIS and the SpaceStat spatial data analysis software are considered in more detail next.

3.3.1 Linking SpaceStat and ArcView

The principle behind the linkage of ArcView (Version 2.1) and SpaceStat (Version 1.80) to facilitate exploratory spatial data analysis is the transfer of *spatial information* from ArcView to SpaceStat for analysis, and the transfer *of location-specific results* from SpaceStat to ArcView for visualization. The spatial information consists of both location and topology (spatial arrangement) of the selected data points (or areal units). The particular ESDA results considered here include spatial lags (for spatial lag pies and spatial lag bar charts), quartiles and outliers (for a box map), the quadrants in a Moran scatterplot (for a Moran scatter map), and the significant LISA and G_i statistics. However, any statistic that has a value assigned to each location in the data set (e.g., observations, regression residuals and predicted values) can be efficiently passed back from SpaceStat to ArcView for visualization using the same principles as outlined in what follows.

The division of labor implemented here is different from the approach taken in Majure et al. (1996), Symanzik et al. (1996) and Cook et al. (1996), who also link ArcView with a statistical package (XGobi). There, ArcView is primarily used to select spatial subsets of locations and the visualization is an output of the XGobi software. The primary aspect of their linkage is to highlight "brushed" data points in both the map (in ArcView) and the various types of scatterplots in XGobi. However, since their ESDA is based on a geostatistical perspective, the topology of the data embedded in the GIS does not need to be exploited. In contrast, this is a central element in the ESDA carried out in SpaceStat, which is based on the use of spatial weights and spatial lags.

The linkage between ArcView and SpaceStat is not dynamic, but based on a loose coupling approach. In part, this is necessitated by the fact that SpaceStat (still) runs under Dos, whereas ArcView is a 32 bit Windows product. While the new MS

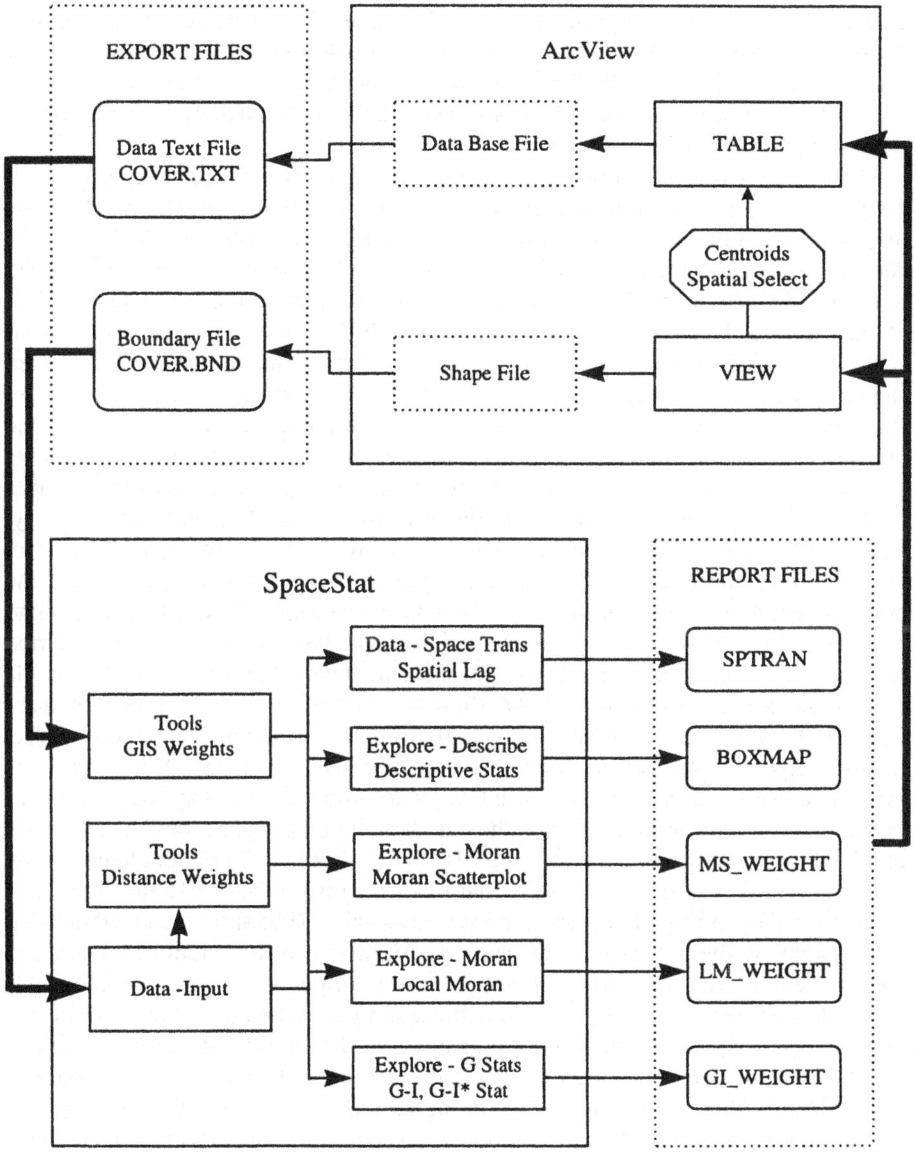

Fig. 3.2. Linkage of SpaceStat and ArcView

Windows platforms (Windows NT and Windows 95) allow SpaceStat to multi-task with ArcView and run "simultaneously" in a separate window, there is no mechanism to call internal SpaceStat functions from within ArcView, the way this is possible through DDE (dynamic data exchange) and/or OLE (object linking and embedding) between true windows programs (in unix, the same effect is achieved by means of remote procedure calls). However, the amount of additional work is minimized by exploiting the structure of the files used in the data exchange between ArcView and SpaceStat, and in most instances it is reduced to a single click on a menu item or windows toolbar button.

A schematic overview of the linkage between ArcView and SpaceStat is given in Figure 3.2. The typical point of departure is ArcView (in the upper right hand corner) in which both a map view (the "View") and a tabular view (the "Table") of spatial data is standard. These standard features can be supplemented by a few specialized functions that operate on the View and extract or construct spatial "variables" (fields) for addition to the attribute table associated with the View. Specifically, "spatial selection" creates an indicator variable that takes on a value of one for the selected spatial units. The selection itself can be carried out interactively with the map, using the standard ArcView selection tools (or some additional tools that can easily be constructed, see section 3.2). The resulting indicator variable can be used to create subsets of data sets and spatial weights in SpaceStat, or form the basis for a spatial analysis of variance or for spatial regimes in the study of spatial heterogeneity. Another pair of important spatial attributes of a view are the x and y coordinates that correspond to the centroids of the (selected) areal units. These coordinates are the point of departure for the construction of distance-based spatial weights in SpaceStat.

Every Table in ArcView is implemented as a data base file that can be converted to an ascii text format in a straightforward manner. This is the first type of export file that will be used to link the data in ArcView to a SpaceStat data set (upper left hand corner in Figure 3.2). The second type contains the information needed to construct the topology or spatial arrangement of the areal units in the View. While ArcView is not a "topological" GIS, in the sense that the left-right polygon topology of the arcs in the map is not recorded explicitly, the data format of the "shape files" is public and can be exploited to build the topology [ESRI (1995c)]. This is implemented by means of an existing SpaceStat utility that reads the binary shape file and converts it to a standard boundary file in ascii format [see Anselin (1995a) for technical details].

In SpaceStat, the two export files from ArcView are converted into data files and spatial weights by means of the functions Data - Input and Tools - GIS Weights (lower left corner of Figure 3.2). Once the x and y coordinates of the centroids are contained in a SpaceStat data set, they can be used to build distance-based spatial weights with the corresponding function in the SpaceStat Tools module. The spatial weights can further be row-standardized, higher order contiguity can be constructed and several other manipulations can be carried out in the SpaceStat Tools module. The data and spatial weights are used in a wide range of exploratory and confirmatory spatial data analyses in SpaceStat [for a detailed description, see Anselin (1992, 1995a)]. Specifically, one data transformation and four types of exploratory analyses can be efficiently linked with ArcView by means of a SpaceStat Report File: spatial lag transformation for visualization of spatial autocorrelation (Data - Space Trans,

Spatial Lag), quartiles with outliers (Explore - Describe, Descriptive Stats), Moran scatterplot (Explore - Moran, Moran Scatterplot), LISA-local Moran (Explore - Moran, Local Moran), and G_i and G_i^* statistics (Explore - G Stats, G_i or G_i^*). The Report File is a comma delimited ascii file with a distinctive file name and containing the values of an indicator variable as its first column (lower right hand corner of Figure 3.2). This indicator variable is exploited to join the Report File with the attribute table of an active View in ArcView, with minimal user input. Once the Report Files are joined, the relevant variables in them can be visualized in the View in a direct manner.

While the data input and report file output are standard features of SpaceStat, the data export and table join aspects of the linkage in ArcView require some customization of the software. We next turn to the specifics of the operational implementation of the interface in ArcView.

3.3.2 Operational Implementation of the Interface in ArcView

As illustrated in Figure 3.2, the essence of the linkage between ArcView and SpaceStat consists of the creation of export files from attribute tables, and the importing and joining of SpaceStat Report files with existing attribute tables. Both of these tasks are carried out within ArcView. They are implemented by means of a collection of customized scripts in the Avenue object oriented macro language [ESRI (1994)], and by the extension of the standard ArcView user interface with two additional menus and a few extra buttons and tools [the collection of scripts is available as the SpaceStat.apr project file from the SpaceStat web site at http://spacestat.rri.wvu.edu; for a technical description, see Anselin and Bao (1996)].

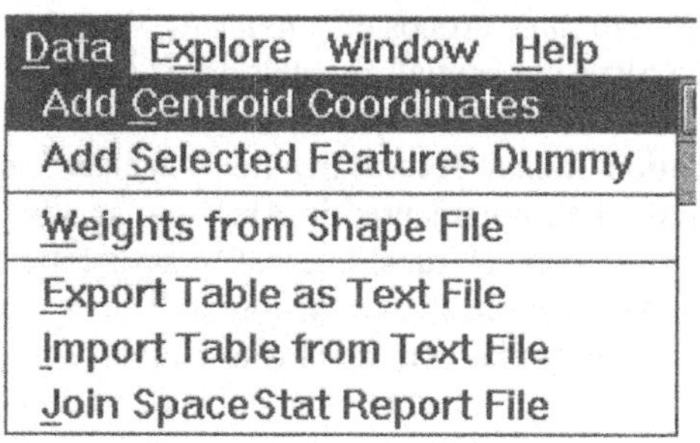

Fig. 3.3. Contents of Data Menu

The menus, buttons and tools are added to the standard View interface in ArcView. The Data menu (Figure 3.3) consists of six commands divided into three categories: (a) the computation of "spatial" variables, i.e., centroid coordinates (Add Centroid Coordinates) and indicator variables for spatial selection (Add Selected Features Dummy); (b) the conversion of shape files to ascii boundary files (Weights from Shape File); and (c) the linking of data tables, either the conversion of attribute tables to text format for later input into SpaceStat (Export Table as Text File), importing any

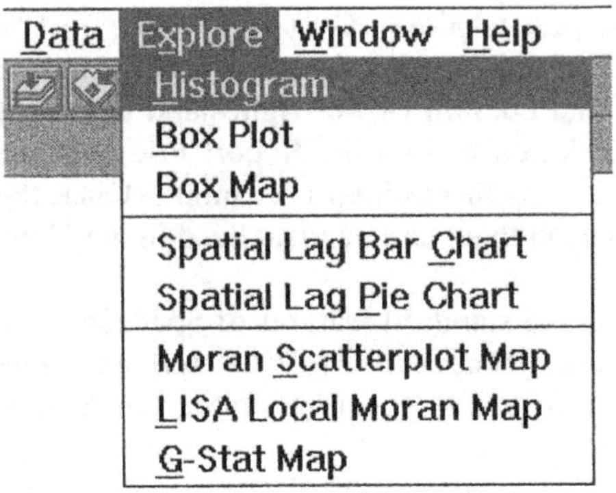

Fig. 3.4. Contents of Explore Menu

text file and adding it as an additional Table (Import Table from Text File), or adding any SpaceStat Report file and joining it with an existing attribute table (Join SpaceStat Report File). The latter is a generic join function for any Report file generated by SpaceStat and is less efficient than the implementations specific to ESDA. Most of the functions in the Data menu are streamlined versions of standard ArcView operations (e.g., adding a Table) or of existing scripts from the Avenue script library (e.g., computing centroids). The streamlining is such that any unnecessary user input is avoided, in the sense that the SpaceStat conventions for file names and file structure are imposed or assumed. The only exception to this is the Weights from Shape File item, which invokes an external utility to carry out the transformation.

The Explore menu (Figure 3.4) is divided into three groups of commands. The first group contains tools to describe and visualize the spatial distribution of the data. Both the Histogram and Box Plot commands do not interact with SpaceStat but create ArcView "Charts" to represent the distribution of data (records in a field) in an active View. The histogram is standard and must be implemented in ArcView as a bar chart. It forms a non-spatial counterpart to a so-called "Equal Interval" classification in the View Legend Editor of ArcView. More precisely, the histogram shows the number of observations that fall in each category that corresponds to a given color (value interval) in an equal interval choropleth map. An additional tool button allows the identification of a selected histogram interval on the View. This is particularly useful when the View does not correspond to an equal interval classification, e.g., when it represents a quantile map. The histogram is only constructed for the selected spatial units in the View, so that it is possible to carry out a somewhat simplistic form of dynamic linking by constructing histograms for different spatial subsets of the data. However, each spatial subset in the View will require the explicit invocation of the histogram command and will result in a new chart (i.e., the selected data are not highlighted on the histogram for the complete data set, in contrast to what would happen in a standard implementation of brushing and linking). Also, while the select tool must be clicked on a given bar of the histogram to highlight the corresponding locations in the map, the bar itself is not highlighted, in contrast to common practice.

In addition, earlier selections of spatial subsets are lost for future comparisons. These non-standard features are due to design constraints within ArcView. While they limit the extent of dynamic linking, they nevertheless provide a very useful way to visualize the spatial distribution of different subsets of the data in an interactive manner.

The Box Plot implementation is somewhat unusual in the sense that the four quartiles and the two sets of outliers (lower and upper) are visualized as a bar chart in ArcView. The height of the bars corresponding to each quartile is the same, except when outliers are present (the number of lower outliers and elements in the first quartile that are not outliers sum to the same total as the second and third quartile, and the same holds for the fourth quartile and upper outliers). The box plot can be constructed for any spatial subset of the data and the areal units corresponding to any bar can be highlighted in the View, in the same manner as for the histogram. In addition, a "graphic" is added to the View that provides a more traditional depiction of the box plot, including both median and mean of the data. This graphic can be moved around on the View in the same way as any other graphic object, but it cannot be used for interactive data analysis due to constraints in the ArcView design.

The Box Map command constructs a box map in the active View based on a joined Report File generated by SpaceStat. Since the file name for a box map Report File is always the same (boxmap.txt), no user intervention is needed to carry out the join. The only query involved is for the name of the variable that must be mapped.

The second group of commands in the Explore menu contains two functions to visualize spatial autocorrelation, a Spatial Lag Bar Chart and a Spatial Lag Pie Chart. Both of these require a Report File from SpaceStat that contains the spatial lags for the variables of interest. As for the box map, this Report File has a fixed file name (sptran.txt) thereby avoiding the need for user interaction to carry out the join. However, the user is queried for the names of the variable and its associated spatial lag, as well as for the colors to represent them. These queries are shortened forms of the generic implementation of pie charts and bar charts as spot symbols on a View in the Avenue script library.

The last group of commands in the Explore menu deals more formally with local spatial association, in the form of a Moran Scatterplot Map (and associated Moran scatterplot as a chart), and maps highlighting the locations with significant values for the local Moran (LISA Local Moran Map) or G statistics (G-Stat Map). Each of these functions requires the input of a SpaceStat Report file with a fixed file name prefix (respectively MS , LM , or GI) followed by the name of the spatial weights file for which the statistics were constructed. Again, user interaction is limited to queries for the file name and the variable of interest. The resulting maps are so-called "unique value" maps in the sense that each color corresponds with a unique value for the variable of interest (a quadrant in the Moran scatterplot or an indicator for the significance level of the LISA statistics). The interpretation of the resulting maps is straightforward.

In addition to the two extra menus, the ArcView interface is also augmented with two buttons: one to invoke SpaceStat (which can run as a true multitasked program in Windows 95 and Windows NT), the other to invoke the Dos command window. Three extra tool buttons are provided as well: one to select bars in the histogram and bar charts as outlined above, the two others to select spatial units within a given circle or within an arbitrary polygon. These tool buttons invoke slightly customized versions of scripts from the Avenue script library and are included to provide some degree of (albeit limited) dynamic linking between the different graphs.

3.4 Illustration: The Spatial Pattern of Housing Values in West Virginia

We illustrate the linked SpaceStat-ArcView environment with an initial exploration of the spatial pattern of housing values in West Virginia counties. The data are the median value of owner-occupied housing from the 1990 U.S. Census [Summary Tape File 1C, contained on the 1994 U.S. Counties CD Rom, U.S. Department of Commerce (1994)].

The spatial distribution of the housing values is illustrated in the four views represented in Figure 3.5. In the upper left corner is a familiar quintile map, to which the two histograms on the right are linked. The histogram on the bottom is for all West Virginia counties, the one on top only for those counties that border another

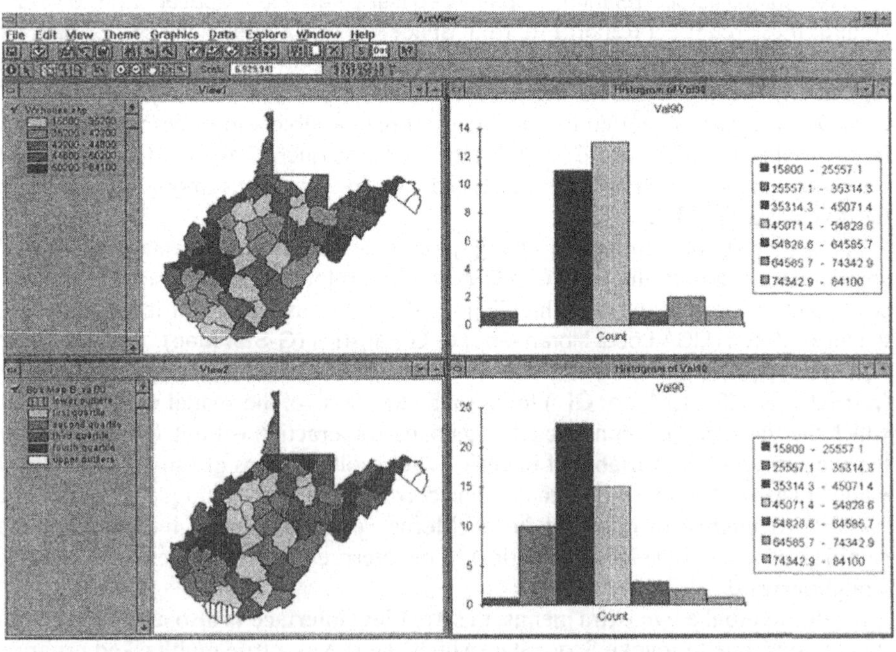

Fig. 3.5. Visualizing the Spatial Distribution of West Virginia Housing Values

state (i.e., the counties on the outer rim of the state). The view in the bottom left is a box map for the housing values. The median value for the whole state is \$44,000, which falls in the interval that corresponds with the third bar in the histogram, illustrating the skewed nature of the distribution. In View 1, the three counties with the highest values are highlighted. This is obtained by clicking on the highest categories in the histogram on the bottom. Two of the counties are in the so-called Eastern Panhandle (Berkeley and Jefferson counties) while the third is Monongalia county, the location of West Virginia University. Interestingly, the highest value (\$84,100) is for the easternmost county. The two panhandle counties are also singled out as upper outliers in the box map of View 2, but Monongalia county is not (its median of \$64,600 is well below 1.5 times the interquartile range of \$11,300 higher than the third quartile — \$49,500). The box map also reveals a lower outlier in the southern part of the state (McDowell county, with a median value of \$15,800). An initial visual inspection of the two maps may suggest a systematic difference between the values in the inner core of the state and those at the outer rim. In order to assess the extent of this (in an exploratory fashion), a histogram is constructed for the "spatial selection" of the outer rim counties (upper chart). A comparison between the

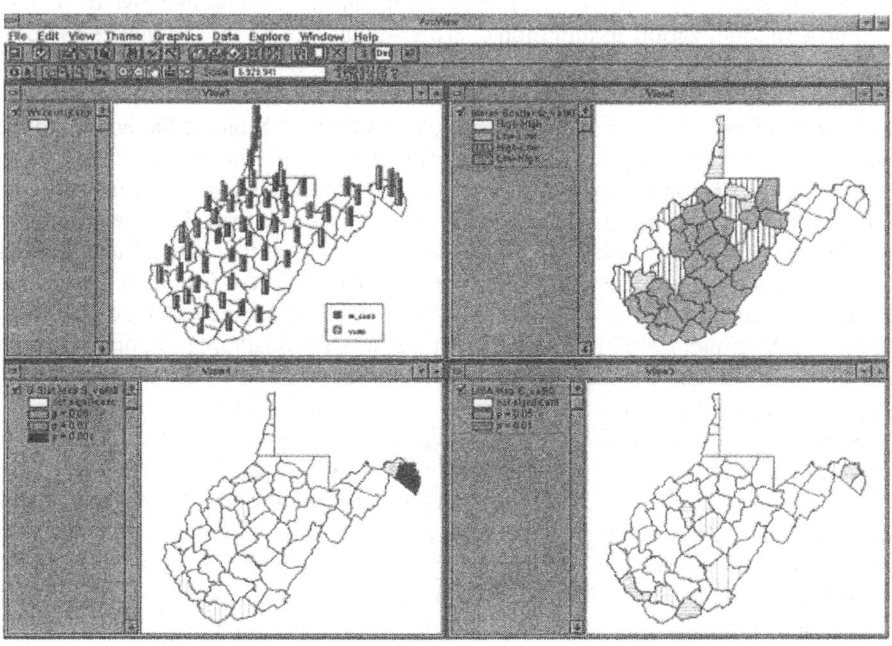

Fig. 3.6. Visualizing Spatial Association in West Virginia Housing Values

two histograms indicates a stronger representation of higher values in the outer rim [note that the vertical scales in the two charts are not the same], with almost all the counties in the highest four histogram bars from the bottom chart included in the upper chart. This overall pattern is countered by the lone outlier at the low end which corresponds with the same county that also was identified in the box map (McDowell county is an outer rim county). Consequently, our initial hypothesis may need to be refined to exclude the southern rim counties from the high "spatial regime." In substantive terms, it turns out that the spatial pattern suggested by the views in Figure 3.5 could be associated with the influence of urbanization. While West Virginia itself is a highly rural state, a number of metropolitan areas are located close to its borders (except for its southern border), the most influential of which may be the Washington D.C. area near the eastern panhandle counties. This hypothesis would need to be assessed more rigorously, for example, by carrying out an adjusted spatial selection (excluding the southern counties from the outer rim regime) and constructing an indicator variable to be used for a spatial analysis of variance in SpaceStat.

Since the results of many statistical analyses (such as an analysis of variance) are affected by the presence of spatial autocorrelation, we assess global and local indicators of spatial association in Figure 3.6. The four views correspond with a spatial lag bar chart (View 1), a Moran scatterplot map (View 2), a map of locations with significant local Moran statistics (View 3) and a map of locations with significant G_i* statistics (View 4). All measures were computed in SpaceStat, using a row-standardized first order contiguity weights matrix constructed from the shape file of the West Virginia counties. The global Moran's *I* statistic for the housing values is 0.460, corresponding to a highly significant standard normal z-value of 5.35 (using a randomization assumption). This global pattern is dominated by local indications of positive spatial association, as illustrated in Views 1 and 2. Overall, the spatial bar charts show very similar heights for each location and its spatial lag, with only a few notable spatial outliers. One of these is Monongalia county in the north, where the third largest median housing value (as indicated by the histogram in Figure 3.5) is surrounded by much lower values for the neighboring counties (i.e., a much shorter bar for the spatial lag in View 1). This is confirmed by the shading in the Moran scattermap (View 2), where Monongalia county is one of ten counties showing this pattern. However, the dominant pattern in the Moran scattermap is clearly associated with positive spatial autocorrelation between counties with low housing values, indicated by 27 counties in the low-low quadrant (13 counties are in the high-high quadrant). The smoothed picture represented by the scatterplot map suggests a cluster of low valued counties in the center and south of the state, with high valued counties in the northeast and along the western border. Only five counties are negative spatial outliers (low-high), located at the fringe of small clusters of higher valued counties. Clearly, not all these associations are "significant" and a query on the scattermap reveals that only three locations are outside the circle with radius equal to two standard deviations on the scatterplot (not illustrated here), the same three as indicated by the box map in Figure 3.5.

A more rigorous assessment of local spatial association is illustrated by the maps in View 3 and View 4 of Figure 3.6. While the local Moran and G_i* statistics measure similar types of association, they are not identical, as indicated by the slight differences in significance between the two maps [see also Anselin (1995b) for further discussion and illustration of the differences and similarities]. All 10 counties with a significant local Moran statistic (based on a conditional permutation) are associated with positive spatial association, but only three of these (the three in the eastern panhandle of the state) indicate a cluster of high housing values, which can be seen from their location in the Moran scatterplot map in View 2. Similarly, two spatial clusters of low values can be distinguished, one in the center of the state (Calhoun, Gilmer and Braxton counties) and a string of three counties in the south (Mingo, Wyoming and Mercer counties). Three fewer counties are indicated with significant G_i* statistics, but except for one difference (McDowell county instead of Wyoming county in the south) all the others overlap with the ones in the LISA map. Again, the three eastern panhandle counties show a cluster of positive association (at a higher level of significance than in the LISA map) while the others show "negative" spatial association, in the sense of indicating clusters of low values.

In sum, the initial exploration of the spatial distribution and patterns of spatial association in West Virginia housing values suggests two main conclusions. One is the potential for the presence of two spatial regimes as mentioned above. The other is the persistent indication of several border counties as "outliers" in the sense that they possibly unduly influence the rest of the analysis. On the one hand, this may suggest that these counties do not fit the same pattern as the rest of the state. On the other hand however, this could be unduly influenced by a misspecification of the spatial weights for the border counties, which in the current example ignore neighbor counties outside the state (i.e., West Virginia is considered to be an "island"). Clearly, counties in the rim outside the state could be included in the analysis, which can easily be implemented in the linked SpaceStat-ArcView framework illustrated here.

3.5 Future Directions

The current implementation of the SpaceStat-ArcView linkage for exploratory spatial data analysis served two primary purposes: (a) to illustrate the types of ESDA techniques that can effectively be integrated within a GIS environment; and (b) to examine the potential and limitations of a loose coupling framework in a realistic setting. In terms of the first objective, we have demonstrated the importance of methods that deal explicitly with both global and local spatial autocorrelation in the data. In terms of the second, the interface between the two modules by means of file import and export is clearly limited and can be much improved by a tighter coupling. Two avenues for further development present themselves. One is to move SpaceStat to a full 32 bit Windows environment and allow spatial data analysis functions to be called from within ArcView (a similar setup could be implemented in a unix environment). Another, perhaps more effective approach would be to move away from a spatial analysis module as a single piece of software and to implement selected methods as spatial data analysis tools in small self-contained software applets, which can be invoked from ArcView using the Windows DLL conventions. Possibly, this is the most effective approach, since it would allow the individual user to customize the spatial data analysis "toolbox" for each application.

ArcView turned out to be a highly effective though limited environment to implement spatial data analysis. The Avenue language scripts, while extremely flexible, result in slow execution which severely limits the scope of analysis than can be carried out. For example, for data sets with hundreds of observations, the time required to compute and draw spatial bar charts or spatial lag pies becomes prohibitive. On the other hand, the linked SpaceStat-ArcView framework applied to small to medium sized examples (a hundred to a few hundreds of observations) is a powerful platform to teach the principles of ESDA within a GIS environment. We hope it will stimulate others to pursue the further integration of *spatial* data analysis techniques into such an environment.

Acknowledgments
The research reported on in this paper was supported in part by Grant SBR-9410612 from the U.S. National Science Foundation. The research was carried out while Shuming Bao was a Visiting Scholar at the Regional Research Institute, West Virginia University. Earlier versions of the paper were presented at the 30th Anniversary Conference of the Regional Research Institute, Morgantown, WV, Nov. 6–8, 1995, and at the 92nd Annual Meeting of the Association of American Geographers, Charlotte, NC, April 9–13, 1996.

The opinions expressed in the paper are solely those of the authors and do not imply an endorsement or any other support by MathSoft, Inc.

References

Anselin L. 1988. *Spatial Econometrics: Methods and Models*, Dordrecht: Kluwer Academic

Anselin L. 1990. What is Special About Spatial Data? Alternative Perspectives on Spatial Data Analysis, in: D.A. Griffith, ed., *Spatial Statistics, Past, Present and Future*, Ann Arbor, MI:Institute of Mathematical Geography, pp. 63–77

Anselin L. 1992. *SpaceStat: A Program for the Analysis of Spatial Data*, National Center for Geographic Information and Analysis, University of California, Santa Barbara, CA

Anselin L. 1994. Exploratory Spatial Data Analysis and Geographic Information Systems, in: M. Painho ed., *New Tools for Spatial Analysis*, Luxembourg: Eurostat, pp. 45–54

Anselin L. 1995a. *SpaceStat Version 1.80 User's Guide*, Regional Research Institute, West Virginia University, Morgantown, WV

Anselin L. 1995b. Local Indicators of Spatial Association — LISA, *Geographical Analysis* **27**:93–115

Anselin L. 1996. The Moran Scatterplot as an ESDA Tool To Assess Local Instability in Spatial Association, in: M. Fischer, H. Scholten and D. Unwin, eds., *Spatial Analytical Perspectives on GIS in Environmental and Socio-Economic Sciences*, London: Taylor & Francis

Anselin L. 1997a. Interactive Techniques and Exploratory Spatial Data Analysis, in: P. Longley, M. Goodchild, D. Maguire and D. Rhind (eds.), *Geographical Information Systems: Principles, Techniques, Management and Applications*, Cambridge: Geoinformation International

Anselin L.1997b. GIS Research Infrastructure for Spatial Analysis of Real Estate Markets. *Journal of Housing Research* **8**

Anselin L. and Bao S. 1996. *SpaceStat.apr User's Guide*, Research Paper 9628, Regional Research Institute, West Virginia University, Morgantown, WV

Anselin L. and Getis A. 1992. Spatial Statistical Analysis and Geographic Information Systems, *The Annals of Regional Science* **26**:19–33

Anselin L., Dodson R. and Hudak S. 1993. Linking GIS and Spatial Data Analysis in Practice, *Geographical Systems* **1**:3–23

Bailey T. C. 1994. A Review of Statistical Spatial Analysis in Geographical Information Systems, in: S. Fotheringham and P. Rogerson (eds.), *Spatial Analysis and GIS*. London: Taylor & Francis, pp. 13–44

Bailey T. C. and Gatrell A.C. 1995. *Interactive Spatial Data Analysis*. Harlow: Longman Scientific and Technical

Bao S. and Henry M. 1996. Heterogeneity Issues in Local Measurements of Spatial Association, *Geographical Systems* **3**:1–13

Bao S., Henry M., Barkley D. and Brooks K. 1995. RAS: A Regional Analysis System Integrated with ARC/INFO, *Computers, Environment and Urban Systems* **18**: 37–56

Barkley D., Henry M., Bao S. and Brooks K. 1995. How Functional are Economic Areas? Tests for Intra-Regional Spatial Association using Spatial Data Analysis, *Papers in Regional Science* **74**:297–316

Batty M. and XieY. 1994. Modelling Inside GIS: Part I. Model Structures, Exploratory Spatial Data Analysis and Aggregation, *International Journal of Geographical Information Systems* **8**:291–307

Becker R. and Cleveland W.S. 1987. Brushing scatterplots, *Technometrics* 29:127–142

Bradley R. and Haslett J. 1992. High Interaction Diagnostics for Geostatistical Models of Spatially Referenced Data, *The Statistician* **41**:371–380

Buja A., Cook D. and Swayne D.F. 1996. Interactive High-Dimensional Data Visualization, *Journal of Computational and Graphical Statistics* **5**:78–99

Buja A., McDonald J.A. , Michalak J. and Stuetzle W. 1991. Interactive Data Visualization using Focusing and Linking, in: G.M. Nielson and L. Rosenblum, eds., *Proceedings of Visualization '91*, Los Alamitos, CA: IEEE Computer Society Press, pp. 155–162

Can A. 1996., Weight Matrices and Spatial Autocorrelation Statistics Using a Topological Vector Data Model, *International Journal of Geographical Information Systems* **10**: 1009–1017

Cleveland W. S. 1993. *Visualizing Data*. Summit, NJ: Hobart Press

Cleveland W.S. and McGill M.E. 1988. *Dynamic Graphics for Statistics*, Pacific Grove, CA: Wadsworth

Cliff A.D. and Ord J.K. 1981. *Spatial Processes: Models and Applications*. London: Pion

Cook D., Majure J., Symanzik J. and Cressie N. 1996. Dynamic Graphics in a GIS: Exploring and Analyzing Multivariate Spatial Data Using Linked Software, *Computational Statistics* **11**:467–480

Cressie N. 1993. *Statistics for Spatial Data*, New York: Wiley

Ding Y. and A. S. Fotheringham The Integration of Spatial Analysis and GIS, *Computers, Environment and Urban Systems* 16 3–19 1992.

ESRI *Avenue, Customization and Application Development for ArcView*, Redlands, CA: Environmental Systems Research Institute, 1994.

ESRI *ArcView 2.1 The Geographic Information System for Everyone* Redlands CA: Environmental Systems Research Institute 1995a.

ESRI *Understanding GIS, The ARC/INFO Method*, Redlands, CA: Environmental Systems Research Institute 1995b.

ESRI *ArcView Version 2 Shapefile Technical Description. White Paper*, Redlands, CA: Environmental Systems Research Institute, 1995c.

Farley J.A., W.F. Limp, and J. Lockhart. The Archeologist's Workbench: Integrating GIS, Remote Sensing, EDA and Database Management, in: K. Allen, F. Green and E. Zubrow

(eds.), *Interpreting Space: GIS and Archaeology*, London: Taylor & Francis, pp. 141–164, 1990.
Fischer M. M. and Nijkamp P. 1993. *Geographic Information Systems, Spatial Modelling and Policy Evaluation*, Berlin: Springer-Verlag
Fischer M.M., Scholten H. and Unwin D. 1996. *Spatial Analytical Perspectives on GIS in Environmental and Socio-Economic Sciences*, London: Taylor & Francis, 1996
Flowerdew R. and Green M. 1991. Data Integration: Statistical Methods for Transferring Data Between Zonal Systems, in: I. Masser and M. Blakemore (eds.), *Handling Geographical Information*, London: Longman, pp. 38–54
Fotheringham A. S. and Charlton M. 1994. GIS and Exploratory Spatial Data Analysis: An Overview of some Research Issues, *Geographical Systems* **1**:315–327
Fotheringham A. S. and Rogerson P. 1994. *Spatial Analysis and GIS*, London: Taylor & Francis
Getis A. and Ord K. 1992. The Analysis of Spatial Association by Use of Distance Statistics, *Geographical Analysis* **24**:189–206
Good I.J. 1983. The Philosophy of Exploratory Data Analysis, *Philosophy of Science* **50**:283–295
Goodchild M. F. 1987. A Spatial Analytical Perspective on Geographical Information Systems, *International Journal of Geographical Information Systems* **1**:327–334
Goodchild M. F. 1992. Geographical Information Science, *International Journal of Geographical Information Systems* **6**:31–45
Goodchild M. F., Haining R.P., Wise S., et al. 1992. Integrating GIS and Spatial Analysis - Problems and Possibilities, *International Journal of Geographical Information Systems* **6**:407–423
Griffith D.A. 1993. Which Spatial Statistics Techniques Should Be Converted to GIS Functions? in: M.M. Fischer and P. Nijkamp (eds.), *Geographic Information Systems, Spatial Modelling and Policy Evaluation*, Berlin: Springer-Verlag, pp. 101–114
Haining R. 1990. *Spatial Data Analysis in the Social and Environmental Sciences*, Cambridge: Cambridge University Press
Haining R. 1994. Designing Spatial Data Analysis Modules for Geographical Information Systems, in: S. Fotheringham and P. Rogerson (eds.), *Spatial Analysis and GIS*, London: Taylor & Francis, pp. 45–63
Haslett J. 1992. Spatial Data Analyis — Challenges, *The Statistician* **41**:271–284
Haslett J. and Power G.M. 1995. Interactive Computer Graphics for a more Open Exploration of Stream Sediment Geochemical Data, *Computers and Geosciences* **21**:77–87
Haslett J., Wills G. and Unwin A. 1990. SPIDER - An Interactive Statistical Tool for the Analysis of Spatially Distributed Data, *International Journal of Geographical Information Systems* **4**:285–296
Haslett J., Bradley R., Craig P., Unwin A. and Wills C. 1991. Dynamic Graphics for Exploring Spatial Data with Applications to Locating Global and Local Anomalies. *The American Statistician* **45**:234–242
Lam N., Fan M. and Liu K-B. 1996. Spatial-Temporal Spread of the AIDS Epidemic, 1982–1990: A Correlogram Analysis of Four Regions of the United States, *Geographical Analysis* **28**:93–107
MacDougall E.B. 1991. A Prototype Interface for Exploratory Analysis of Geographic Data, *Proceedings of the Eleventh Annual ESRI User Conference, vol. 2,*. Redlands, CA: Environmental Systems Research Institute, Inc., pp. 547–553
Majure J. and Cressie N. 1997. Dynamic Graphics for Exploring Spatial Dependence in Multivariate Spatial Data, *Geographical Systems* (forthcoming)
Majure J., Cook D., Cressie N., Kaiser M., Lahiri S. and Symanzik J. 1996. Spatial CDF

Estimation and Visualization with Applications to Forest Health Monitoring, *Computing Science and Statistics* **27**:93–101

Majure J., Cressie N., Cook D. and Symanzik J. 1996. *GIS*, Spatial Statistical Graphics, and Forest Health, in: *Proceedings, Third International Conference/Workshop on Integrating GIS and Environmental Modeling, Santa Fe, NM, January 21–26, 1996*, Santa Barbara, CA, National Center for Geographic Information and Analysis (CD ROM)

MathSoft. 1996a. *S+Gislink*, Seattle: MathSoft, Inc.

MathSoft. 1996b. *S+Spatialstats User's Manual, Version 1.0*, Seattle: MathSoft, Inc.

Monmonier M. 1989. Geographic Brushing: Enhancing Exploratory Analysis of the Scatterplot Matrix, *Geographical Analysis* **21**:81–84

Oden N.L. 1984., Assessing the Significance of a Spatial Correlogram, *Geographical Analysis* **16**:1–16

O'Loughlin J. and Anselin L. 1996. Geo-Economic Competition and Bloc Formation: U.S., German and Japanese Trade Development, 1968–1992, *Economic Geography* **72**:131–160

Openshaw S. 1991. Developing Appropriate Spatial Analysis Methods for GIS, in: D. Maguire, M.F. Goodchild and D. Rhind (eds.), *Geographical Information Systems: Principles and Applications*, Vol 1, London: Longman, pp. 389–402

Openshaw S. and Fischer M.M. 1995. A Framework for Research on Spatial Analysis Relevant to Geo-Statistical Information Systems in Europe, *Geographical Systems* **2**:325–337

Ord J. K. and Getis A. 1995. Local Spatial Autocorrelation Statistics: Distributional Issues and Applications, *Geographical Analysis* **27**:286–306

Painho M. 1994. *New Tools for Spatial Analysis*, Luxembourg: Eurostat

Stuetzle W. 1987. Plot windows, *Journal of the American Statistical Association* **82**:466–475

Symanzik J., Majure J., Cook D. and Cressie N. 1994. Dynamic Graphics in a GIS: A Link between Arc/Info and XGobi, *Computing Science and Statistics* **26**:431–435

Symanzik J., Majure J. and Cook D. 1996. Dynamic Graphics in a GIS; A Bidirectional Link between ArcView 2.0 and XGobi, *Computing Science and Statistics* **27**:299–303

Talen E. 1997. Visualizing Fairness: Equity Maps for Planners, *Journal of the American Planning Association* (forthcoming)

Talen E. and Anselin L. 1997. Assessing Spatial Equity: The Role of Access Measures, *Environment and Planning A* (forthcoming)

Tukey J.W. 1977. *Exploratory Data Analysis*, Reading MA: Addison-Wesley

Unwin A. 1994. REGARDing Geographic Data, in: P. Dirschedl and R. Osterman (eds.), *Computational Statistics*, Heidelberg: Physica Verlag, pp. 345–354

Unwin A. 1996. Exploratory Spatial Analysis and Local Statistics, *Computational Statistics* **11**:387–400

Upton G. J. and Fingleton B. 1985. *Spatial Data Analysis by Example*, New York: Wiley

U.S. Department of Commerce.1994. *USA Counties 1994 CD-ROM*, Washington, D.C.: Bureau of the Census

Venables W. N and Ripley B.D. 1994. *Modern Applied Statistics with S-Plus*, New York: Springer-Verlag

Williams I., Limp W., Briuer F. 1990. Using Geographic Information Systems and Exploratory Data Analysis for Archeological Site Classification and Analysis, in: K. Allen, F. Green and E. Zubrow (eds.), *Interpreting Space: GIS and Archaeology*, London: Taylor & Francis, pp. 239–273

Zhang A., Yu H. and Huang S. 1994. Bringing Spatial Analysis Techniques Closer to GIS Users: A User-Friendly Integrated Environment for Statistical Analysis of Spatial Data, in: T. C. Waugh and R. G. Healy (eds.), *Advances in GIS Research*, London: Taylor & Francis, pp. 297–313

4 Measuring Spatial Variations in Relationships with Geographically Weighted Regression

A. Stewart Fotheringham,[1] Martin Charlton[1] and Chris Brunsdon[2]

[1] Department of Geography, University of Newcastle, Newcastle-upon-Tyne, NE1 7RU, UK

[2] Department of Town and Country Planning , University of Newcastle, Newcastle-upon-Tyne, NE1 7RU, UK

4.1 Spatial Non-Stationarity

A frequent aim of data analysis is to identify relationships between pairs of variables, often after negating the effects of other variables. By far the most common type of analysis used to achieve this aim is that of regression in which relationships between one or more independent variables and a single dependent variable are estimated. In spatial analysis, the data are drawn from geographical units and a single regression equation is estimated. This has the effect of producing 'average' or 'global' parameter estimates which are assumed to apply equally over the whole region. That is, the relationships being measured are assumed to be *stationary* over space. Relationships which are not stationary, and which are said to exhibit *spatial non-stationarity,* create problems for the interpretation of parameter estimates from a regression model. It is the intention of this paper to describe a statistical technique, which we refer to as *Geographically Weighted Regression (GWR),* which can be used both to account for and to examine the presence of spatial non-stationarity in relationships.

It would seem reasonable to assume that relationships might vary over space and that parameter estimates might exhibit significant spatial variation in some cases. Indeed, the assumption that such events do *not* occur, which until recently has been relatively unchallenged, seems rather suspect. There are three reasons why parameter estimates from a regression model might exhibit spatial variation: that is, why we might expect parameters to be different if we calibrated the same models from data drawn from different parts of the region (as shown by Fotheringham et al, 1996a; 1996b). The first and simplest is that parameter estimates will vary due to random sampling variations in the data used to calibrate the model. The contribution of this source of variation is not of interest here but needs to be eliminated by significance testing. That is, in this paper we want to concentrate on large-scale, statistically significant variations in parameter estimates over space, the source of which cannot be attributed solely to sampling. The second is that, for whatever reason, some relationships are intrinsically different across space. Perhaps, for example, there are spatial variations in people's tastes or attitudes or there are different administrative, political or other contextual issues that produce differing responses to the same stimuli across space. In which case, it is clearly useful to have a technique that can identify the nature of these variations in relationships over space; without such a technique only a global or average relationship can be estimated and this may bear little resemblance to particular local relationships. This is a situation where we throw away a great of interesting spatial detail in relationships.

The third reason why some relationships might exhibit spatial variation is that the model from which the relationships are being estimated is a gross misspecification of reality and that one or more relevant variables have either been omitted from the model or represented by an incorrect functional form and are making their presence felt through the parameter estimates. Given that all models, by their nature, are likely to be misspecifications of reality, the potential for this misspecification to be sufficiently gross as to cause spatial non-stationarity in parameter estimates would seem quite high. If misspecification *is* the cause of spatial non-stationarity, GWR has two roles to play. The first is in an exploratory mode when it can help to identify the nature of the misspecification by an examination of the spatial pattern of the localised parameter estimates. This would be particularly important when a global estimate is insignificant, and when the variable associated with this estimate is often excluded from the analysis, but when some of the localised estimates from GWR are statistically significant. The second would be as a means of incorporating otherwise 'unmeasurable' effects. Suppose, for example, the omitted variable causing the gross misspecification were a function of individuals' attitudes or tastes which could not be measured, the localised parameter estimates from GWR would provide a means of both incorporating this effect into the regression framework and of measuring its intensity.

The latter two reasons for thinking that spatial non-stationarity might be a possibility in regression modelling are interesting in that they reflect opposite views of modelling. The belief that some relationships are intrinsically different across space is consistent with a post-modernist view of spatial analysis where striving to identify global relationships is seen as having little relevance to real-world situations where relationships are very complex and are likely to be highly contextual. On the contrary, the belief that global models of human behaviour do exist if only we could find the correct set of explanatory variables is consistent with a 'spatial analysis is physics' approach in which the search for general 'laws' is paramount. We do not take sides on this debate in this paper; we merely note that GWR is consistent with either view and that it can be seen as a possible bridge between them.

4.2 Previous Attempts at Measuring Spatial Non-Stationarity

Several techniques aimed at measuring and incorporating spatial non-stationarity already exist in the literature. Perhaps the most well-known is that of the Expansion Method (Casetti 1972; Casetti and Jones 1992) which attempts to measure parameter 'drift'. In this framework, parameters of a global model can be made functions of other attributes including geographic space so that *trends* in parameter estimates over space can be measured (Fotheringham and Pitts 1995; Eldridge and Jones 1991). While this is a useful and easy-to-apply framework in which improved models can be developed, it is essentially a trend-fitting exercise in which complex patterns of relationships will be hidden. The output from the expansion model is essentially a second order set of relationships describing how the first order relationships vary across space. The output from GWR, on the other hand, is a set of parameter estimates which describe the first order relationships. These parameter estimates can be mapped to show their exact spatial distributions rather than just any trends that might exist in them.

At least three other statistical methods to handle varying parameter estimates have been proposed but which are of limited use in a spatial context: spatial adaptive filtering (SAF) (Foster and Gorr 1986; Gorr and Olligschlaeger 1994); the random coefficients model (Aitken 1996); and multilevel modelling (Goldstein 1987). The first incorporates spatial relationships in a rather *ad hoc* manner and produces parameter estimates which cannot be tested statistically and so has found very limited applicability. In the latter two techniques the parameter estimates in a regression model are assumed to be random variables. In multilevel modelling the distribution of the estimates is assumed to be Gaussian, while in the random coefficients model, the parameter estimates are modelled as finite mixture distributions. In either case, by using Bayes' theorem it is possible to obtain local parameter estimates although in neither case is any spatial dependency assumed in the estimation of the parameters which seems unrealistic in models of spatial phenomena. Although geographical variations of multilevel modelling have been attempted (Jones 1991) they rely heavily on a pre-defined hierarchy of spatial units which again probably represents an unrealistically discrete view of spatial dependency. In contrast, the GWR procedure described below utilises a distance-decay-based spatial weighting structure which is probably more reasonable in most spatial applications.

4.3 Geographically Weighted Regression

Consider a standard linear regression model in which a dependent variable, y, is represented as an additive linear combination of independent variables, x_k. Each y and each x_k can be measured for a point in space i so that the equation can be written as:

$$y_i = a_0 + \Sigma_k \; a_k x_{ik} + \epsilon_i \tag{4.1}$$

where a denotes a parameter to be estimated and the ϵ_is are independent normally distributed error terms with mean zero. Usually the least squares method is used to estimate the a_ks and using matrix notation the estimator can be expressed as:

$$\mathbf{\hat{a}} = (\mathbf{x}'\mathbf{x})^{-1}\mathbf{x}'\mathrm{y} \tag{4.2}$$

where **x** is an n by (k+1) matrix of independent observations and **y** is an n by 1 matrix of dependent observations where n is the number of observations. The resulting matrix of estimated parameters, $\hat{\mathbf{a}}$, has dimensions (k+1) by 1. It is important to note that this matrix is constant with respect to i and that only one set of parameter estimates is obtained for all points i. That is, the effect of any x on y is assumed constant across the space.

GWR is a relatively simple technique that extends the traditional regression framework of equations (4.1) and (4.2) by allowing local rather than global parameters to be estimated. That is, the parameter estimates become specific to location i and the model is rewritten as:

$$y_i = a_{i0} + \Sigma_k a_{ik} x_{ik} + \epsilon_i \tag{4.3}$$

where a_{ik} is the value of the kth parameter at location i. Note that equation (4.1) is a special case of equation (4.3) in which the parameters are constant over space. The GWR equation in (4.3) allows spatial variations in relationships to be measured. As is shown in the empirical example below, the point i in equation (4.3) is now completely generalizable and whereas in equation (4.1) it is restricted to a point at which data are collected, in equation (4.3) parameter estimates can be obtained for any point in space even if data on the dependent and independent variables is not collected for that point. This makes it possible to produce parameter estimates for a sample set of points defined by the user as a precursor to mapping. As it stands, equation (4.3) cannot be calibrated. There are more unknowns than observed variables. However, many models of this kind have been proposed before and they are reviewed by Rosenberg (1973), and Spjotvoll (1977). More recent work has been carried out by Hastie and Tibshirani (1990). Our approach borrows from the latter particularly in the fact that we do not assume the coefficients to be random, but rather that they are deterministic functions of some other variables - in our case location in space. The general approach when handling such models is to note that although an *unbiased* estimate is not possible, estimates with a small amount of bias can be provided. We argue here that the estimation process in GWR can be thought of as a trade-off between bias and standard error. Assuming the parameters exhibit some degree of spatial consistency then values near to the one being estimated should have relatively similar magnitudes and signs. Thus, when estimating a parameter for a given point i, one can approximate (4.3) in the region of i by (4.1), and perform an OLS regression using a subset of the points in the data set that are close to i. Thus, the aij's are estimated for i in the usual way and for the next i, a new subset of 'nearby' points is used and so on. These estimates will have some degree of bias, since the coefficients of (4.3) will exhibit some drift across the local calibration subset. However, if the local sample is large enough, this will allow a calibration to take place - albeit a biased one. The greater the size of the local calibration subset the lower the standard errors of the coefficient estimates; but this must be offset against the fact that enlarging this subset increases the chance that the coefficient 'drift' introduces bias. To reduce this effect one final adjustment to this approach may also be made. Assuming that points in the calibration subset further from i are more likely to to have differing coefficients, a weighted OLS calibration is used, so that more influence in the calibration is attributable to the points closer to i.

As noted above, the calibration of equation (4.3) assumes implicitly that observed data near to point i has more of an influence in the estimation of the a_{ik}s that do data located farther from i. In essence, the equation measures the relationships inherent in the model *around each point i.* Hence weighted least squares provides a basis for understanding how GWR operates. In ordinary least squares, the sum of squared differences between predicted and observed y_is is minimized by the parameter estimates. In weighted least squares a weighting factor w is applied to each squared difference before minimizing so that the inaccuracy of some predictions carries more of a penalty than others. The weighted least squares estimator for equation (4.1) can be written as:

$$\hat{\mathbf{a}} = (\mathbf{x}^t \mathbf{w}\, \mathbf{x})^{-1} \mathbf{x}^t \mathbf{w} \mathbf{y} \tag{4.4}$$

where $\mathbf{w}$ is an n by n matrix whose off-diagonal elements are zero and whose diagonal elements are the weights of each observation. Note that the weight of each observation is a constant and that only one set of parameter estimates is obtained for all points in the space.

In GWR the weighted least squares approach is taken one step further by weighting an observation in accordance with its proximity to point i so that the weighting of an observation is no longer constant in the calibration but varies with i. Data from observations close to i are weighted more than data from observations farther away. In this way, the estimator for the parameters in equation (4.3) is similar to that in equation (4.4) but with the important difference that the weighting matrix depends on i. That is,

$$\hat{\mathbf{a}}_i = (\mathbf{x}^t \mathbf{w}_i \mathbf{x})^{-1} \mathbf{x}^t \mathbf{w}_i \mathbf{y} \tag{4.5}$$

where $\mathbf{w}_i$ is an n by n matrix whose off-diagonal elements are zero and whose diagonal elements denote the geographical weighting of observed data *for point i.* There are parallels between GWR and that of kernel regression and kernel density estimation (Parzen 1962; Cleveland 1979; Cleveland and Devlin 1988; Silverman 1986; Brunsdon 1991, 1995; Wand and Jones 1995). In kernel regression, $\mathbf{y}$ is modeled as a non-linear function of $\mathbf{x}$ by weighting data in attribute space rather than geographic space. That is data points closer to x_i are weighted more heavily than data points farther away and the output is a set of localized parameter estimates in x space. It should be noted that as well as producing localized parameter estimates, the GWR technique described above will produce localized versions of all standard regression diagnostics including goodness-of-fit measures such as r-squared. The latter can be particularly informative in understanding the application of the model being calibrated and for exploring the possibility of adding additional explanatory variables to the model.

4.4 Choice of Spatial Weighting Function

Until this point, it has merely been stated that $\mathbf{w}_i$ is a weighting scheme based on the proximity of i to the sampling locations around i without an explicit relationship being stated. The choice of such a relationship will be considered here. Firstly, consider the implicit weighting scheme of the OLS framework in equations (4.1) and (4.2). Here

$$w_{ij} = 1 \;\; \forall \, i,j \tag{4.6}$$

where j represents a specific point in space at which data are observed and i represents any point in space for which parameters are estimated. That is, in the global model each observation has a weight of unity. An initial step towards weighting based on locality might be to exclude from the model calibration observations that are further than some distance d from the locality. This would be equivalent to setting their weights to zero, giving a weighting function of

$$\begin{aligned} w_{ij} &= 1 \text{ if } d_{ij} < d \\ w_{ij} &= 0 \text{ otherwise} \end{aligned} \tag{4.7}$$

The use of equation (4.7) would simplify the calibration procedure since for every point for which coefficients are to be computed only a subset of the sample points need to be included in the regression model. However, the spatial weighting function in (4.7) suffers the problem of discontinuity. As i varies around the study area, the regression coefficients could change drastically as one sample point moves in to or out of the circular buffer around i and which defines the data to be included in the calibration for location i. Although sudden changes in the parameters over space might genuinely occur, in this case changes in their estimates would be artefacts of the arrangement of sample points, rather than any underlying process in the phenomena under investigation. One way to combat this is to specify w_{ij} as a continuous function of d_{ij}, the distance between i and j. One obvious choice is:

$$w_{ij} = \exp(-\beta \, d_{ij}^{\,2}) \tag{4.8}$$

where β is a distance-decay parameter so that if i is a point in space at which data are observed, the weighting of that data will be unity and the weighting of other data will decrease according to a Gaussian curve as the distance between i and j increases. In the latter case the inclusion of data in the calibration procedure becomes 'fractional'. For example, in the calibration of a model for point i, if $w_{ij} = 0.5$ then data at point j contribute only half the weight in the calibration procedure as data at point i itself. For data a long way from i the weighting will fall to virtually zero, effectively excluding these observations from the estimation of parameters for location i.

Whatever the specific weighting function employed, the essential idea of GWR is that for each point i there is a 'bump of influence' around i corresponding to the weighting function such that sampled observations near to i have more influence in the estimation of i's parameters than do sampled observations farther away. A problem common to most weighting functions and all distance-based weighting functions is determining the steepness of the decay curve, β, around each point and this topic is now addressed.

4.5 Calibrating the Spatial Weighting Function

One difficulty with GWR is that the estimated parameters are, in part, functions of the weighting function or kernel selected in the method. In (4.7), for example, as d becomes larger, the closer will be the model solution to that of OLS and when d is equal to the maximum distance between points in the system, the two models will be equal. Equivalently, in (4.8) as β tends to zero, the weights tend to one for all pairs of points so that the estimated parameters become uniform and GWR becomes equivalent to OLS. Conversely, as the distance-decay becomes greater, the parameter estimates will increasingly depend on observations in close proximity to i and hence will have increased variance. The problem is therefore how to select an appropriate decay function in GWR.

Consider the selection of β in equation (4.8). One possibility is to choose β on a 'least squares' criteria. One way to proceed would be to minimise the quantity

$$\sum_{i=1,n} [y_i - y_i^*(\beta)]^2 \tag{4.9}$$

where $y_i^*(\beta)$ is the fitted value of y_i using a distance-decay of β. In order to find the fitted value of y_i it is necessary to estimate the a_{ik}s at each of the sample points and then combine these with the *x*-values at these points. However, when minimising the sum of squared errors suggested above, a problem is encountered. Suppose β is made very large so that the weighting of all points except for i itself become negligible. Then the fitted values at the sampled points will tend to the *actual* values, so that the value of (4.9) becomes zero. This suggests that under such an optimising criterion the value of β tends to infinity but clearly this degenerate case is not helpful. Firstly, the parameters of such a model are not defined in this limiting case and secondly, the estimates will fluctuate wildly throughout space in order to give locally good fitted values at each i.

A solution to this problem is a *cross-validation* (CV) approach suggested for local regression by Cleveland (1979) and for kernel density estimation by Bowman (1984). Here, a score of the form

$$\sum_{i=1,n} [y_i - y_{\neq i}^*(\beta)]^2 \tag{4.10}$$

is used where $y_{\neq i}(\beta)$ is the fitted value of y_i with the observations for point i omitted from the calibration process. This approach has the desirable property of countering the 'wrap-around' effect, since when β becomes very large, the model is calibrated only on samples near to i and not at i itself. Plotting the CV score against the required parameter of whatever weighting function is selected will therefore provide guidance on selecting an appropriate value of that parameter. If it is desired to automate this process, then the CV score could be maximised using an optimisation technique such as a Golden Section search (Greig, 1980).

4.6 Testing for Spatial Non-Stationarity

Until this point, the techniques associated with GWR have been predominantly descriptive. However, it is useful to assess the question: "Does the set of a_{ik} parameter estimates exhibit significant spatial variation?" The variability of a_{ik} can be used to describe the plausibility of a constant coefficient. In general terms, this could be thought of as a variance measure. For a given k suppose a_{ik}^* is the GWR estimate of a_{ik}. As there are n values of this parameter estimate (one for each point i within the region), an estimate of variability is given by the standard deviation of these values. This statistic will be referred to as s_k.

The next stage is to determine the sampling distribution of s_k under the null hypothesis that the global model in equation (4.1) holds. Although it is proposed to consider theoretical properties of this distribution in the future, for the time being a Monte Carlo approach will be adopted. Under the null hypothesis, any permutation of $(\mathbf{x}_i, y_i)$ pairs amongst the geographical sampling points i are equally likely to occur. Thus, the observed value of s_k could be compared to the values obtained from randomly rearranging the data in space and repeating the GWR procedure. The comparison between the observed s_k value and those obtained from a large number (99 in this case) of randomisation distributions forms the basis of the significance test. Making use of the Monte Carlo approach, it is also the case that selecting a subset of random permutations of $(\mathbf{x}_i, y_i)$ pairs amongst the i and computing s_k will also give a significance test when compared with the observed statistics.

An Empirical Example:
The Distribution of Limiting Long-Term Illness in The North-East of England

The application and potential uses of GWR are described in an empirical examination of the spatial distribution of Limiting Long-Term Illness (LLTI) which is a self-reported variable asked in the UK Census of Population. It encompasses a variety of severe illnesses such as respiratory diseases, MS, heart disease, severe arthritis as well as physical disabilities which prevent people from being in the labour market. The study area encompasses 605 census wards in four administrative counties in

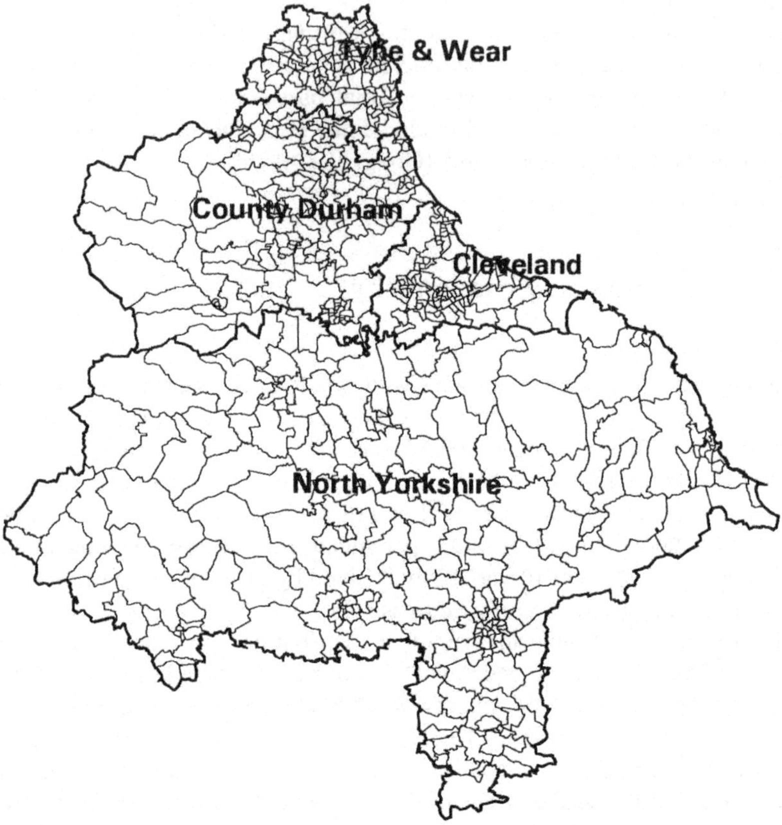

Fig. 4.1. Ward Boundaries in the Study Area

North-East England: Tyne and Wear, Durham, Cleveland and North Yorkshire as shown in Figure 4.1.

Tyne and Wear is a heavily populated service and industrial conurbation in the northern part of the study area and is centered on the city of Newcastle. To the south, Durham has been heavily dependent on coal mining in the eastern half of the county with the western half being predominately rural. Cleveland, to the southeast, is a largely urban, industrial area with heavy petrochemical and engineering works clustered around the Tees estuary and centered on Middlesbrough. North Yorkshire, to the south, is a predominantly rural and fairly wealthy county with few urban areas. The distribution of urban areas throughout the region is shown in Figure 4.2.

The spatial distribution of a standardised measure of LLTI (defined as the percentage of individuals aged 45-65 living in a household where LLTI is reported) throughout the study region is shown in Figure 4.3. As one might expect, the LLTI variable

Fig. 4.2. Urban Areas

tends to be higher in the industrial regions of Tyne and Wear, east Durham and Cleveland and lower in the rural areas of west Durham and North Yorkshire. To model this distribution, the following global regression model was constructed:

$$LLTI_i = a_0 + a_1 UNEM_i + a_2 CROW_i + a_3 SPF_i + a_4 SC1_i + a_5 DENS_i \tag{4.11}$$

where LLTI is the age-standardised measure of LLTI described above; UNEM is the proportion of economically active males and females who are unemployed (the denominator in this variable does not include those with LLTI who are not classed as being economically active); CROW is the proportion of households whose inhabitants

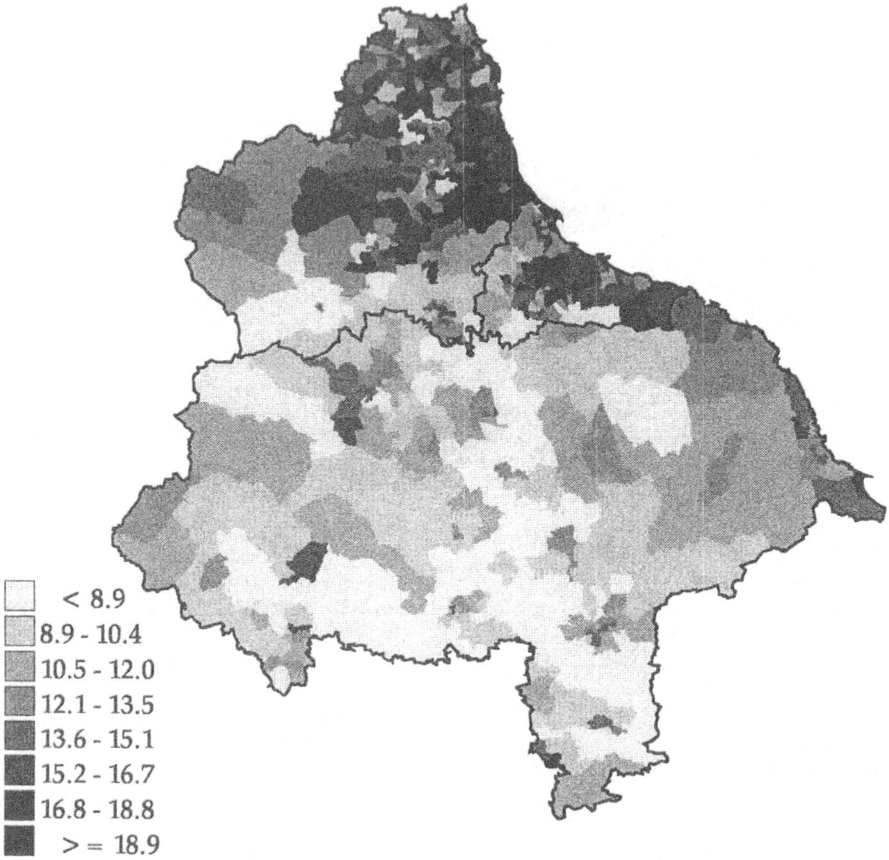

Fig. 4.3. Spatial Distribution of LLTI

are living at a density of over 1 person per room; SPF is the proportion of households with single parents and children under 5; SC1 is the proportion of residents living in households with the head of household in Social Class I (employed in professional, non-managerial occupations); and DENS is the density of population in millions per square kilometre. This latter variable discriminates particularly well between urban and rural areas. The model is guided by the findings of Rees (1995) in his examination of LLTI at a much coarser spatial resolution (English and Welsh counties and Scottish regions). The data are extracted from the 1991 UK Census of Population Local Base Statistics. The areal units used are census wards which contain on average approximately 200 households per ward. Using these data, the calibrated form of the global model is:

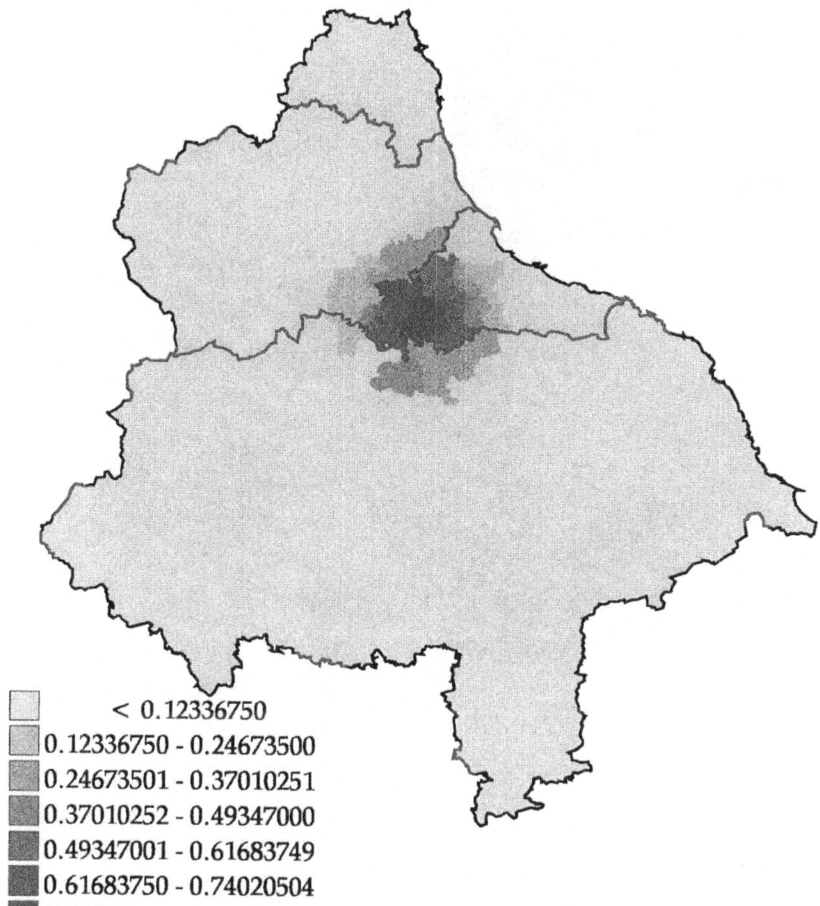

Fig. 4.4. An Example of Spatial Weighting Function Applied to a Single Point

$$LLTI_i = \underset{[.8]}{13.5} + \underset{[2.4]}{46.1UNEM_i} - \underset{[2.5]}{7.6CROW_i} - \underset{[1.5]}{1.7SPF_i} - \underset{[2.6]}{17.4SC1_i} - \underset{[1.6]}{4.5DENS_i} \quad (4.12)$$

where the numbers in brackets represent t-statistics and the r-squared value associated with the regression is .55. The results suggest that across the study region LLTI is positively related to unemployment levels and crowding. The former relationship reflecting perhaps that the incidence of LLTI is related to both social and

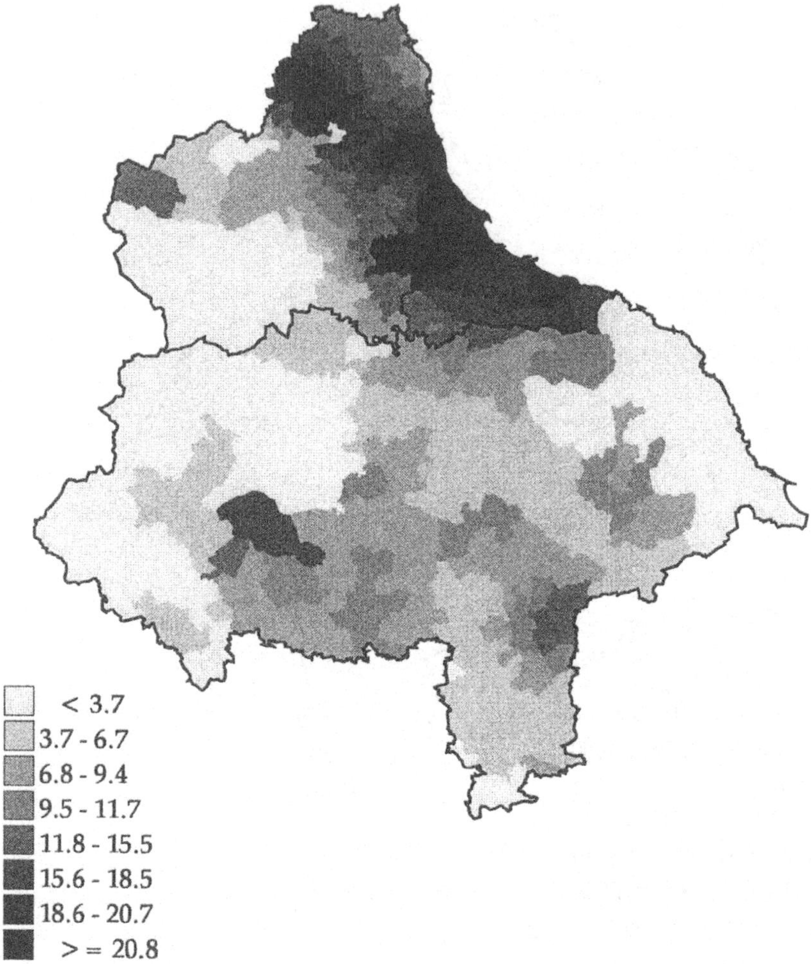

Fig. 4.5. Spatial Distribution of the Regression Constant

employment conditions in that the areas of higher unemployment tend to be the poorer wards with declining heavy industries. The latter relationship suggests a link between LLTI and social conditions with levels of LLTI being higher in areas with high levels of overcrowding. LLTI is negatively related to the proportion of professionally employed people in a ward and to population density. The negative relationship between LLTI and SC1 supports the hypothesis that LLTI is more prevalent in poorer areas which have fewer people in professional occupations. It also reflects the fact that industrial hazards which are a factor in the incidence of LLTI are less likely to occur to people in professional occupations. The negative relationship between LLTI and DENS is somewhat counter-intuitive in that it suggests that LLTI is greater in less densely populated areas, *ceteris paribus*. The nature of this latter relationship is explored in greater detail below in the discussion of the GWR results. Only the single parent family variable is not significant (at 95%) in the global model although it is locally significant.

To this point, the empirical results and their interpretations are typical of those found in standard regression applications: parameter estimates are obtained which are assumed to describe relationships that are invariant across the study region. We now describe the application of GWR which examines the validity of this assumption and explores in greater detail spatial variations in the relationships described above. Figures 4.5-4.10 describe the spatial variation in the parameter estimates obtained from the calibration of equation (4.3) using the estimator in equation (4.5). The Gaussian weighting function in equation (4.8) was calibrated by the cross-validation technique described above and the estimated value of β resulted in a weighting function that tends to zero at a distance of approximately 19 km from a point at which parameter estimates are obtained. An example of the spatial weighting function is shown centered on one ward in Figure 4.4. Each parameter estimate is obtained by weighting the data according to this function around each point. Hence, an interpretation of each of the spatial estimates depicted in Figures 4.5-4.10 is that each estimate reflects a particular relationship, *ceteris paribus, in the vicinity of that point in space.* It is evident from the spatial distributions of the parameter estimates in Figures 4.5-4.10 that there appears to be considerable variation in relationships across space and we now consider these results in greater detail.

Figure 4.5 shows the spatial variation in the estimated constant term. This is an interesting map because it shows the extent of LLTI after the spatial variations in the explanatory variables have been taken into account. The high values which occur primarily in the industrial areas of Cleveland and east Durham suggest a raised incidence of LLTI in these areas even when the relatively high levels of unemployment and low levels of employment in professional occupations are accounted for. The Monte Carlo test of significance for the spatial variation in these estimates described above indicates that the spatial variation is significant. This latter test might form a useful basis for testing for model specification. Presumably there are other attributes that might be added to the model that would reduce the spatial variation in the constant term. One could be satisfied with the specification of the model once the spatial variation in the constant term fails to be significant. The map in Figure 4.5 also acts as a useful guide as to what these attributes should be. The model apparently still does not adequately account for the raised incidence of LLTI in the mainly industrial areas of the North-East and perhaps other employment or social factors would account for this.

The spatial variation in the unemployment parameter shown in Figure 4.6 depicts the differing effects of unemployment on LLTI across the study area. All the parameters are significantly positive but are smaller in magnitude in the urbanised wards centred on Cleveland and Tyne and Wear. Again, the spatial variation in these parameter estimates is significant. The results suggest a possible link to environmental causes of LLTI with levels of LLTI being high regardless of

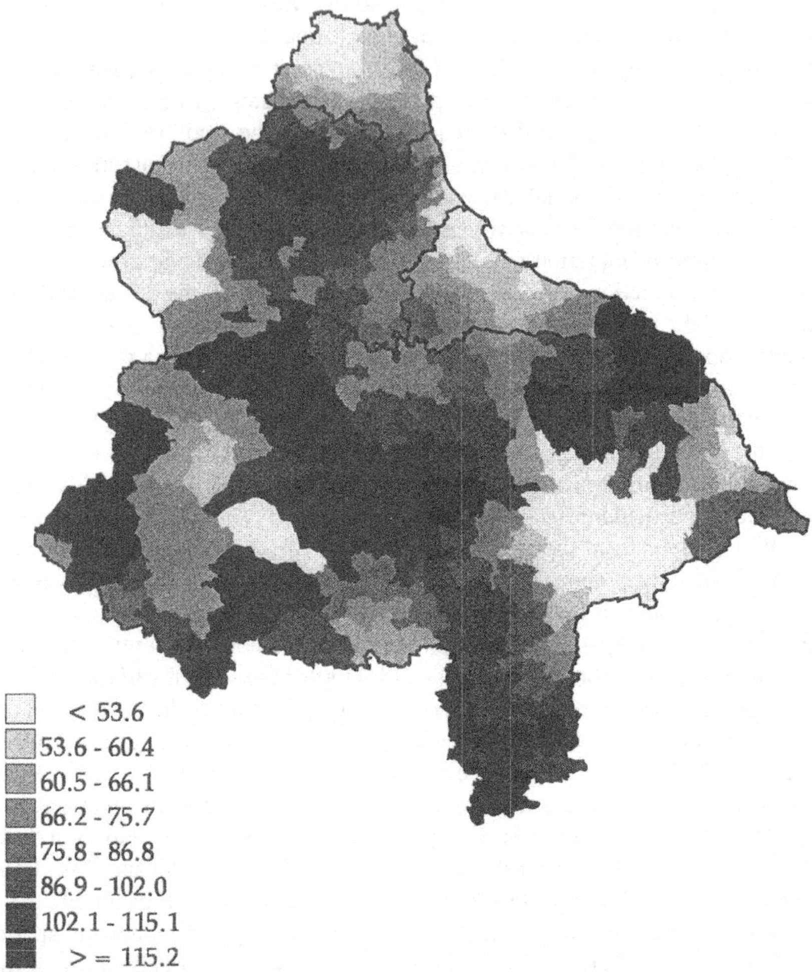

Fig. 4.6. Spatial Distribution of the Unemployment Parameter

employment status in Cleveland which has large concentrations of chemical processing plants and, until recently,steelworks. Another possibility is that levels of LLTI are high regardless of employment status in these areas because employment is concentrated in declining heavy industries and a large proportion of the unemployed were probably formerly employed in such industries which are associated with high levels of LLTI.

The overcrowding parameter in Figure 4.7 is negative in some parts of the region, albeit only marginally so, and positive in other parts across the region although the spatial variation is not significant. The parameter estimate for the single parent

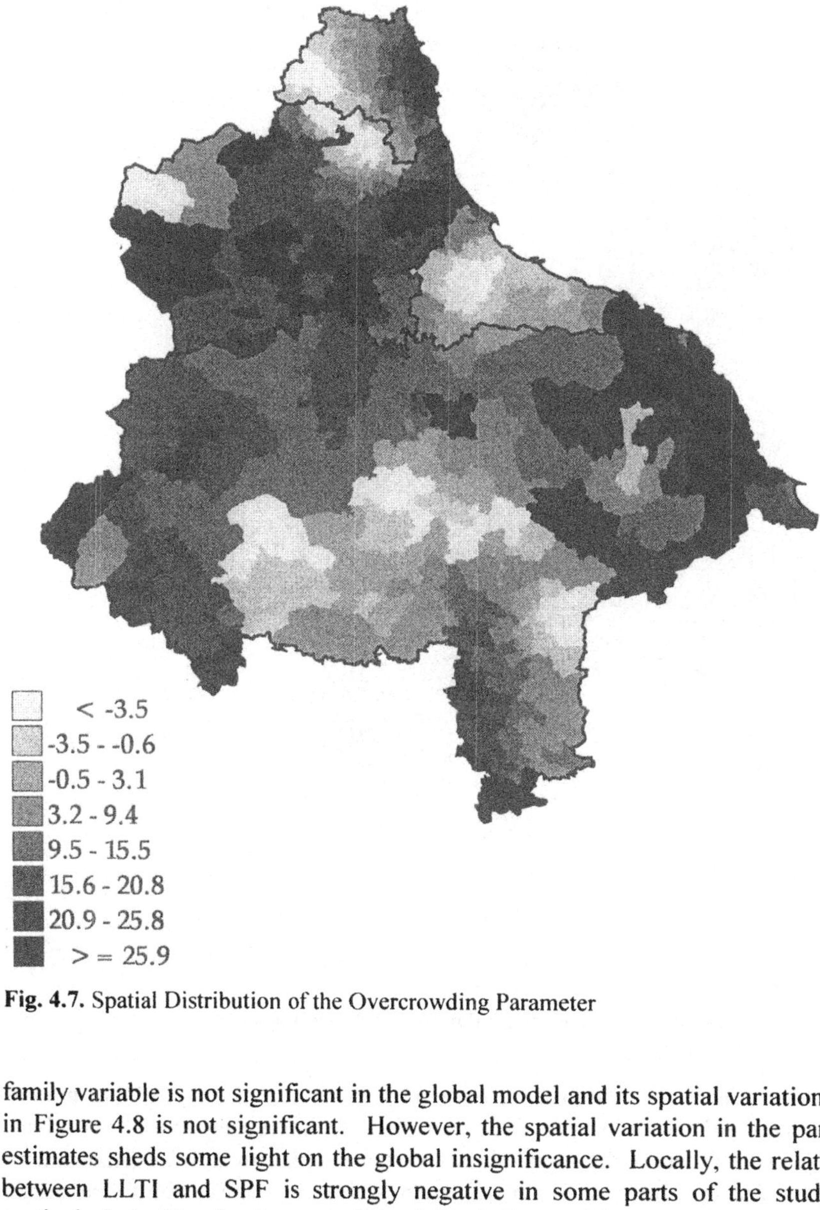

Fig. 4.7. Spatial Distribution of the Overcrowding Parameter

family variable is not significant in the global model and its spatial variation shown in Figure 4.8 is not significant. However, the spatial variation in the parameter estimates sheds some light on the global insignificance. Locally, the relationship between LLTI and SPF is strongly negative in some parts of the study area, particularly in Cleveland, parts of North Yorkshire and West Durham, and strongly postive in other parts, notably central and east Durham. These results perhaps relate to peculiar demographic compositions in some areas. In east Durham, for example, which is a coalfield area, there might be a low proportion of oung families and SPF will be low although levels of LLTI are high. Whatever the reason, the net result in

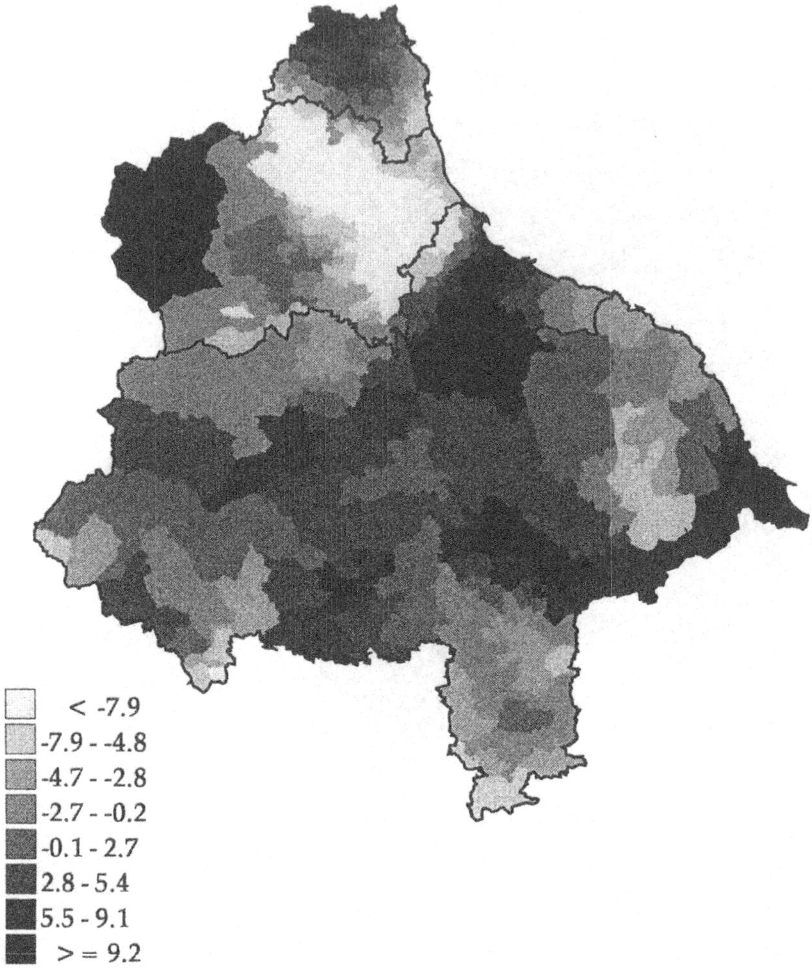

Fig. 4.8. Spatial Distribution of the Single Parent Family Parameter

the global model is an insignificant parameter estimate. Clearly, global models indicate 'average' relationships which can disguise interesting spatial variations.

The global estimate of the social class 1 variable is significantly negative and all the spatial estimates, shown in Figure 4.9, are negative and exhibit significant spatial variation. The more negative estimates are concentrated along the industrial parts of Cleveland, east Durham and Tyne and Wear indicating that levels of LLTI are more sensitive to variations in social class in urban areas than in rural areas. Within urban areas LLTI is presumably linked to blue-collar occupations whilst in rural areas, the incidence of LLTI is more evenly distributed across types of employment.

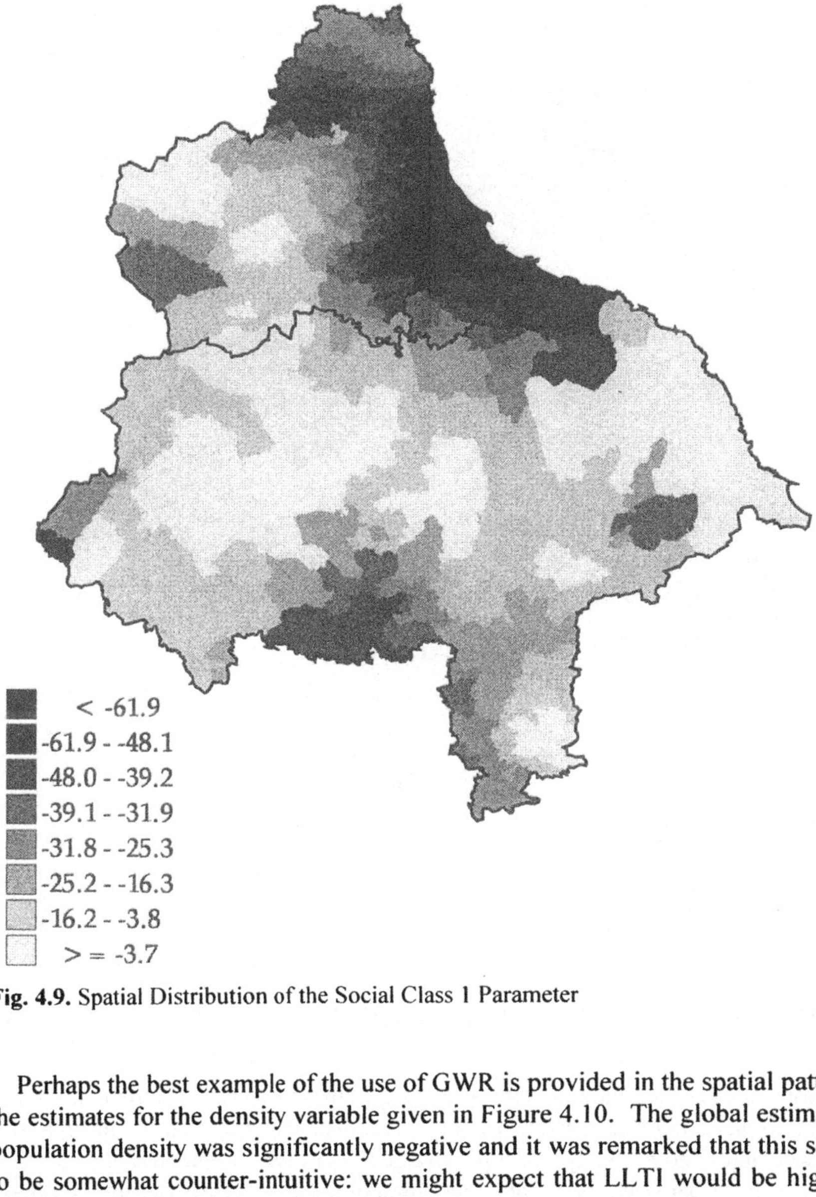

Fig. 4.9. Spatial Distribution of the Social Class 1 Parameter

Perhaps the best example of the use of GWR is provided in the spatial pattern of the estimates for the density variable given in Figure 4.10. The global estimate for population density was significantly negative and it was remarked that this seemed to be somewhat counter-intuitive: we might expect that LLTI would be higher in more densely populated urban wards than in sparsely populated rural wards, *ceteris paribus*. The spatial variation of this parameter estimate shown in Figure 4.10 and the pattern indicates that the most negative parameter estimates are those for wards centered on the coalfields of Eastern Durham. The probably explanation for this is that LLTI is closely linked to employment in coal mining (pneumoconiosis and other

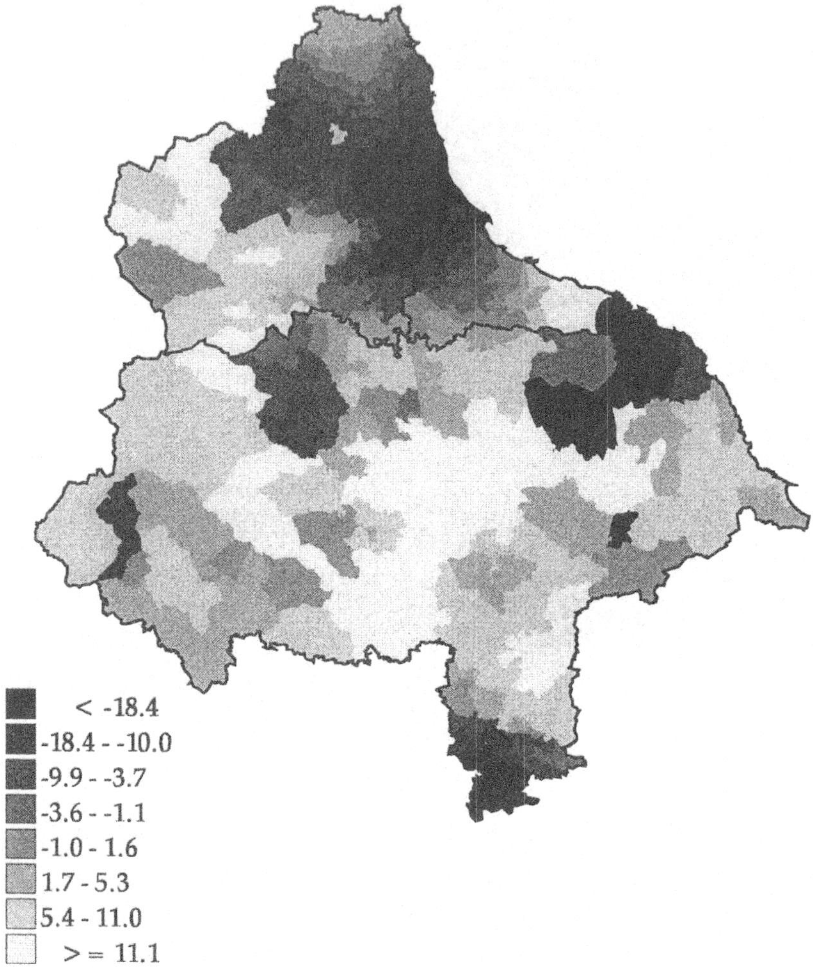

Fig. 4.10. Spatial Distribution of the Density Parameter

respiratory diseases being particularly prevalent in miners) but that settlements based on coal mining do not have particularly high densities of population in this area - the area is characterised by many small pit villages. However, population density rises rapidly in the urbanised areas both immediately south and north of the coalfields where employment is less prone to LLTI. Hence, *within the locality of east Durham* it is clear that the relationship between LLTI and density is significantly negative. In the more rural parts of the study area, particularly in west Durham and North Yorkshire, the relationship is positive with t values in excess of 2 in many places. Hence the more intuitive relationship, where LLTI increases in more densely populated areas, does exist in much of the study region but this information is completely hidden in the global estimation of the model and is only seen through GWR. The different relationships between LLTI and population density that exist across the region and which are depicted in Figure 4.10 highlight the value of GWR as an analytical tool.

Although they are not reported here because of space limitations, it is quite easy to produce maps of t values for each parameter estimate from GWR. These maps depict each spatially weighted parameter estimate divided by its spatially weighted standard error. Generally, the patterns in these maps are very similar to those depicted in the maps of the parameter estimates but occasionally some minor differences can occur because of spatial variations in standard errors which are based on data points weighted by their proximity to each point for which the model is calibrated. The statistics are useful, however, for assessing variations in the strengths of relationships across space.

One further spatial distribution from the GWR analysis is that of the spatially varying goodness-of-fit statistic, r-squared, shown in Figure 4.11. These values depict the accuracy with which the model replicates the observed values of LLTI in the vicinity of the point for which the model is calibrated. The global value of this goodness-of-fit statistic is .75 but it can be seen that there are large variations in the performance of the model across space ranging from a low of .23 to a high of .99 for the local model. In particular, the model explains observed values of LLTI well in a large group of wards in south Cleveland and the northern extremity of North Yorkshire and also in a group of wards in the southern and westerly extremes of the study region. The model appears to replicate the observed values of LLTI less well in parts of North Yorkshire and parts of Durham. The distribution of r-squared values in Figure 4.11 can also used to develop the model framework if the areas of poorer replication suggest the addition of a variable that is well represented in such areas and less well-represented in areas where the model already works well. For instance, there is evidently a coalfield effect missing from the model and the low values of r-squared in North Yorkshire suggest the model still fails to account adequately for rural varaitions in LLTI.

4.7 Summary

GWR is an analytical technique for producing local or 'mappable' statistics from a regression framework. It is a response to calls such as those of Fotheringham (1992, 1994), Fotheringham and Rogerson (1993) and Openshaw (1993) for a move away from 'whole-map' or global statistics which merely present averages across space and which therefore discard large amounts of potentially interesting information on spatial variations of relationships and model performance. The output from GWR, as shown in an empirical example using the spatial distribution of limiting long-term illness in

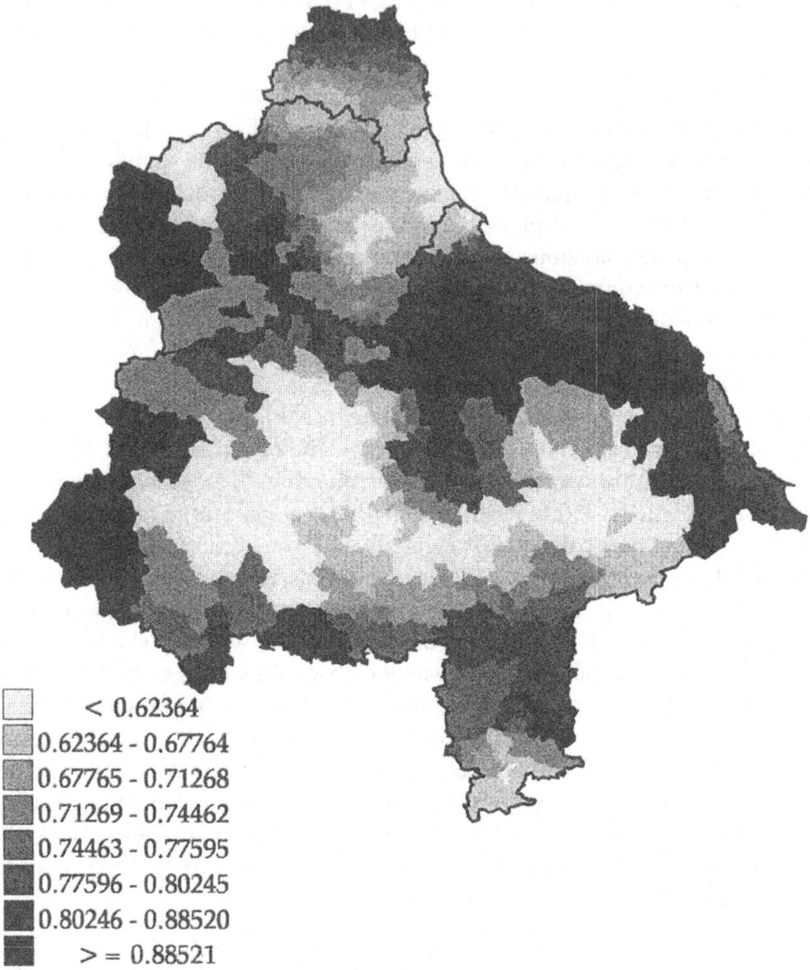

Fig. 4.11. Spatial Distribution of Goodness-of-Fit

four counties in the United Kingdom, is a set of local indicators of spatial relationships which inform on the spatial nature of relationships and on the spatial consequences of modelling such relationships. The technique can also be used in an exploratory mode to assist in model building. The spatial distributions of the regression constant and the goodness-of-fit statistic are seen as particularly important in this context. The spatial distributions of the other regression parameters indicate the degree of spatial non-stationarity in relationships and describe the interesting way in which relationships vary over space.

The concept of measuring spatial variations in parameter estimates within a GWR framework lends itself to some obvious developments beyond those described above. It would be interesting and useful for example to show what, if any, relationships exist between GWR results and those from more traditional regression diagnostics such as the battery of leverage statistics available. Similarly, it would be interesting to compare GWR results with those from a spatial variant of the expansion method, the latter presumably producing a simpler trend surface of spatially varying relationships.

Perhaps the most obvious development of the GWR is to experiment with various ways of making the spatial weighting function adapt to local environments. As defined above, the weighting function is a global one. Local versions could be produced with the aim of producing larger kernels in more sparsely populated rural areas and smaller kernels in more densely populated urban areas. Another avenue of research that could be pursued is to see how well various forms of GWR recover attributes of a simulated data set with known properties. Again, the GWR results could be compared to other methods that produce spatially varying results.

It is felt that GWR provides a significant advance in spatial analysis and that global regression applications with spatial data will be seen as rather limited. The application of GWR is not difficult and code will soon be available from the authors. It is hoped that GWR will promote interest in more genuinely geographical approaches to spatial analysis.

References

Aitken, M. 1996. A General Maximum Likelihood Analysis of Overdispersion in Generalized Linear Models, *Statistics and Computing* (forthcoming).

Bowman, A.W. 1984. An Alternative Method of Cross-Validation for the Smoothing of Density Estimates, *Biometrika* **71**: 353-360

Brunsdon, C.F. 1991. Estimating Probability Surfaces in GIS: an Adaptive Technique, in *Proceedings of the Second European Conference on Geographical Information Systems*, eds Harts, J., Ottens H.F. and Scholten, H.J., 155-163, Utrecht : EGIS Foundation.

Brunsdon, C.F. 1995. Estimating Probability Surfaces for Geographical Point Data: An Adaptive Kernel Algorithm, *Computers and Geosciences*, **21**: 877-894.

Casetti, E. 1972. Generating Models by the Expansion Method: Applications to Geographic Research, *Geographical Analysis* **4**: 81-91.

Casetti, E. and J.P. Jones III 1992. *Applications of the Expansion Method* London: Routledge.

Cleveland, W. S. 1979. Robust Locally Weighted Regression and Smoothing Scatterplots, *Journal of the American Statistical Association* **74**: 829-836.

Cleveland, W.S. and S.J. Devlin 1988. Locally Weighted Regression: An Approach to Regression Analysis by Local Fitting, *Journal of the American statistical Association* 83: 596-610.

Eldridge, J.D. and J.P. Jones III 1991. Warped Space: a Geography of Distance Decay, *Professional Geographer* **43**: 500-511.

Foster, S.A. and W.L. Gorr 1986. An Adaptive Filter for Estimating Spatially Varying

Parameters: Application to Modeling Police Hours Spent in Response to Calls for Service, *Management Science* **32**: 878-889.

Fotheringham, A.S. 1992. Exploratory Spatial Data Analysis and GIS, *Environment and Planning A* **25**: 156-158.

Fotheringham, A.S. 1994. On the Future of Spatial Analysis: The Role of GIS, *Environment and Planning A* Anniversary Issue: 30-34.

Fotheringham, A.S. and T.C. Pitts 1995. Directional Variation in Distance-Decay, *Environment and Planning A* **27**: 715-729.

Fotheringham, A.S. and P.A. Rogerson 1993. GIS and Spatial Analytical Problems, *International Journal of Geographic Information Systems* **7**: 3-19.

Fotheringham, A.S., Charlton, M. and Brunsdon C.F. 1996a. The Geography of Parameter Space: An Investigation into Spatial Non-Stationarity, *International Journal of Geographical Information Systems* **10**: 605-627.

Fotheringham, A.S., Charlton, M and Brunsdon C.F. 1996b. Two Techniques for Exploring Non-Stationarity in Geographical Data, *Geographical Systems* (forthcoming).

Goldstein, H. 1987. *Multilevel Models in Educational and Social Research*, London: Oxford University Press.

Gorr, W.L. and A.M. Olligschlaeger 1994. Weighted Spatial Adaptive Filtering: Monte Carlo Studies and Application to Illicit Drug Market Modeling, *Geographical Analysis* **26**: 67-87.

Greig, D.M., 1980. *Optimisation*, Longman:London.

Hastie T. and Tibshirani, R. 1990. *Generalized Additive Models*, Chapman and Hall, London.

Jones, K 1991. Specifying and Estimating Multilevel Models for Geographical Research, *Transactions of The Institute of British Geographers* **16**: 148-159.

Openshaw, S. 1993. Exploratory Space-Time-Attribute Pattern Analysers, in *Spatial Analysis and GIS* , eds. Fotheringham A.S. and Rogerson, P.A, 147-163, London: Taylor and Francis.

Parzen, E. 1962. On the Estimation of a Probability Density Function and the Mode, *Annals of Mathematical Statistics* **33**: 1065-1076.

Rees, P. 1995. Putting the Census on the Researcher's Desk, Chapter 2 in *The Census Users' Handbook*, ed. S. Openshaw, 27-81, GeoInformation International: Cambridge.

Rosenberg, B. 1973. A Survey of Stochastic Parameter Regression, *Annals of Economic and Social Measurement* **1**:381-397.

Silverman, B.W. 1986. *Density Estimation for Statistics and Data Analysis*, London:Chapman and Hall

Spjotvoll, E. 1977. Random Coefficients Regression Models, A Review, *Mathematische Operationsforschung und Statistik* **8**: 69-93.

Wand, M.P. and Jones, M.C. 1995. *Kernel Smoothing*, London:Chapman and Hall.

5 Regionalisation Tools for the Exploratory Spatial Analysis of Health Data

Steve Wise, Robert Haining and Jingsheng Ma

Sheffield Centre for Geographic Information and Spatial Analysis, Department of Geography, University of Sheffield, Sheffield S10 2TN, UK.

5.1 Introduction

This paper considers issues associated with the construction of regions as part of a programme of exploratory spatial data analysis in the case of what Cressie (1991) refers to as "lattice data". Lattice data arise where a study area has been partitioned into a set of zones or regions attached to each of which is a vector that describes the set of attributes for that zone. The focus of this paper will be the analysis of health data so the attributes in question may be health related but may also include demographic, socio-economic and environmental attributes.

Exploratory spatial data analysis (ESDA) comprises a set of statistically robust techniques that can be used to identify different forms of spatial variation in spatial data. ESDA represents an extension of exploratory data analysis (EDA) to handle spatially referenced data where in addition to the need for techniques to identify distributional properties of a set of data there is also a need for techniques to identify spatial distributional properties of the data. Typically these techniques comprise numerical summaries (e.g. measures of central tendency, measures of dispersion, regression) and graphic displays (e.g. boxplots, histograms, scatterplots) but in the case of ESDA cartographic displays make a vital and distinctive additional contribution enabling the analyst to see each attribute value in its geographical context relative to other attribute values. Like EDA therefore, ESDA exploits different methods of visualisation. Moreover like EDA, ESDA is not associated with any one stage in the process of data analysis for the techniques can be appropriate both for preliminary stages of analysis (pattern detection; hypothesis formulation) and later stages (model assessment).

EDA is underpinned by a conceptual data model in which data values comprise an element that is "smooth" (sometimes called the "fit") and an element that is "rough" (sometimes called the "residual"). In the case of ESDA the "smooth" is often associated with large scale patterns (such as spatial trends, patterns of autocovariation or concentration) whilst the "rough" may be outliers, that is individual areas with values that are higher ("hot") or lower ("cold") than found in the neighbouring areas. To identify these data properties in the case of ESDA a number of techniques have been brought together, either adapted from EDA (e.g. median polish to detect trends (Cressie 1984)) or custom developed for the identification of spatial properties (e.g.

spatial autocorrelation tests (Cliff and Ord 1981)). A distinction is also drawn between "whole map" techniques which generate aggregate measures (e.g. the Cliff and Ord statistics for spatial autocorrelation) and "focused" or "local" tests which only treat subsets of the data (e.g. the G statistics of Getis and Ord (1992), Ord and Getis (1995) and others (Anselin, 1995). For a longer review of these techniques and underlying models see Haining (1996).

There are several fundamental properties of the spatial system (the zones) that will influence the results of the application of ESDA techniques. One of these is the definition of the assumed spatial relationships between the zones. This is represented by the so-called "connectivity" or "weights" matrix and represents the analyst's chosen definition of inter-zonal relationships. The analyst may well wish to explore the robustness of ESDA findings to alternative definitions and hence alternative constructions of the weights matrix. At least as fundamental as this, however, is the construction of the zones themselves - the regionalisation (or spatial filter) through which events are observed. An important feature of any system that purports to offer spatial data analysts ESDA facilities ought to be a capacity to look at the spatial distribution of values in many different ways and to observe the robustness, or conversely the sensitivity, of findings to alternative (but equally plausible) regional frameworks. Regionalising by aggregating smaller spatial units is not a sine qua non for ESDA since a variety of smoothing techniques (e.g. kernel smoothing (Bithell 1990)) might be preferable, since they will allow the analyst to stay close to the original data whilst still making it possible to detect spatial properties, but, as will be argued in section 5.2, there are times when aggregation is essential. However in those situations where regionalisation through aggregation is part of the analysis there is a fundamental problem. Whilst the analyst engaged in ESDA may need to be able to construct alternative regionalisations under defined criteria, the analyst will also want to obtain different regionalisations reasonably quickly. This will require trade offs between many things but particularly (at the present time on most generally available hardware platforms) between optimality according to the specified criteria and the time taken to arrive at a an acceptable solution.

The spatial analysis of health data is normally undertaken with one of two objectives in mind:

a. To detect or describe patterns in the location of health-related events such as incidence of a disease or the uptake rate of a service. The focus here is on seeking explanations or causative factors - spatial patterns can rarely provide direct evidence for either but can be suggestive of particular mechanisms (Elliott et al 1992)

b. To assist in the targeted delivery of health services, by identifying areas of need. Here the emphasis is not on explanation but on producing a reliable picture of the variation in health needs across an area, allowing for the variations in other factors such as population age structure.

Both types of analysis require information about the population in the area of interest, whether this is simply a count of people for calculating incidence or mortality rates, or more detailed information relating to the age structure or levels of economic deprivation. Such data is usually available for small areas - in Great Britain for example the reporting areas, known as Enumeration Districts (EDs), normally contain about 150 households. This very detailed level of spatial resolution can cause problems for many types of analysis, as described in detail below, and it is common practice to undertake analysis at more aggregated spatial scales. However, larger standard statistical areas may also be unsuitable because they are too large or are too inhomogeneous. There is therefore a need to be able to merge the basic spatial units into tailor-made groups by means of a classification or regionalisation procedure.

As Openshaw and Rao (1994) point out, the availability of digital boundaries and Geographical Information System software capable of manipulating them has made this task far easier than before, and opened up the possibility of making regionalisation an intrinsic part of the analysis of spatial data, rather than a one-off, time-consuming exercise.

This paper describes work to develop a suite of classification and regionalisation tools suitable for use in the analysis of health data. The research forms part of a larger project which is developing a software system for the analysis of health data, with the emphasis on rapid interactive visualisation of data supported by a suite of cartographical, graphical and analytical tools. The system uses a GIS for the storage, manipulation and cartographical display of the spatial data and is called SAGE - Spatial Analysis in a GIS Environment. The architecture of the whole system is described elsewhere (Haining et al, 1996) but two of the key design elements of SAGE are important in what follows:

a. The focus of the software is on the provision of tools for data visualisation. A range of display techniques are provided including map display, histograms, box plots and scatter plots. More importantly, these different views of the data are linked so that highlighting an area on the map will also cause the equivalent point on the scatter plot to be identified. This is in keeping with much of the software that has been developed following the pioneering work of Haslett et al (1990,1991).

b. The emphasis is on building a system by utilising existing software and techniques wherever possible, rather than writing it from scratch. One of the important consequences of this is that users can explore data which they already hold in a GIS without having to export it to another software package (although the facility to export to some specialised packages is being provided). It also means that all the standard tools of data input and management are provided by the GIS.

The structure of this paper is as follows. We firstly elaborate on the importance of classification and regionalisation in the analytical process, and then describe the criteria governing the design of the classification and regionalisation tools in SAGE. This is followed by a brief description of the methods used, some initial results and a discussion of future research directions.

5.2 The Importance of Classification and Regionalisation in Analysis

A common problem with area-based analyses is that the results of the analysis may be sensitive to the choice of spatial unit. It is well known that census EDs are designed to minimise enumerator's workloads, and to nest within higher order areas such as wards and districts. As a result these basic spatial units are quite varied in terms of areal extent, population composition and population size and do not reflect the underlying variation in socio-economic conditions. If analysis reveals a link between levels of deprivation and levels of ill-health or leads to the identification of disease "hot-spots" there remains the possibility that if the boundaries of the areas were drawn differently, then different results might be found. This is the well-known modifiable areal unit problem or MAUP (Openshaw 1978, Fotheringham and Wong 1991) although we would emphasise that in a well defined analytical context only a subset of possible zoning systems would be considered a plausible basis for analysis.

Despite these problems, there are a number of reasons why it is often beneficial to group the basic spatial units to form larger regions as the framework for the analysis:

a. To increase the base population in each area, so that incidence rates for example will be based upon a larger sample size and hence more robust to small random variations in the number of cases. This will be particularly important when the events being studied are rare (such as incidences of a rare cancer).

b. To reduce the effect of suspected inaccuracies in the data. When dealing with small areas, an error of one or two in the count of health events could have a large effect on the calculated rates, whereas such errors will have far less effect for larger areas (assuming the positive and negative errors for each ED tend to cancel out when the areas are aggregated - any systematic bias will simply carry forward into the aggregated data). In the UK census, counts are routinely modified by the random addition of -1,0 or +1 as a further protection of confidentiality thus producing a level of 'error' in the population data which could similarly affect the calculation of expected rates in small areas.

c. To reduce the effect of suspected inaccuracies in the location of the health events. A common form of locational reference for health data in the UK is the unit postcode, which is shared by 15 addresses on average (Raper et al 1992) and therefore has an inherent precision of the order of 100 m. This problem is compounded by the fact that the file which is widely used to assign UK National Grid references to postcodes (the Postzon file) gives the location of the first address in the postcode to the nearest 100m, and is known to contain errors. In a study in Cumbria, Gatrell et al (1989) found that using the Postzon file to allocate postcodes to EDs (using a point in polygon operation) resulted in the incorrect assignment of 39% at the ED level, but only 3% at the ward level (the first level of aggregation of EDs to standard statistical units). (This is quoted as an example of the effects of scale on the problem - for the 1991 UK census more accurate data is available than was the case for the 1981 census). Using larger areal units essentially means that less of the cases are near the edges of units, and hence less are likely to be assigned to the wrong unit.

d. To make the analysis computationally tractable. Some spatial analytical techniques require the manipulation of an NxN connectivity matrix, where N is the number of regions in the analysis (and element (ij) in the matrix defines the spatial relationship between regions I and j). In the study of Sheffield by Haining et al (1994) the initial dataset was based on 1000 EDs, which would give a contiguity matrix with a million elements.

e. To facilitate visualisation. An important element of visualisation techniques is that they must operate quickly enough that the user is encouraged to explore different views of the data and use different techniques. Dealing with large numbers of small areas may cause the system to be too slow. In addition, it may be difficult for a user to see trends and patterns in the data when confronted by maps and graphs relating to large numbers of areas (although techniques such as kernel smoothing may also be employed to deal with this problem).

All these are situations where regionalisation might be performed prior to some other form of analysis, such as fitting a model to the data. However, as noted in the introduction, regionalisation can also be seen as an intrinsic part of the analytical process, in the sense that it can be considered as a means of visualising broad scale patterns in the data, and as a technique for testing the robustness of model results.

The construction of new regions can be achieved by classifying the basic spatial units, and merging together those that fall into the same class. A map of these new regions will often appear very fragmented however, and it is often desirable to use a regionalisation approach which will ensure that the regions form contiguous areas on the map.

5.3 The Design of a Regionalisation System for the Analysis of Health Data

Both classification and regionalisation are topics of long standing in many areas of science, and a large number of methods now exist for both. The intention was to make use of existing techniques wherever possible, but before reviewing the main approaches described in the literature we describe the design decisions which governed the choice of methods for SAGE.

Different applications of regionalisation will have different criteria. Cliff et al (1975) suggested that, in general, an 'optimal' regionalisation should simultaneously satisfy the following criteria:

a. It should be simple, in the sense that a solution which produces few regions is better than one which produces many.

b. Regions should be homogeneous in terms of the characteristics of the zones which comprise them.
c. Regions should be compact.

As they point out, these criteria are competitive. For example, the simplest regionalisation would group all zones into one region, but this would be the least homogeneous. These are not universal criteria - in political redistricting for example homogeneity may well be very undesirable (since it smacks of gerrymandering) and compactness plus equality in terms of population are more likely to be important (Horn 1995, Sammons 1978).

In terms of the analysis of health data, the following three criteria seem most appropriate as objectives:

Homogeneity. The new regions should be made up of zones which are as similar as possible in terms of some characteristic which is relevant to the analysis. In a study relating ill health and deprivation, for example, it clearly makes sense to use regions in which deprivation levels are relatively uniform.

Equality. In some cases it may also be important that the new regions are similar to one another in certain respects. The most obvious example of this is in terms of their population sizes. It is known that there are problems in comparing incidence rates which have been calculated on the basis of different base populations, and adjustment techniques have been developed to allow for this (e.g. Bayes adjustment (Clayton and Kaldor 1987). One way to minimise this problem is to ensure that the populations in the new regions are of sufficient size and as similar as possible.

Contiguity. There are two aspects to this. Firstly there is the distinction between classification, in which the location of the original zones is not considered, and regionalisation, in which only neighbouring zones may be merged to form the regions. Secondly, in the case of regionalisation there is the question of the shape of the regions, since merely enforcing contiguity may lead to long thin regions rather than compact ones. This is sometimes regarded as merely a cosmetic aim (e.g. Openshaw 1994) but Horn (1995) argues that it accords with our intuitive understanding that regions within cities for example, form tight units of economic and social activity and hence are spatially compact.

It is not necessarily the case that all three of these will be important in all applications. For example, in the design of regions for the administration of health delivery homogeneity may be unimportant, whereas equality of population size and compactness would be very desirable. Conversely for any statistical analysis homogeneity is almost certain to be desirable, and perhaps equality, but contiguity may be undesirable, especially since enforcing it will probably reduce the level of homogeneity and equality. This implies that it should be possible to specify any combination of these three criteria in the classification or regionalisation process, and since the objectives may be competitive, it must be possible to decide on the balance between them.

5.4 Classification and Regionalisation Methods in SAGE

The aim of both classification and regionalisation is to group N initial observations into M classes (where M∴N) in a way which is optimal according to one or more objective criteria. The terminology varies depending on the discipline, but here we will use the term zones to refer to the initial small areas and groups or regions to refer to the larger units. One of the key problems is that the number of possible solutions with even modest numbers of zones is astronomical (Cliff et al 1975), so that enumerating all possible solutions to find the optimal one is not possible. It is often not even clear what the number of classes or regions should be, unless this is dictated by external requirements (as is often the case with political redistricting for example). Most methods are therefore heuristics in the sense that they cannot guarantee to find a global optimum solution, but employ a range of techniques to attempt to find a good local optimum.

Methods are generally based upon some measure of similarity (or dissimilarity) between the zones - the simplest is Euclidean distance between the values of one or more attributes measured for each zone, although other metrics have also been used. Early approaches used a hierarchical (Lankford 1969) in which zones are merged one at a time. The distance metric is calculated for all possible pairs of zones, and held in a dissimilarity matrix. The two zones which are most similar are then merged to form a group, and the matrix updated by calculating the distance between this group and the remaining zones. This process is repeated until all the original zones have been allocated to a group.

The earliest attempts to produce contiguous regions simply included the coordinates of the zone centroids as additional variables. This will tend to merge adjacent zones, but cannot guarantee to produce M regions when grouping to M classes. This can be guaranteed by building in a contiguity constraint such that only distances between neighbouring zones are considered in deciding which zones to add in next.

Hierarchical methods are very fast since each zone is only considered once, although storage of the whole matrix means memory requirements can be high. However, because each zone is only considered once it is known that these methods produce sub-optimal solutions (Semple and Green 1984).

More recent work has therefore concentrated on iterative techniques, of which the best known is the k means method McQueen 1967, Anderberg 1973). The number of classes (k) must first be specified, and the first k observations are assigned to a class each. All the other observations are then assigned to the class to which they are closest and the class mean is calculated for all classes. On each iteration, each case is examined to see if it is closer to the mean of different class than its current one - if it is then it is swapped. This process continues until the process converges on a solution (i.e. no more swaps are possible). As with the hierarchical methods, the k-means method can be used for regionalisation by building in a contiguity constraint, and this is essentially the method used by Openshaw (1978) in his Automatic Zoning Procedure (AZP).

The k-means approach is rapid and will often converge on a solution in a few iterations (Spath 1980). The main problem is that this solution is often only a local optimum, and so the method is sensitive to the initial allocation of observations to classes - subsequent runs can produce quite different results. One solution to this is simply to run the algorithm several times and save the best result. This seems inelegant and potentially rather slow and cumbersome, and a number of other techniques have been tried to improve the method. In a recent development of the AZP approach, Openshaw (1994) has studied techniques such as simulated annealing and tabu which are designed to prevent the search getting stuck in a local optimum. Simulated annealing does this by using the analogy of a material slowly cooling. In the initial stages when the temperature is high, swaps are allowed even when they make the objective function worse. As the temperature is cooled, the probability of allowing this is reduced until by the end only improving exchanges are allowed. It has been shown that as the rate of cooling approaches zero, the method will find the global optimum. However all these methods result in much slower execution, as far more iterations are necessary to converge to a solution.

An alternative approach to regionalisation is to modify the original dissimilarity matrix to take account of the relative positions of pairs of zones. The idea is to weight the differences such that ones which are close in space are allowed to differ more in terms of their attributes than zones which are distant in space. Some workers have used a simple exponential distance decay but Oliver and Webster (1989) used the estimated semi-variogram to base the weightings on the spatial autocorrelation present in the data. The modified matrix can then be used in a normal classification such as k-means or even a hierarchical approach. This will produce a rapid result, but the initial stage of fitting a variogram adds an extra complication and it is difficult to see how it would be possible to allow the user to control the degree of importance to be attached to the compactness criterion.

It was therefore decided to base the regionalisation tools in SAGE on a k-means classification because this is relatively rapid, and can be adapted to deal with more than one objective function. Its main weakness is that it is known to be sensitive to the initial allocation of zones to regions, and a series of methods have been developed to try and reduce this problem as described in the next section.

5.5 Implementation

The heart of the SAGE regionalisation tools is a k-means based classification procedure which can classify zones according to any of the following functions:

- Homogeneity - the objective is to minimise the within group variance of one or more attributes.

- Equality - the objective is to minimise the difference between the total value of an attribute for all zones in a region and the mean total of this value for all regions.

•Compactness - this is currently achieved simply by treating the X and Y coordinates of the zone centroids as two variables. Although this will tend to create compact regions, it is not ideal because the location of the centroid is to some extent arbitrary, and especially in large zones, different positions for the centroid could lead to quite different results with no obvious way to select which is the 'right' one. A better alternative may be to base the measure on the length of perimeter between regions (Horn 1995) or to use some measure of area shape such as the perimeter/area ratio.

In the case of trying to satisfy competing objectives, an interesting issue is what to do when a swap improves one objective but makes another one worse. The problem is complicated by the fact that the objective functions are measured in different units as described above.

Cliff et al (1975) used two objective functions which took values from 0 (the worst solution for that objective) to 1 (the best). The optimum regionalisation was therefore the one which maximised the sum of these functions. This would also be amenable to weighting, although it relies on finding suitable objective functions which can be scaled in this way.

A alternative solution to the problem was used by Sammons (1987) who found that insisting that all objective functions must improve with every swap lead to poor results because of premature convergence i.e. the method tends to stabilise quickly but at a local optimum which gives a poor result. He therefore adopted a scheme whereby a swap was accepted if it improved any of the functions.

In SAGE the problem is tackled by expressing the changes in the objective functions due to a swap in percentage rather than absolute terms. The percentage changes are summed, and the swap accepted if this sum is positive (i.e. the gains outweigh the losses). One advantage of this scheme is that it is then easy to weight each objective function in calculating the sum. This provides an intuitive means for the user to control the balance of objectives for the regionalisation - any objective can be weighted from 0 (i.e. not considered) to 100%. One option would be to make 100% mean that the objective must be maximised at all costs i.e. no swap to be accepted unless it improves this objective function. However, this would certainly constrain the regionalisation to certain parts of the solution space and may lead to poor solutions (even in terms of the objective set at 100%).

As mentioned above the k-means method is sensitive to the initial allocation of zones into regions. When this is done randomly, different runs of the method will produce very different results. Our solution to this problem is to regard the k means regionalisation as simply one in a set of procedures which a user may use so that the process of regionalisation in SAGE can be visualised as shown in Figure 5.1.

The k means regionalisation forms stage 2 in the process. Prior to this the user may decide how to make the initial allocation of zones to regions. The default is a random allocation, but Sammons (1978) found it improved the results of his political redistricting algorithm if this was started with a 'good' allocation of zones to regions. In the context of political re-districting the main objective was of equal population, and hence the initial allocation could be based on the population distribution in the area.

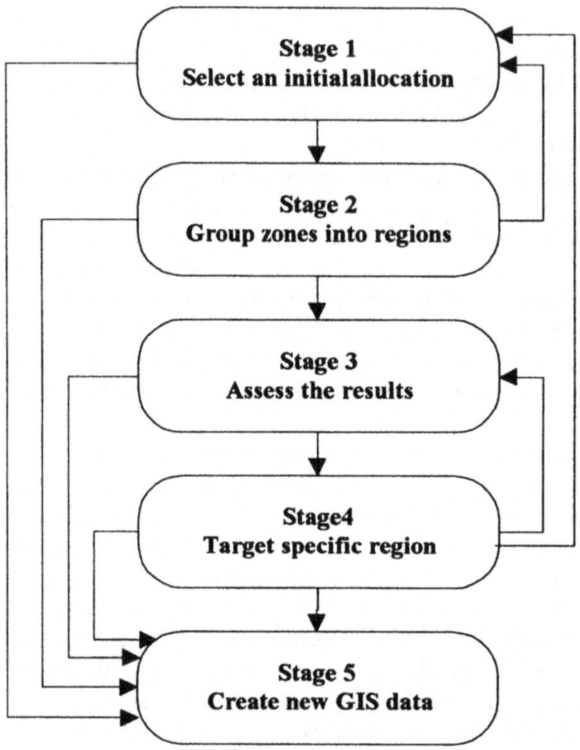

Fig. 5.1. Sequence of Steps for Performing Regionalisation in SAGE

For a more general solution, SAGE employs an adaptation of a method devised by Taylor (1969) in which a regionalisation is based on first identifying 'nodes' - zones which are typical of their local area and which can be regarded as the centre of uniform regions. In Taylor's method each zone was visited in turn, and a count made of the number of zones within a certain distance (D_g) whose attributes differed by less than a specified threshold (D_s). Given the number of regions required (M), the M zones which had the greatest number of similar neighbours were selected as region seeds. In successive steps the remaining zones were added on to the region which they bordered and which they most resembled. As a regionalisation method this has a number of drawbacks - it is not clear how to select suitable values for D_g and D_s, and it is a hierarchical method. What is interesting though is the idea of identifying relatively uniform areas of the map from which to grow regions, an idea which is similar to some of the methods used in image recognition to identify objects in images (Rosenfeld and Kak 1982). In the initial version of SAGE, we have simply implemented the first step of Taylor's method as a means of identifying a starting point for the regionalisation.

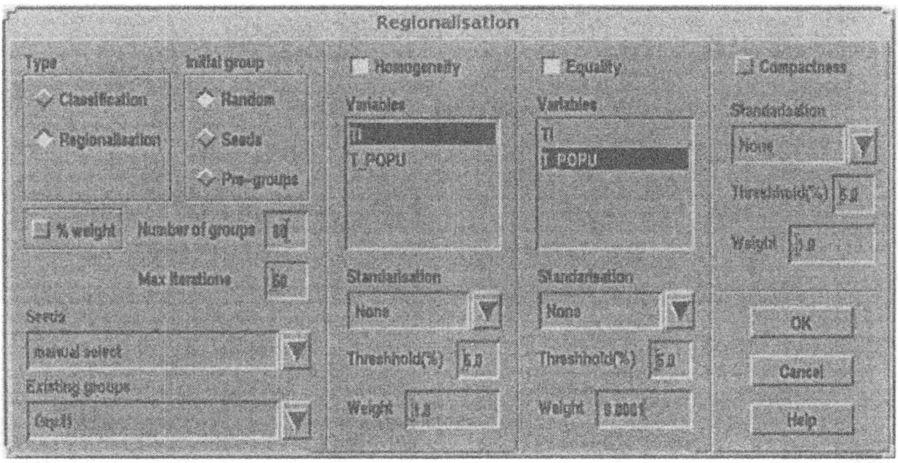

Fig. 5.2. User Interface for Stage 1 and 2

Future work will focus on better methods for identifying uniform regions in which to locate seed points, possibly using some of the LISA statistics which are local measures of spatial association which do not require the choice of threshold values (Getis and Ord 1992, Barkley et al 1995).

The selection of the initial allocation of zones is done on the main menu screen for the regionalisation package which is shown in Figure 5.2. The type of initial group is chosen from the following options:

•Random - this is the default.

•Seeds - selecting this option activates the seeds option lower down, giving a choice on how the seed points are to be chosen. Currently the options are for manual selection and Taylor's method.

•Pre-groups - the starting point for regionalisation can be an existing regionalisation, where each zone has already been allocated to a region, but the boundaries within the regions have not been removed. This could have been produced by an earlier run of SAGE, or by some other software package, with the results imported into an ARC/INFO coverage.

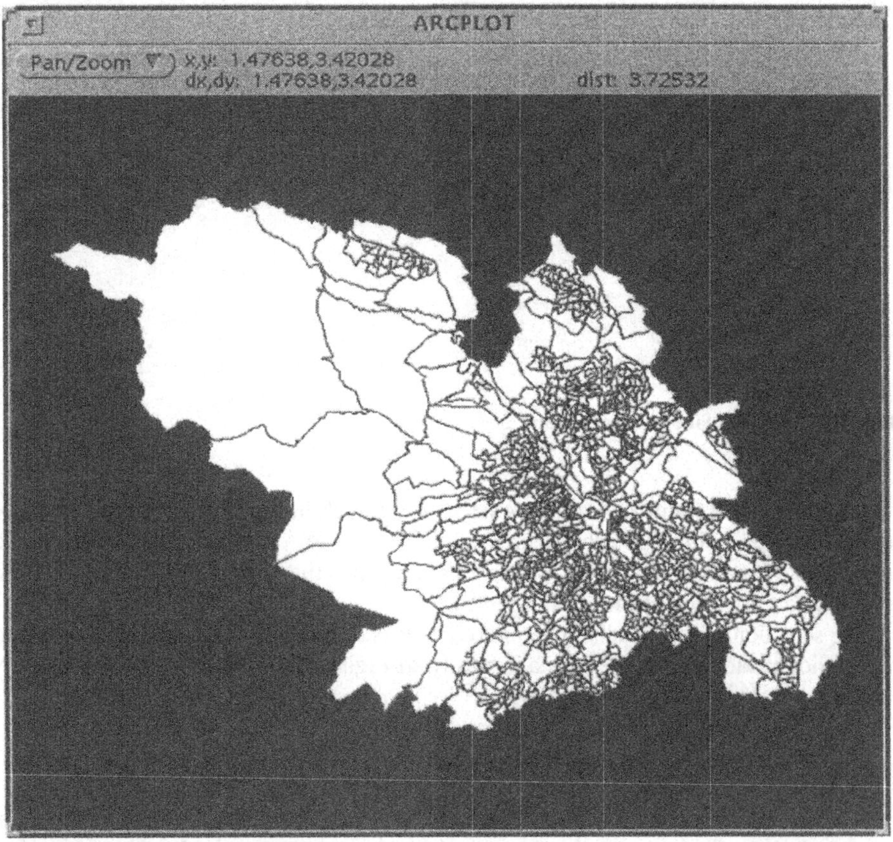

Fig. 5.3. 1991 Enumeration districts of Sheffield

This menu also allows the user to decide whether to perform a classification or a regionalisation and how many groups or regions are to be created. The other menu items allow the user to select which of the three objectives are to be considered, and where more than one is chosen, what weighting is to be attached to each. Two types of weighting are supported, selected by means of the % weight option on the left hand side.

- When %weight is selected, the weights under each of the objective functions are used to weight the percentage change in the objective functions when a

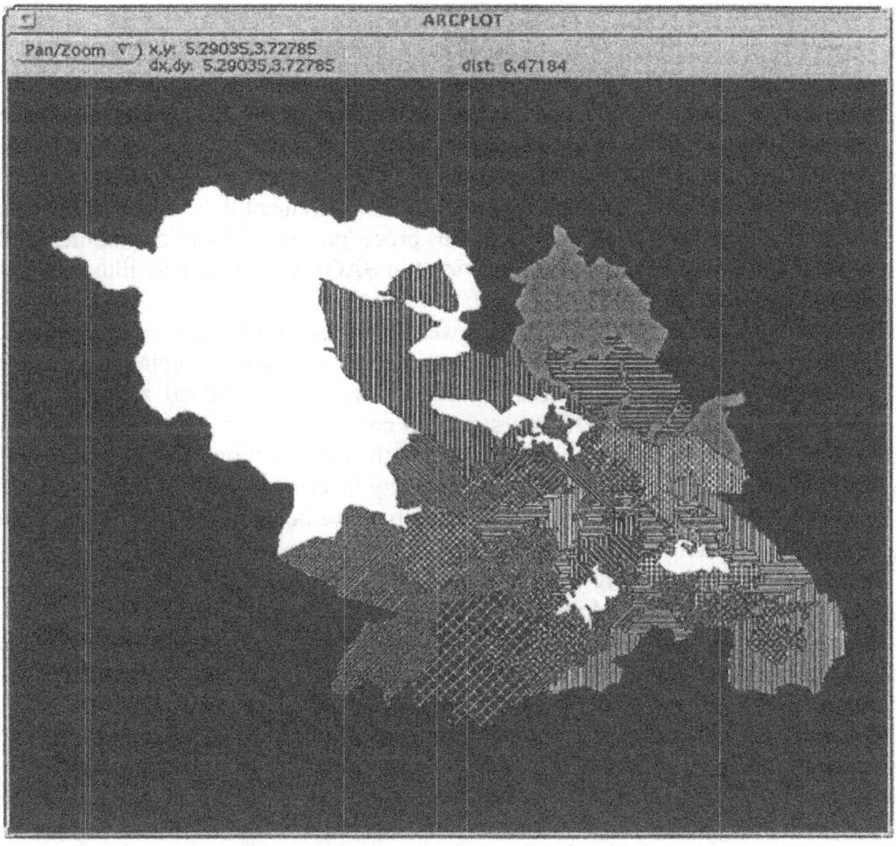

Fig. 5.4. Regions produced from the EDs using the options indicated on Figure 5.1.

swap is being considered.

•When %weight is not selected the weights under the objective functions are used to weight the objective functions directly. Since these are measured in different units, the user has the option of either allowing for this in assigning values to the weights, or standardising all the objective functions using the appropriate options for that function - in the case of the homogeneity variables for example, the values can be converted to z scores.

There is an overall weighting set between 1 and 100%, which is used to weight the percentage changes in each function when a swap is considered. This will allow some functions to get worse while others improve as long as there is an overall improvement. However, a threshold can also be set, such that if a function is made worse by this amount (also expressed as a percentage) the swap will not be accepted, no matter how great the improvement in the other functions.

A key element of the interactive approach taken in SAGE is the importance attached to feedback to the user on the performance of the regionalisation, which takes two forms. As the k means procedure is running, a graph is displayed showing the improvement in the objective function - this allows the user to terminate the process if repeated iterations are making little improvement for example. The main tools for assessing the output of the k means procedure are the range of graphical and statistical visualisation techniques provided in SAGE which are best illustrated by means of a simple example.

Figure 5.3 shows the 1159 EDs which were used for the 1991 census in Sheffield. Associated with the boundary data is a table of attributes, which contains one row for each of the 1159 zones. This table is initially stored as an internal attribute table within ARC/INFO, but SAGE takes a local copy to use for display and analysis purposes. These EDs were regionalised using the options shown in Figure 5.2. The objectives for the new regions were homogeneity in terms of the Townsend index of deprivation and equality of population, each objective being given equal importance (weights of 1). Each was also allowed to become worse by 5% on any given swap. The map of the resulting regions is shown in Figure 5.4, as it would appear in the ARCPLOT window when drawn from SAGE. The results of the regionalisation are stored as a new column in the attribute table, each zone being given the number of the region to which it belongs. Note that at this stage the ED boundaries are not altered - there are still 1159 zones in the coverage.

In order to assess the success of the method in achieving the objectives, the graphical capabilities of SAGE can be used to draw the two histograms shown in Figure 5.5. The lower graph is a histogram of the interquartile ranges of the Townsend Index values for the zones within each region. The interquartile range of values for the Townsend Index for the original EDs was 4.66 , whereas many of the new regions have interquartile ranges of around 1, although some still have high values . The upper graph shows a histogram of total population in the new regions, and shows that most of the new regions have very similar values. Again though one or two regions have populations which are far larger than the rest.

If there are problems with the regions which are constructed, one option is clearly to go back and re-run the regionalisation with modified objectives. Since the results will be saved into the same table, it is relatively easy to compare the effects of

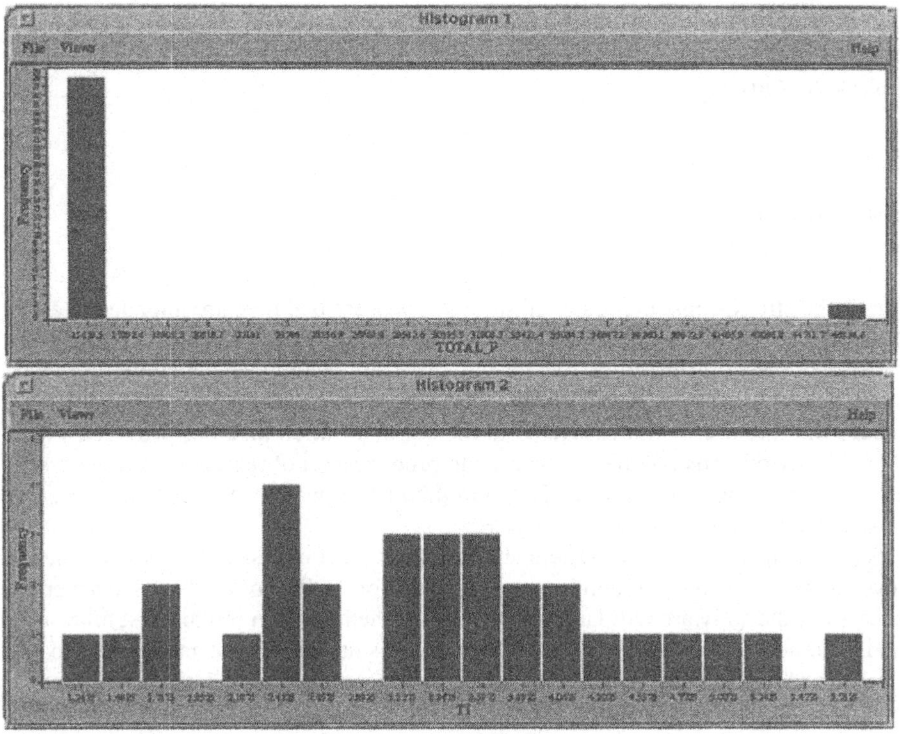

Fig 5.5. Graphical Assessment of Regionalisation Results.
The upper graph is a histogram of the total population in each region, the lower one is a histogram of the interquartile range of Townsend deprivation index values for zones making up each region.

different runs.

Alternatively, it is possible to move on to step 4 in Figure 5.1, and use a series of tools which will split large regions or merge small ones in order to improve the equality or homogeneity of a set of regions. In performing these split and merge operations, constraints can also be built in to avoid making the overall objective function much poorer - in trying to equalise population for example a threshold can be set on how much this is allowed to reduce the performance in terms of homogeneity.

These procedures can result in a change in the number of regions. If the best result seems to be to split one large region no attempt is currently made to merge two others to compensate. It is felt that this is justified by the fact the number of regions is often a relatively arbitrary decision, possibly guided by some idea of what the minimum population should be in the regions for the calculation of reliable rates, rather than by some fixed notion of a number of regions.

The final stage of the process (stage 5 on Figure 5.1) is to create a new set of boundaries by removing the boundaries between zones in the same region and merging their attribute values together, which in a GIS such as ARC/INFO is a very simple operation.

5.6 Conclusions

It must be stressed that the regionalisation tools described here are intended to assist with the interactive spatial analysis of data. Speed has therefore been one of the key criteria in selecting methods, which has ruled out techniques which may produce superior results. However, the design of SAGE, as a system linked to a GIS package means that the user is not restricted to the regionalisation methods provided within SAGE. Any other technique can be used to produce a set of region boundaries which can then be imported into ARC/INFO in the normal way, and used as the basis for analysis using the other tools of SAGE.

Some of the details of the system are still the subject of research - for example the investigation of ways of identifying nodal zones for the Taylor method. However the success of the software will largely stand upon whether it can produce useful results, and whether SAGE provides an environment suitable for the interactive spatial analysis of health data, which will only become apparent after it has been tested by others, which is the next stage of this research project.

Acknowledgments

The authors acknowledge receipt of ESRC grant R000234470 which has made the work reported here possible.

References

Anderberg M.R. 1973. *Cluster analysis for applications.* Academic Press. New York

Anselin L. 1995. Local indicators of spatial association - LISA. *Geographical Analysis,* **27**:93-115.

Barkley D.L., Henry M.S., Bao S. and Brooks K.R. 1995. How functional are economic areas? Tests for intra-regional spatial association using spatial data analysis. *Papers in Regional Science* **74**:297-316.

Bithell J.F. 1990. An application of density estimation to geographical epidemiology. *Statistics in Medicine* **9**:691-701.

Clayton D. and Kaldor J. 1987. Empirical Bayes estimates of age-standardized relative risks for use in disease mapping. *Biometrics* **43**:671-681.

Cliff, Haggett, Ord, Bassett and Davies. 1975. *Elements of spatial structure*. Cambridge University Press.

Cliff A.D. and Ord J.K. 1981. *Spatial processes : models and applications*. Pion, London.

Cressie N. 1984. Towards resistant geostatistics. In G.Verly, et al (eds) *Geo-statistics for natural resources characterization*, 21-44. Reidel, Dordrecht.

Cressie N.A.C. 1991. *Statistics for spatial analysis*. John Wiley and Sons, New York.

Elliott P., Cuzick J., English D. and Stern R. 1992. *Geographical and environmental epidemiology : methods for small area studies*. Oxford University Press.

Fotheringham A.S. and Wong D.W.S. 1991. The modifiable areal unit problem in multivariate statistical analysis. *Environment and Planning A*, 23:1025-1044.

Gatrell A.C. 1989. On the spatial representation and accuracy of address-based data in the United Kingdom. *International Journal of Geographical Information Systems* **3**:335-48.

Getis A. and Ord J.K. 1992. The analysis of spatial association by use of distance statistics. *Geographical Analysis* **24**:75-95.

Haining R.P., Wise S.M. and Blake M. 1994. Constructing regions for small area analysis: material deprivation and colorectal cancer. *Journal of Public Health Medicine* **16**:429-438.

Haining R.P, Wise S.M. and Ma J. 1996. The design of a software system for interactive spatial statistical analysis linked to a GIS. *Computational Statistics* (in press).

Haining R.P. 1996. *Spatial statistics and the analysis of health data*. Paper presented to the GISDATA workshop on GIS and health. Helsinki, June 1996.

Haslett J., Wills G. and Unwin A.R. 1990. SPIDER - an interactive statistical tool for the analysis of spatially distributed data. *International Journal of Geographical Information Systems* **4**:285-296.

Haslett J., Bradley R., Craig P.S.,Wills G. and Unwin A.R. 1991. Dynamic graphics for exploring spatial data with application to locating global and local anomalies. *American Statistician*, **45**:234-42.

Horn M.E.T. 1995. Solution techniques for large regional partitioning problems. *Geographical Analysis* **27**:230-248.

Lankford P.M. 1969. Regionalisation: Theory and Alternative Algorithms. *Geographical Analysis* **1**:196-212.

McQueen J. 1967. Some Methods for Classification and Analysis of Multivariate Observations. *Proceedings of the 5th Berkeley Symposium on Mathematical Statistics and Probability*, **1**:281-297.

Oliver M.A. and Webster R. 1989. A geostatistical basis for spatial weighting in multivariate classification. *Mathematical Geology* **21**:15-35.

Openshaw S. 1978. An optimal zoning approach to the study of spatially aggregated data. in Masser I. and Brown P.J. (eds) *Spatial Representation and Spatial Interaction*. Martinus Nijhoff,Leiden, 95-113.

Openshaw S. 1984. *The modifiable areal unit problem*. Concepts and Techniques in Modern Geography 38. GeoAbstracts, Norwich.

Openshaw S. and Rao L. 1994. Re-engineering 1991 census geography: serial and parallel algorithms for unconstrained zone design.

Ord J.K. and Getis A. 1995. Local spatial autocorrelation statistics : distributional issues and an application. *Geographical Analysis* **27**:286-306.

Raper J., Rhind D.W and Shepherd J.W. 1990. *Postcodes: the new geography*. Longman, Harlow.

Rosenfeld A. and Kak A. 1982. *Digital picture processing*. Academic Press, London.

Sammons R. 1978. A simplistic approach to the redistricting problem. in Masser I. and Brown P.J. (eds) *Spatial Reprsentation and Spatial Interaction.* Martinus Nijhoff, Leiden71-94.

Semple R.K. and Green M.B. 1984. Classification in Human Geography. in G.L.Gaile and C.J.Wilmott (eds) *Spatial statistics and models*, Reidel, Dordrecht. 55-79.

Spath H. 1980. *Cluster Analysis Algorithms*. John Wiley and Sons, New York.

Taylor P.J. 1969. The location variable in taxonomy. *Geographical Analysis* **1**:181-195.

6 A Structural Time Series Approach to Forecasting the Space-Time Incidence of Infectious Diseases: Post-War Measles Elimination Programmes in the United States and Iceland

J.D. Logan[1] and A.D. Cliff[2]

[1]Statistical Laboratory, University of Cambridge, Mill Lane, Cambridge, CB21SB, UK
[2]Department of Geography, University of Cambridge, Downing Place, Cambridge, CB23EN, UK

6.1 Introduction

When the epidemiological history of the present century comes to be written, the outstanding success that historians will be able to record is the global eradication of smallpox. The complex story, which culminated in the last recorded natural case in October 1977, has been told by Fenner, Henderson, Arita, Jezek and Ladnyi (1988). That success has inevitably raised questions as to whether other infectious diseases can also be eradicated[1]. At the present time, WHO has eight diseases so targeted by the millenium – an original list of six (diphtheria, measles, poliomyelitis, whooping cough, neo-natal tetanus and tuberculosis) and two recent additions, hepatitis B and yellow fever.

The main vehicle for eradication is vaccination. The ten-year smallpox eradication campaign, launched in 1967, gives an idea of the costs involved. Quite apart from the disease and death from smallpox itself, the cost of vaccination, plus that of maintaining quarantine barriers, is calculated to have been about $US1,000 million per annum in the last years of the existence of the virus in the wild. These costs disappear completely only if *global* eradication is achieved. Eradication is therefore an expensive business for which efficient vaccination strategies need to be found if it is to be a global possibility for target diseases.

One approach to the problem of reducing the costs of eradication would be to devise models to forecast the space-time incidence of disease. If this were possible, spatially targeted vaccination programmes might be devised to replace near blanket coverage. Indeed WHO recognised the importance of selective spatial control in the

---

[1] We follow Fenner in confining the term *eradication* to the total global removal of the infectious agent (except, as with smallpox, for preserved laboratory samples). *Elimination* is used to refer to the stamping out of the disease in a particular country or region, but it leaves open the possibility of reinfection from another part of the world.

early years of the smallpox campaign when mass vaccination was reinforced by contact tracing and vaccination in locations where cases had occurred.

There have been many attempts to devise forecasting models using both *SIR* (*S*usceptible–*I*nfective–*R*emoved) and *ARMA* (*A*uto*R*egressive–*M*oving *A*verage) time series methodologies. These are reviewed in Cliff, Haggett and Smallman-Raynor (1993, pp. 359-410) and in a series of papers produced as part of the proceedings of the NATO Advanced Research Workshop on Epidemic Models held at the Newton Institute in Cambridge in 1993 (Mollison, 1995). When applied to the space-time forecasting problem, these models suffer from one of two defects: they either miss the start of epidemics or, if they detect epidemic starts, they seriously over-estimate the total case load. If forecasting is seen as an aid to the development of spatially-targeted vaccination, the first of these defects implies that epidemics are likely to be well under way before effective vaccine-based intervention can be mounted; the second implies vaccine stocks required are likely to be over-estimated.

In this paper, we return to the age-old question of forecasting and develop a structural time series approach to epidemic projecting. The models were originally proposed by Harvey and Durbin (1986) in a very different context – to study the impact that legislation to make the wearing of car seat belts compulsory in the UK had upon the incidence of injuries and deaths among car drivers in road traffic accidents. In this context, the introduction of the new legislation was represented as an intervention in a time series of injuries/deaths that had trend, seasonal and irregular features. There are direct parallels here with the assessment of the impact of vaccination (the intervention) upon the form of epidemic time series of infectious diseases such as diphtheria, measles and whooping cough on WHO's eradication list. The model is illustrated by application to the quarterly time series of reported measles cases in Iceland, 1945-85, at the geographical scale of six standard regions.

The outline of the paper is as follows. In Section 6.2, we use the example of the campaign to eliminate measles from the United States to show both the strengths and weaknesses of aspatial mass vaccination in reducing the incidence of disease. In Section 6.3, the characteristics of the Icelandic data set are described. Section 6.4 defines the structural model. In Section 6.5, a spatial version of the model is fitted to the Icelandic data. The use of temporally-lagged explanatory variables, based upon events in the epidemic lead area of Iceland (the capital, Reykjavík) to act as triggers to predict epidemics in other regions of Iceland is discussed and illustrated in Section 6.6. The paper is concluded in Section 6.7.

6.2 Measles Elimination in the United States

At the start of the twentieth century, thousands of deaths were caused by measles each year in the United States, and an annual average of more than half a million measles cases and nearly 500 deaths were still reported in the decade from 1950–59. It was against this background that the Centers for Disease Control and Prevention evolved in the United States a programme for the elimination of indigenous measles once a safe and effective vaccine was licensed in 1963. In 1966, Senser (Senser, Dull and Langmuir, 1967) announced that the epidemiological basis existed for the eradication of measles from the United States using a programme with four tactical elements: (a) routine immunisation of infants at one year of age; (b) immunisation at school entry of children not previously immunised (catch-up immunisation); (c) surveillance; and (d) epidemic control. The immunisation target aimed for was 90–95 percent of the childhood population – effectively blanket vaccination.

Following the announcement of possible measles eradication, considerable effort was put into mass measles immunisation programmes throughout the United States. Federal funds were appropriated and, over the next three years, an estimated 19.5 million doses of vaccine were administered. This caused a significant reduction in the time series of reported cases (Figure 6.1), with the reported incidence plummeting in 1968 to around 5 percent of its 1962 value. Despite this dramatic success, total elimination in the United States has still not been achieved. Many reasons have been adduced. These include the emergence of new diseases in the US, such as AIDS, Legionnaires' and Lyme, which have shifted the focus of attention, and a general reduction in vaccination levels against measles as public awareness fell with the decline of measles as a public health problem.

But one other factor played its part, namely the aspatial nature of the vaccination programme. Many of the 5% or so of individuals who escaped vaccination each year were either spatially concentrated or became so at certain points in time. Examples are provided by geographically concentrated groups such as the Amish of Pennsylvania who were exempt on religious grounds, and a miscellany of unvaccinated childen who eventually coalesced on American university campuses in 1985, 20 years on from the start of mass vaccination in 1965. Each group experienced its own spatially confined epidemic when the local susceptible population built up to critical levels, documented in Sutter, Markowitz, Bennetch, Morris, Zell and Preblud (1991) for the Amish of Pennsylvania, and in Frank, Orenstein, Bart, Bart, El-Tantawy, David and Hinman (1985) for the campus outbreaks.

Figure 6.2 makes the point in a different way. This shows the distribution of counties in the United States reporting measles early in the elimination campaign in 1978, and five years later in 1983. The contraction of infection from most settled

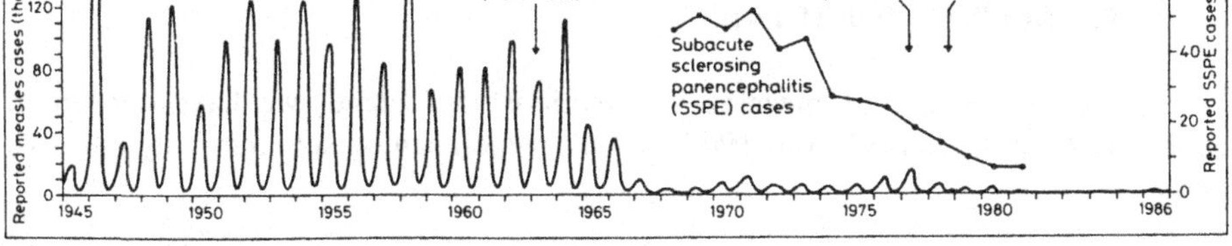

Fig. 6.1. United States: Reported Measles Cases, 1968-1986, showing also the concomitant reduction in a complication of measles (SSPE). Source: Cliff and Haggett (1988, Figure 4.9A, p164).

areas of the United States to restricted parts of the Pacific Northwest, California, Florida, the northeastern seaboard and the Midwest is pronounced. The persistence of indigenous measles in these regions has a strong geographical component, related to the movement of seasonal workers and migration from Mexico and the Caribbean in the case of Florida and California, and to cross border movements between Canada and the US in the case of the northern foci; for many years Canada lacked systematic measles vaccination programmes.

Such geographically-concentrated epidemics make the case for a spatial component to vaccination campaigns, and with it the advantages that would accrue if spatial forecasting of such occurrences were possible. Against this background, we consider spatial forecasting models for measles in the remainder of this paper, but for the much more isolated geographical environment of Iceland.

6.3 The Icelandic Data Set

6.3.1 Measles and Vaccination Data

The modelling to be described in subsequent sections is based upon reported cases of measles and vaccinations against the disease in the period, 1945-85. In Iceland, such vaccinations commenced in 1965 and so a time period symmetric about this date has been selected. In this section, we briefly describe the characteristics of the Icelandic data as a backcloth to the modelling. A fuller account is provided in Cliff, Haggett, Ord and Versey (1981, pp. 46-91).

Monthly reported measles data are available for 47 Icelandic medical districts from the latter years of the nineteenth century. Figure 6.3 plots the national annual number of recorded cases and cumulative vaccinations since 1945. The diagram shows that, as vaccination cover has built up, measles epidemics in Iceland have behaved like those in the United States. Levels have fallen sharply since 1977 and cases now occur in isolated, small outbreaks. To provide a manageable spatial data set for modelling, the monthly data for the 1945-85 period relating to the 47 districts were combined into quarterly time series for six standard administrative regions – the capital, Reykjavík (*R*), the Northwest (*NW*), the North (*N*), the East (*E*), the South (*S*) and the Southwest (*SW*). The locations of these regions appear later in Figure 6.5.

6.3.2 The Study Period, 1945-85

To model the impact of measles vaccinations in Iceland, we have chosen to concentrate on the period from 1945 to 1985 which can be split naturally into two parts:

a. 1945-65, when disease transmission was unaffected by mass vaccination;

b. 1966-85, when transmission was increasingly controlled by vaccination.

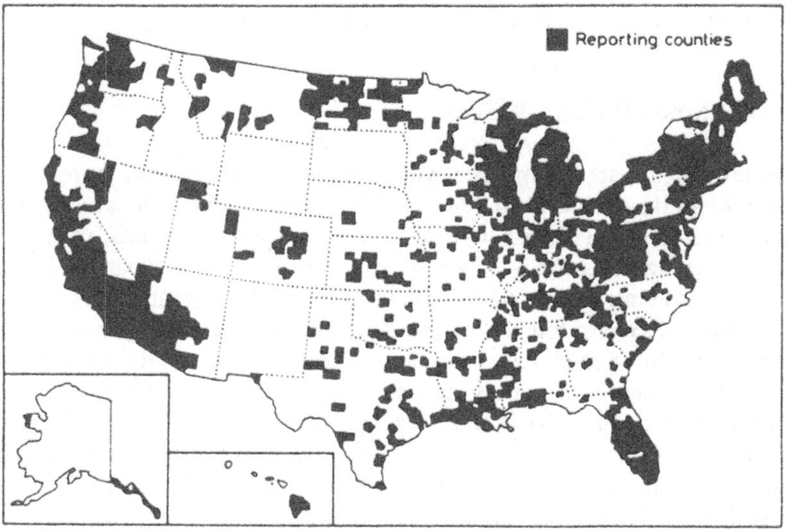

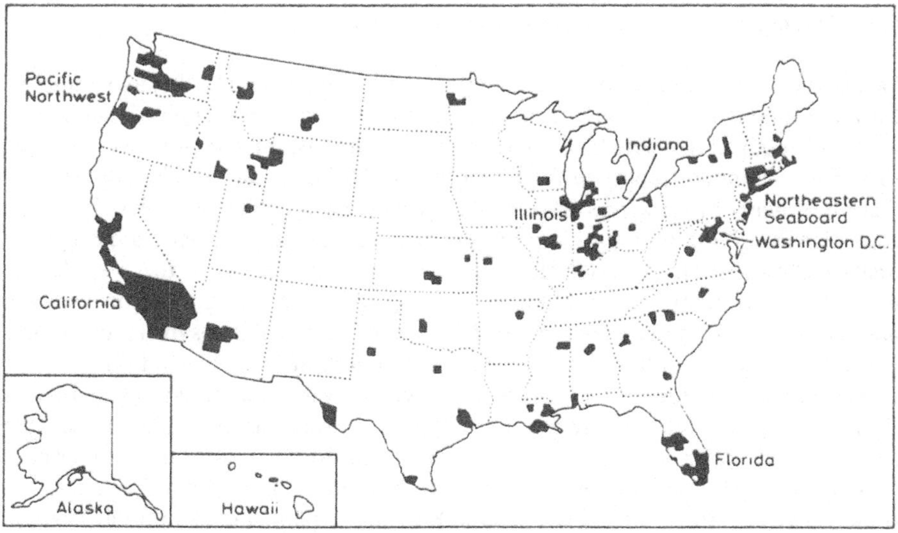

Fig.6.2. United States Counties Reporting Measles. (Upper) 1978. (Lower) 1983. Counties reporting measles during any week of the year are shown in black. Source: Cliff and Haggett (1988, Figures 4.9C and D, p165).

This period was itself divided into two: 1965-75 was used for model estimation and 1976-85 for the generation of *ex-post* forecasts. In Iceland, mass vaccination against measles was introduced in 1965-6 in all regions except the Northwest where the starting date was 1975. To give a common starting date across all regions for model comparisons, it was decided to estimate models on data through 1975, and to check forecasts against the actual values for the period, 1976-85. Thus each of the regional series included a section for both pre- and post-vaccination so that the effect of this intervention could be included in the estimation process.

6.3.3 The Study Period, 1945-85

To model the impact of measles vaccinations in Iceland, we have chosen to concentrate on the period from 1945 to 1985 which can be split naturally into two parts: 1. 1945-65, when disease transmission was unaffected by mass vaccination; 2. 1966-85, when transmission was increasingly controlled by vaccination. This period was itself divided into two: 1965-75 was used for model estimation and 1976-85 for the generation of *ex-post* forecasts. In Iceland, mass vaccination against measles was introduced in 1965-6 in all regions except the Northwest where the starting date was 1975. To give a common starting date across all regions for model comparisons, it was decided to estimate models on data through 1975, and to check forecasts against the actual values for the period, 1976-85. Thus each of the regional series included a section for both pre- and post-vaccination so that the effect of this intervention could be included in the estimation process.

6.3.4 Geographical Corridors of Measles Spread

One of the crucial features of the Icelandic measles data that make them so attractive for modelling is that, to support the quantitative record, the chief physician in each of the 47 medical districts that make up the six administrative regions was required to submit an annual report to the Chief Health Officer in Reykjavík. An essential feature of many of these reports is contact tracing information for each epidemic in each district, giving both the origin of the outbreak and its local spread through the district. Thus, in all essential elements, the pattern of spread of these epidemics through the Icelandic communities is actually available for comparison with our attempts to identify spread patterns from the data. Figure 6.4 maps the results of this contact tracing for all the measles epidemics to have affected Iceland this century. Geographically, epidemics of infectious diseases in Iceland are propagated partly through the network of towns and villages – from larger places to smaller (*hierarchical diffusion*) – and partly from individual towns out into their hinterlands (*contagious diffusion*). These features are to be seen in Figure 6.4 for measles. The principal vectors fan out from the capital, Reykjavík (population 95,000 in 1990) to the three main regional towns in Iceland of Akureyri (population 15,000) in the North, Ísafjörður (population 3,600) in the Northwest and Seyðisfjörður (population 1,000) in the East. Local spread from these centres into their hinterlands is illustrated

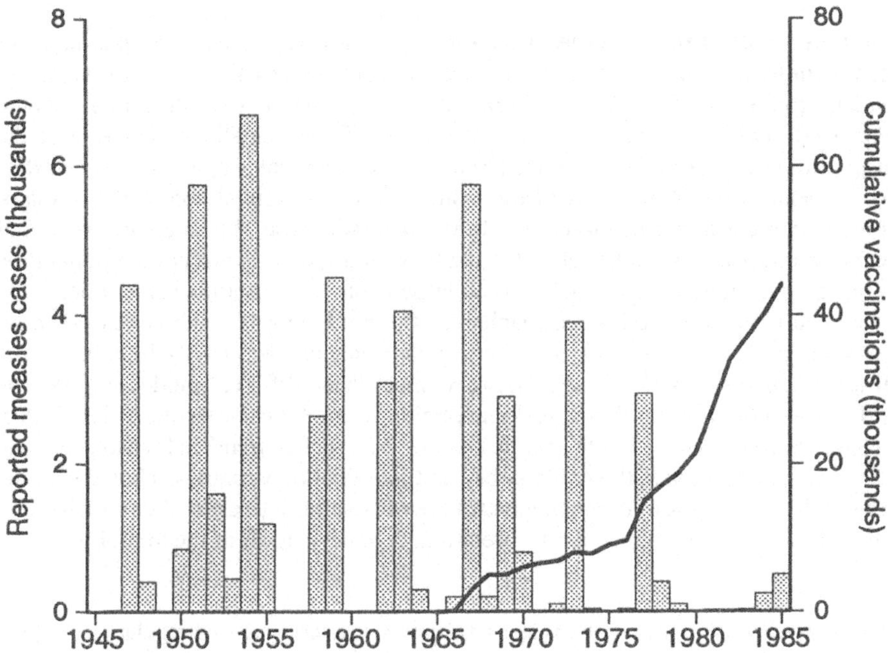

Fig 6.3. Icelandic measles epidemics and vaccinations. Annual reported cases (stippled) and vaccinations (line trace), 1945-1985.

by the vectors within the stippled areas on the map, and is most clearly seen around Akureyri.

6.3.5 Modelling Issues

The modelling to be described focuses upon two issues which spring from the preceding sections.

a. Can geographical components be added to structural time-series models in such a way as to avoid the characteristic problems of existing spatial epidemic forecasting models: namely, missing epidemic starts or overestimating the case load?

b. How can vaccination data be used to study the impact of immunisation upon the size and spacing of measles epidemics?

6.4 The Structural Time Series Model

The time series of reported measles cases and case rates for Iceland, like those for most countries of the world, exhibits three main characteristics, some of which are evident from Figure 6.3: (i) a long-term falling trend that may be attributed partly to improved health care and partly to intervention by systematic vaccination of the susceptible population since 1965; (ii) a seasonal element – epidemics, when they occur, peak in the winter part of the year; and (iii) a cyclical component – major epidemics recur at regular intervals. In the Icelandic case, this is about every four years, an interval that can be related to the time required for a susceptible population to build up through births to a threshold large enough to sustain an epidemic.

Conventional time series approaches to the modelling of such series begin by removing trend and seasonality by filtering (Box and Jenkins, 1970). In contrast, the structural time series model (*STSM*) recognises these different model components and, instead of filtering them, seeks explicitly to model each element. To do this, *STSMs* are composed of *unobservable components*, such as trend and seasonality, and *observable components*, the explanatory and intervention variables. Thus the model is formulated in terms of components that are of specific interest to the modeller and which have a direct empirical interpretation. A generic form of the model is:

$$\text{Observed} = \text{Trend} + \text{Seasonal} + \text{Cycle} + \text{Explanatory} + \text{Intervention} + \text{Irregular} \qquad (6.1)$$

This type of model then serves two purposes. Not only does it provide a basis for making predictions of future observations, but it also provides a description of the relative importance of significant features of a time series. A further advantage of the *STSM* is that it can be estimated using the Kalman filter (Kalman, 1960).

6.4.1 The Basic Structural Model

Table 6.1 shows how trend, cyclical and seasonal components (the unobserved components of a time series) can be combined to yield model (F), the basic structural model (*BSM*). This is the traditional representation of a time series in structural form as the sum of trend, seasonal and irregular components, and it is the starting point for the construction of structural models.

The seasonal component, γ_t The *BSM* handles seasonality by including either dummy variables or through trigonometric terms. Both involve a summation where s is the number of seasons per year; in the dummy variable seasonal, ω_t is simply a white noise term.

The irregular component, ε_t The irregular component, ε_t, is white noise. In the *BSM*, it is assumed to be independent of the two other white noise terms, η_t and ζ_t, which come from the stochastic trend component.

Table 6.1. Structural Time Series Analysis. Main components and models

Model	Component	Specification
	1a Random walk	$\mu_t = \mu_{t-1} + \eta_t$
	1b Random walk with drift	$\mu_t = \mu_{t-1} + \beta + \eta_t$
A *Local level / random walk plus noise*		$y_t = \mu_t + \varepsilon_t$ with μ_t as in (1a)
	2 Stochastic trend	$\mu_t = \mu_{t-1} + \beta_{t-1} + \eta_t$ $\beta_t = \beta_{t-1} + \zeta_t$
B *Local linear trend*		$y_t = \mu_t + \varepsilon_t$ with μ_t as in (2)
	3 Stochastic cycle	$\begin{bmatrix} \psi_t \\ \psi_t^* \end{bmatrix} = \varrho \begin{bmatrix} \cos\lambda_c & \sin\lambda_c \\ -\sin\lambda_c & \cos\lambda_c \end{bmatrix} \begin{bmatrix} \psi_{t-1} \\ \psi_{t-1}^* \end{bmatrix} + \begin{bmatrix} \varkappa_t \\ \varkappa_t^* \end{bmatrix}$ where ψ_t is the cycle, $0 \le \varrho < 1$, $0 \le \lambda_c \le \pi$
C *Cycle plus noise model*		$y_t = \mu + \psi_t + \varepsilon_t$ where $0 \le \varrho < 1$
D *Trend plus cycle*		$y_t = \mu_t + \psi_t + \varepsilon_t$ with μ_t as in (2)
E *Cyclical trend*		$y_t = \mu_t + \varepsilon_t$ $\mu_t = \mu_{t-1} + \psi_{t-1} + \beta_{t-1} + \eta_t$ with β_t as in (2)
	4 Non-stationary cycle	as (3) but $\varrho = 1$
	5a Dummy variable seasonality	$\gamma_t = \sum_{j=1}^{s-1} \gamma_{t-j} + \omega_t$
	5b Trigonometric seasonality	$\gamma_t = \sum_{j=1}^{[s/2]} \gamma'_{t,j}$; $\gamma'_{t,j}$ is a non-stationary cycle, (4), with $\lambda_c = \lambda_j = 2\pi j/s, j = 1,2,\cdots,[s/2]$
F *Basic structural model (BSM)*		$y_t = \mu_t + \gamma_t + \varepsilon_t$ where μ_t is as in (2) and γ_t is as in (5a) or (5b)

Source: Harvey (1989, Appendix 1, p. 510).

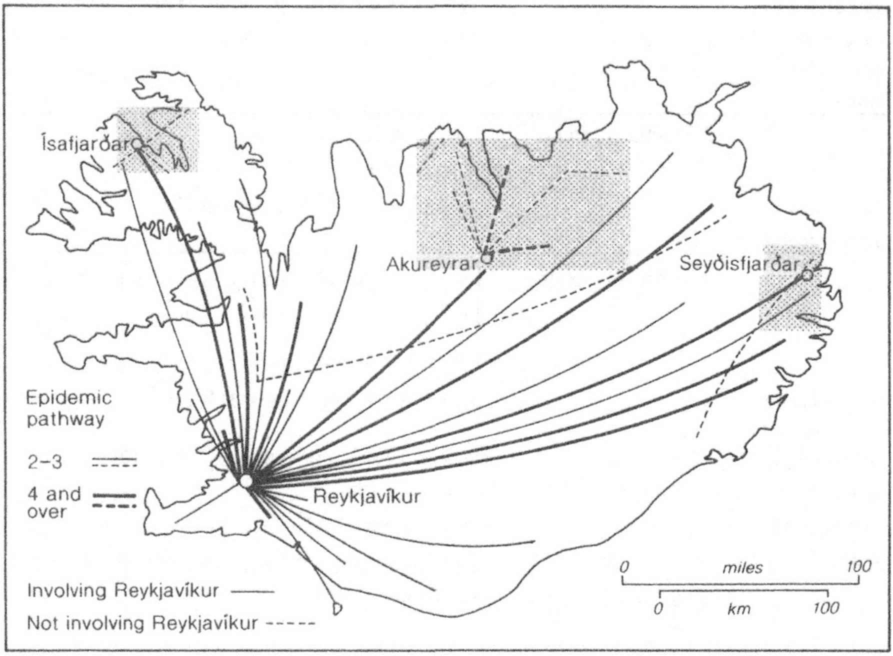

Fig. 6.4. Observed Epidemic Corridors for Measles Transmission in Iceland, 1900-90.

These four components may be combined to produce a generalised version of the *BSM* which includes a direct component for the cycle, for example $y_t = \mu_t + \gamma_t + \psi_t + \varepsilon_t$. However we shall consider only the *BSM* itself as our general underlying model when we add explanatory and intervention terms as described in the next subsection. Thus our model of unobserved components is given by model (F), namely

$$y_t = \mu_t + \gamma_t + \varepsilon_t\,, \tag{6.2}$$

where the trend is stochastic and the seasonal component may be either dummy variable or trigonometric.

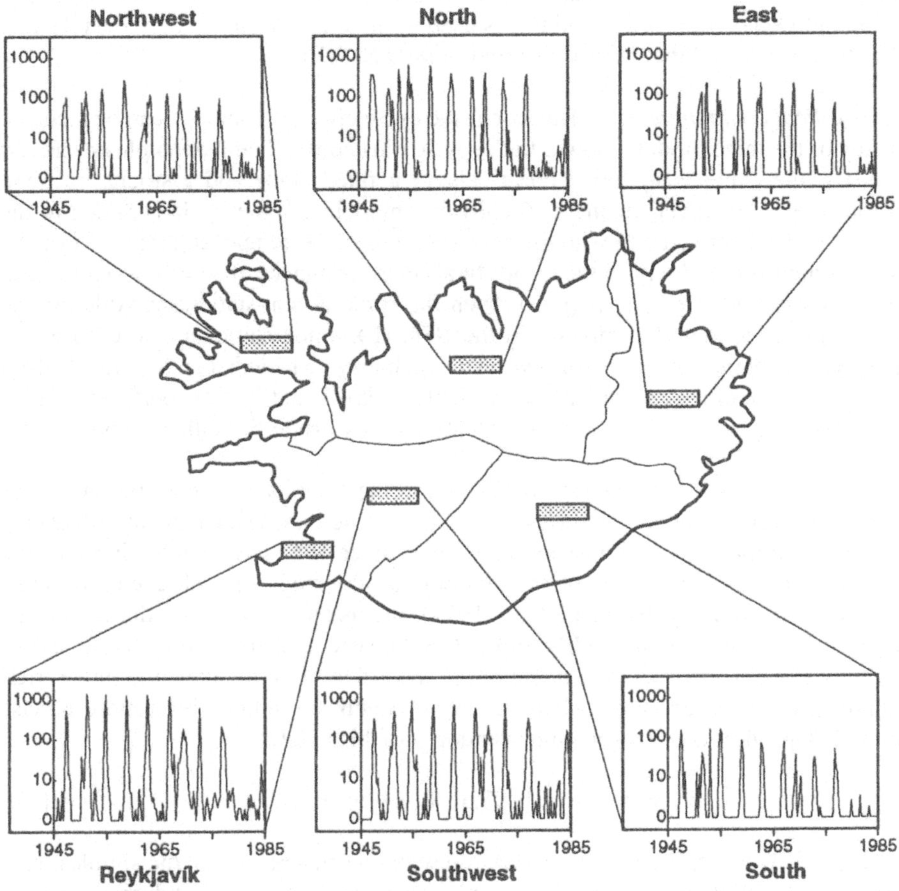

Fig. 6.5. Icelandic Regional Boundaries and Quarterly Time Series of Reported Measles Cases, 1945-85 (logarithmic scale).

6.4.2 Explanatory Variables and Interventions

The principal structural time series models can be regarded as regression models in which the explanatory variables are functions of time and the parameters are time-varying. With this interpretation, the addition of p observable explanatory variables to the right hand side of the *BSM* (F) is a natural progression, yielding

$$y_t = \mu_t + \gamma_t + \Sigma\delta_j x_{jt} + \varepsilon_t \qquad (6.3)$$

The trend component, μ_t The trend is divided into two terms, the level, μ_t, and the slope, β_t. In the random walk model, (1a), the level evolves over time but no slope term is included; in a random walk with drift, (1b), a fixed slope is included. In a stochastic trend, the slope is also allowed to vary over time; that is, both level and slope follow a random walk, and this is used in the local linear trend model (B). It is the stochastic trend that is incorporated in the *BSM* (F).

The cyclical component, ψ_t The stochastic cycle given in Table 6.1 can be included in any of the three models shown. In the cycle plus noise (C) and trend plus cycle (D) models, the cycle is included additively to a fixed slope and a stochastic trend respectively. However, in the cyclical trend model, (E), the cycle is added to the stochastic level term, along with the corresponding slope component; thus the cyclical term is incorporated within the trend. In all these instances, the cycle is assumed to be stationary, with $\varrho < 1$. If $\varrho = 1$, then the cycle is non-stationary. Although no cyclical term is included directly in the *BSM* (F), a non-stationary cycle forms an integral part of the trigonometric seasonal which may be included in this model where x_{jt} is the value of the *j*th explanatory variable at time *t*, and δ_j is its coefficient. There is no necessity for the $\{x_{jt}\}$ to have any stationarity properties either before or after differencing.

One of the other key features of *STSM*s is their ability to handle the impact of particular events upon a time series. As we have noted, the critical event influencing the shape of the Icelandic measles time series since 1965 has been the introduction of mass vaccination. Within *ARMA* modelling, the study of such events is called *intervention analysis* (Box and Tiao, 1975); the intervention is introduced at some time, $t = \tau$. This idea is readily applied to the structural framework by the use of dummy explanatory variables to represent the effect of events upon the time series. Dummy variables are used when the event cannot be defined by a more precise model. The addition of q such intervention variables yields

$$y_t = \mu_t + \gamma_t + \Sigma\delta_j x_{jt} + \Sigma\lambda_i w_{it} + \varepsilon_t \tag{6.4}$$

where w_{it} is the *i*th intervention variable with coefficient λ_i. In the simplest case when $q = 1$ and the response is instantaneous and constant, w_{it} is a dummy variable defined by

$$w_{it} = \begin{cases} 0, & t < \tau \\ 1, & t \geq \tau \end{cases} \tag{6.5}$$

Alternatively there can be an instantaneous response at time τ, which then either increases or decreases, or the response may build up gradually following the intervention. Other more complicated intervention variables may be used and these are described in the next section.

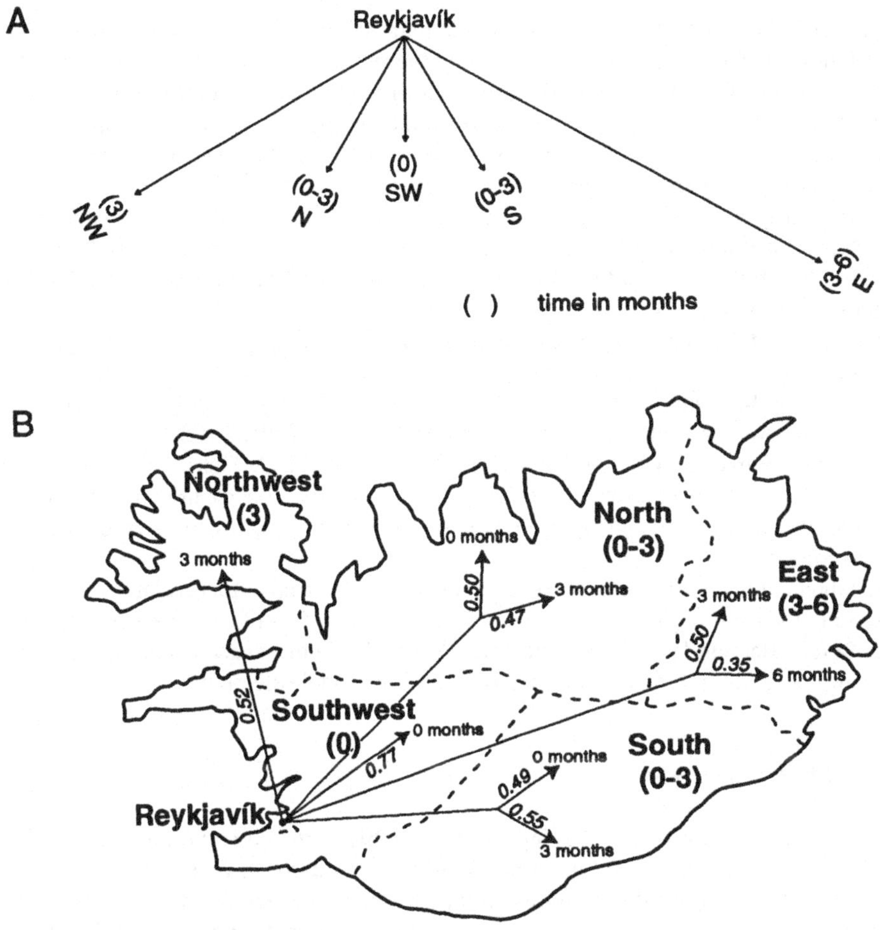

Fig. 6.6. Inter-Regional Propagation Corridors for Measles in Iceland, 1945-85. (A)Schematic representation of all significant cross-correlations between regions with lags. (B)Geographical linkages implied by (A). Regional boundaries are shown by pecked lines and cross-correlations are given in italics on vectors.

6.5 Application to Iceland

6.5.1 Spatial Structure

Figure 6.5 graphs the quarterly time series of reported measles cases for each of the six standard regions. Mirroring the national series of Figure 6.3, the introduction of vaccinations in 1965 caused a change in the trend of the series, with epidemic magnitude collapsing rapidly after 1975. To determine the patterns of inter-regional transmission of measles, cross-correlations (*CCF*) were calculated between all pairs of regional time series; the data were detrended and deseasonalised prior to analysis.

Figure 6.6 translates these *CCF*s into a map by plotting the maximum cross-correlations between Reykjavík at time t and the standard regions at time $t + k$, where k is the value of the lag in months at which the maximum occurred. Figure 6.6 is consistent with Figure 6.4. The cross-correlations imply that, on average, epidemics occur first in Reykjavík and then fan out across the country. They reach the geographically proximate South and Southwest regions in under 3 months; the North is also rapidly reached, linked as it is to Reykjavík by a major air link to Akureyri, the second largest town in Iceland. Epidemics take slightly longer to reach the more isolated Northwest and up to 6 months to reach the East region, geographically the farthest from Reykjavík.

The spatial dependence of regions away from Reykjavík upon events in the capital revealed by the cross-correlation analysis suggests that temporally-lagged terms involving the Reykjavík series could be used as explanatory variables in specifying appropriate structural models for these regions. Such a spatial impulse may be viewed as a trigger for events in dependent regions. In addition, the *CCF* analysis implies that Reykjavík acts as a measles diffusion pole whose reservoir of infection must be isolated by appropriate control strategies to reduce the risk of measles epidemics in other regions.

We then calculated the autocorrelation (*ACF*) and partial autocorrelation (*PACF*) functions for each regional time series. The most significant lags were used later as criteria for selecting lagged *AR* explanatory variables in the structural models tested. Bearing these comments in mind, we now turn to model selection.

6.5.2 Model Selection

For each of the six regions, the dependent variable in the corresponding structural model was the reported number of measles cases in quarter t; we use R_t, NW_t, N_t, E_t, and SW_t to denote these values. Initially, untransformed variables were tried. However, experience soon showed that log (base 10) transformations stabilized the extreme contrasts in means and variances between epidemic and non-epidemic phases in each of the districts, and so these were employed exclusively in the analysis to be described.

Unobserved Components. For the unobserved components of the model, we began by using the Basic Structural Model (F) described in Section 6.4.1. Best fits were obtained from a model that involved only a stochastic trend and an irregular term (i.e., no seasonal or cyclical component), namely:

$$y_t = \mu_t + \varepsilon_t \tag{6.6}$$

where $\mu_t = \mu_{t-1} + \beta_{t-1} + \eta_t$, $\beta_t = \beta_{t-1} + \zeta_t$, and ε_t, η_t, and ζ_t are random disturbances.

Explanatory Variables. a. The number of reported cases of a disease in a given area at time t will reflect in part the past history of the generating process in that area. Consequently, for the first of our explanatory variables, we used temporally-lagged versions of the dependent variables. As described in Section 6.5.1, the lags to include were identified from the *ACF*s and *PACF*s of the regional time series.

b. To handle the inter-regional transmission of measles virus from Reykjavík to other areas implied by Figure 6.6, R_t and R_{t-1} were added to the explanatory variables used for regions other than Reykjavík. Measles has a serial interval of about two weeks, so that new cases 'batch' through the population at fortnightly intervals (Burnet and White, 1972, pp. 15-16). The implication is that, given quarterly data, a term in R_t must be included to allow for transmissions that appear as simultaneous events simply because quarterly data can mask some six two-weekly crops of infections. Such 'simultaneous event' terms do not help with forecasting, and so the lagged-in-time term, R_{t-1}, was included to handle this feature of the model.

c. To incorporate into the model a simple measure of the size of the susceptible population, each region's quarterly population total was added to its list of explanatory variables. By analogy with the notation used for the number of cases of measles reported in each region, we use $R_{P,t}$ to denote the population of Reykjavík at time t, with similar notation for the other regions.

The exact forms of the explanatory variables used are summarized in Table 6.2.

Table 6.2. Forecasting Regional Variations in Measles Cases in Iceland, 1945–85, using a structural time series model. Dependent variables form column headings; explanatory variables appear in body of table.

Reykjavík	North	Northwest	South	Southwest	East
	R_{t-1}	R_{t-1}	R_{t-1}	R_{t-1}	R_{t-1}
	R_t	R_t	R_t	R_t	R_t
R_{t-1}	N_{t-1}	NW_{t-1}	S_{t-1}	SW_{t-1}	E_{t-1}
R_{t-16}				SW_{t-16}	
	N_{t-17}	NW_{t-17}			E_{t-17}
$R_{P,t}$	$N_{P,t}$	$NW_{P,t}$	$S_{P,t}$	$SW_{P,t}$	$E_{P,t}$

Intervention Variables. Following the licensing of measles vaccine in 1963, a large number of countries, Iceland included, embarked upon mass vaccination campaigns with a view to eliminating measles. The impact of the Icelandic campaign is described Cliff, Haggett and Smallman-Raynor (1993, pp. 256–59). In the structural time series approach, vaccination represents the intervention described in Section 6.4.2. Several forms of the intervention variable were tried, and these are now described. The interventions are shown schematically in Figure 6.7.

Region-independent variables Three standard intervention variables, V, reflecting the introduction of measles vaccination in Iceland were tried: a level (step) intervention (ℓ), a slope intervention (s) and a dynamic response intervention (d), given at time t by:

$$V_{l,t} = \begin{cases} 0, & t < \tau \\ 1, & t \geq \tau \end{cases}$$

$$V_{s,t} = \begin{cases} 0, & t < \tau \\ t-\tau, & t \geq \tau \end{cases} \tag{6.7}$$

$$V_{d,t} = \begin{cases} 0, & t < \tau \\ t-\varphi^{t-\tau}, & t \geq \tau \end{cases}$$

Here t indexes the quarters in the time series, and τ is the quarter in which the intervention occurred. In terms of the generic equation, (6.4), $V \equiv w$. As described in Section 6.3.1, measles vaccination was introduced in Iceland in 1965, so that we took τ to correspond to 1966, quarter 1 in the 1945-85 regional time series of reported measles cases. In addition, we chose $\varphi = 0.95$ so that $V_{d,t}$ approached an asymptote of one over the 20-year period, 1966–85.

Vaccination-based variables The impact of vaccination can be explored more fully by working directly with the time series of recorded vaccinations, v_t, rather than the simple dummy variables described above. The number of people vaccinated against measles in each of the six regions is recorded by the Icelandic authorities on a quarterly basis. For each region, a quarterly time series, $v_{c,t}$, of cumulative vaccinations administered was generated. Two variables were created by taking the logs of both the original and cumulated data so that, adjusting for zeros, we have:

$$l_t = \log(v_t + 1); \tag{6.8}$$

$$l_{c,t} = \log(v_{c,t} + 1). \tag{6.9}$$

Population size and vaccination effects can be combined to produce variables representing the percentage of people vaccinated by quarter t. Let $p_{v,t}$ denote the proportion of the population vaccinated against measles by quarter t, so that

$$p_{v,t} = 100 \left| \frac{v_{c,t}}{P} \right| \tag{6.10}$$

where P_t is the population of a given region at time t.

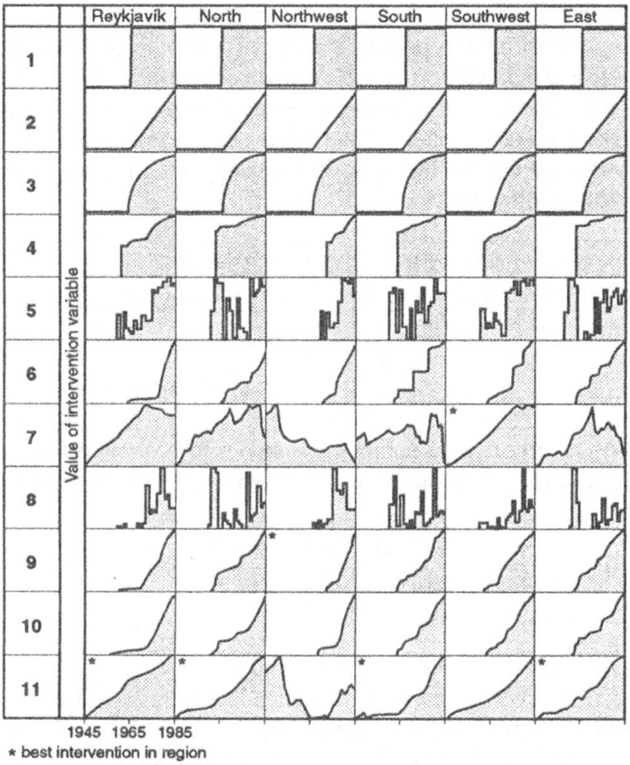

Fig. 6.7. Icelandic Measles Data, 1945-85. Form of intervention variables used in each region. Best-fit versions are astericked and models are as follows:

1 $V_{I,t}=\begin{cases}0 & t<\tau\\ 1 & t\geq\tau\end{cases}$

2 $V_{s,t}=\begin{cases}0 & t<\tau\\ t-\tau & t\geq\tau\end{cases}$

3 $V_{d,t}=\begin{cases}0 & t<\tau\\ 1-\varphi^{t-\tau} & t\geq\tau\end{cases},\ \varphi=0.95$

4 $l_{c,t}=\log(v_{c,t}+1)$

5 $l_t=\log(v_t+1)$

6 $p_{v,t}=100(\frac{v_{c,t}}{P_t})$

7 $G_{1,t}=\dfrac{P_t - v_{c,t}}{1000[1-(v_t/P_t)]^2}$

8 $G_{2,t}=\dfrac{1}{[1-(v_t/P_t)]^2}$

9 $G_{3,t}=\dfrac{1}{[1-(v_{c,t}/P_t)]^2}$

10 $G_{4,t}=100(G_{3,t}-1)$

11 $G_{5,t}=\dfrac{P_t - v_{c,t}}{1000[1-(v_{c,t}/P_t)]^2}$

where τ is the start of the intervention period, v_t is the number of recorded vaccinations, $v_{c,t}$ is the cumulative number of recorded vaccinations and P_t is the population at time t.

Griffiths factor variables The minimum population size of a community required to sustain an infectious disease endemically is called the *critical community size*. For measles, this is conventionally taken as c.250,000 in an unvaccinated population; see Bartlett (1957, 1960), Black (1966) and Schenzle and Dietz (1987). Griffiths (1973) has studied the impact of vaccination upon the critical community size. To follow his arguments, we define the following terms:

S	the number of individuals susceptible to the measles virus;
I	the number of individuals actively infective with measles;
ν	the birth rate;
λ	the infection rate;
μ	the recovery rate.

In the absence of vaccination, equilibrium is achieved when $S = \mu/\lambda = n$, and $I = \nu/\mu = m$, and the expected time, T, to the fade-out of infection, with $n >> m$, is given by

$$T_{m,n} \sim \frac{(2\pi n)^{1/2}}{\mu m} \exp\left\{ \frac{\left(M + \frac{n}{m} \right)^2}{2n} \right\} \tag{6.11}$$

Now suppose that a proportion (1-x) of people is vaccinated each quarter. Then $m' = xm$ is the number of people infective post-vaccination and $n' = n$ is the number of people susceptible post-vaccination. This causes a change in the fade-out time, T, such that

$$T(m,n) \rightarrow T(m',n') = T(xm,n) \sim T(m,n/x^2) \tag{6.12}$$

Thus, the effect on the critical community size is to multiply it by a factor of $1/x^2$.

Based on Griffiths' analysis, the five intervention variables shown in Table 6.3 were defined to reflect the n/x^2 relationship where n is the number of susceptibles post-vaccination. The variables were scaled to correspond with the scales of the dependent variables. $G_{4,t}$ is a rescaled version of $G_{3,t}$ to create a variable with zeros prior to the start of vaccination in Iceland in 1966. The quantity, $x = 1 - (v_t/P_t)$, should be used to articulate the Griffiths formula. However, we also tried $x' = 1 - (v_{c,t}/P_t)$. In the event (see next subsection) this modification improved the goodness-of-fit of the model.

Intervention vs explanatory variables In the structural model framework, the explanatory variables are observable exogenous variables that help to account for variation in the time series. If the series is affected by a once-only event, such as the introduction of mass vaccination, this effect is included in the model using an intervention variable. In their simplest form, intervention variables are just dummy explanatory variables and are estimated as such. The only difference between explanatory and intervention variables is that the latter are not included in the state vector until the intervention takes place. Diagnostic and goodness-of-fit tests are based on the generalised recursive residuals if either explanatory or intervention variables are included in the model.

Intervention analysis can be performed with or without the presence of other explanatory variables; in pure intervention analysis, only a single intervention variable is included. If the intervention takes place near the end of the series, the model should be constructed from the observations prior to the intervention and only then should the intervention be added and tested. However, if the intervention occurs early in the series, the specification of the model and the intervention should be carried out jointly. This is the more appropriate method for the Icelandic data analysed here because it allows the build-up of vaccinations over time to be modelled (by the Griffiths factor interventions). Consequently, we placed our intervention approximately midway through the time series.

Ordinarily the impact of an intervention cannot be explicitly defined because of the nature of the event it is designed to represent (the introduction of seat belt legislation in the Harvey and Durbin study, for example). Then it is natural to try to model the effect of the intervention using simple variables that consist of zeros up to the point of intervention, followed by a level, slope or 'dynamic response' change respectively. However, in the present study, we wanted to model the effect of the introduction of vaccinations, an intervention that opens itself up to explicit modelling using the number of vaccinations administered. While it would be possible to treat the Griffiths factors given in Table 6.3 as conventional explanatory variables, they have been handled as interventions because they are based on the number of vaccinations; this links them directly to the idea of intervention. Thus, although the variables are non-zero prior to intervention, the formulae in Table 6.3 make clear that this variation is due only to temporal changes in regional populations, and that such variation is independent of the intervention.

Table 6.3. Icelandic measles data, 1945-85. Definition of Griffiths intervention variables.

Variable identifier	Griffiths definition	Definition in terms of Icelandic variables
$G_{1,t}$	$\frac{n}{x^2}$	$\frac{P_t - v_{c,t}}{1000[1-(v_t/P_t)]^2}$
$G_{2,t}$	$\frac{1}{x^2}$	$\frac{1}{[1-(v_t/P_t)]^2}$
$G_{3,t}$	$\frac{1}{(x')^2}$	$\frac{1}{[1-(v_{c,t}/P_t)]^2}$
$G_{4,t}$		$100(G_{3,t}-1)$
$G_{5,t}$	$\frac{n}{(x')^2}$	$\frac{P_t - v_{c,t}}{1000[1-(v_{c,t}/P_t)]^2}$

6.5.3 Results

The model was fitted to each of the six regions, and all combinations of the variables defined above were tested on data for the period, 1945-75. *Ex-post* forecasts for 1976-85 were then generated using the parameters calculated for the estimation period. A number of measures of goodness-of-fit for structural time series models have been proposed by Harvey (1989). In the present study, goodness-of-fit was determined visually from a *CUSUM* plot of residuals; the smaller the *CUSUM*, the better the fit. The following diagnostics were also computed: the prediction error variance (σ^2), the coefficient of determination (R^2) and the post-sample predictive test statistic (ξ). Figure 6.8 illustrates the best-fit models for each of the six regions, while Table 6.4 gives the statistical details of the models. Thus, for example, the best model for the Reykjavík region is defined by $R_t = \mu_t + R_{t-1} + G_{5,t} + \varepsilon_t$, where μ_t and ε_t are terms from the basic structural model (F) defined in Section 6.4.1. At the 95% significance level, the post-sample predictive test statistic, $\xi(40)$ in Table 6.4, is not significant for any of the models listed. This means that the fitted series was consistent between the estimation and forecast periods, thus ensuring that the model did not give a spurious fit in the forecast period. In regions 2-6, R^2 was more significant for regions which included a Reykjavík explanatory variable than for those without such a term (but see the discussion, later, of the so-called *phase-shift problem*). In these regions, the prediction error variance was also lower, indicating a closer match between the fitted and the observed series. The *CUSUM* plots for all six models lay easily within their 10% confidence bands, implying that there was no model breakdown in the forecast period.

All of the explanatory variables except E_{t-17} were significant at the 5% level and, of those, only R_{t-1} from region 1 was not significant at the 1% level. Except for the East and Southwest, the best model for each region contained an explanatory variable from the Reykjavík region. In the East and Southwest, the poorer fit was caused by the fact that the Reykjavík variables did not allow the magnitude of epidemic peaks to be predicted accurately. This was a special problem in the East which is considered further in Section 6.6.

Table 6.4. Icelandic Measles Data, 1945–85. Explanatory and intervention variables in best fit structural time series models for each region and goodness-of-fit statistics.

Region	Explanatory variable	Intervention variable	$\tilde{\sigma}^2$	R^2	$\xi(40)$	*CUSUM*
1. Reykjavík	R_{t-1}^*	$G_{5,t}$	0.54	0.05	0.01	0.51
2. *N*	R_{t-1}^*	$G_{2,t}$	0.43	0.28	0.57	1.61
3. *NW*	R_t^*	$G_{3,t}$	0.36	0.15	0.11	1.12
4. *S*	R_t^*	$G_{5,t}$	0.23	0.49	0.30	1.61
5. *SW*	SW_{t-16}^*	$G_{1,t}$	0.45	0.14	0.04	0.83
6. *E*	E_{t-17}	$G_{5,t}$	0.53	0.03	0.10	0.98

* Significant at α=0.05 level (2-tailed test).

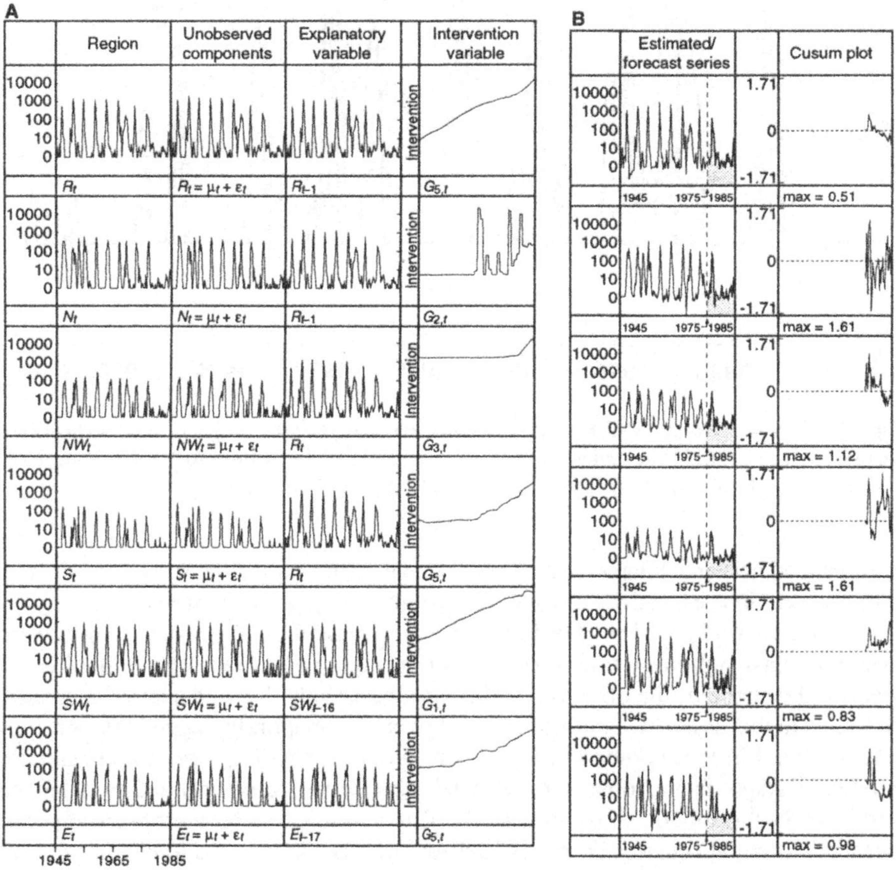

Fig.6.8. Icelandic Measles Data, 1945-85. (A) Schematic representation of the best-fit structural time series models for each of the six regions. (B) Estimated and forecast series

None of the intervention variables was statistically significant at the 5% level, yet they improved fits noticeably and this warranted their inclusion from a forecasting viewpoint. In three of the six regions, $G_{5,t}$ was the best of the intervention variables. In the Northwest, $G_{3,t}$ was the best, but this might be because mass vaccination was

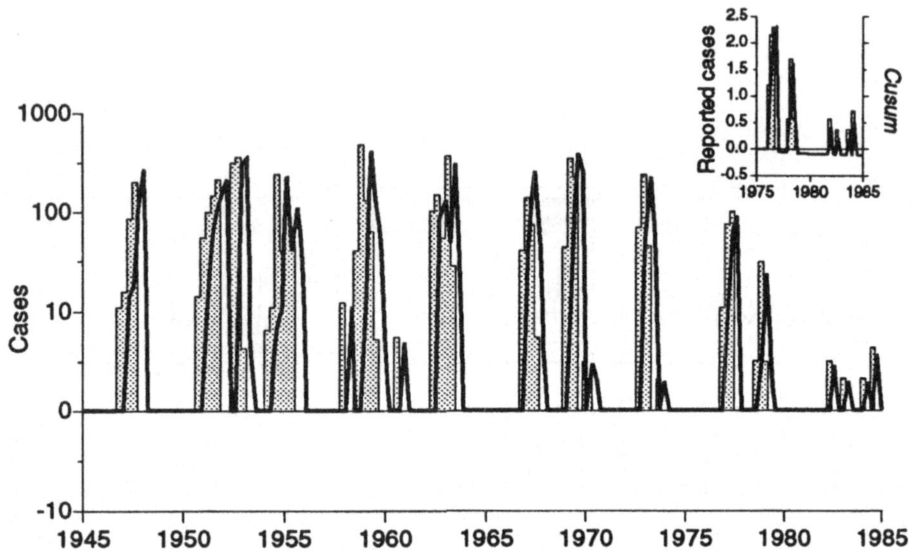

Fig. 6.9. Eastern Iceland, 1945-85. Reported measles cases (histogram) and fitted series (line trace) from a structural time series model to illustrate the phase shift problem. *CUSUM* plot of residuals highlighting the offset nature of the fitted series is inset.

commenced here much later (in January 1975) than in the other regions. The only real surprise was in the Southwest where $G_{I,t}$ produced by far the best results.

The phase-shift problem Deciding what is 'a good model' is not straightforward with the quasi-stationary time series shown in Figure 6.5. Many of the fitted series tracked the shape of the observed series very accurately, but one quarter in arrears. We refer to this as a *phase-shift problem*. Given the temporally-lagged relationships encapsulated in the models used here, the sudden switching from no epidemic to epidemic states in the observed regional time series commonly led to a lagged response in the model, thus causing the fitted series to echo the observed series accurately, but one quarter in arrears. Because of this effect, it was found that several models matched the shape of the observed series accurately, but produced goodness-of-fit statistics that were noticeably inferior to those from models which simple visual inspection showed were much poorer fits over the series as a whole.

The nature of this problem is illustrated in Figure 6.9, where the model used is given by $E_t = \mu_t + V_{s,t} + \varepsilon_t$. The epidemics in the fitted series match the observed almost perfectly except for a lag of about one quarter. This effect can be seen in the *CUSUM* plot which exhibits lagged peaks that correspond to the region's epidemic peaks. Because of these peaks in the *CUSUM*, this model gives a high *CUSUM* value of 1.76 and, since the fitted series is completely offset, the value for R^2 is zero. Thus the model is excluded from statistical consideration, although the fitted series, ignoring the lag, is a very good match to the observed series.

6.6 Triggering Strategies

The phase shift problem was recorded in all the regions for all combinations of local explanatory and intervention variables. However when the (lagged) number of measles cases in Reykjavík, R_t and R_{t-1}, were used as explanatory variables for the other regions, this lag of one quarter was reduced. Indeed, the inclusion of these lagged terms produced the best results in three of the regions, but this was frequently achieved at the expense of the accuracy with which epidemic magnitudes were estimated.

If time-lagged Reykjavík explanatory variables are used to tackle the phase shift problem, the deleterious impact of the *size* of epidemics in Reykjavík may be counteracted by *clipping* the actual series to a square waveform; this is done by replacing the actual values with ones or zeros if the values in the observed series exceed (1) or equal (0) some constant, c. For example, for the Reykjavík time series, we set $R_t = 1$ if cases were reported in month t and $R_t = 0$ otherwise. This is a 'natural' form of clipping on a presence/absence basis, with $c = 0$. Similar definitions were made for the regional series. Taking the Eastern region as an example, let $C^R_{\bullet\bullet}$ be the clipped variables for Reykjavík region and $C^E_{\bullet\bullet}$ be the clipped variables for the East region. The idea is to use these variables to predict the start of the regional epidemics and to combine them with a regional explanatory variable which would then estimate the size of the epidemics. In the variables below, the definitions only apply in the period immediately before epidemics, so ensuring the variables have a trigger effect. Occasionally the Reykjavík epidemic started in the same quarter as the regional epidemic. The most successful trigger, T, tried was defined as follows:

$$T_t = \begin{cases} 0 \text{ } if \text{ } C^R_{1,t} = 0 \text{ } and \text{ } C^E_{1,t} = 0 \\ 1 \text{ } if \text{ } C^R_{1,t} = 1 \text{ } and \text{ } C^E_{1,t} = 0 \\ 2 \text{ } if \text{ } C^R_{1,t} = 1 \text{ } and \text{ } C^E_{1,t} = 1 \\ or \text{ } if \text{ } C^R_{2,t} = 1 \text{ } and \text{ } C^E_{2,t} = 0 \\ 3 \text{ } if \text{ } C^R_{2,t} = 1 \text{ } and \text{ } C^E_{2,t} = 1 \end{cases} \tag{6.13}$$

Here, we define

$$C_{1,t} = \begin{cases} 1 \; \textit{if} \; R_t > 0 \\ 0 \; \textit{otherwise} \end{cases}$$

$$C_{1,t-1} = \begin{cases} 1 \; \textit{if} \; R_{t-1} > 0 \\ 0 \; \textit{otherwise} \end{cases}$$

$$C_{2,t} = \begin{cases} 1 \; \textit{if} \; R_t > 0 \\ 0 \; \textit{otherwise} \end{cases} \tag{6.14}$$

$$C_{2,t-1} = \begin{cases} 1 \; \textit{if} \; R_{t-1} > 0 \\ 0 \; \textit{otherwise} \end{cases}$$

The best model fit model was

$$E_t = \mu_t + T_t + E_{t-1} + V_{s,t} + \varepsilon_t \tag{6.15}$$

Figure 6.10 plots the observed and fitted series and shows the way in which this modification successfully reduced both the phase offset and matched epidemic magnitudes while maintaining a low *CUSUM*.

6.7 Conclusion

This paper has explored the use of a structural time series approach to model the regional incidence of reported measles cases in Iceland for the period, 1945-85. One of the cardinal advantages of this approach is its ability to model the impact of interventions in the time series process – here, the introduction of mass vaccination programmes after 1965. The methodology also allows a wide range of explanatory and intervention variables to be explored. One of the main modelling problems encountered, apart from the effect of vaccination, was handling the transition between non-epidemic and epidemic phases. The rapid switching between these states in all the regional time series was difficult to model. But the use of clipped time series to design triggering mechanisms shows how the problem might be tackled. The following main conclusions were reached:

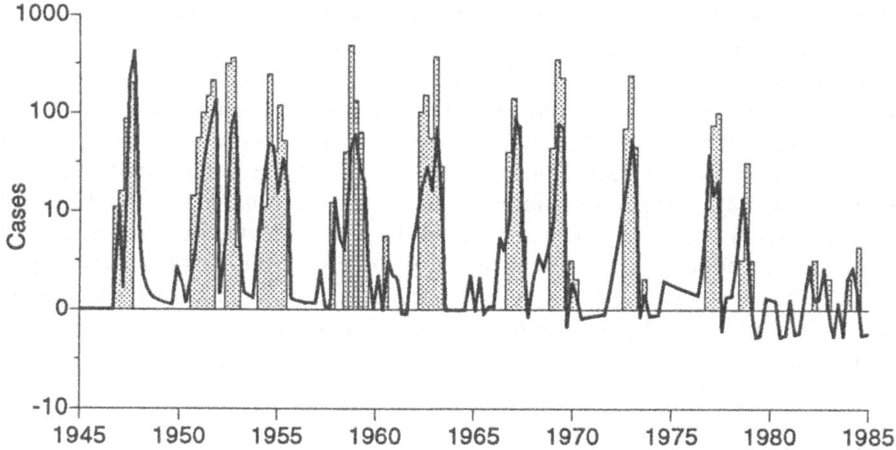

Fig. 6.10. Eastern Iceland, 1945-85. Reported measles cases (histogram) and fitted series (line trace) for structural time series model given by equation (6.15), using trigger variables based on Reykjavík time series of reported measles cases to model epidemic starts in the East.

a. Measles spreads hierarchically in Iceland from Reykjavík to other regions of the country. The use of temporally lagged variables based on the reported number of measles cases in Reykjavík as explanatory variables in other regions produced significant improvements in model fits as compared with single region models that excluded such a spatial transmission term.
b. The use of intervention variables confirmed the importance of Griffiths' (1973) work in reducing the amplitude of epidemics, and again produced real gains in reducing residual variance in the models.
c. Capturing epidemic starts remains difficult. Spatial triggers based on clipped variables are one way of avoiding the phase shift problem. Further work is required to develop triggers that are more sophisticated than the simple example used in Section 6.6 to illustrate the principles involved.

Our view is that the *STSM* approach to epidemic forecasting is capable of further development. It is much simpler to implement for forecasting than the more usual *SIR* models, and yet it does not make the heroic assumptions of standard *ARMA* models. We can only speculate about the portability of the model to other geographical areas and, in particular, how it would translate from an island setting to continental areas. But, given the flexibility of the *STSM* framework, we do not anticipate any significant problems. A pointer in this direction is provided in a recent review paper by Harvey and Koopman (1996). One of the case studies described by these authors is a brief examination of the time series of monthly reported measles cases in New York City, 1928-73. The model selected was based on the *BSM* (see Table 6.1), although the authors used a fixed seasonal and two additional cyclical terms. As in the Icelandic case studied here, the New York series showed a sharp decline from the mid 1960s due to the introduction of mass vaccinations in the United States (see Section 6.2). No attempt was made to model this feature of the data.

So far as we are aware, no sensitivity analysis of *STSMs* has been undertaken to determine their utility from a forecasting rather than a modelling viewpoint. It will be important to determine the robustness of forecasts as the model is applied at different spatial scales, and to time series in which regular sequences of epidemic waves give way to separate outbreaks, and then to isolated cases. As the World Health Organisation continues to extend its mass vaccination programmes to eliminate infectious diseases from areas of the globe, the continued time–space disintegration of epidemic waves will proceed apace: the answers provided by sensitivity studies will be of special interest to those responsible for articulating control programmes.

References

Bartlett M.S. 1957. Measles periodicity and community size. *Journal of the Royal Statistical Society A*, **120**:48-70.

Bartlett M.S. 1960. The critical community size for measles in the United States. *Journal of the Royal Statistical Society A*, **123**:37-44.

Black F.L. 1966. Measles endemicity in insular populations; critical community size and its evolutionary implication. *Journal of Theoretical Biology*, **11**:207-11.

Box G.E.P. and Jenkins, G.M. 1970. *Time series analysis, forecasting and control*. San Francisco: Holden-Day.

Box G.E.P. and Tiao G.C. 1975. Intervention analysis with applications to economic and environmental problems. *Journal of the American Statistical Assosciation*, **70:**70-79.

Burnet M. and White D.O. 1972. *Natural History of Infectious Diseases* (Fourth Edition). Cambridge: Cambridge University Press.

Cliff A.D. and Haggett P. 1988. *Atlas of Disease Distributions*. Oxford: Blackwell Reference Books.

Cliff A.D., Haggett P., Ord J.K. and Versey G.R. 1981. *Spatial diffusion: an historical geography of measles epidemics in an island community*. Cambridge: Cambridge University Press.

Cliff A.D., Haggett P. and Smallman-Raynor M.R. 1993. *Measles: an historical geography of a major human viral disease from global expansion to local retreat, 1840-1990*. Oxford: Blackwell Reference Books.

Fenner F., Henderson D.A., Arita I., Jezek Z. and Ladnyi I.D. 1988. *Smallpox and its eradication*. Geneva: World Health Organization.

Frank J.A., Orenstein W.A., Bart K.J., Bart S.W., El-Tantawy N., David R.M. and Hinman A.R. 1985. Major impediments to measles elimination. *American Journal of Diseases of Children*, **139**:881-8.

Griffiths D.A. 1973. The effect of measles vaccination on the incidence of measles in the community. *Journal of the Royal Statistical Society A*, **136**:441-49.

Harvey A.C. 1989. *Forecasting, structural time series models and the Kalman filter*. Cambridge: Cambridge University Press.

Harvey A.C. and Durbin J. 1986. The effects of seat belt legislation on British road casualties: a case study in structural time series modelling. *Journal of the Royal Statistical Society A*, **149**:187-227.

Kalman R.E. 1960. A new approach to linear filtering and prediction problems. *Transactions of the ASME Journal of Basic Engineering*, **82**:34-45.

Mollison D. (ed). 1995. *Epidemic models: their structure and relation to data*. Publications of the Isaac Newton Institute, 5. Cambridge: Cambridge University Press.

Schenzle D. and Dietz K. 1987. Critical population sizes for endemic virus transmission. In Fricke W. and Hinz E. (eds) *Räumliche Persistenz und Diffusion von Krankheiten*, **83**. Heidelberg: Heidelberg Geographical Studies, 31-42.

Senser D.J., Dull H.B. and Langmuir A.D. 1967. Epidemiological basis for the eradication of measles. *Public Health Reports*, **82**:253-6.

Sutter R.W., Markowitz S.E., Bennetch J.M., Morris W., Zell E.R. and Preblud S.R. 1991. Measles among the Amish: a comparative study in primary and secondary cases in households. *Journal of Infectious Diseases*, **163**:12-16.

7 Estimating the Impact of Preventative Action on the Space-Time Incidence of HIV/AIDS

Richard Thomas and Fiona Smyth

School of Geography, University of Manchester, Manchester, M13 9PL, UK

7.1 Introduction

Ever since the isolation of HIV (Barré-Sinoussi et al., 1983; Gallo et al., 1984), debates about the appropriate methods to curtail the transmission of the virus have been fiercely contested (Smyth and Thomas, 1996a). In the continued absence of an effective vaccine (Cease and Berzofsky, 1988), these preventative actions have either entailed implementing medical interventions linked to the serological test or social measures, like community orientated education, intended to modify high-risk behaviours. Medical opinion, for example, has often justified contacting and testing the partners of those who are known to be HIV+ on the grounds that it is unethical to deny them the opportunity to take precautions to prevent further passage of the virus (May et al., 1989; Knox et al., 1993). In contrast, those most at risk to infection have been quick to counter that such public interventions are an extreme invasion of privacy (Krieger and Appleman, 1994) and, instead, argue for the adoption of socially constructed measures made sensitive to their specific needs (Kirp and Bayer, 1992).

Despite this contention about their legitimacy, the successful implementation of all these interventions requires *direct* action to be taken by either voluntary or government agencies. Alternatively, prevention may also be construed as a *passive* activity which confers protection on a community through their collective attributes. A low rate of acquisition of new sexual partners, for example, might preclude the circulation of HIV, while both social and geographical distance from partnership networks where infection is present offer a more temporary degree of protection. This paper, therefore, draws upon results obtained from various versions of the recurrent epidemic model to evaluate the possible effects of both direct and passive prevention on the transmission of HIV. The next section considers the implications of estimates for the effects of direct actions on the recorded epidemics of HIV/AIDS among gay men and intravenous drug users (IVDUs) in Ireland (Smyth and Thomas, 1996b). Then, the possible effects of passive protection are explored through modelling systems (Thomas, 1994; 1996a; 1996b) which take account of estimated variations in the main transmission parameters (Anderson and May, 1988). The discussion sets these findings within the debate about the appropriateness of the various interventions against HIV/AIDS.

7.2 Direct Actions

7.2.1 Methods

The most recent advances in the theory of infectious disease are reviewed in Mollison (1995) and Isham and Medley (1996). We begin, however, with the adaptions to the recurrent epidemic model (Bailey, 1975, Haggett, 1994) that are necessary to mimic the transmission of HIV are due to Anderson (1988) and Isham (1988). They both describe a model that is derived for a community of gay men and may easily be reinterpreted to match other epidemic settings. Suppose that, prior to the first infection, this community numbers n individuals. Then, let x denote the number who are susceptible, y the number who have antigen levels that are sufficient for them to transmit HIV, and w the number of seropositive uninfectives, who carry antibodies to the virus but have ceased to be infectious because their antigen levels are too low to be detectable (May et al., 1989). Accordingly, at any point in time, t, the transmission of HIV is represented by the following state diagram.

$$\begin{array}{l} \mu n \rightarrow [x] \rightarrow \beta rxy/n \rightarrow [y] \rightarrow y/D \rightarrow [w] \rightarrow w/A \\ \qquad\;\; \downarrow \\ \qquad\;\; \theta x \end{array} \tag{7.1}$$

Here, is the rate of entry of men into the gay population and is their death rate from causes other than AIDS. The incidence of HIV is given by the expression rxy/n which has the following derivation. Let r denote the average number of sexual partners per unit of time, then y/n is the probability a new partner is infectious and rxy/n is the total number of partnerships between susceptibles and infectives. The number of new infections (the incidence) is obtained by multiplying this total by β, which is the probability that such a partnership results in the transmission of HIV. The removal of infectives from active circulation depends upon D, which is the average duration of the period of communicability for HIV. Then, y/D counts the number of these individuals who are removed in each unit of time. Similarly, people are uninfectious seropositive for an average duration A, such that w/A counts their incidence as new cases of AIDS.

A crucial property of this modelling system is revealed at the start of the simulated epidemic when a single infection is presumed to occur in an otherwise wholly susceptible community. Then, it can be shown that the reproduction rate (R) of HIV infection is given by

$$R = \beta r D \tag{7.2}$$

Here, R is seen to be the average number of susceptibles who acquire HIV from the initial infective during the period of communicability (Anderson and May, 1991). Therefore, if R is less than unity, an epidemic cannot begin because the first infective will fail to reproduce himself prior to his removal. Similarly, the critical partner acquisition rate for an epidemic to start in a community is derived as

$$r' = 1/\beta D \tag{7.3}$$

The modelling system described by the state diagram (7.1) may be made operational by using finite difference methods (Thomas, 1993) to simulate the deterministic time paths of the epidemic variables (x, y, w) and the incidence of HIV ($\beta rxy/n$) and AIDS (w/A). This simulation procedure adopts constant parameter estimates and, therefore, takes no account of direct actions taken by the community during the course of the epidemic to prevent the transmission of HIV. Such actions, however, may be represented in time dependent framework where a relevant parameter estimate is reduced during the simulation in response to a particular community initiative. In this respect, the voluntary adoption of most safer behaviours and practices correspond to reductions to particular parameter estimates (Knox et al., 1993). Both increased condom use and needle cleaning by IVDUs, for example, will be equated with reductions to the transmission probability, β. Similarly, decreased promiscuity and needle sharing are synonymous with a lower rate of partner acquisition, r. Medical interventions are not so easily represented, although contact tracing or quarantining those testing HIV+ are both intended to terminate the epidemic eventually by reducing the number of circulating infectives (y) to zero.

In practice, the legacy of AIDS prevention suggests these relationships between specific interventions and a particular parameter can usefully be generalised. With the exceptions of quarantining in Cuba (Santana et al., 1991) and contact tracing in Sweden (Hendriksson and Ytterberg, 1992), sustained and mandatory medical interventions have been rare. Therefore, it might be expected that any influence of prevention on observed AIDS incidence would be attributable largely to voluntary actions and reflected in changes to β and r. Moreover, because these parameters combine to form the rate of infection (βr), an assessment of the general effects of voluntary interventions can be made through an analysis of either of their values.

One method for making this assessment is described in Smyth and Thomas (1996b). Here, the model is calibrated with all the parameters given epidemiological estimates except one which is free to vary. Usually, this parameter will be the partnership rate which is notoriously difficult to measure reliably (Gagnon and Simon, 1974; Knox, 1986). The start of the epidemic is taken to be the time of the first diagnosis of HIV+. Then, by linear interpolation, the model is run with successive values of r until a specified AIDS incidence in the observed series is predicted exactly. This procedure may be repeated for each observed AIDS incidence. Accordingly, any reductions to these fitted r-values through time should reflect the extent to which voluntary actions have managed to check the initial force of HIV infection.

It should be noted, however, that these fitted r-values are averages obtained from a deterministic simulation procedure and, therefore, the differences between them have no formal statistical significance. To make this distinction requires the model to be expressed in a stochastic format where the deterministic input and output terms are recast as transition probabilities. The incidence of HIV, for example, is expressed as

$$p[(x,y,w) \rightarrow (x-1,y+1,w)] = \beta rxy/n\Delta t, \tag{7.4}$$

and is the probability of a specified individual making the transition from susceptible to infective during the interval Δt. The occurrence of such events are estimated by Monte Carlo methods to generate time series describing random variations around the deterministic average (Bartlett, 1960; Murray and Cliff, 1977; Thomas, 1988). Although not formally demonstrated in this paper, such stochastic variation enables some tests for the significance of pairs of fitted r-values to be posited. First, successive simulations with a given r-value would generate an error distribution around the time of the fit to the observed AIDS incidence. Alternatively, it may be preferable not to assume the other epidemiological parameters (β and D) are constants but, instead, regard them as random variables drawn, as in Bayesian analysis, from some appropriate prior distribution. Then, the process of drawing parameter values from these distributions and running a series of stochastic simulations would generate error distributions which not only reflected uncertainty in the progression of infection, but also in the values of the population parameters.

So far, the statistical analysis of temporal variations in the model parameters has provided a retrospective assessment of the possible impacts of preventative actions. In contrast, prospective appraisals can be made by adopting modelling strategies which imitate one of the interventions against HIV by altering one of the parameter values during the course of the simulation (May et al., 1989; Knox et al., 1993). Suppose the parameter in question is again the partnership rate. Then, any simulated intervention which subsequently alters the starting estimate of r will be represented by two statistics (Smyth and Thomas, 1996b): first,

$$\Delta r = r - r^*, \tag{7.5}$$

which is the decrement to the initial rate induced by the new modified rate, r*; and second, $t(\Delta r)$ is the time when this intervention was implemented. The impact of a particular intervention denoted by the pair, $[\Delta r, t(\Delta r)]$, may be measured either by the size (cumulative AIDS incidence) of the modified epidemic or the duration of the epidemic cycle. Variations in these impact indicators may be obtained by running successive simulations for alternative $[\Delta r, t(\Delta r)]$ pairs with the values of $t(\Delta r)$ spanning the initial epidemic cycle and those of Δr the range $r^* = r$ to $r^* = 0$. The results of this procedure may then be converted into control charts detailing the impact of each possible intervention.

7.2.2 Irish Evidence

The results reported here are derived from the following estimates of the epidemiological parameters. The transmission probability is taken to be $\beta = 0.1$, which is an estimate derived from a large sample of partners of seropositive gay men in the US (May et al., 1989). Similarly, estimated periods of communicability, D = 2 years, and seropositive uninfectiousness, A = 4 years, have been derived from research into viral abundance and antigen concentration after seroconversion (Pederson et al., 1987; Anderson and May, 1988; Nowak et al., 1991). Together,

Table 7.1. Fitted annual rates of partner acquisition (r) by risk behaviour and country of origin

		r		AIDS incidence	
Risk Group	Population(n)[1]	1987-fit	1991-fit	1987	1991
IVDUs, Ireland	1289	16.00	11.65	9	31
IVDUs, UK	11513	16.00	13.18	10	84
Gay Men, Ireland	31732	9.10	8.90	6	23
Gay Men, UK	506370	15.61	12.45	113	306

[1]These are minimum estimates of the initial population. Sources are given in Smyth and Thomas (1996b)

these values are consistent with a critical partnership rate (equation 7.3) of $r' = 5$ per year for an initial infection to be self-sustaining. The demographic rates for Ireland are $\mu = 0.016$ and $\theta = 0.016$ persons per year and are adjusted for the effects of net migration (see Smyth and Thomas, 1996b).

These values were applied to the procedure for deriving fitted values for the unknown partnership rate. Table 7.1 summarises the results obtained from this procedure for recorded AIDS incidence among IVDUs and gay men in Ireland and the UK between 1987 and 1991. Despite the lack of formal statistical testing, the variations observed in the fitted r-values are subject to a variety of interpretations. The partnership rates for IVDUs, for example, decline consistently between 1987 and 1991 and the Irish rate is lower in 1991 than the UK rate. A number of factors may have contributed to the more favourable Irish outcome: first, the more widespread adoption of actions that reduce the risk of needle sharing; and second, the improved recording of IVDU AIDS incidence in the UK consequent upon belated official awareness of the magnitude of this risk. Moreover, the evolution of AIDS campaigning in both countries lends support to each of these factors. In particular, the illegality of anal sex in Ireland until 1993 stopped the government initiating campaigns targeted specifically at gay men and, instead, the limited resources that were available were most likely to be directed towards drug users. In the UK, this orientation was the reverse. HIV had become established earlier in the gay community and the official response to IVDU infection was slower.

The fitted r-values for gay men display more pronounced national contrasts. The Irish rates are consistently less than their UK equivalents, but only the latter exhibit a marked decline between 1987-91. This aspect of the UK experience might be accounted for by more preventative action, like increased condom use. This explanation, however, would seem inappropriate for Catholic Ireland where condoms might be expected to be underused by gay men because of their association with contraception and, therefore, heterosexuality. Alternatively, the lower Irish rates might be the result of the under-reporting of diagnoses in gay men as a consequence of the illegality of homosexual acts. A similar outcome is associated with the reported emigration of gay men for HIV testing or the treatment of AIDS, which has been prompted either by the desire to maintain anonymity or the unavailability of such services in Ireland (Smyth, 1995).

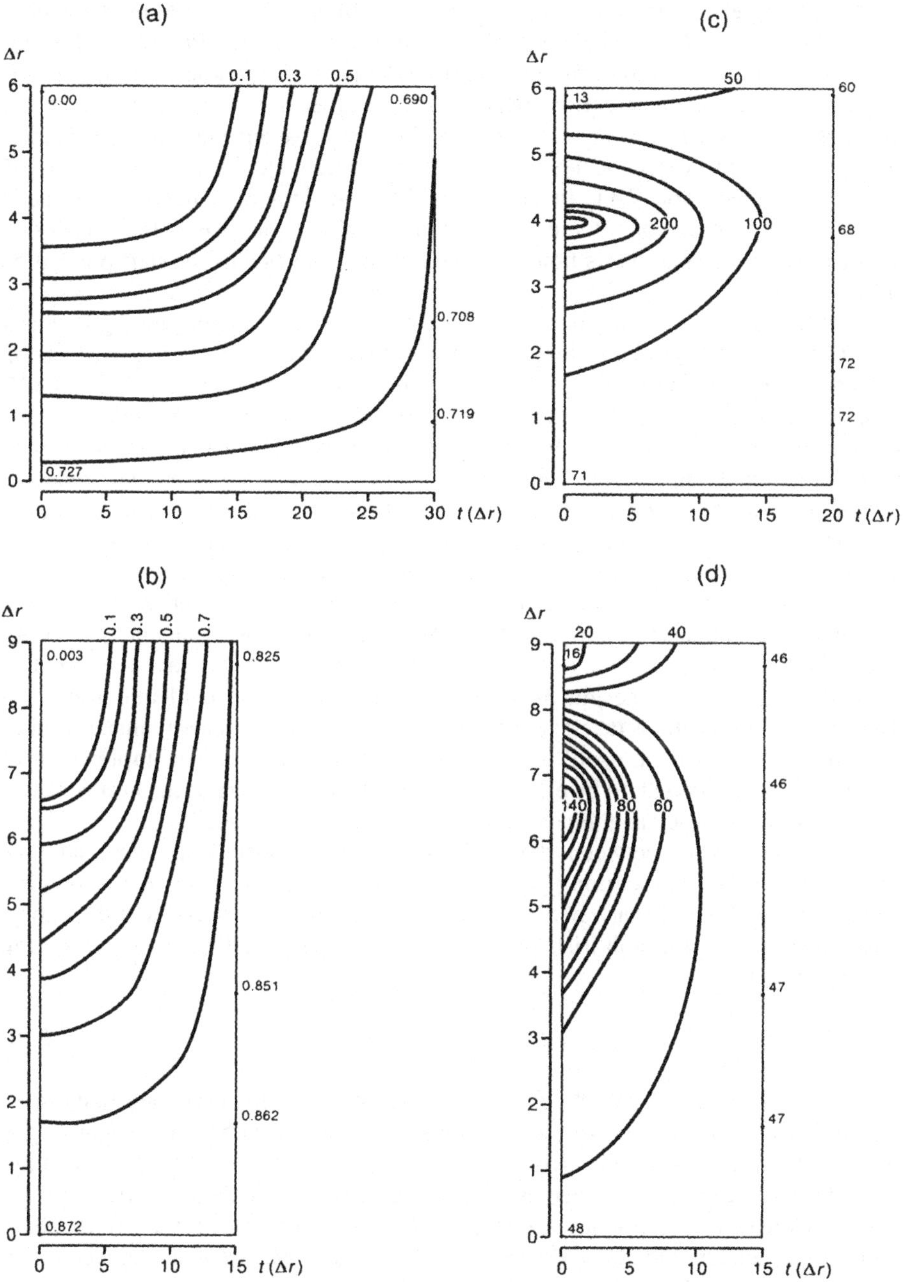

Fig. 7.1. Control Charts for the Irish AIDS Epidemic for its simulated size among (a) gay men (r = 8.90. t(Δr) = 0 = 1982) and (b) IVDUs (r = 11.65. t(Δr) = 0 = 1985) and its duration (years) among (c) gay men and (d) IVDUs. t(Δr) = 0 is the year of the first recorded AIDS case and Δr is the simulated reduction to the estimated partnership rate (r) at t.

A second perspective on AIDS prevention in Ireland is provided by the control charts (Figure 7.1) describing how different interventions of the form [Δ, t(Δ)] impact upon the size and duration of the forecast epidemic. The contours on these charts describe sets of such interventions which generate either an epidemic of the same size or duration. These results are based on simulations that assume the partnership rates for gay men and IVDUs at the start of the epidemic were r = 8.90 and r = 11.65 respectively (see Table 7.1). In each simulation, the duration is the time elapsed in years to the termination of the first AIDS cycle and the size is the number of simulated AIDS cases to this time expressed as a proportion of the start population (n).

Figures 7.1a and 7.1b are the charts for the size of the epidemic among Irish gay men and IVDUs. When no intervention is simulated and the starting partnership rates are maintained throughout, the respective epidemic sizes are 0.727 and 0.872. These maximum, unaltered sizes plot at the origin [Δ r = t(Δr)] of each chart. Then, both charts possess the property that interventions taken early and with great magnitude make the largest reductions to the size of the unaltered epidemic. Moreover, a comparison of the charts for gay men and IVDUs reveals how their different starting partnership rates influence the likely success of preventative actions. The slower progress of the forecast HIV epidemic predicted by the lower value of r = 8.90 for gay men provides extra time to mount effective interventions. For example, for unit changes to Δr, the largest decrements to epidemic size occur immediately prior to the time of peak HIV prevalence which, for gay men, is at t(0) = 22 (the year 2004). The equivalent peak for the faster IVDU epidemic is at t(0) = 10 (1995). Interventions taken after these dates are predicted modest impacts against epidemic activity which, itself, is in a state of natural decline. The implications of these peak times are that, to be effective, strong interventions are needed immediately for Irish IVDUs, whereas the position for gay men is less critical.

The relationship between the duration of the simulated epidemic and the magnitude of the intervention (Figures 7.1c and 7.1d) is more complex. Its form depends upon the difference between the starting partnership rate(r) and the critical rate, which is r' = 5 for both Irish epidemics. Then, defining the critical partnership rate reduction as

$$\Delta r' = r - r' \qquad (7.6)$$

gives a value of Δr' = 3.90 (8.90 - 5.00) as the rate reduction necessary to meet the start criterion. Inspection of the durations of the predicted epidemics among gay men (Figure 7.1c) reveals they lengthen over the interval Δr = 0 (no action taken) to Δr' = 3.90 as the reduced partnership rate slows down the simulation of transmission events. Conversely, interventions that are in excess of Δr' create a parameter setting with a reproduction rate of less than unity and a characteristic simulated epidemic that is not naturally self-sustaining. This instability causes interventions taken in this interval to shorten the duration the more Δr exceeds Δr'. Therefore, only very large interventions can be expected to advantage temporal control.

The simulated epidemics for Irish IVDUs are characterised by the values r = 11.65 and Δr' = 6.65, which are higher than those estimated for gay men. Accordingly, these simulated epidemics progress more quickly than those for gay men and the duration times associated with equivalent valued interventions are typically shorter (Figure 7.1d). Again, these predictions reinforce the need for prompt action to curtail transmission among Irish IVDUs.

7.3 Passive Protection

7.3.1 Interlocking Transmission Rates

It should not pass without mention that our analysis of HIV/AIDS in Ireland treats the epidemics among IVDUs and gay men as separate entities. In this respect, our research is subject to the criticism of stigmatisation (Brown, 1995) that has been levelled against other geographical analyses which have categorised AIDS incidence by the population at risk to infection (Smallman-Raynor and Cliff, 1990; Löytönen, 1991). Foremost, this issue has concerned the stereotyping of gay men (Altman, 1988). Brown (1995), for example, argues that such categorisation maps sexual practices over social identity and, thereby, obscures the fact that a high risk activity, like anal intercourse, may be undertaken by both heterosexual and homosexual couples. Moreover, Grover (1988) has noted how, in much political debate, the notion of a risk group has been used to stigmatise people who are already seen to be outside the moral and economic parameters of the general population.

Leaving aside, for the moment, their moral dimension, these arguments carry important implications for the ways populations and their variable infection risks are represented in epidemiological models. The basic recurrent model, for example, adopts a constant average partnership rate which presumes each member of the community behaves identically. In contrast, observations made on gay men in San Francisco during the early phases of this epidemic found behaviours ranging from celibacy to extreme promiscuity (Grant et al., 1987). The frequency distribution was similar to that observed for other STDs, where small core groups of individuals who acquire numerous sexual partners play a disproportionate role in the maintenance and transmission of infection (Yorke et al., 1978). To imitate this variation, Anderson and May (1991) have described an extension to the basic model where the epidemic variables are disaggregated according to the number of sexual partners each individual acquires. In comparison to the basic model, this specification showed that the activities of those with frequent partner changes raised the prevalence of HIV during the early phases of the simulated epidemic. However, because these individuals are also removed from circulation with AIDS earlier in the epidemic, they play little part in the later phase when transmission relies increasingly on the less sexually active and when HIV prevalence is less than that predicted by the basic model. Consequently, the decline in the rate of seroconversion observed among gay men in San Francisco in the mid-1980s (Winkelstein et al., 1987; Evans et al., 1988) could have been more the consequence of the early onset of AIDS among those with most partners than the effect of preventative action.

This framework, therefore, adjusts for variations in the frequency with which a risk activity associated with a *specific* behaviour is undertaken. However, it does not take account of the interactions between different risk populations that are facilitated by intermediate behaviours like prostitution, bisexuality and the sexual partners of IVDUs. This task requires a compartmental design where each behaviour is specified separately and the HIV incidence terms contain cross-infection mechanisms for feasible intermediate behaviours. Knox (1986) was one of the first to develop a model of this type. Unfortunately, the absence of precise data for many of the relevant parameters and population sizes forced him to adopt an approximate calibration procedure which, in hindsight, has been found to have over-estimated the size of the epidemic in the UK. Moreover, the complex designs of many compartment models (see Jacquez et al., 1988; Van Druten et al., 1990) have led some to argue that beyond a certain refinement threshold they add little to the predictive capabilities of less elaborate systems (Kaplan, 1989; Kaplan and Lee, 1989).

Despite this criticism, compartment models do begin to address important hypotheses concerning the potential for HIV transmission within the entire population. A simplified framework for the direct investigation of this problem is described in Thomas (1994; 1996b). The specification of this model presumes that, at the start of the epidemic, every individual in the population can be assigned to either a low (A) or high (B) risk category according to the number of partners with whom they undertake risk activities. Then, the membership criterion for A is that the number of such partners be less than or equal to the critical rate (r') for an epidemic to begin. Conversely, the number of such partners in B is always in excess of r'.

Let the average partnership rate in each group be denoted by r^A and r^B, then the state diagram representing the various HIV transmission routes between these populations(n^A, n^B) takes the following form.

$$\begin{array}{ccccc} & & \beta r^A x^A p^{AA} & & \\ [x^A] & \rightrightarrows & & \rightrightarrows & [y^A] \\ & & \beta r^A x^A p^{AB} & & \\ & & & & \\ & & \beta r^B x^B p^{BA} & & \\ [x^B] & \rightrightarrows & & \rightrightarrows & [y^B] \\ & & \beta r^B x^B p^{BB} & & \end{array} \tag{7.7}$$

Here, notation of the form AB denotes the incidence of HIV being passed from infecteds in B (or more generally the second letter in the sequence) to susceptibles in A (the first letter in the sequence). Moreover, define q^{AA} as the probability of a susceptible in A acquiring a partner from A together with the complementary probability $q^{AB} = 1 - q^{AA}$ and the group B equivalents q^{BB} and q^{BA}. Then, the terms of the form $p^{\cdot\cdot}$ are defined by

$$p^{AA} = q^{AA} y^A / n^A, \tag{7.8}$$

$$p^{AB} = q^{AB}y^B/n^B, \tag{7.9}$$

$$p^{BA} = q^{BA}y^A/n^A, \text{ and} \tag{7.10}$$

$$p^{BB} = q^{BB}y^B/n^B, \tag{7.11}$$

The expressions labelled AA and BB are indicative of the degree of simulated infection within the populations and those labelled AB and BA denote cross-infections. The term p^{AB}, for example, is the probability that a high risk partner from B of a susceptible in low risk A is an infective. Similarly, p^{BA} is the probability that a partner from A of a susceptible in B is an infective.

The calibration of this system requires estimates of partnership rates and choice probabilities that are not generally available from official sources. For this reason, the following illustrative application adopts fictive estimates based on some fragmentary sources of data for the UK given in Knox (1986) and Knox et al. (1993). In this setting, those in A are taken to comprise 90% of the population who acquire partners at the rate of $r^A = 1.11$ per year, while the respective values for high risk B are 10% and $r^B = 10$. The partner choice probabilities are taken to be $q^{AB} = 0.05$ ($q^{AA} = 0.95$) and $q^{BA} = 0.05$ ($q^{BB} = 0.95$), while the transmission parameters are given the previously justified estimates of $\beta = 0.1$ and $D = 2$ years. Together, this configuration defines a critical partnership rate for the entire population of $r' = 5$ per year to indicate that a simulated epidemic can begin only with an infection in B. Consequently, the transmission of HIV among those at low risk will be dependent on cross-infections from those in B. This relationship is illustrated in Figure 7.2 which plots the simulated incidence of HIV according to each type of partnership formation (AA, AB, BA, BB). If the discussion is couched in terms of sexual activity, then the low incidence of HIV from BA mixing is synonymous with the activity of a small number of group B prostitutes, who generate a much larger incidence (AB) among their more numerous group A clients. In turn, because of the low rate of partner acquisition in A, these clients infect only a small number of their partners (AA) in A. This type of chaining conforms closely with the observed incidence of AIDS in most developed countries (Kirp and Bayer, 1992), where occurrences of the direct transmission between low risk heterosexual partners have been extremely rare (Knox et al., 1993).These outcomes suggest that a low rate of partner acquisition will confer a high degree of natural protection on such a community.

In contrast to the Irish analysis, where the different risk behaviours were simulated distinctively timed epidemics, the interlocking transmission model predicts HIV prevalences(y^A, y^B) where the minor epidemic in A is synchronised entirely within the longer duration of the major cycle in B. This dependency, together with the simulated chaining of transmission, has some important implications for the targeting of preventative actions. Persuading those in B to reduce partner acquisition or increase condom use, for example, is expected to have a much greater impact on HIV incidence than equivalent actions taken by those in A. Moreover, the continued modification of high risk activities should eventually 'switch off' the dependency relations which maintain the epidemic in A. Knox et al. (1993) report similar outcomes from a detailed application of more segmented models to sexual

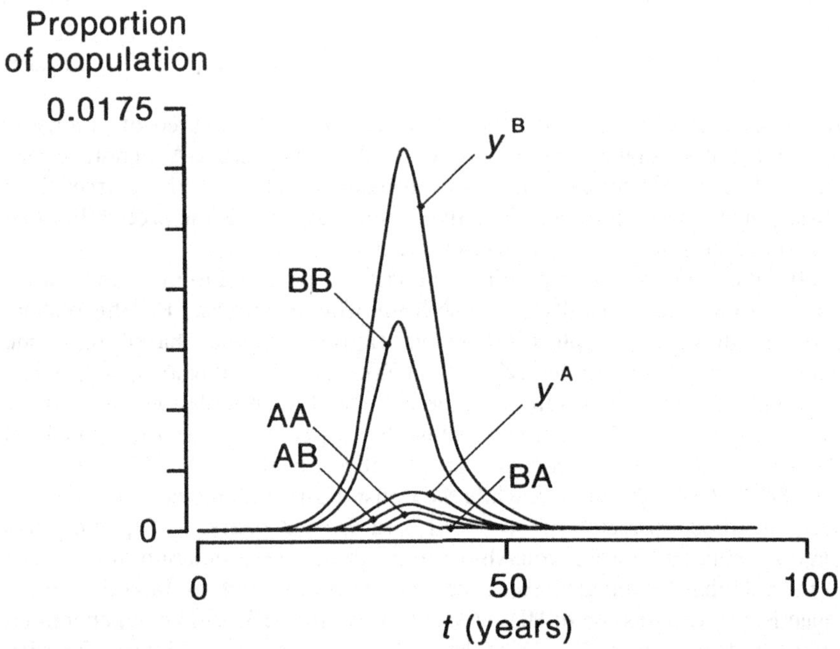

Fig. 7.2. Simulated Series of HIV Incidence by Source of Infection (AA, AB, BA, BB) and HIV prevalence (y^A, y^B) from a model imitating transmission between a high (B) and a low (A) risk population

survey data collected in Birmingham. In particular, they propose the bulk of public health investment should be directed towards the prevention of transmission among gay men, to curtail the main epidemic, and bisexuals, to limit the heterosexual transmission of HIV. Their research, however, does not take account of the transmission activities of IVDUs.

Up to this point, the preventative implications of the interlocking model have been discussed in relation to particular risk behaviours. In practice, however, the adoption of the critical partnership rate(r') to distinguish group membership defines risk according to the frequency with which any unsafe activity is taken and irrespective of its associated behaviour. Therefore, group B will contain some with a stereotypically low risk behaviour but a high activity rate and vice versa for group A.

This type of distinction may also be applied to the long experience of AIDS in central Africa (Earickson, 1990; Barnett and Blaikie, 1992), where the predominantly heterosexual mode of transmission has often been hastily attributed to high promiscuity rates among this population. An alternative explanation is related to the higher transmission probability of $\beta = 0.4$, which has been estimated from samples of African patients (Peterman et al., 1988) and is thought to be linked to the high prevalence of genital cuts and ulcers consequent upon prior infection with other STDs (Latif et al., 1989; Bassett and Mhloyi, 1991). This revised probability defines a much more conservative critical partnership rate of $r' = 1.25$ per year which, together with the positively skewed partner acquisition frequency distribution, is likely to place many more African heterosexuals at high risk than those elsewhere.

7.3.2 Starts, Pathways and Warning Times

All the models described so far represent HIV incidence as a temporal cycle(s) which, in certain communities, may fail to materialise if local activity rates pre-empt the starting conditions. This framework, however, takes no account of the geographical diffusion of infection which has often been visualised as a wave rippling outwards from the initial point of infection (Cliff and Haggett, 1989; Haggett, 1990; 1992; Thomas, 1992). Following this analogy, the most distant place from the source will be afforded a long waiting time prior to the arrival of the wave, which may be utilised to plan and implement interventions. By contrast, many people living close to the origin might experience a full epidemic cycle before the disease agent is isolated (Thomas, 1993). In this respect, distance from the source may be thought of as a form of spatial protection which might be enhanced if local risk activity is sufficiently infrequent to either deflect, or decelerate, the passage of the wave.

One method for testing such eventualities is to convert the recurrent model into a multiregion format (Thomas, 1994). In the version of this system for a high risk population, the distribution of these individuals (n_i) is counted for a set of m regions and the epidemic variables for the ith region are subscripted accordingly. The contact rules in this model draw upon spatial interaction theory (Wilson and Bennett, 1985) such that these regional populations are assumed to mix in a manner that is consistent with a negative exponential function of the form $\exp(-\lambda d_{ij})$. Here, d_{ij} is the distance between the centres of regions i and j, and λ is a parameter calibrated to represent the degree to which distance curtails contact frequencies in the study area. Then, the incidence of HIV in any region i is described by the following state diagram.

$$[x_i] \rightarrow \beta r x_i W_i \, [\Sigma\, y_j \exp(\lambda d_{ij})] \rightarrow [y_i], \quad \forall\, i. \tag{7.12}$$

Here, W_i is a scalar given by

$$W_i = [\Sigma\, n_j \exp(-\lambda\, d_{ij})]^{-1}, \quad \forall\, I, \tag{7.13}$$

which ensures the frequency of simulated contact in each region is consistent with its population potential (see Thomas, 1994).

The start criterion for a single infection in *any* region i is that the reproduction rate (equation 7.2) must be greater than unity. This same condition applies to each region because the epidemic parameters(β, r, D) are assumed to be invariant among the regional populations. In this representation, therefore, the epidemic parameters determine a constant duration for the cycle of HIV incidence *within* each region and act independently of the spatial parameter(λ), which interacts with distance to control the time elapsed *between* first infection in one region and another.

This multiregion modelling format provides the opportunity to analyse the structure of the international contact networks that support the transmission of HIV as a pandemic infection. To this end, the following simulation of the pathway of HIV/AIDS adopts the previously justified disease parameters($\beta = 0.1$, $r = 10$, $D = 2$) and uses a system of $m = 16$ cities to represent the demography of the globe in miniature. These cities (see Figure 7.3) were chosen to make their number and population sum in each continent proportional to the continent's share of the world population in 1959. This particular date was picked to coincide with one possible estimate of the time of first human infection with HIV in Zaire (Gotlieb et al., 1981).

The key parameter in this system is the exponential decay parameter, which is fitted to the observed progression of HIV. In this last respect, evidence from the serological dating of strains of HIV-1 (Li et al., 1988; Haggett, 1992) suggest that, after its likely African origins, the virus spread through the Caribbean to the US by the mid-1970s and then to Europe about 1980. Figure 7.3 compares this route with that predicted by the multi-region model with $\lambda = 2$, which is the value of the exponential decay parameter that most closely matches the dated arrival of HIV in the US. This fit corresponds to a simulated first infection in Chicago in 1978. Unfortunately, the model also predicts a full scale epidemic in Eurasia well before this date. Serological evidence, for example, has adjudged HIV to have been in circulation in India since 1982 (Banerjee, 1989), whereas the model predicts the time of entry into Bombay as 1972. Moreover, the forecast indicates that, according to the relationship between the rate of diffusion and population potential embedded in spatial interaction models, the most likely natural pathways out of Africa are through the Middle East and the Mediterranean. One explanation for this divergence might be that the serological pathway is subject to bias, because the early concentration of HIV research overlooked other contingencies (Chirimuuta and Chirimuuta, 1987). Alternatively, the predominantly sexual partnership networks, which facilitated the transfer of infection to the US, might have been of sufficient density to break the constraint of population potential.

A more extended application of the multiregion format is to simulate pandemics started at different locations and record the time elapsed to first infection in each of the other cities (Thomas, 1993). This procedure yields a frequency distribution for each city of the duration to first infection of a pandemic begun in every other city. In effect, these distributions serve as warning times which may be exploited to plan interventions against a disease agent arriving from a known source. This idea is germane to the control of the influenza virus which periodically mutates its antigenic character and, thereby, erodes the acquired immunity to infection of much of the world's population (Cliff et al., 1986; Pyle, 1986). Accordingly, the warning time for a given source is the episode granted to a place to acquire vaccine and immunise those at risk against the new pandemic strain.

These simple space-time structures, however, seem unlikely to be so applicable to the control of HIV. In this respect, it is important to note that HIV was isolated only in 1983 (Barré-Sinoussi et al., 1983) which, according to the timing invoked here, implies the virus circulated unnoticed for at least twenty-four years prior to this date.

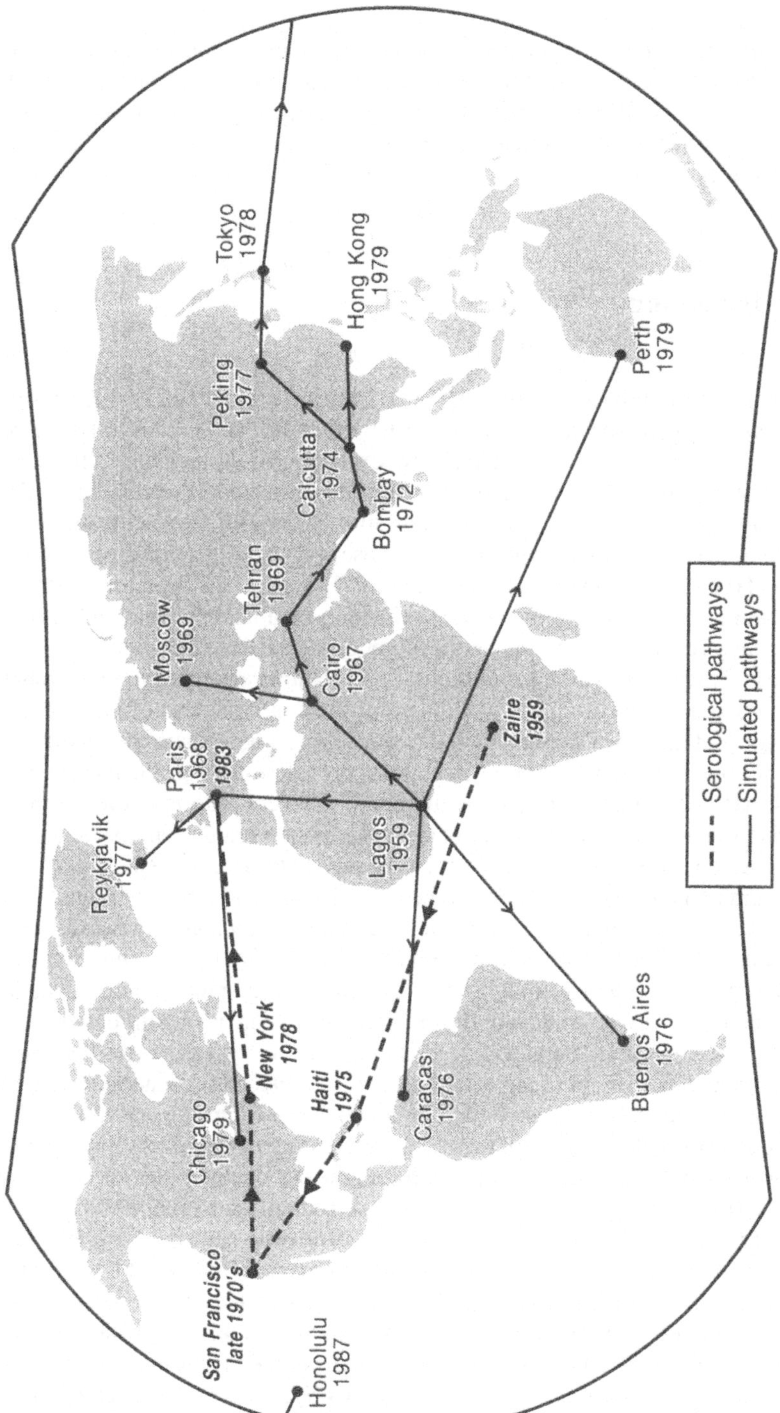

Fig. 7.3. HIV Pandemic Pathways. (a) Serological (Li et al. 1988) and (b) Simulated with $\lambda=2$ (Thomas. 1994)

During this interval, HIV managed to gain entry to most countries outside Asia and, thereby, exhausted many potentially long warning times. Moreover, the continued lack of a vaccine implies that those places, where warning times are yet to lapse, will be unlikely to utilise their outstanding durations to mount effective immunisation programmes.

7.4 Discussion

The epidemic modelling systems described in this paper have enabled a series of hypotheses to be elaborated about the distribution and prevention of HIV infection in space and time. The investigation of Irish AIDS incidence, for example, demonstrated how temporal variations in the fitted partnership rate might reflect the beneficial consequences of past interventions. Generally, this assessment yielded modest impacts probably because most of those with AIDS recorded in the series had acquired HIV infection prior to its isolation in 1983, when the need for protection was not apparent. This outcome is related to the long incubation period and implies that only recently recorded AIDS incidence might be expected to be modified by changed activities. Alternatively, the simulation of preventative action within the community model generates control charts which reveal the consequences of variations in the magnitude and timing of hypothetical interventions. These charts stress the need for strong actions prior to the time of peak HIV prevalence if significant reductions to forecast AIDS incidence are to be anticipated. Present reports of the US incidence appearing to reach a plateau (Rosenberg et al., 1992) would seem to suggest that, in many other countries, this critical time may also soon be imminent.

In contrast to the Irish analysis by types of risk behaviour, the representation of the interlocking model construed risk as the frequency with which any unsafe activity is undertaken. This last notion was dichotomised according to the value of the critical partnership rate. Then, through cross-infections, the simulated epidemic among those at high risk 'drove' a small and synchronised HIV incidence among those at low risk which was not self-sustaining. This particular scenario stresses the reductions to AIDS incidence in both populations that are the likely outcome of targeting preventative measures at those with a high frequency of unsafe activity.

By comparison, the outputs of the multiregion model engage less directly with the prevention debate. These results do highlight the complexities of administering immunisation campaigns the longer a vaccine remains unavailable. However, the failure of this model to replicate the observed pandemic pathway is indicative of the need for a more geographically sensitive specification. One possibility would be to make one, or all, of the disease parameters region specific to reflect spatial variations in their estimated values, such as the higher transmission probability reported for central Africa. This adjustment would serve to quicken and intensify the forecast epidemic within the affected area. Alternatively, making the exponential decay parameter vary by region would presume differences in the ease of contact between these places. A relatively low local estimate, for example, would shorten the duration of infection transfer times out of this region in recognition of the increased rate of contact.

Finally, mention should be made about some of the criticisms that have been levelled against the application of epidemic modelling to the understanding of preventative action. Brown (1995, p.168), for example, argues such applications are intended to benefit only those who are not infected and, further, that the constant stress on the prevention of diffusion reinforces a dichotomy founded on fear and homophobia where gay bodies are seen as a threat to the general public. These separate distinctions by behaviour and infection status, however, are not mutually exclusive and, like analyses of incidence by risk group (Dyck, 1990; Kearns, 1996), imply a fragmentation of the response to HIV/AIDS. In contrast, we have explored model formulations where, initially, everyone is presumed susceptible and the infection risk is then ubiquitous but variable in space and time. The corresponding simulations stress the important role of public health for promoting comprehensive strategies which co-ordinate the diverse efforts of communities where HIV/AIDS happens to be prevalent. Without such co-operation, the prevention of infection seems likely to remain more a consequence of passive protection than belated direct actions.

Acknowledgements
We would like to thank Graham Bowden and Nick Scarle for preparing the figures. Funding for part of this work has been provided by a postgraduate studentship from the Economic and Social Research Council, whose support we would like to acknowledge.

References

Altman D. 1988. Legitimation through disaster: AIDS and the gay movement in *AIDS: the burden of history.* Eds E Fee & D M Fox (Berkeley: University of California Press) pp 301-15

Anderson R.M. 1988. The epidemiology of HIV infection: variable incubation plus infectious periods and heterogeneity in sexual activity. *Journal of the Royal Statistical Society A* **151**: 66-93.

Anderson R.M. and May R.M. 1988. The epidemiological parameters of HIV transmission. *Nature* **333**:314-9.

Anderson R.M. and May R.M. 1991. *Infectious diseases of humans dynamics and control.* Oxford: Oxford University Press.

Bailey N.J.T. 1975. *The mathematical theory of infectious diseases and its applications.* High Wycombe: Charles Griffin.

Banerjee K. 1989. *Rising prevalence of antibodies against human immunodeficiency virus (HIV-1) in Western Maharashtra, India.* Abstracts of the Fifth International Conference on AIDS, Montreal, Abstract TG022.

Barnett T. and Blaikie P. 1992. *AIDS in Africa: its present and future impact.* London: Belhaven Press.

Barré-Sinoussi, F. Chermann, J.C., Rey F., Nugeyre M.T., Charnaret S., Gruest J., Danduet C., Axler-Blin C., Vézinet-Brun F., Rouzioux C., Rozenbaum W. and Montagnier L. 1983. Isolating of a T-lymphotropic retro-virus from a patient at risk from acquired immune deficiency syndrome. *Science* **220**:865-71.

Bartlett M.S. 1960. *Stochastic population models in ecology and epidemiology.* London: Methuen.

Bassett M.T. and Mhloyi M. 1991. Women and AIDS in Zimbabwe: the making of an epidemic. *International Journal of Health Services* **21**:143-56.

Brown M 1995. Ironies of distance: an ongoing critique of the geographies of AIDS *Environment and Planning D* **13**:159-83

Cease K.B. and Berzofsky J.A. 1988. Antigenic structures recognised by T-cells: towards the rational design of an AIDS vaccine. *AIDS* **2(1)**:95-101.

Chirimuuta R.C. and Chirimuuta R.J. 1987. *AIDS, Africa and racism* Bretby, Derbyshire: Chirimuuta.

Cliff A.D., Haggett P., and Ord J. K.1986. *Spatial aspects of influenza epidemics* London: Pion.

Cliff A.D. & Haggett P. 1989 Spatial aspects of epidemic control Progress in Human Geography, **13**, 315-47

Dyck I. 1990. Context, culture and client: geography and the health for all strategy *Canadian Geographer* **34**:338-41

Earickson, R.J. 1990. International behavioural responses to a health hazard: AIDS. *Social Science and Medicine* **31**:951-62.

Evans P.E., Rutherford G.W.. Amory J.W., Ilessol N.A., Bolan G.A., Herring M. and Werdegar D. 1988. *Does health education work? Publically funded health education in San Francisco, 1982-1986.* Abstracts of the Fourth International Conference on AIDS, Stockholm, Abstract 6044.

Gagnon J.H. & Simon W. 1974. *Sexual contacts: the social sources of human sexuality* Chicago: Aldine.

Gallo R.C., Salahuddin S.Z., Popvic M., Shearer G.M., Kaplan M., Haynes B.F., Palker T.J., Redfield R., Oleske J., Safai B., White G., Foster P. and Markham P.D. 1984. Frequent detection and isolation of cytopathic retroviruses (HTLV III) from patients with AIDS and at risk of AIDS. *Science* **224**:500-3.

Gotlieb M.S., Schroff R., Schanker H.M., Weisman J.D., Fan P.T., Wolf R.A. and Saxon A.S. 1981. Pneumocystis carinii pneumonia and mucosal candidiasis in previously healthy homosexual men: evidence of a new acquired cellular immunodeficiency, *New England Journal of Medicine*, **305**:1425-31.

Grant R.M.. Wiley J.A. and Winkelstein W. 1987. Infectivity of the Human Immunodeficiency Virus of homosexual men. *Journal of Infectious Diseases* **156**:189-93.

Grover J. Z., 1988. AIDS, keywords and cultural work in *Cultural studies* Eds L. Grossberq, C. Nelson, and P. Treichler. Cambridge: MIT Press, pp 295-337

Haggett P. 1990. *The geographer's art*. Oxford: Blackwell.

Haggett P. 1992. Sauer's 'Origins and dispersals': its implications for the geography of disease. *Transactions of the Institute of British Geographers* **17**:387-98.

Haggett P. 1994. Prediction and predictability in geographical systems. *Transactions of the Institute of British Geographers* **19**:6-20.

Hendriksson B. and Ytterberg H. 1992. Sweden: the power of the moral(istic) left in *AIDS in the industrialised democracies: passions, politics and policies,* Eds D.L. Kirp and R. Bayer, New Brunswick, New Jersey: Rutgers University Press, pp 317-38.

Isham V. 1988. Mathematical modelling of the transmission dynamics of HIV infection and AIDS. *Journal of the Royal Statistical Society A* **151**:5-30.

Isham V. and Medley G. 1996. *Models for infectious human diseases*, Cambridge: Cambridge University Press

Jacquez J.A., Simon C.P., Koopman J., Sattenspiel L. and Perry T. 1988. Modelling and analyzing HIV transmission: the effect of contact patterns. *Mathematical Biosciences* **92**: 119-99.

Kaplan E.H. 1989. Can bad models suggest good policies? Sexual mixing and the AIDS epidemic. *The Journal of Sex Research* **26**:301-14.

Kaplan E.H. and Lee Y.S. 1989. How bad can it get? Bounding worst case endemic heterogeneous mixing models of HIV/AIDS. *Mathematical Biosciences* **99**:157-80.

Kearns R.A. 1996. AIDS and medical geography: embracing the Other. *Progress in Human Geography* **20**:123-31

Kirp D.L. and Bayer R. 1992. *AIDS in the industrialized democracies: passions, politics and policies*. New Brunswick, New Jersey: Rutgers University Press.

Knox E.G. 1986. A transmission model for AIDS. *European Journal of Epidemiology* **2**:165-77.

Knox E.G., MacArthur C. and Simons K.J. 1993. *Sexual behaviour and AIDS in Great Britain*. London: HMSO.

Krieger N. and Appleman R. 1994. The politics of AIDS, in *AIDS: the politics of survival*, Eds N. Krieger and G. Margo, New York: Baywood, pp 3-54

Krieger N. and Margo G. 1994. *AIDS: the politics of survival*, New York: Baywood

Latif A.S., Bassett M.T. and Mhloyi M. 1989. Genital ulcers and transmission of HIV among couples in Zimbabwe. *AIDS* **3**:519-23.

Li W.H., Tanimura M. and Sharp P.M. 1988. Rates and dates of divergence between AIDS virus nucleotide sequences. *Molecular Biology and Evolution* **54**:313-30.

Löytönen M. 1991. The spatial diffusion of human immunodeficiency virus type 1 in Finland, 1982-1997. *Annals of the Association of American Geographers* **81**:127-51.

May R.M., Anderson R.M. and Blower S.M. 1989. The epidemiology and transmission dynamics of HIV/AIDS. *Daedalus* **118**:163-201.

Mollison D. 1995. *Epidemic models*, Cambridge: Cambridge University Press

Murray G.D. and Cliff A.D. 1977. A stochastic model for measles epidemics in a multi-region setting, *Transactions of the Institute of British Geographers*, **2**:158-74.

Nowak M., Anderson R.M., McLean A.R., Wolfs T., Goudsmit J. and May R.M. 1991. Antigenic diversity thresholds and the development of AIDS. *Science* **254**:963-9.

Pederson C., Nielsen C.M., Vestergaard B.F., Gerstoft N., Krogsgaard K. and Nielsen J.O. 1987. Temporal relation of antigenaemia and loss of antibodies to core antigens to development of clinical disease in HIV infection. *British Medical Journal*, **295**:567-569.

Peterman T.A., Stoneburner R.L., Allen J.R., Jaffe H.W. and Curran J.W. 1988. Risk of HIV transmission from heterosexual adults with transfusion-associated infections. *Journal of the American Medical Association* **259**:53-63.

Pyle G. 1986. *The diffusion of influenza: patterns and paradigms* Totowa: Rowan and Littlefield.

Rosenberg P.S., Gail M.H. and Carroll R.J. 1992. Estimating HIV prevalence and projecting AIDS incidence in the United States: a model that accounts for therapy and changes in the surveillance definitions of AIDS. *Statistics in Medicine* **11**:1633-55.

Santana S., Faas L. and Wald K. 1991. Human immunodeficiency virus in Cuba: the public response of a third world country. *International Journal of Health Services* **21**:511-37.

Smallman-Raynor M.R. and Cliff A.D. 1990. Acquired Immunodeficiency Syndrome (AIDS): literature, geographical origins and global patterns. *Progress in Human Geography* **14**: 157-213.

Smyth F.M. 1995. *Social and epidemiological constructions of HIV/AIDS in Ireland*, Unpublished PhD thesis, University of Manchester.

Smyth F.M. and Thomas R.W. 1996a. Preventative action and the diffusion of HIV/AIDS. *Progress in Human Geography* **20**:1-22.

Smyth F.M. and Thomas R.W. 1996b. Controlling HIV/AIDS in Ireland: the implications for health policy of some epidemic forecasts. *Environment and Planning A* **27**:99-118.

Thomas R.W. 1988. Stochastic carrier models for the simulation of Hodgkin's disease in a system of regions. *Environment and Planning A* **20**:1575-601

Thomas R.W. 1992. *Geomedical systems: intervention and control*. London: Routledge.

Thomas R.W. 1993. Source region effects in epidemic disease modelling: comparisons between influenza and HIV. *Papers in Regional Science* **72**:257-82

Thomas R.W. 1994. Forecasting global HIV/AIDS dynamics: modelling strategies and preliminary simulations. *Environment and Planning A* **26**:1147-66

Thomas R.W. 1996a. Alternative population dynamics in selected HIV/AIDS modelling systems: some cross-national comparisons. *Geographical Analysis* 28 :108-25

Thomas R.W. 1996b. Modelling space-time HIV/AIDS dynamics: applications to disease control. *Social Science and Medicine* **41**(in press)

Van Druten J.A.M., Reintjes A.G.M., Jager J.C., Heisterkamp S.H., Poos M.J.J.C., Coutinho R.A., Dijkgraaf M.G.W. and Ruitenberg E.J. 1990. Infection dynamics and intervention experiments in linked risk groups. *Statistics in Medicine* **9**:721-36.

Wilson A.G. and Bennett R.J. 1985. *Mathematical models in human geography and planning*. Chichester: Wiley.

Winkelstein W. Jr., Samuel M., Padain N.S., Wiley J.A., Lane W., Anderson R.B. and Levy J.A. 1987. The San Francisco men's health study III: reduction in human immunodeficiency virus transmission among homosexual/bisexual men, 1982-1986. *American Journal of Public Health* **76**:685-9.

Yorke J.A., Hethcote H.W. and Nold A. 1978. Dynamics and control of the transmission of gonorrhoea. *Sexually Transmitted Diseases* **5**:51-156.

Part B

Behavioural Modelling

8 Longitudinal Approaches to Analysing Migration Behaviour in the Context of Personal Histories

John Odland

Department of Geography, Indiana University, Bloomington IN 47405 USA

8.1 Introduction

Longitudinal approaches to analyzing migration behaviour are reviewed in this paper and used to investigate interdependencies between the migration histories of individuals and their histories of participation in employment. Precise information about the timing and sequencing of events can be especially useful in the analysis of relations between employment and migration and the empirical analysis is based on some longitudinal data that are especially detailed in this respect: The Survey of Income and Program Participation (SIPP) for the United States. Relations between these migration histories and the histories of other aspects of individual lives, including employment histories, can be investigated within a general framework in which discrete-state continuous-time stochastic processes serve as general models for the lifetime migration behaviour of individuals and observed migration histories are treated as particular realizations of these processes. The processes are formally defined, in a general and abstract way, by a set of discrete states and a set of functions that describe the chances that an individual will make a transition between any pair of states at any time. In the case of migration histories the states of the model correspond to particular localities where an individual might reside and the transition functions summarize the chances that an individual will, at any time, move from one locality to another. The observable phenomena that correspond to the operation of these processes are, for any individual, a set of migration events that occur on particular dates during the individual's lifetime and a corresponding series of episodes of residence in particular localities, with each episode beginning and ending on a particular date.

Investigations of individual migration behaviour typically center on hypotheses about how the chances of these migration events are determined. The approaches that have been used to examine these hypotheses, and the specifications of the hypotheses themselves, have been conditioned by the nature of the data about migration behaviour that have been available. Most analyses of individual migration behaviour have been based on cross-sectional data that provide very little information about the locational histories of individuals. Cross-sectional data are limited to information on the whereabouts of individuals on only two dates, and reveal whether an individual migrated at least one time during the interval between the dates. Many of the analyses of migration behaviour that have used cross-sectional data have been carried out by applying discrete-choice models, which have a solid basis in utility theory (see Anderson and Papageorgiou, 1994; Liaw, 1990; Liaw and Ledent, 1987; and Moss, 1979). Cross-sectional data also are typically available for very large samples, since they generally come from a national census. These large sample sizes have made it possible for migration research to emphasize the effects of localized variables, such local labour market characteristics, on migration behaviour. The samples are usually large enough to detect differences in migration behaviour between the populations of fairly small subregions or to carry out separate analyses of destination choice for the populations of different subregions. Cross-sectional data limit investigations to a particular format, however, in which hypotheses are formulated in terms of covariations between the frequencies of migration events and the values of exogenous variables.

A much broader array of hypotheses can be formulated on the basis of the longitudinal models reviewed in this paper. It is possible to use these longitudinal models to formulate and test hypotheses that correspond to the hypotheses that have been traditionally been investigated with cross-sectional methods. Since event history data provide more information about the number and timing of migration events than cross-sectional data it may not be surprising that the longitudinal models can resolve many of the problems of measurement and inference that occur when the same hypotheses are investigated with cross-sectional approaches (Odland, 1996). The major contributions of longitudinal approaches will probably come, however, from the capacity to support the formulation and testing of new hypotheses rather than the reformulation of traditional ones. Extensions of migration analysis based on longitudinal approaches are discussed in this paper and some empirical analyses based on event history data are presented in the later sections of the paper.

Empirical investigations based on longitudinal models depend on the availability of longitudinal data. In the case of migration these data should provide continuous records for the whereabouts of individuals over major portions of their lifetimes and should also identify the timing of events, including migration events, with precision. Longitudinal data are expensive to gather and longitudinal samples are much smaller than the cross-sectional samples that have dominated research on migration behaviour. Consequently, even though some cross-sectional hypotheses may have longitudinal counterparts, the sizes of longitudinal samples may be insufficient to allow thorough testing, especially if testing depends on detecting differences in migration behaviour across subregions. Longitudinal data also vary considerably in the precision with which the dates of migration are identified. The SIPP data used in the empirical analysis in this paper provides very precise information about the dates of events, but for a sample that is much smaller than the cross-sectional samples that have been used to investigate migration in the U.S. The information about the timing of events in these data makes them especially useful for analyzing how migration is related to an individual's on-going history of employment.

The traditional hypotheses about migration, those that center on covariations between migration and the status of exogenous variables, can be reformulated, in terms of longitudinal models, in ways that make use of the additional information in longitudinal data . They have generally been formulated as hypotheses about covariations between the durations of episodes of residence and the values of exogenous variables (Bailey and Ellis, 1993; Davies and Flowerdew, 1992; Sandefur and Scott, 1981; Waldorf, 1994; Waldorf and Esparza, 1991). These hypotheses have generally been tested by means of proportional hazards models (Kalbfleisch and Prentice, 1980; Lawless, 1982).

The detailed information on timing that is available in some event history data also makes it possible to formulate and test hypotheses about interdependencies between migration events and other events that amount to important transitions in the personal histories of individuals, such as getting married or divorced, losing jobs and finding others, or entering and leaving the labour force. These events correspond to changes in the status of exogenous variables but if the change in status corresponds to an important transition in an individual's life then migration may be associated with the event itself, as well as having different associations with different values for the status of the variable. The distinction is made clear in Mulder and Wagner's (1993) analysis of migration and marriage. The status of the variable "marriage" covarys with migration so that migration rates differ between married and unmarried persons. Substantial amounts of migration are also associated, however, with the marriage event itself (the transition from "unmarried" to "married") as people migrate in order to form a conjugal household.[1]

8.2 Migration and Employment

Information about the precise timing of events is especially useful in analyzing relations between migration and employment. Employment histories for individuals are characterized by frequent changes of status, short-lived episodes of employment and unemployment, and jobs that are held for brief periods; especially among the young people who account for most migration. Consequently, the timing of events has to be known with some precision in order to ascertain an individual's employment status at the time of migration. Further, the relations between migration and employment may involve interdependencies with transitions in employment status as well as interdependencies with the status of employment at the time of migration. For example, individuals who are not involved in the labour force may be more likely, or less likely, to migrate than those who are in the labour force; but migration may also be associated with transitions into or out of the labour force, as individuals become full-time workers or retire from participation in employment. Similarly, migration can be associated with episodes of unemployment and job search but migration may be triggered by a particular event that is part of those episodes, such as entry into a new job.

[1]Estimates of the difference in migration rates between married and unmarried people are likely to be misspecified if the interdependence between the events is neglected because the difference depends in part on unmarried persons migrating in connection with the marriage event. Mulder and Wagner (1993) find that most of the difference in migration rates between married and unmarried persons is accounted for by unmarried persons moving in connection with their own marriages.

Employment has, in fact, been a major component of theorizing about migration behaviour and the most general explanations for migration have either treated it as a strategy that individuals can use to enhance their returns from labour force activity (Sjaastad, 1962; Molho, 1986); or as the consequence of job searches that extend beyond the migrant's home community (Bailey, 1994; Herzog, Schlottman and Boehm, 1993). The cross-sectional approaches that have been used to carry out most empirical investigations of the behaviours of individual migrants are poorly suited, however, for analyzing relations between migration and employment because those methods accommodate only limited information about the timing of events. Longitudinal methods have recently been applied to analyze relations between migration and some aspects of work histories (Bailey, 1994; Shumway, 1993, Yoon, 1995) and longitudinal analyses of interdependencies between migration and employment can be organized within a general framework where migration and employment are analyzed in terms of interdependent histories that unfold simultaneously; with the progress of each history potentially affecting the progress of the others. This framework of interacting histories provides a very general approach for the analysis of life histories; one that treats different aspects of life histories as interdependent domains that interact with one another as they develop (Willekens, 1991). This framework can also accommodate analyses of the relations between migration and the histories of other aspects of life besides employment, such as the development of households and families (Courgeau, 1990; Courgeau and Lelièvre, 1992a; Mulder and Wagner, 1993).

8.3 Migration Behaviour in the Context of Interdependent Histories

A comprehensive framework for analyzing interdependent histories can be constructed by treating these histories as realizations of a set of stochastic processes that unfold simultaneously and may influence one another as they develop. Locational histories, which are logically the focus of migration research, can be formalized as realizations of discrete-state continuous-time processes in which each of the discrete states corresponds to a region or location that an individual might occupy, and the individual's history amounts to an accounting of the dates when the individual resided in particular regions. Migration events, which may occur at any time, correspond to state-to-state transitions and the periods of time between succeeding migration events define the lengths of time that the individual resides in various regions. Other aspects of individual lives can also be organized as realizations of stochastic processes. An individual's employment history can be described by a process in which the states correspond to the possible values of an employment variable. A relatively simple set of states might correspond to general categories of labour force participation, such as "employed," "unemployed," and "not in the labour force," but a more detailed set of states could correspond to particular jobs or particular employers. Individuals spend varying lengths of time in each of these states and may make transitions between pairs of states at any time. Interdependencies between locational histories and employment histories may occur in a variety of ways. Migration may depend on employment: if the chances of migration events depend on the current status of the employment variable (people may be more likely to migrate when they are not employed), or migration events may be associated with changes in employment status (migration may be associated with entering or leaving employment). The development of an individual's employment history may also depend on his or her locational history. The chances of making a transition into unemployment may vary with an individual's current location, if labour market conditions vary across locations, and individuals may also relinquish employment in order to migrate. Married women in particular may be likely to relinquish a job in order to migrate in the company of a spouse (Lichter, 1980).

Other personal histories, such as marital histories can also be formalized as realizations of discrete-state continuous-time process. Constraints on state-to-state transitions can be imposed on these processes in order to preserve the logical characteristics of different histories. For example, locational histories may have no restrictions on state-to-state transitions, so that an individual may migrate from his or her current location to any other location and may do so at any time. Marital histories, on the other hand, may be characterized by constraints on transitions between certain pairs of states. An individual who leaves the state "never married" can never return to that state and cannot enter the state "divorced" without making an intermediate transition through the state "married." These constraints can be imposed by specifying prior values of zero on the hazards or probabilities for certain state-to-state transitions.[2]

The framework of interdependent histories can also accommodate factors beyond the personal characteristics of individuals. Place-related variables, such as the characteristics of local labour markets, can be treated as variables that have their own histories and affect the migration decisions made by residents of particular places. The marital and family relations that cause individuals to share common locational histories for parts of their lifetimes can also be represented in the framework of interdependent histories, if constraints are imposed that extend across individuals. The locational effects of marriage amount to constraints that dictate that two individuals share a common locational history for the duration of the marriage and an individual's status as a dependent child means that his or her locational history is determined by the locational history of his or her parents as long as he or she retains the status of a dependent child.

[2] Constraints on state-to-state transitions make it possible to include variables that usually are not treated in historical terms within the formalism of a set of parallel histories. For example, an individual's age in years can be treated as a variable whose value advances exactly one year in each calendar year in a completely predictable way because the only transitions that are allowed are transitions to the succeeding value of age and those are allowed to occur only on certain calendar dates.

8.3.1 Migration and the Life Course

The framework of interdependent histories can accommodate, in at least a conceptual way, the effects on migration behaviour of changing personal characteristics, interdependencies with the locational decisions of other persons, and changing economic conditions. These features make it possible to use the framework of interdependent histories to analyze migration within the general perspective of the life course. The life course perspective, which has found widespread application in social science in recent years, emphasizes the integration of research on individual histories with the social structures and historical circumstances that prevail over an individual's life span (Elder, 1985; Mayer and Tuma, 1990). The life course perspective has been applied to migration research by Courgeau (1990) and Mulder and Wagner (1993) as well as to the closely related area of residential mobility by Clark, Deurloo and Dieleman (1994).

Perhaps the most important implication of the life course perspective for migration research may be drawn from the notion that individuals pass through a series of fairly distinct social roles over the period of their lifetimes and the transitions between these social roles are occasions when several aspects of the individual's life are reorganized. The times when an individual assumes these social roles, or retires from them, correspond to major turning points in individual lives; such as departing from a parental home, beginning or ending a career as a full-time worker, and forming or dissolving marriages and households. These major milestones are likely to involve transitions in several aspects of the individual's life, and those transitions will often include migration. For example, the completion of an individual's formal education may coincide with release from the authority of his or her parents; entry into the full-time labour force; and the beginning of a period of eligibility for marriage. These major turning points in individual lives, when many aspects of an individual's circumstances change substantially, are also likely to be periods when the individual is especially liable to evaluate his or her locational *status quo* and migrate in response to the changes in circumstances.

8.3.2 Interdependencies between Migration Events and Status Variables

Statistical methods for the analysis of event histories have developed rapidly in recent years and most of these developments have centered on models for testing hypotheses about how the occurrence of an event or transition in one variable is associated with the values of exogenous variables (Lancaster, 1990; Yamaguchi, 1991; Petersen, 1995). These models, which are generally known as hazard models or models for failure times, are constructed for the duration of a period of risk that elapses before an event occurs. In the case of migration a period of risk for leaving an origin begins when an individual begins a period of residence at the origin and ends with migration to another locality. That is, models would typically be constructed for the duration of an individual's period of residence in a particular locality (Bailey, 1993; Odland and Bailey, 1990).

Models for covariations between a set of durations and the values of exogenous variables differ from ordinary regression models because data for durations typically include some that are incomplete, or censored, at the time when they are observed; and because the values of the exogenous variables may change over the corresponding durations of time. Models based on hazard functions can accommodate both censored intervals and time-varying covariates. The hazard is a function of elapsed time since the beginning of a period of risk and gives the rate at which durations end at any value of elapsed time, t:

$$h(t) = \lim_{\Delta t \to 0} \frac{(\Pr(t \le T < t + \Delta t \mid T \ge t))}{\Delta t}. \tag{8.1}$$

The proportional hazards models is a regression-type model that can be used to investigate how observed durations of time covary with the values of a set of exogenous variables, **z**;

$$h(t|\mathbf{z}) = h_0(t)g(\mathbf{z}) \tag{8.2}$$

In this formulation $h_0(t)$ is a baseline hazard function that specifies a general functional form for the hazard and $g(\mathbf{z})$ is a function of the exogenous variables. The functional form for $g(\mathbf{z})$ is usually specified as an exponential, which yields

$$h_0(t|\mathbf{z}) = h_0 \exp(\mathbf{z}\beta), \tag{8.3}$$

where β is a set of parameters that measure covariation between the hazard and the values of exogenous variables. The exponential form produces a model in which the covariation between **z** and the hazard for ending a duration is independent of the time elapsed since the beginning of the duration. The baseline hazard may be specified as a parametric functional form but the class of semi-parametric models introduced by Cox (1972) makes it possible for inferences about the effects of **z** to be made on the basis of non-parametric forms for the hazard.

The parameters of the proportional hazards model can be fitted by likelihood methods (Kalbfleish and Prentice, 1980; Lawless, 1982) and these parameters indicate how the hazard responds to variation in the values of exogenous variables. Frequency distributions or survivor distributions that correspond to any hazard function are readily derived once values for the parameters have been obtained (Lawless, 1982).

Proportional hazards models can be used to investigate the same kinds of hypotheses that have traditionally been the focus of cross-sectional migration research. That is, they address questions about how the chances of migration events are conditioned by exogenous variables, albeit in a slightly different form. Hypotheses in cross-sectional analysis center on the chances that an individual will migrate within a fixed period, and the way that exogenous variables condition those chances; while the equivalent hypotheses in the proportional hazards models centers on the durations of time that elapse before migration and the way that exogenous variables condition those durations. The longitudinal formulation makes use, however, of the greater amounts of information in event histories in order to resolve inferential and measurement problems that are associated with cross-sectional analysis (Odland, 1996). The parameter estimates for cross-sectional models may be unreliable because they omit information about locational histories that may lead to correlations between the included variables and the error terms for the models, (Davies and Pickles, 1985) but that information is included directly in longitudinal models. Proportional hazards models can also be fitted for time-varying covariates, so that the values of exogenous variables need not be invariant over the period of the analysis, and the competing risks version of the model (Hachen, 1988) makes it possible to enlarge the model to include choices among alternative destinations.

If proportional hazards models for the relations between migration and employment were to parallel cross-sectional models in a simple way employment would be treated as an exogenous variable that could affect the hazard for a migration event. More complex patterns of dependencies become evident when the two histories are examined in a longitudinal framework, however, and the direction of dependence cannot be treated as a forgone conclusion. In fact, some recent research examines the effects of migration on employment histories; specifically the effects of migration on the lengths of periods of unemployment (Bailey, 1994; Shumway, 1993). Complex relationships, in which employment (or other variables) may both influence migration and be influenced by migration are certainly possible. Courgeau and Lelièvre (1992a; 1992b) have developed a method for classifying processes that involve two or more histories by examining the differences in the hazard rates for transitions in each variable before and after the occurrence of a transition in the value of the other variable. The possibilities include reciprocal dependence, if the hazard of each event changes after the occurrence of an event in the other, as well as unilateral dependence, in which transitions in one variable affect the hazard of transitions in a second but the hazards for the first variable are unaffected by transitions in the second. The approach depends very heavily on information about the sequencing of events, and even longitudinal data sets often do not provide information about timing that is precise enough to identify the sequencing of events.

8.3.3 Synchronization between Migration and Other Transitions

Proportional hazards models provide a means for investigating some of the possible hypotheses about interdependencies among a set of histories that unfold simultaneously over time; particularly hypotheses that center on covariation between the chances of migration events and the characteristics of individuals. A wider range of interdependencies becomes evident when processes are treated as simultaneous histories, including interdependencies between migration events and transitions in other histories. These interdependencies between events are distinct from interdependencies with the status of the same variables and their importance has been recognized in the context of several kinds of processes. Willekens (1991) has discussed the interdependencies between women's work histories and their fertility histories within a general framework that shows how the relations between these histories include dependencies between events as well as dependencies between events and status variables. Mulder and Wagner (1993) refer to marriage and migration as "synchronized events" in order to distinguish dependencies between these events from the effects of the status of "married" or "unmarried." Deurloo, Clark and Dieleman (1994) recognize a similar distinction in their analysis of tenure change, where they analyze not only the effect of family composition (a status variable) on the chances of a change in housing tenure but also the effect of *changes* in family composition, an event which they investigate for its possible "triggering effect" on tenure change.

Migration events and transitions in employment histories are likely to occur simultaneously (or nearly simultaneously) if both types of events are involved in major turning points or discontinuities in the life courses of individuals; if migration is the result of a successful job search that extends beyond a local area; or if migration is an event that disrupts an individual's work history. The major turning points that are emphasized in the life course perspective include, among others, an individual's embarkation on a career as a full-time worker, a change that may involve reorganization in several aspects of an individual's life. That is, the individual may relinquish social roles that are preliminary to that of full-time worker, including the role of student, and become eligible for new roles that are associated with adult status. Individuals who embark on new and different phases of their lives are likely to reevaluate their current location according to new criteria that reflect their new circumstances and, in many cases, the transition to a new phase of life will involve migration to a new location.

Events in migration and work histories may be synchronized even if the events are not part of a major life course transition. The search for a new job is likely to be associated with migration, at least for individuals who are willing to extend their job searches beyond their local area. Even though migration is usually treated as a strategy that can be used to enhance returns from working migration may also disrupt work histories, especially when an individual's migration is not directly motivated by considerations of his or her own labour force activity. Many individuals move as part of a family so that the migration of a family can affect the work histories of several members. In these cases the same move that enhances returns from labour force activity for one family member may cause a disruption in the work history of another. In particular, women may relinquish employment in order to migrate with their husbands and enter a period of unemployment or leave the labour force in conjunction with migration.

There have been only a few analyses of the synchronization of events but two general approaches to hypothesis testing have emerged. The first is to model the effects of an event by using the date of event to define a time-varying covariate that may affect the hazard of migration, but only for a fixed period following the event. Deurloo, Clark, and Dieleman (1994) take this approach in their analysis of changes in housing tenure where they analyze the effects of changes in family status as possible triggering events for a tenure change. They define time-varying variables such as "change from couple to family in the previous year" and fit proportional hazards models. This approach has the advantage that the effects of events can be integrated into the same proportional hazards models that are used to analyze the effects of status variables so that the model can include both an estimate of the effect of the status of a variable as well as an estimate for the effect of a recent change in the status of the same variable. Mulder and Wagner (1993) use a similar approach but theirs is a model of occurrence rates rather than a hazard model. The period of time in which an event exerts an effect on the hazard has to be determined outside the model, however.

A second approach, based on the intervals between two different kinds of events, has been used by Odland and Shumway (1993) who test for synchronization by treating the interval between migration and another event as a random variable. The two events tend to be synchronized if the durations of time that intervene between them are generally smaller than what would be expected if the timing of the two events were independent. Odland and Shumway define an expected survivor function for the durations of these intervals, based on the observed timing of events in each history and an assumption of independence in the timing of events between the histories. They then compare this expectation with the observed set of intervals between migration and marriage events.

Neither of these approaches is entirely satisfactory as a means of identifying the relations among interacting histories because the temporal ordering of events in these histories does not necessarily correspond to their causal priority (Willekens, 1991). That is, decisions may be make in anticipation of future events so that events which occur earlier are not necessarily exogenous to events that occur later. In the case of hazard models that use time-varying variables to capture the effect of a triggering event, a triggering event may actually occur later than the "triggered" event if an action is taken in anticipation of the triggering event. Inferences based on hazard models that include a time-varying covariate to indicate that a triggering event had occurred in the recent past would be likely to underestimate the effects of triggering events, since the triggering event would actually be recorded after the "triggered" event in some cases. Hypotheses about the synchronization of events can be tested without treating either event as exogenous if the time intervals between events are recorded without regard to their temporal ordering. For example, Odland and Shumway (1993) analyze the synchronization of marriage and migration by examining the intervals between the date of marriage and the nearest migration event for the same individual, regardless of which event occurs first. The usual methods for analyzing censored data cannot be applied in this case, however, because the "nearest migration event" cannot always be identified on the basis of censored data.[3]

8.4 Longitudinal Models and the Scope of Migration Research

Longitudinal frameworks, such as the general model of discrete-space continuous-time processes described above, provide comprehensive perspectives on migration behaviour that are useful because they clarify the ways that migration decisions are embedded in both the personal history of a decision-maker and the changing social and economic circumstances that surround an individual's decisions. It is unlikely, however, that any single empirical analysis can incorporate all of relations that are implied by these models, if only because no single data source provides sufficient information to investigate all the important aspects of migration behaviour.

[3] This occurs if a marriage event ends an interval that begins with migration but begins a second interval that is censored while it is still shorter than the first interval.

Longitudinal analyses have to be based on longitudinal data and while longitudinal data provide much more information about the timing of migration events they typically provide much less information about the locational context of migration than cross-sectional samples. The problem is sample size. The major longitudinal surveys for the United States, the Survey of Income and Program Participation (SIPP), the Panel Survey of Income Dynamics (PSID), and the National Longitudinal Surveys (NLS) gather information for several thousands of individuals for the entire country; and those sample sizes provide only a few individuals in any locality. The special requirements of migration research were not a factor in the design of these surveys and sample sizes were chosen to meet the needs of research agendas that did not involve analyzing problems in the contexts of particular localities. Comparisons across localities, and analyses that account for the contextual effects of particular localities, have been a major component in the design of research on migration, and those comparisons require sizable samples not just for the overall population but for the population of each locality. Cross-sectional data have provided very large samples that are sufficient to test hypotheses about differences in migration behaviour that are associated with the contexts of different localities. These samples are even large enough to support a notable emphasis on destination choice in migration research by providing sizable samples of migrants for particular origin-destination pairs (Fotheringham, 1986). There is no prospect that longitudinal samples large enough to replicate cross-sectional studies will be available, although comparisons between some regional contexts may be possible with existing longitudinal samples.[4]

Longitudinal data are likely to be especially important in analyses where the developmental or historical contexts of decision-making are predominant concerns, or where inferences about decision-making can be derived from precise information about the timing of events. Developmental or historical contexts are, of course, important in virtually all analyses of decision-making and it is the absence of this kind of information that leads to serious inferential problems in analyses of cross-sectional migration data (Davies and Pickles, 1985; Odland, 1996).[5] In fact, the inferential problems associated with cross-sectional modelling of on-going processes are potentially very serious, although Clark (1992) has shown that longitudinal and cross-sectional models for the same migration data are likely to yield similar results and Dieleman, Clark, and Deurloo (1994) make a case for the complementarity of cross-sectional and longitudinal approaches. A focus on those developmental or historical contexts is likely to be available only through the use of longitudinal data that provide generally insufficient information about locational contexts, however. A number of recent studies have demonstrated the value of longitudinal approaches

[4]For example, Odland and Bailey (1990) are able to make comparisons of migration behaviour in the two largest metropolitan areas in the U.S., New York and Los Angeles, on the basis of data from the National Longitudinal Surveys.

[5]The omission of historical information, especially about the persistence of values over time. leads to interdependencies between the covariates of a model and the error term. Davies and Pickles (1985) show how this may lead to serious problems in the estimates of parameters for cross-sectional models of on-going processes.

in investigations of the effects of the personal histories of migrants, especially their migration histories, and several of these studies have dealt with the problem of locational contexts by focusing on migration systems that involve relatively limited sets of locations (Bailey, 1993; Bailey and Ellis, 1993; Odland and Bailey, 1990). Other studies have used longitudinal data to analyze the interdependencies between migration and major transitions in personal histories (Courgeau and Lelièvre, 1992b; Lelièvre and Bonvalet, 1994) or on interdependencies which involve the effects of aggregate economic conditions as well as the effects of transitions in the histories of persons or households (Clark, Deurloo and Dieleman, 1994; Deurloo, Clark, and Dieleman, 1994). Analyses that focus on personal or household histories have to be based on longitudinal data that contain information for substantial portions of individual lifetimes and records for extensive segments of individual lives are available for the United States in the PSID and NLS surveys. Migration can be dated only to the nearest year in the PSID and NLS data, however, so there is considerable uncertainty about the timing of migration with respect to other events, such as episodes of unemployment. Migration can be identified to the nearest week in the SIPP data, but the SIPP surveys cover only brief segments of individual lives. The precise information about timing of events in the SIPP data are especially valuable for analyzing relations between migration and employment, however, and an investigation of interdependencies between migration events and employment histories is presented in the next section.

8.5 Empirical Analyses of Timing in Migration and Employment Histories

The employment status of individuals is reported on a weekly basis in the SIPP and migration events can also be identified to the nearest week. This makes it possible to identify synchronization of migration with employment histories with much greater precision than is possible with other data sources and two types of questions about the relations between the timing of migration and employment are investigated here. First, the SIPP data are used to investigate whether migration is associated with particular states in employment histories, where employment histories can be in three states in any week: "with a job," "without a job, looking for a job," and "without a job, not looking for a job." Economic explanations for migration rely heavily on employment and job search as motives for migration so it is important to ascertain how much migration occurs when people are neither employed nor seeking a job. The importance of job search in migration can be evaluated by investigating how frequently persons are in the state "without a job, looking for a job" in the weeks near the migration event. The second set of questions is concerned with the possible synchronization of migration with changes in employment status. That is, migration may be associated not only with the current status of an individual's work history but the timing of the migration event may be associated with the timing of transitions in work histories.

8.5.1 Migration Histories in the Survey of Income and Program Participation

The temporal detail in the SIPP data makes it possible to investigate these questions but these data provide individual work and locational histories for only relatively brief periods. The 1990 panel of SIPP, which is used in the analysis reported here, provides locational and work histories for a maximum of 28 months for a sample of the non-institutionalized resident population of the United States.[6] These data were gathered in a series of interviews and record the activities of individuals in the sample between October, 1989 and September, 1992.[7]

The analyses of interdependencies in this research centers on the behaviour of work histories at the time of migration and hypotheses are tested by comparing the behaviour of work histories during brief periods that coincide with the time of migration events to the general patterns of behaviour in work histories over all times. This approach makes it possible to determine whether work histories at the time of migration are more likely to be in particular states, such as "without a job- looking for a job" than at other times. That is, persons may be more likely to be spending time unemployed or out of the labour force at the time of migration. Hypotheses about the timing of work histories with respect to the timing of migration are also examined by comparing behaviour near the time of migration to behaviour at other times. Tests are carried out by comparing frequencies for particular week-to-week transitions in the labour force status of individuals for weeks that coincide with migration to frequencies of the same transitions for all pairs of weeks.

This approach involves comparing selected segments of work histories (those that coincide with migration events) to work histories in general and it is necessary to choose some particular length for these segments. There are no *a priori* reasons to select a particular length of time, although an "ideal" period would be long enough to exclude any work history events that are unrelated to the migration event. An admittedly arbitrary but reasonable length of five weeks is chosen in this investigation. That is, hypotheses are tested on the basis of segments of work histories five weeks long, and centered on the week of the migration event so that the segments include the week of migration as well as the two weeks preceding and two weeks following the migration event.

[6]Work histories are actually available for up to 32 months but locational histories are available for only 28 months. Members of the SIPP Panel are interviewed every four months, with each interview covering their activities in the preceding four months. Information about migration is obtained from questions about changes of address "since the last interview" so this information is not available from the first interview, which covers the first four months of the panel. Histories for some individuals in the panel are shorter than 32 months (for employment) or 28 months (for location) because individuals enter and leave the SIPP panels over the course of the study period.

[7]Although interviews extend over a 36 month period information for any individual was gathered for no more than 32 months. The sample consists of four "rotation groups" and each rotation group was interviewed every four months. The sample period for the first rotation group covers a 32 month period that is timed four calendar months before the 32 month period for the fourth rotation group.

The definition of migration events depends on the level of locational detail in the SIPP data. It is possible to identify all changes of address made by members of the panel while they are in the panel, but migration events are defined for this analysis as "interregional" moves, with local moves within the same community being excluded as much as the definition of regions in the SIPP data allows. Migration events are defined as moves that cross either the boundary of a Metropolitan Statistical Area and/or a state boundary.[8] This definition produces a set of 2738 migration events over the 28 months when migration events can be observed among members of the 1990 SIPP panel. These migration events are generated by a sample population whose size changes over the survey period, because of the loss of some respondents and the addition of others, but the total size of the panel varies from 56631 persons for the initial set of interviews to 54887 for the final set of interviews.

The analysis of interdependencies between migration and work histories in this research is carried out separately for different categories of the sample population, with these categories defined on the basis of age, gender, and marital status. This categorization is necessary because migrants are usually younger than the population as a whole. Hypotheses are tested in this research by comparing the behaviours of work histories at the time of migration to their behaviours at other times and a large proportion of the migration events are generated by younger people. Most of the five-week periods of work histories associated with migration are generated by young people and comparing those periods with work histories in general would confuse age differences in work histories with differences associated with migration.

The categorization of the population also makes it possible to examine the hypothesis that relations between migration and work histories differ between different subgroups of the population. In particular, the behaviour of employment histories at the time of migration may differ between married women and other categories of persons. Unmarried individuals are much more likely than married individuals to make locational decisions autonomously, by calculating the costs and benefits only for themselves. Consequently their migration decisions are likely to be closely related to their career development; provided that career development is the major element in their calculations. Migration decisions for married persons are more likely to involve compromises with respect to the work careers of two individuals so the relations between migration and work careers may be determined in more complex ways for a married person. It is possible, however, that family migration decisions are made largely in response to the careers of husbands, with wives more likely to make career sacrifices in conjunction with migration. The work histories of married women at the time of migration would differ from the histories of other subgroups if that is the case.

[8] This makes it possible for some short distance moves to be included, so long as they cross one of these boundaries. Moves over relatively long distances may be excluded if they occur within a single state and do not cross the boundaries of an MSA. The alternative would be to confine the analysis to interstate migration.

8.5.2 The Status of Work Histories at the Time of Migration

The status of work histories at the time of migration can be summarized, and compared to work histories at other times, by calculating the percentages of person-weeks that migrants spend in each of the three work history states at the time of migration. The "time of migration" is the five-week interval centered on the migration event for an individual. The distribution over work history states for migrants at the time of migration can then be compared to the distribution of person-weeks over work history states for all persons in all weeks. The results of these calculations are shown in Table 8.1.1 (for men) and Table 8.1.2 (for women) for categories formed on the basis of age, gender, and marital status.

There are consistent differences in the ways that migrants spend their time in the weeks near to migration events, compared to the general distribution of person-weeks over work-history categories. Migrants are more likely to be out of the labour force (without a job- not looking) or looking for a job in the weeks surrounding the migration event (or equivalently, they are less likely to be employed in those weeks). That pattern is consistent with the notion that migration events are associated with periods of interruption in work histories, including periods of unemployment and periods of time spent outside the labour force. The only exceptions are young unmarried persons (ages 15-24) and women over 65. Migrants in these groups are more likely to spend at least some of the weeks near the migration event with a job. Differences in the time spent in labour force activity between the migration period and other periods are greater for persons in late middle age, when rates of labour force participation are diminishing. Unmarried persons generally spend more weeks in the category "looking for a job" than their married counterparts, but the difference is greater for migrants in the weeks near to the migration event. This may indicate that unmarried persons are more likely than married persons to cope with unemployment by migrating.

8.5.3 Transitions in Work Histories near the Time of Migration

The general hypothesis that the timing of transitions in individual migration histories coincides with the timing of migration events can be tested by comparing the frequencies for work history transitions that occur among migrants at the time of migration with the frequencies for such transitions among the sample population at other times. It is sufficient to examine the frequencies with which individual migration histories leave their current states and the SIPP data makes it possible to enumerate these events on a week-to-week basis. Each migration event is associated with four "trials" when the migrant's employment status may leave its current state (given the five-week time period that is associated with migration), and the total number of such trials is four times the number of migrants. Pairs of adjacent weeks which are not within the five-week intervals associated with migration provide a much larger set of trials that can be used to estimate the probability that an individual employment history leaves its current state at times that are not associated with migration.

Table 8.1.1 Percentages of Person-Weeks for Men Person-Weeks Spent in Categories of Labour Force Activity Near the Time of Migration and by all Men at All Times

Age		Married Men		Unmarried Men	
		all times	migration	all times	migration
	with a job	.879	.764	.558	.622
15-24	looking for a job	.072	.103	.098	.086
	without a job, not looking	.049	.134	.345	.291
	with a job	.933	.831	.814	.747
25-34	looking for a job	.038	.052	.072	.113
	without a job, not looking	.029	.116	.114	.140
	with a job	.934	.842	.815	.645
35-44	looking for a job	.029	.031	.070	.175
	without a job, not looking	.037	.128	.115	.180
	with a job	.901	.646	.753	.506
45-54	looking for a job	.030	.116	.053	.062
	without a job, not looking	.069	.238	.194	.431
	with a job	.623	.321	.451	.153
55-64	looking for a job	.018	.072	.034	.001
	without a job, not looking	.359	.607	.515	.847
	with a job	.154	.082	.127	.110
> 65	looking for a job	.002	.001	.004	.001
	without a job, not looking	.843	.917	.868	.890

Let $p_{x,M}$ be an estimate of the probability that an employment history for an individual leaves state x between week t and week t+1; given that weeks t and t+1 are within five weeks of a migration event for the same individual. Let p_{x,M^*} be an estimate of the probability of leaving the same state between pairs of weeks that are not associated with migration. Then the quantity

$$z = \frac{p_{x,M} - p_{x,M^*}}{[p_x(1-p_x)(n_M^{-1} + n_{M^*}^{-1})]^{1/2}} \tag{8.4}$$

has approximately a normal distribution where p_x is an estimate based on all the trials and n_M and n_{M^*} are the numbers of trials of each type.

Results for this test are shown in Tables 8.2.1 -8.2.3 for categories of persons based on age, gender, and marital status. The analysis is carried out for only three age groups, 15-24, 25-34, and 35-44, because the numbers of migrants over age 44 in the sample is too small to calculate reliable estimates of the probabilities for transitions in work histories.

Table 8.1.2 Percentages of Person-Weeks for Women Person-Weeks Spent in Categories of Labour Force Activity Near the Time of Migration and by all Women at All Times

Age		Married Women		Unmarried Women	
		all times	migration	all times	migration
	with a job	.623	.550	.548	.594
15-24	looking for a job	.048	.051	.060	.102
	without a job, not looking	.329	.399	392	.303
	with a job	.679	.520	.728	.632
25-34	looking for a job	.025	.052	.052	.091
	without a job, not looking	.300	.428	.220	.287
	with a job	.723	.595	.787	.666
35-44	looking for a job	.023	.047	.039	.105
	without a job, not looking	.254	.359	.174	.230
	with a job	.661	.484	.733	.501
45-54	looking for a job	.016	.069	.033	.021
	without a job, not looking	.323	.447	.234	.292
	with a job	.362	.388	.444	.200
55-64	looking for a job	.007	.000	.019	.000
	without a job, not looking	.030	.012	.537	.800
	with a job	.068	.094	.084	.203
> 65	looking for a job	.001	.014	.003	.013
	without a job, not looking	.931	.891	.913	.784

Table 8.2.1 Estimated Probabilities for Remaining in the State "with a job" from Week to Week

Age	All Weeks	Weeks near Migration	Test for Difference
Age	Married Men		
15-24	0.9926	0.9707	4.325**
25-34	0.9944	0.9668	9.327**
35-44	0.9929	0.9823	2.291*
	Unmarried Men		
15-24	0.9928	0.9762	4.645**
25-34	0.9945	0.9665	7.811**
35-44	0.9930	0.9612	4.411**
	Married Women		
15-24	0.9935	0.9319	12.027**
25-34	0.9971	0.9393	22.171**
35-44	0.9972	0.9395	16.958**
	Unmarried Women		
15-24	0.9936	0.9322	18.405**
25-34	0.9960	0.9598	10.305**
35-44	0.9964	0.9916	0.935

* indicates significance at .05 level ** indicates significance at .01 level

Table 8.2.2 Estimated Probabilities for Remaining in the State "looking for a job" from Week to Week

Age	All Weeks	Weeks near Migration	Test for Difference
	Married Men		
15-24	0.9132	0.8699	2.319*
25-34	0.9198	0.8607	5.155**
35-44	0.9056	0.7071	11.909**
	Unmarried Men		
15-24	0.9172	0.7411	14.713**
25-34	0.9150	0.8088	7.633**
35-44	0.9137	0.8098	4.240**
	Married Women		
15-24	0.9044	0.8732	1.601
25-34	0.9317	0.8653	5.210**
35-44	0.9371	0.9500	-0.814
	Unmarried Women		
15-24	0.9163	0.8190	8.206**
25-34	0.9241	0.8608	4.202**
35-44	0.9294	0.7470	8.289**

* indicates significance at .05 level ** indicates significance at .01 level

Table 8.2.3 Estimated Probabilities for Remaining in the State "without a job-not looking for a job" from Week to Week

Age	All Weeks	Weeks near Migration	Test for Difference
	Married Men		
15-24	0.9787	0.8611	5.915**
25-34	0.9821	0.8117	11.121**
35-44	0.9835	0.8789	5.691**
	Unmarried Men		
15-24	0.9780	0.8754	10.760**
25-34	0.9829	0.8530	8.388**
35-44	0.9843	0.6761	15.268**
	Married Women		
15-24	0.9654	0.9378	1.967*
25-34	0.9695	0.9219	5.098**
35-44	0.9769	0.9171	5.075**
	Unmarried Women		
15-24	0.9773	0.9225	6.357**
25-34	0.9810	0.8748	10.258**
35-44	0.9815	0.8357	7.791**

* indicates significance at .05 level ** indicates significance at .01 level

Table 8.3 Estimated Probabilities that a Married Woman enter the state "looking for a job" given that she was in the state "with a job" in the Previous Week

Age	All Weeks	Weeks Near Migration	Test for Difference
15-24	0.6769	0.1118	5.082**
25-34	0.6145	0.0386	5.413**
35-44	0.6088	0.0841	4.573**

** indicates significance at .01 confidence level

The differences in estimated probabilities indicate, in general, that the weeks near to migration activity are associated with greater frequencies of week-to-week transitions in employment histories. That is, persons are more likely to make transistions out of their current employment status in those weeks than during other times. Persons who are "with a job" during one of the five weeks associated with migration are more likely to make a transition to one of the other two states in the next week than are persons at other times. The differences in the estimated transition probabilities are numerically small but statistically significant (except for unmarried women aged 35-44). The difference is somewhat larger for married women than for others. That is, the migration event seems to increase the chances for leaving employment to a greater degree for married women than for other subgroups.

Migration also increases the probabilities for ending a job search by leaving the state "looking for a job," at least for sub-groups other than married women. Persons other than married women who are looking for a job are more likely to leave that state (by finding a job or ceasing to look) in the five weeks surrounding migration than persons whose job search is not associated with migration. Married women who are looking for a job may not experience increased chances of ending their searches at the time of migration, although the difference of proportions is significant for married women in one age group: those aged 25-34. Persons who are not employed or looking for work are also more likely to enter the labour force during the weeks surrounding migration.

Persons who leave one of the three work history states also enter one or the other of the remaining two and it is possible that the probabilities for entering one or the other of the remaining states, given the person's previous work history state, differ between events at the time of migration and events at other times. Transitions out of any state are not frequent events and the samples of migrants who leave each state and enter each of the other two are generally too small to support reliable inferences. The samples of married women who leave the state "with a job" are larger than for other categories of migrants, however, and will support some cautious inferences about differences in the behaviour of women who leave a job in conjunction with migration versus those who leave the state "with a job" at other times. The information in Table 3 indicates that married women who leave the state "with a job" in conjunction with migration are much less likely to embark immediately on a search for a new job than women who leave "with a job" at other times. That is, married women who leave a job in conjunction with migration are much more likely to enter a period when they are neither working nor looking for a job than married women who leave a job at other times.

8.6 Summary

Longitudinal approaches provide a comprehensive framework where migration and mobility behaviour can be analyzed in the context of individual locational histories that develop as the individual migrates from place-to-place and spends segments of his or her lifetime residing in different localities. These locational histories develop simultaneously with other histories, such as the individual's own histories of work activity, marriage, and family formation as well as the economic histories of particular regions where the individual might reside. Investigations of individual migration behaviour can be organized around the general question of how locational histories interact with these other histories as they develop. Discrete-state continuous-time stochastic models provide a general framework for modelling individual migration histories and statistical methods for analyzing duration data make it possible to use the additional information in longitudinal data to avoid inferential and measurement problems that are characteristic of cross-sectional analyses of migration.

Longitudinal analyses are not likely, however, to replace cross-sectional approaches in all areas of migration and mobility research. Longitudinal data are not easy to manage (Dieleman, 1992; 1995) and longitudinal samples usually are not sufficient to provide the kind of information about the locational contexts of migration decisions that have been available from cross-sectional data. Longitudinal data are especially useful for analyzing the effects of personal histories on migration behaviour and for investigating the relations between migration and employment.

The SIPP data used here provide precise information about the timing of events and have been used to reveal what is happening in the lives of migrants at the times when they migrate. These comparisons indicate that migrants spend their time, in the weeks near to migration events, in ways that do not differ greatly rom the activities of other persons in the same categories of age, gender, and marital status. Migrants do spend more time looking for jobs and more time out of the labour force in the weeks surrounding migration but these differences are not large enough to indicate that migration is selective of persons who are much less likely to be employed that other members of the population; nor are the differences large enough to indicate that migration is usually associated with drastic interruptions of work histories.

The time of migration is, however, a time when people experience increased rates of transition into and out of employment. The times when people migrate are also times when they are more likely than usual to leave employment, end a period of job search, or shift into the active labour force. These increased frequencies of transition are consistent with the general hypothesis that migration histories and work histories are interdependent and that events in the two histories are either directly related, as when a successful job search results in migration; or are components of major episodes of reorganization in people's lives.

Acknowledgment
This research was partly supported by National Science Foundation grant SBR9514971.

References

Anderson W. P. and Papageorgiou Y.Y. 1994. An analysis of migration streams for the Canadian regional system 1952-1983 1. Migration Probabilities. *Geographical Analysis* **26**:15-36.

Bailey A. J. 1993. Migration history migration behavior, and electivity. *The Annals of Regional Science* **27**:315-326.

Bailey A.J. 1993. Migration and unemployment duration among young adults. *Papers in Regional Science* **73**:289-307.

Clark W.A.V. 1992. Comparing cross-sectional and longitudinal analyses of residential mobility and migration.*Environment and Planning A* **45** 1291-1302.

Clark W.A.V. Deurloo M.C. and Dieleman F.M. 1994. Tenure change in the context of micro-level family and macro-level economic shifts. *Urban Studies* **31** 137-154.

Courgeau D. 1990. Migration, family and careet: A life course approach, in: *Life Span Development and Behavior*, eds. P.B. Baltes, D.L. Featherman and R. Lerner Hillsdale, N.J.: Lawrence Erlbaum Associates. pp. 219-255.

Courgeau D. and Lelièvre E. 1992a. *Event History Analysis in Demography* Oxford: Clarendon Press..

Courgeau D. and Lelièvre E. 1992b. Interrelations between first home-ownership, constitution of the family, and professional occupation in France, in: *Demographic Applications of Event History Analysis*, eds. J. Trussell, R. Hankinson, and J. Tilton Oxford: Clarendon Press. pp. 120-140.

Cox D.R. 1972. Regression models and life tables. *Journal of the Royal Statistical Society Series B* **34**, 187-220.

Davies, R.B. and Flowerdew, R. 1992. Modelling migration careers using data from a British survey. *Geographical Analysis* **24**, 35-57.

Davies, R.B. and Pickles, A.R. 1985. Longitudinal versus cross-sectional methods for behavioral research: A first round knock out. *Environment and Planning A* **17**, 1315-1329.

Deurloo, M.C., Clark, W.A.V. and Dieleman, F.M. 1994. The move to housing ownership in temporal and regional contexts. *Environment and Planning A* **26**, 1659-1670.

Dieleman, F.M. 1992. Struggling with longitudinal data and modelling in the analysis of residential mobility. *Environment and Planning A* **24**, 1527-1530.

Dieleman, F.M. 1995. Using panel data: Much effort, little reward? *Environment and Planning A* **27**, 676-682.

Dieleman, F.M., Clark, W.A.V., and Deurloo, M.C. 1994. Tenure choice: Cross-sectional and longitudinal analyses. *Netherlands Journal of Housing and the Built Environment* **9**, 229-246.

Elder, G.H. 1985. *Life Course Dynamics.* Ithaca, N.Y. Cornell University Press..

Fotheringham, A.J. 1986. Modelling hierarchical destination choice. *Environment and Planning A* **18**, 401-418.

Hachen, D.S. 1988. The competing risks model: A method for analyzing processes with multiple types of events. *Sociological Methods and Research* **17**, 21-54.

Herzog, H.W., Schlottmann, A.M. and Boehm, T.P. 1993. Migration as spatial job-search: A survey of empirical findings, *Regional Studies* **37**, 327-340.

Kalbfleisch, J.D. and Prentice, J.L. 1980. *The Statistical Analysis of Failure Time Data.* New York: John Wiley..

Lancaster, T. 1990. *The Econometric Analysis of Duration Data.* Cambridge: Cambridge University Press..

Lawless, J.E. 1982. *Statistical Models and Methods for Lifetime Data.* New York: John Wiley..

Lelièvre, E., and Bonvalet, C. 1994. A compared cohort history of residential mobility, social change and home-ownership in Paris and the est of France. *Urban Studies* **31**, 1647-1665.

Liaw, K-L. 1990. Joint effects of personal factors and ecological variables in the interprovincial migration pattern of young adults in Canada: A nested logit analysis. *Geographical Analysis* **22**, 189-208.

Liaw, K-L, and Ledent, J. 1987. Nested logit model and maximum quasi-likelihood methods: A flexible methodology for analyzing interregional migration patterns. *Regional Science and Urban Economics* **17**, 67-88.

Lichter, D.T. 1980. Household migration and the labor market position of married women. *Social Science Research* **9**, 83-97.

Molho, I. 1986. Theories of migration: A review. *Scotttish Journal of Political Economy* **33**, 396-419.

Moss, W.G. 1979. A note on individual choice models of migration. *Regional Science and Urban Economics* **9**, 333-343.

Mulder, C.H., and Wagner, M. 1993. Migration and marriage in the life course: A method for studying synchronized events, *European Journal of Population* **9**, 55-76.

Odland, J. 1996. Longitudinal analysis of migration and mobility: Spatial behavior in explicitly temporal contexts. forthcoming in: *Spatial and Temporal Reasoning* eds. M.J. Egenhofer and R.G. Golledge.

Odland, J. and Bailey, A.J. 1990. Regional out-migration rates and migration histories: A longitudinal analysis. *Geographical Analysis* **22**, 158-170.

Odland, J. and Shumway, J.M. 1993. Interdependencies in the timing of migration and mobility events, *Papers in Regional Science* **72**, 221-237.

Petersen, T. 1995. Analysis of event histories, in: *Handbook of Statistical Modeling for the Social and Behavioral Sciences* eds. G.Arminger, C. C. Clogg, and M.E. Sobel New York, Plenum Press. pp. 453-517.

Sandefur, G.S. and Scott, W. 1981. A dynamic analysis of migration: An assessment of the effects of age, family and career variables. *Demography* **18**, 355-368.

Shumway, J.M. 1993. Factors influencing unemployment duration with a special emphasis on migration: An investigation using SIPP data and event history methods. *Papers in Regional Science* **72**, 159-176.

Sjaastad, L. 1962. The costs and returns of human migration, *Journal of Political Economy* **70**, 80-93.

Waldorf, B.S. 1994. Assimilation and attachment in the context of international migration: The case of guest workers in Germany, *Papers in Regional Science* **73**, 241-266.

Waldorf, B.S., and Esparza, A. 1991. A parametric failure time model of international return migration. *Papers in Regional Science* **70**, 419-438.

Willekens, F. J. 1991. Understanding the interdependence between parallel careers, in: *Female Labor Market Behavior and Fertility* eds. J.J. Siegers, J. de Jong-Gierveld, and E. Van Imhoff Berlin: Springer-Verlag. pp. 2-31.

Yamaguchi, K. 1991. *Event History Analysis* Newbury Park, CA: Sage Publications..

Yoon, B. 1995. An estimation of the returns to migration of male youth in the United States: A longitudinal analysis. Ph.D. dissertation, Indiana University.

9 Computational Process Modelling of Disaggregate Travel Behaviour

Mei-Po Kwan[1] and Reginald G. Golledge[2]

[1] Department of Geography, The Ohio State University, Columbus, OH 43210-1361, USA
[2] Department of Geography & Research Unit on Spatial Cognition and Choice, University of California Santa Barbara, Santa Barbara, CA 93106-4060

9.1 Introduction

It is common for researchers adopting an activity based approach to travel behaviour to differentiate between behaviours that are routinized and behaviours that result from deliberate choice. For example, a significantly large part of work-trip behaviour is routinized; individuals tend to use the same mode for each trip, to leave their home base at approximately the same time, to aim at arriving at approximately the same time at their work place, and to follow the same route and the same path segments that make up that route. Some other trip purposes are similarly routinized such as trips for religious purposes, and trips for medical or health related reasons - routinized in the sense of using the same mode and following the same path, even though the times at which travel may be undertaken might vary because of temporal differences in the scheduling of appointments by health professionals. Other trips such as grocery or food shopping may be routinized to a lesser extent. Instead of choosing a single destination and following a repetitive path to that destination, several alternative destinations may be kept as part of a feasible alternative set. Trip making on any given day then becomes more of a deliberate choice both in terms of selecting a specific destination and in terms of selecting the travel path. Variation can also occur in terms of travel mode. Many other trip purposes fall within the deliberated choice purview. In particular trips for social or recreational purposes, trips to meet with friends, trips undertaken for the purpose of dining away from home, business trips, and so on, all may be scheduled with different episodic intervals or frequencies, different lengths or durations, different destinations, different temporal units, different priorities, different sequences, as well as being undertaken either as single purpose or multiple purpose trips with single stop or multiple stop destinations.

While the modelling of the routinized choices and the prediction of consequent travel behaviour has been achieved with a considerable degree of success using discrete choice models, dynamic Markov models, and even via variations of the fundamental spatial interaction or gravity type models, less success has traditionally accrued when trying to model behaviours resulting from deliberated choice. As part of the effort to model, explain, and predict trip making, geographers and transportation scientists generally have developed or adopted a number of strategies that focus either on network characteristics (shortest path models), aggregate behaviours (spatial interaction model and entropy models), individual preferences (compositional and decompositional preference modes), and choice models (discrete choice models, models of variety seeking behaviour, compositional and decompositional choice models) (Timmermans and Golledge, 1990). Discrete choice models have also been used in transportation science for the modelling of choices of modes, departure times, or other characteristics relating to how single trip choices and choice alternative characteristics match up, or to the extent to which individual trip making behaviour matches the behaviour of a group to which they are assumed to belong.

The seminal work of Jones, Koppelman and Orfeuil (1990) firmly established a mutual dependency between travel choices and household or individual agendas of activities. Previously, Root and Recker (1983) had suggested that choices of destinations, departure times, and frequency and duration of activity participation should all be treated in a single conceptual framework that entails behavioural assumptions accounting for the process of making these interrelated choices. In other words, they developed the idea of focusing on the activity scheduling process and defining the type of model whose input consisted of components of this scheduling process. While this approach was conceptually and theoretically appealing, it proved difficult to implement within the context of the existing transportation and behaviour models, particularly the dominant discrete choice modelling framework that existed at that time.

As an alternative, a new form of modelling of travel behaviour began to develop based on the idea of a set of interacting computer programs which would relate elements of real and perceived environments, factors influencing choice of destination, household preferences for scheduling activity sequences, and a variety of authority, coupling, and capability constraints that had been offered as part of the emerging field of time geography (Hägerstrand, 1970; Carlstein, Parkes and Thrift, 1978). Simultaneously, awareness of the limitations of simple discrete choice models encouraged the development of tools suitable to model interdependent or joint choices, which include the nested logit (Ben-Akiva and Lerman, 1985; McFadden, 1979) or structural equations models (Golob and Meurs, 1988) evolved. Axhausen and Gärling (1992) have summarized other attempts at estimating discrete choice models in which activities are important components and include summaries of the works of Kitamura (1988), Thill and Thomas (1987), and the trip-chaining models of Damm and Lerman (1981), Kitamura, Kazuo and Goulias (1990), and activity choice and activity duration (Kitamura, 1984). Following the lead of Root and Recker (1983) activity based choice models emerged (Recker, McNally and Root, 1986a and 1986b), along with some econometric research on time allocation (Winston, 1987). Much of this work, however, invariably rested on utility maximizing assumptions. Questioning of this assumption had been extensive in psychology (Simon, 1955 and 1990; Tversky and Kahneman, 1991 and 1992) as well as in transportation research (Supernak, 1992). In general, these models were limited in that they specified the factors affecting final choice but neglected the processes resulting in these choices. Obviously if the primary aim is to forecast travel choices, this criticism is not an important one; but if the goal is to understand the entire process and to develop appropriate relevant theory, then the shortcoming does become significant.

In a style similar to the STARCHILD model developed by Recker, et al. (1986a and 1986b) another alternative model format called Computational Process Modelling (CPM) emerged. In an activity scheduling context, these Computational Process

Models (CPM) focused on interdependent choices where choice involved acquisition, storage and retrieval of information, including retrieval from long-term memory, tradeoffs between accuracy of recalled information concerning locations, hours of business, and remembered paths, in terms of effort (time or distance) expended in order to make the tradeoff and achieve a goal. It also included the possibility for a conflict resolution where uncertainty may exist in terms of competition for a travel mode (e.g., who gets the household car), which activities are considered primary and which secondary and therefore take precedence in scheduling, and which destination choices provide the greatest flexibility and judge success in terms of completing a planned activity schedule. These models are built on some of the seminal ideas of Hayes-Roth and Hayes-Roth (1979) who produced a production system model which was accepted as a feasible alternative to existing discrete choice models for travel behaviour analysis.

Production systems were initially developed by Newell and Simon (1972) as elaborations of how people think when they solve problems. They are frequently used in theories involving the higher cognitive processes (Anderson, 1990; Newell, 1992). Essentially a production system is a set of rules in the form of condition-action pairs that specify how a task can be solved. If the task requires an individual to choose one alternative in a choice set, the rules may specify what information is searched under different conditions, how the information is evaluated, how the evaluation or judgments are integrated. The system is usually realized in terms of a cognitive architecture comprising a perceptual parser, a limited capacity working memory, a permanent long-term memory, and a system for effecting a behaviour. An operational CPM is a production system implemented as a computer program. The resulting CPM offers a testbed for assessing the consequences of different policy measures, or as a mechanism for facilitating the development of different testable hypotheses. One may also incorporate different testable assumptions into the model to examine their effects on potential choice and consequent behaviour.

Essentially a CPM was assumed to be capable of providing a detailed description of the individual choice process, but there were some questions as to whether or not it was amenable to travel diary data input which by the late 1980's was becoming accepted as a dominant and detailed form of obtaining information for models of activity scheduling and choice behaviour (Gärling, Kwan and Golledge, 1994). The emphasis on this highly disaggregate base also raised questions as to whether or not output could be aggregated in order to provide some reasonable basis for forecasting and policy development. Two parallel alternative approaches have been suggested: microsimulation has been developed for forecasting from systems of disaggregate discrete choice models (e.g., Kitamura and Goulias, 1989) and combining CPM and discrete choice models in a single complementary context (Ettema, Borgers and Timmermans, in press). In either case, one can argue that the intrinsic value of the CPM or production system approach would be to provide the theoretical basis for the microsimulation or CPM/discrete choice model approach. In the following sections, therefore, we will review a selection of Computational Process Models, discuss a recent contribution from geography which combines the Computational Process Model idea with a Geographical Information System (GISICAS) (Kwan, 1994, 1995), and comment on both the microsimulation and CPM plus discrete choice model combination approach.

9.2 Review of Computational Process Models

One of the first attempts at developing a CPM of travel choices was that offered by Kuipers (1978) - the TOUR model. This focused on an individual's memory representation of the environment (i.e., cognitive map), the acquisition of environmental information through search and exploration, and the use of experience and stored memories for making route choices. The model was developed in an artificial simulated environment and lacked empirical application using actual examples of cognitive maps, spatial orientation capabilities, and wayfinding procedures.

A similar type of model (NAVIGATOR) (Gopal, Klatzky and Smith, 1989; Gopal and Smith, 1990), is based more on the principles uncovered during empirical research on the spatial knowledge acquisition and wayfinding abilities of children travelling through a well known neighborhood (Golledge, Smith, Pellegrino, Doherty and Marshall, 1985). Using this practical knowledge base the route planning procedure in NAVIGATOR is modeled by various choice heuristics. When information for making this particular segment choice is lacking, a general route selection criteria such as "moving in the same general heading" or "make a random turn at an intersection" represent options for next segment selection in the path following process. Again, however, there was considerable input from prior empirical testing of human behaviour in route selection tasks in a real environment, the model was still developed in a small hypothetical space.

Route following in a static environment is also the focus of another CPM, TRAVELLER (Leiser and Zilberschatz, 1989). An equivalent type of model in a dynamic environment was labeled ELMER (McCalla, Reid and Schneider, 1982). TRAVELLER simplifies the route selection problem by assuming that the relative locations of origins and destinations are known. This, of course, is perfectly reasonable in most routinized travel activities; it may not be quite so acceptable when we look at the question of deliberated choice where choosing a destination from a set of feasible options is part of the travel planning process. TRAVELLER then constructs a route from origin to destination via a process of search. In this case the production system consists of a set of rules which constrain how search and exploration can take place. In comparison to this the ELMER model conceives of routes as sequences of instructions for how to travel (e.g., go ahead 200 yards, turn right at the intersection). These instructions are retrieved when a particular need arises - for example when one must make a decision about a turn that could result in heading towards or away from a potential destination. Thus, route following is seen as a dynamic decision making process in which choices for segment selection and turns are made en route upon recall of appropriate constraining rules.

The above models, however, did not stress the dependence between travel choice and activity choice. CARLA (Jones, Dix, Clarke and Heggie, 1983) and STARCHILD (Recker et al., 1986a, 1986b) do attempt to address this problem. CARLA in general has the fewest behavioural assumptions but is tied more strongly to various time geographic concepts (e.g., Lenntorp, 1978). In particular it incorporates a variety of time geographic interaction constraints including capability and coupling constraints. Its output consists of sets of feasible activity schedules and sets of possible activity patterns.

STARCHILD is a very comprehensive CPM and emphasizes modelling of the choice between activity schedules. It incorporates a conventional discrete choice model to make such selections, but the authors agree that other theoretically sound choice models could be appropriately substituted. The emphasis in STARCHILD is on the utilities associated with each activity and the sum of the utilities that comprises a particular schedule. Thus, utilities of waiting time and travel time are important features to consider. Perhaps the conceptually weakest part of STARCHILD is its acceptance of utility maximizing assumptions and its use of combinatorics to evaluate all possible feasible choice patterns. In practice, of course, people have limited capability both for considering a finite number of options and for accessing what is truly optimal. A boundedly rational selection mechanism could however be incorporated into STARCHILD thus bringing it much closer to the realities of human decision making.

In an attempt to incorporate more realistic behavioural assumptions and to begin the process of including a perceptually valid environment in the model, Gärling, Brännäs, Garvill, Golledge, Gopal, Holm and Lindberg (1989) outlined a conceptual framework which could be implemented into a model they called SCHEDULER. This model focuses on an individual's choice of activities, selects from feasible set of destinations, examines possible departure times which are critical in forming a travel agenda for a particular time period. Activities were generally stored in a long-term memory system called "the long term calendar." Associated with each activity is a priority for waiting time and a maximum duration for completing the activity. A specific activity is retrieved from long-term memory and scheduled on the basis of its relative priority weight and the expected duration required for its completion. Spatio use and temporal constraints, including the hours of business and the determination of sets of reachable locations, are retrieved from a memory representation or stored cognitive map of the environment. This is obtained a priori and stored in long-term memory. Choice of the location of a feasible destination and a potential departure time is then determined by the SCHEDULER as it works in a top-down fashion scheduling the highest priority and most repetitive needs first. A possible activity schedule is stored in a short-term memory (the short-term calendar) for possible later execution, depending on whether or not critical input variables such as the duration of time allocated to the activity remain constant or are changed. In this model members of the feasible alternative set are evaluated according to a nearest neighbor heuristic (Hirtle and Gärling, 1992; Gärling and Gärling, 1988). Since in the SCHEDULER the constraints on when activities can be performed (e.g., open hours for business purposes) are part of the initial filtering process, the choice of a feasible location is at least partly determined by the duration allocated to the activity and business open hours during which the activity can be performed. If a possible sequence of activities is defined but cannot be executed because of a temporal overlap, the activity sequence is redefined by resolving conflict among the competing activities. Here, higher priorities are given precedence. This conceptual model has been expanded by Kwan (1994, 1995) by hosting the scheduling module in a GIS context. We now explore this in more detail.

GISICAS represents an attempt to overcome some of the limitations of existing CPMs in terms of their lack of geoprocessing capabilities to handle the vast amount of real-world location and route data, and to perform spatial search using information about the objective and cognitive environments of the traveler. Although the scheduling algorithm of GISICAS benefited considerably from that of the SCHEDULER, it extended the framework of the SCHEDULER in several ways. Besides an individual's home and work locations, and the priority, duration and timing of activities, a person's preferred and fixed destinations were also included in GISICAS as important elements of the cognitive environment. The procedures of activity scheduling were integrated with a comprehensive geographic database of a real urban environment and interact dynamically with the module of spatial search heuristics in relation to that environment. Realizing the simplicity of the nearest neighbor heuristic used by the SCHEDULER, several new spatial search heuristics were developed in GISICAS for handling the effect of locational preference and the binding effect of fixed destination on an individual's travel behaviour. Basically, they were high-order spatial search heuristics which, instead of looking for the nearest activity locations locally, search for globally satisficing locations in relations to the next fixed destination to which an individual has to travel to. GISICAS's procedure for delimiting the choice set in explicit spatial terms also represents a departure from the largely aspatial or pseudo-spatial method of the SCHEDULER. The concept of feasible opportunity set was first formulated by Golledge, Kwan and Gärling (1994) for expressing the effect of bounded rationality on the choice set and implementing the satisficing principle in explicit spatial terms. It was defined with respect to a person's home and work locations. In GISICAS the feasible opportunity set was defined dynamically with respect to the current location and the immediate spatio-temporal constraints of the traveler (e.g. time allowed for the activity and travel to the destination, distance willing to travel, etc.). This sequential identification of the choice set regarding feasible destinations enables the dynamic interaction of the planning and execution of the activity agenda. The focus of GISICAS on the spatial dimension of activity scheduling and travel decision making suggests an alternative way of modelling travel behaviour at the individual level. With the data handling and geoprocessing capabilities of GIS, modelling and analysis of travel behaviour does not depend on any prior schema of zonal division of the study area. This may open up a new arena for future research on the use of CPM in transportation science.

9.3 Micro-Simulation Models

A modified form of the CPM has recently been developed by Ettema, et al. (in press) and Ettema, Borgers and Timmermans (1995). Like SCHEDULER, SMASH emulates the scheduling process by computing utilities for choices that result in inclusion, deletions, or substitution of activities. In this model, as the number of potential choices increase, disutility occurs. The decision making process is cumulative and it terminates when no choice results in a positive utility. The utilities associated with the choice of activities depend on the value of priorities for activities in the schedule, travel distance or travel time, the attractiveness of possible locations, the pressure on the individual to complete the activity within a certain time range, and the waiting time before the activity can be implemented and completed. Once a schedule is determined, its realism is evaluated. The SMASH model includes factors that can be assumed to affect activity scheduling and in some respects may be said to have similarities with STARCHILD. It appears to be a more complete model than SCHEDULER. However, whereas in STARCHILD utility maximization is an end product, in SMASH utilities are maximized at each step in the scheduling process. Thus, as each activity is evaluated for possible inclusion, deletion, or substitution in a schedule, the utilities associated with the appropriate process are assessed and ordered. It is thus a more disaggregate and implemented model than is STARCHILD and it raises the question as to global versus local maximization. Once again the original SMASH model required all possible choices to be evaluated at each scheduling step which would in reality place a considerable burden on the human traveler. Substituting a feasible opportunity set reduces the magnitude of this presumed computational component.

A significant added strength of SMASH, however, is that the authors have envisaged a way of empirically testing the model using mathematical statistical modelling (Ettema and Timmermans, 1993). The fundamental characteristics of this empirical testing focuses on the choice processes of including, deleting, or substituting activities. These actions are predicted from variables describing the current state of the scheduling process such as the number of activities already scheduled and attributes of the activities that still remain to be scheduled. While some inferences about these actions can be made from examining detailed travel diary records, the authors point out that there still remains much to be learned about the scheduling process itself. Particular items suggested for inclusion in future versions of the model include incorporation of mode choice, the planning of time spent at home, the addition of constraints and the sequence of activities so that some activities are not planned before undertaking others, and the linking of a production system to observed behaviour so that specific parameter values can derived for the model and tested in different environmental, socio-economic, demographic, and mode choice mix environments. This would allow comparison of the outcomes of simulations of a selection of activity schedules using different parameter settings, with the observed scheduling behaviour of humans. Extensive testing in this way should be able to provide a range of "best specification" for parameter values in different environments. As an alternative, parameter values for the model could be collected during interactive simulations in which subjects would be required to complete a task consisting of a number of clearly specified steps that are part of the scheduling process. However, modelling each step separately could provide substantial problems in terms of integrating the results into a single final comprehensive model. At this stage, however, it appears that SMASH, STARCHILD, and the SCHEDULER/GISICAS alternatives are equally feasible alternatives to further examine the process of activity scheduling and the act of tying together scheduling and travel behaviour.

Epstein (1996) has built a pragmatic navigation model. This model acquires facts about a two-dimensional environment in order to travel through it without an explicit map in its memory prior to travel. Spatial information is accumulated during travel which can be described as sequential trip-making through a fixed maze with specific barriers. Insights about basic points (landmarks), jagged or smooth barriers, bottlenecks, and other obstructions are learned during a sequence of trials. What is learned on one trial is stored and used to help make decisions on the next trial.

As with the other computational process type models previously discussed this device is tested in a simulated environment consisting of cells of a specific size (i.e., a grid network). The argument is that during experimentation individuals do not build detailed maps of the environment through which they wander but rather encode pragmatic representations of environmental features that assist in path-finding. In this type of world a robot or other travelling agent can learn it's way around in an efficient and effective manner. While doing so it creates a spatial representation consisting of selected parts of the environment and in some ways can be regarded as similar to the cognitive maps that humans would build in following the same set of tasks. These cognitive maps would be incomplete, in parts fuzzy, and may not lend themselves to exact solution procedures such as minimizing distance or time. In some sense, therefore, the solution is a satisficing or perhaps boundedly rational one. A critical feature, however, is the limited nature of the space and it's dimensionality, but it is interesting to note that navigation can eventually take place in a rapid and effective manner even though the environment through which travel takes place is always a partially obstructed space. In the solution process, Epstein uses a "FORR" ("FOr the Right Reasons") process. This type of program gradually acquires useful knowledge using a hierarchical reasoning process. It relies on different tiers of "Advisors" which are domain-specific but problem-class independent. They provide rational or support criteria for decision making (e.g., get closer to your destination). Each Advisor is implemented as a time-limited procedure and must operate within a constraint set which limits the number of permissible actions. Advisor recommendations are in the form of comments each of which has a weight, salience, or strength attached to it (i.e., an integer from 0 to 10 that measures the intensity and direction of the Advisor's opinion). Advisors usually do not recommend extensive search activities. Many of the Advisors embody within them the results of comprehensive spatial cognition work from recent decades. This includes advice on alignment, direction of travel, orientation, navigating around obstacles, and neighborhood definition. The results of the application of the model would appear to have relevance for human navigation in that in general there is a tendency to select routes such that there are no substantial changes of direction during the wayfinding process; initially the strongest desire is to move toward the goal; travel is preferred along the main rather than orthogonal axes; there is a tendency to avoid neighborhoods with limited entrance and exit possibilities. On the whole, however, as with many of the other simulation models that could be found in the artificial intelligence literature, this is more a model for navigation and wayfinding than a model that explicitly attacks the question of activity scheduling and route selection in a spatial and temporal context that facilitates successful completion of an activity schedule.

Another computational process model recently developed is that by Chown, Kaplan, and Cortenkamp (1995). Chown, et al., offer a model called PLAN (Prototype, Location, and Associative Network model) which is an integrated representation of large scale space which they claim is commensurate with a cognitive map. PLAN is used as a means for including wayfinding as a process which is concurrent with the rest of cognition not apart from it. The model is seeing-based because it takes advantage of the properties of the human visual system which provides continuous answers to the spatial updating or "where" problem. It synthesizes parts of prototype theory, associate network theory, and locational principles together in a connectionist system. The views in the PLAN model are not from an aerial or survey prospective but reflect what an observer would see from different head positions at a single location. This is similar in some ways to the model suggested by Golledge, Smith, Pellegrino, Gopal, Doherty, and Marshall (1985) which differentiates the environment into view (straight ahead linear observation) and scenes (more detailed local observations that might occur with head movements to the side as one is walking along a path). In the latter, views represent what might be seen when looking down the street into the distance, while scenes might be the explicit characteristics and features of a single house that would be observed if one turned their head from a facing direction while travelling. In the PLAN model, path overlap provides the mechanisms for developing new path segments which combine sections of previously experienced routes, and also provide the appropriate geometric base for integrating separate but partially overlapping paths into a more general knowledge of spatial layout (see also Golledge, Ruggles, Pellegrino, and Gale, 1993). Chown argues that a significant advantage of PLAN is that it only stores a fraction of the available information internally and relies on the perceptual system to fill in gaps when PLAN goes into action. This appears to be a reasonable way to operate a robot travelling through a learning environment but it also has some direct similarities concerning the way humans collect encode, store, and use information as they travel through complex environments. Particularly the idea of storing a minimal spatial representation as a cognitive map and perceptually updating one's location as travel takes place, represents a reasonable wayfinding and search strategy for human behaviour. PLAN thus responds to the criticisms of computational process models offered by Gärling, et al. (1989) who complained that an appropriate mix of attention paid to the processes of spatial cognition used by humans to extract and process information about environments as well as encapturing the essential components, barriers, and paths of the environment itself is required. The model GISICAS (Kwan, 1995) also takes these suggestions seriously and integrates cognitive processes, real-world environmental systems, and the intervening mechanisms associated with Geographic Information Systems (GIS) as a way of handling both wayfinding and activity scheduling problems in complex real-world systems.

9.4 Multi-Criterion Equilibrium Traffic Assignment Models

Multi-criterion equilibrium traffic assignment models developed in part because of a lack of attention to any realistic interpretation of the value of time problem in traffic assignment. Virtually every transportation model is ultimately evaluated in terms of how users interpret the value of time (VOT) that the model requires them to expend.

This is true for mode choice models, congestion pricing models, and traffic assignment models. It is generally recognized that each individual has a different VOT depending on factors such as the person's economic resource base or the time that one is willing or able to spend on a trip. For the most part, transportation planning models acknowledge this by developing an average view of time figure and as a result they invariably produce large estimation errors and inaccurate forecasts. Recently, Ben-Akiva and Bowman (1995) developed a logit mode choice model by assuming that there was a distribution of the value of time characteristics rather than assuming a similar VOT for all users. The result was a significant improvement of goodness-of-fit between predicted choices and actual choices. Dial (1995) consequently proposed a similar remedy for traffic assignment models. His model admits that VOT is best captured by a distribution and it uses a bi-criteria user optimal equilibrium traffic assignment model which generalizes classic traffic assignment by relaxing the VOT parameter in the generalized cross function from a constant to a random variable with an arbitrary probability density function (Dial, 1995). His model, called T2, is said to respond to a variety of difficult existing problems including the mode/route choice problem, parking policy, and congestion pricing. T2 models mode choice by assigning trips to paths in a multi-modal or hyper network. The latter combines walking, riding, transit and highway links. It is able to selectively route auto trips to parking lots that have a specific range of charges associated with them (cheaper lots that may require a longer walk to a destination), or to other higher priced lots that reduce the walking component of a multi-modal trip. It is also touted as being an appropriate model for determining where to place toll booths and what prices to levy in order to reduce congestion. In discussing his model, Dial captures some of the time/cost tradeoffs of a variety of different forms of transportation, but it satisfies none of the behavioural criticisms levied against econometric and mathematically optimizing traffic assignment models.

T2 works as follows. Assume it is necessary to make a trip from an origin O to a destination D. The problem is to determine the mode choice for the trip. Dial first assumes that it is possible to enumerate all feasible paths for this trip and to know the time and cost of each path. Each path is then plotted at a point in a graph according to its time and cost. One might show fifteen feasible "paths" in terms of time and cost between a given O-D pair. Today, a helicopter is often the fastest mode and the most expensive while walking is the slowest and cheapest. Dial then examine the process of selecting among possible mode/cost/time possibilities. It develops a path-finding algorithm which examines all possible combinations of mode/time/cost and determines an optimum. In an argument similar to that used to determine feasible alternative destination sets from among all possible sets that was developed in SCHEDULER and GISICAS, Dial differentiates between mode/time/cost combinations by determining a likelihood that a particular combination will be chosen. In this way many potential combinations are eliminated and only those few feasible paths connecting a given origin and destination are examined. This final traffic assignment model accumulates trips for all O-D pairs and defines a user optimal traffic assignment. This solution is termed a traffic equilibrium, and focuses on the relationship between travel times, cost, and volume of flow along arcs of a given path.

This model is a return to the classical mathematical/econometric model of traffic assignment, and does not deal with travel behaviour. It does make some concessions to the CPM and discrete choice related research which emphasized the significance of disaggregate units and individual differences in evaluating activity schedules and travel paths by including a value of time distribution characteristic rather than a single VOT estimate across an entire population. The question remains, however, whether it is suitable primarily for the routinized travel behaviours or whether it is robust and versatile enough to also incorporate path selection, mode choice, and travel behaviour for the other activities we have previously described as resulting from deliberated considered choice, with conflicts being resolved in a dynamic on-route environment.

9.5 Summary and Discussion

In this paper we have reviewed and assessed a variety of approaches to the activity scheduling and travel behaviour problem. Among these were a range of computational process models and some recently emerging alternatives including microsimulation, multicriterion traffic assignment models, and combined CPM/discrete choice models. Of this set we have argued that the CPM approach allows one to move closest to the real world decision making and choice situation and allows us to incorporate elements of both objective and cognitive environments as the matrix on which activity schedules and travel choices are made. For traffic assignment in a multimode environment, the T2 model appears to have significant promise. Of the various CPMs reviewed, SMASH and GISICAS both appear to be flexible, expandable, developed at the individual level but capable of aggregation, and suited for testing in real world environments. STARCHILD also has been tested on real world travel diary data with considerable success. It is quite possible that with some modifications such as the inclusion of a procedure for determining feasible alternative sets rather than going through complete enumeration of all possible activities, embedding the model in a Geographic Information System in a real-world environment, and allowing a boundedly rational selection criteria to replace the simple maximizing utility assumption, then STARCHILD could be expected to achieve a considerable success at predicting and forecasting travel behaviour. Of the models examined, SMASH attempts to integrate the recent CPM approach with the more traditional discrete choice model approach and by combining the most powerful aspects of each, provides significant hope for successful application - which appears to have recently been completed (Ettema, 1995).

In order to continue evaluating and assessing the different avenues of current research, specific empirical testing will be required. This requires explicit testing of the behavioural assumptions entailed by the different models as well as by assessing their ability to consider different mode combinations, and different combinations of activities in a schedule. GISICAS argues that scheduling must take place over a longer period than a day for some high priority activities (e.g., food shopping) may only be undertaken every second or third day. At that time, however, their importance is extremely great and scheduling must be adapted to allow the activity to take place.

The increasing volume of travel diary, panel, and survey data is providing more and more insights into the process of travel behaviour and mode choice. However, detailed testing of how activities are selected and how the selections are combined into schedules is still a critical point for future research. GISICAS is the only model so far to explicitly incorporate a Geographic Information System (GIS) into its structure. Obviously the potential for GIS must be further examined. This examination should include the suitability of GIS as a host for recording diary, panel or survey data, as well as its potential for developing an interactive framework for the assessment of priorities associated with activities and the choice of an appropriate path selection model. As far as the latter is concerned, more work needs to be done on the criteria that can conceivably be used in the path selection process. All too often assumptions are readily accepted that minimizing distance or minimizing time or cost are the only criteria worth considering. Recent research (Golledge, 1995a and 1995b; Kwan, 1994) has indicated that there are a number of other feasible path selection criteria and spatial search heuristics that may have to be made available in a group of models that could potentially be used in successful path selection and travel behaviour prediction. A GIS seems to be a reasonable host for incorporating a variety of models which can satisfy criteria such as minimizing turns, always heading in the direction of your destination, selecting the longest or shortest leg first, maximizing the aesthetic value of the route, minimizing perceived or actual costs, minimizing perceived or actual distance, minimizing perceived or actual time. Most model frameworks adopt one or another of these path selection criteria in their trip behaviour phase. It appears that different criteria may well be used for different trip purposes. If this is the case then standard models based on single criteria cannot possibly hope to satisfactorily forecast travel behaviour. A GIS that includes a set of path selection algorithms which could be initiated by a predisposition of a traveler to select certain criteria for certain purposes, could add significantly to our ability to understand and perhaps even to forecast the complex set of trips that make up the activity patterns of population aggregates. It is our intention in the future to continue working on these problems and, in particular, to determine the type of GIS (e.g., object oriented or relational) that will best lend itself to the procedures defined above.

References

Anderson, J.R. 1990. *The adaptive character of thought.* Hillsdale, NJ: Erlbaum.

Axhausen, K.W., and Gärling, T. 1992. Activity-based approaches to travel analysis: Conceptual frameworks, models, and research problems. *Transport Reviews,* **12**, 4: 323-341.

Ben-Akiva, M., and Bowman, J.L. 1995. Activity based disaggregate travel demand model system with daily activity schedules. Paper presented at the International Conference on Activity based Approaches: Activity Scheduling and the Analysis of Activity Patterns, May 25-28. Eindhoven University of Technology, The Netherlands.

Ben-Akiva, M., and Lerman, S. 1985. *Discrete choice analysis.* Cambridge, MA: MIT Press.

Carlstein, T., Parkes, D., and Thrift, N. 1978. *Timing space and spacing time, 3 Volumes.* London: Edward Arnold.

Chown, E., Kaplan, S., and Cortenkamp, D. 1995. Prototypes, location, and associative networks (PLAN): Towards a unified theory of cognitive mapping. *Cognitive Science,* **19**,

1-51.

Damm, D., and Lerman, S. 1981. A theory of activity scheduling behavior. *Environment and Planning A,* **13**, 703-718.

Dial, R.B. 1996. Multicriterion equilibrium traffic assignment: Basic theory and elementary algorithms. Part I - T2: The bicriterion model. *Transportation Science* accepted..

Epstein, S.L. 1996. *Spatial representation for pragmatic navigation.* Personal Communication, Department of Computer Science, Hunter College and the Graduate School of The City University of New York.

Ettema, D. 1995. SMASH: A model of activity scheduling and travel behavior. Unpublished manuscript, Eindhoven University of Technology, The Netherlands.

Ettema, D., and Timmermans, H.P.J. 1993. *Using interactive experiments for investigating activity scheduling behavior.* Proceedings of the PTRC 21st Summer Annual Meeting, Manchester, England, Vol. P366, 267-281.

Ettema, D., Borgers, A., and Timmermans, H.P.J. in press. A simulation model of activity scheduling behavior. Transportation Research Record.

Ettema, D., Borgers, A., and Timmermans, H.P.J. 1995. SMASH (Simulation Model of Activity Scheduling Heuristics): Empirical test and simulation issues. Paper presented at the International Conference on Activity based Approaches: Activity Scheduling and the Analysis of Activity Patterns, May 25-28. Eindhoven University of Technology, The Netherlands.

Gärling, T., and Gärling, E. 1988. Distance minimization in downtown pedestrian shopping behavior. *Environment and Planning A,* **20**, 547-554.

Gärling, T., Kwan, M-P., and Golledge, R.G. 1994. Computational-process modelling of household activity scheduling. *Transportation Research B,* **28**B, 5: 355-364.

Gärling, T., Brännäs, K., Garvill, J., Golledge, R.G., Gopal, S., Holm, E., and Lindberg, E. 1989. *Household activity scheduling.* In: Transport policy, management and technology towards 2001: Selected proceedings of the fifth world conference on transport research, Volume IV. Ventura, CA: Western Periodicals, pp. 235-248.

Golledge, R.G. 1995a. Defining the criteria used in path selection. Paper presented at the International Conference on Activity based Approaches: Activity Scheduling and the Analysis of Activity Patterns, May 25-28. Eindhoven University of Technology, The Netherlands.

Golledge, R.G. 1995b. Path selection and route preference in human navigation: A progress report. In A.U. Frank and W. Kuhn Eds.., *Spatial information theory: A theoretical basis for GIS.* Proceedings, International Conference COSIT '95. Semmering, Austria, September. Berlin: Springer-Verlag, pp. 207-222.

Golledge, R.G., Kwan, M-P., and Gärling, T. 1994. Computational process modeling of household travel decisions using a geographic information system. *Papers in Regional Science,* **73**, 2:99-117.

Golledge, R.G., Ruggles, A.J., Pellegrino, J.W., and Gale, N.D. 1993. Integrating route knowledge in an unfamiliar neighborhood: Along and across route experiments. *Journal of Environmental Psychology,* **13**, 4: 293-307.

Golledge, R.G., Smith, T.R., Pellegrino, J.W., Doherty, S., and Marshall, S.P. 1985. A conceptual model and empirical analysis of children's acquisition of spatial knowledge. *Journal of Environmental Psychology,* **5**, 125-152.

Golob, T.F., and Meurs, H. 1988. Development of structural equations models of the dynamics of passenger travel demand. *Environment and Planning A,* **20**, 1197-1218.

Gopal, S., and Smith, T.R. 1990. NAVIGATOR: An AI-based model of human way-finding in an urban environment. In M.M. Fischer and Y.Y. Papageorgiou Eds.., *Spatial choices and processes.* North-Holland: Elsevier Science Publishers, pp. 168-200.

Gopal, S., Klatzky, R.L. and Smith, T.R. 1989. NAVIGATOR: A psychologically based model of environmental learning through navigation. *Journal of Environmental Psychology,* **9**, 4: 309-332.

Hägerstrand, T. 1970. What about people in regional science? *Papers and Proceedings, North American Regional Science Association,* 24, 7-21.

Hayes-Roth, B., and Hayes-Roth, F. 1979. A cognitive model for planning. *Cognitive Science,* **3**, 275-310.

Hirtle, S.C., and Gärling, T. 1992. Heuristic rules for sequential spatial decisions. *Geoforum,* 23, 2: 227-238.

Jones, P., Koppelman, F., and Orfeuil, J.-P. 1990. Activity analysis: State-of-the-art and future directions. In P. Jones (Ed.), *Developments in dynamic and activity-based approaches to travel analysis.* Aldershot: Avebury, pp. 34-55.

Jones, P.M., Dix, M.C., Clarke, M.I., and Heggie, I.G. 1983. *Understanding travel behavior.* Aldershot: Gower.

Kitamura, R. 1984. A model of daily time allocation to discretionary out-of-home activities and trips. *Transportation Research B,* **18**, 255-266.

Kitamura, R. 1988. An evaluation of activity-based travel analysis. *Transportation,* **15**, 9-34.

Kitamura, R., and Goulias, K.G. 1989. MIDAS: A travel demand forecasting tool based on dynamic model system of household car ownership and mobility. Unpublished manuscript.

Kitamura, R., Kazuo, N., and Goulias, K. 1990. Trip chaining behavior by central city commuters: A causal analysis of time-space constraints. In P. Jones Ed.., *Developments in dynamic and activity-based approaches to travel analysis.* Aldershot: Avebury, pp. 145-170.

Kuipers, B.J. 1978. Modelling spatial knowledge. *Cognitive Science,* **2**, 129-153.

Kwan, M-P. 1995. GISICAS: An activity-based travel decision support system using a GIS-interfaced computational-process model. Paper presented at the International Conference on Activity based Approaches: Activity Scheduling and the Analysis of Activity Patterns, May 25-28. Eindhoven University of Technology, The Netherlands.

Kwan, M.-P. 1994. GISICAS: A GIS-interfaced Computational-Process Model for Activity Scheduling in Advanced Traveler Information Systems. Unpublished Ph.D. dissertation, University of California, Santa Barbara.

Leiser, D., and Zilberschatz, A. 1989. The TRAVELLER: A computational model of spatial network learning. *Environment and Behavior,* **21**, 4: 435-463.

Lenntorp, B. 1978. A time-geographic simulation model of individual activity programmes. In T. Carlstein, D. Parkes and N. Thrift (Eds.), *Human activity and time geography.* London: Edward Arnold, pp. 162-180.

McCalla, G.I., Reid, L. and Schneider, P.K. 1982. Plan creation, plan execution, and knowledge execution in a dynamic micro-world. *International Journal of Man-Machine Studies,* **16**, 89-112.

McFadden, D. 1979. Quantitative methods for analyzing travel behavior of individuals: Some recent developments. In D. Hensher and P. Stopher (Eds.), *Behavioural travel modelling.* London: Croom Helm, pp. 279-319.

Newell, A. 1992. *Unified theories of cognition.* Cambridge, MA: Harvard University Press.

Newell, A., and Simon, H.A. 1972. *Human problem solving.* Englewood Cliffs, NJ: Prentice-Hall.

Recker, W.W., McNally, M., and Root, G.S. 1986a. A model of complex travel behavior: Part I: Theoretical development. *Transportation Research A,* **20**, 4: 307-318.

Recker, W.W., McNally, M., and Root, G.S. 1986b. A model of complex travel behavior: Part II: An operational model. *Transportation Research A,* **20**, 4: 319-330.

Root, G.S., and Recker, W.W. 1983. Towards a dynamic model of individual activity pattern formation. In S. Carpenter and P. Jones (Eds.), *Recent advances in travel demand analysis.* Aldershot: Gower.

Simon, H.A. 1955. A behavioral model of rational choice. *Quarterly Journal of Economics,* **69**, 99-118.

Simon, H.A. 1990. Invariants of human behavior. *Annual Review of Psychology,* **41**, 1-19.

Supernak, J. 1992. Temporal utility profiles of activities and travel: Uncertainty and decision making. *Transportation Research B,* **26**, 61-76.

Thill, J-C., and Thomas, I. 1987. Towards conceptualizing trip-chaining behavior: A review. *Geographical Analysis,* **19**, 1-17.

Timmermans, H.J.P., and Golledge, R.G. 1990. Applications of behavioral research on spatial problems II: Preference and choice. *Progress in Human Geography,* **14**, 311-354.

Tversky, A., and Kahneman, D. 1991. Loss aversion in riskless choice: A reference-dependent model. *The Quarterly Journal of Economics,* **106**, 1039-1061.

Tversky, A., and Kahneman, D. 1992. Advances in prospect theory: Cumulative representation of uncertainty. *Journal of Risk and Uncertainty,* **5**, 4: 297-323.

Winston, G.C., 1987, Activity choice: A new approach to economic behavior. *Journal of Economic Behavior and Organization,* **8**, 567-585.

10 Modelling Non-Work Destination Choices with Choice Sets Defined by Travel-Time Constraints

Jean-Claude Thill[1] and Joel L. Horowitz[2]

[1]Department of Geography and National Center for Geographic Information and Analysis, State University of New York, Amherst NY 14261, USA
[2]Department of Economics, University of Iowa, Iowa City IA 52242, USA

10.1 Introduction

In the general theory of choice behaviour generally, a common modelling approach postulates that individuals follow a two-stage decision process leading to the choice of one alternative over all other alternatives that could be picked at the time of decision. The universe of alternatives is reduced in the first stage to a smaller set called the *choice set* or *consideration set* whose construction hinges upon one's knowledge and awareness of choice alternatives, the feasibility and accessibility of alternatives as well as their perceived saliency at decision time (Shocker *et al.* 1991). In the second stage, only the choice set alternatives are analyzed and a final selection is operated. [1]

Because choice sets are not directly observable, the conceptual two-stage decision model is frequently simplified in the revealed-preferences approach to spatial choice by *ad hoc*, deterministic, and sometimes arbitrary, assumptions on the size, composition, and interpersonal variability of the choice set of alternatives. For example, in a problem of grocery shopping destination choice, the individual choice set may be composed of all supermarkets within a 10-mile radius of home or within a 1-mile buffer along the shortest commuting route. Even if this rule of thumb can be validated empirically in the aggregate, significant incongruities are bound to exist for individual decision-makers. Choice sets can hardly be imputed with certainty to individuals.

In fact, several authors have argued that the generation of the choice set is itself a complex and hierarchical process along which the universe of choice alternatives (universal choice set) is incrementally weeded out of options deemed unworthy of further consideration in accordance with specific cognitive, social and contextual mechanisms. Desbarats' (1983) conceptual framework stresses the "interplay of individual free will and of social and environmental contingencies over which individuals have little control and of which they may not even be aware" (p. 346).

[1] It is not universally accepted that choice takes place in two stages. In fact, empirical evidence has recently been provided by Horowitz and Louviere (1995) against two-stage choice in certain choice settings.

It is believed that, in first instance, structural constraints on the structure and supply of opportunities are produced by the institutional context and geographic accessibility and shape an objective choice set. Personal characteristics may enhance or diminish the impact of structural constraints. The decision-maker's social conditioning to conform to social norm and their familiarity with the opportunities included in the objective choice set further contract this set to an effective choice set. Burnett (1980) and Burnett and Hanson (1979, 1982) espoused a very similar approach framing spatial choice in a constrained context. This characterization of decision-making is a natural extension of Hägerstrand's (1970) time geography theory whose tenet is that an individual's path of daily activities in space and through time is confined to a feasible "prism" shaped by three main forces defined as "capacity," "coupling," and "authority."

Another hierarchical model of decision-making choice was recently articulated by Shocker *et al.* (1991) out of the vast applied marketing literature on brand choice. Their characterization traces its roots in the seminal contribution of Howard (1963) and Howard and Seth (1969). They describe a funneling process consisting of three stages: development of an *awareness or knowledge set*, which includes all items in the universal set of which the decision-maker is aware; retention of those items that show sufficient consistency with the expected outcome of the decision, are accessible and come to mind at that time (consideration set); iterative weeding-out of alternatives to produce a final choice set that is more differentiated and whose size allows for more effective processing of information by the decision-maker. The same structure is advocated by Crompton (1992) and Crompton and Ankomah (1993) to explain how individuals select vacation destinations.

Interest for choice set definition is motivated by errors that may result in case choice sets are mis-specified. The inconsistency of parameter estimates of the utility function in random utility models was studied by Manski (1977), McFadden (1978), Williams and Ortuzar (1982), and others. The failure to endogenize processes of choice set generation and formation brings estimation biases [2] and renders more elusive the goal of many disaggregate modelling efforts to *explain* choice among geographically-referenced alternatives and separate it from contextual constraints. With a crude delineation of the choice set, respective factors of choice set generation and of alternative selection may not be identifiable unambiguously. This was illustrated by Stetzer and Phipps (1977) who established that, while the well-known distance decay in observed spatial interaction patterns is usually interpreted in terms of preference for closer choice opportunities, it may just as well be the result of maximal distance restrictions on the choice sets. The same concern for tampering with the concept of individual preferences and free will is echoed by Sheppard (1980, 1984).

Finally, the mis-specification of choice sets may lead to erroneous choice predictions by assigning non-trivial probabilities to alternatives that are out of the true choice set, or *vice versa*. Simulations on constrained destination choice behaviour by Thill and Horowitz (1996) indicate that the accuracy of predictions that incorporate mis-specified choice sets is poor to fair. In product or service marketing,

[2] Pellegrini *et al.* (1997) illustrate from an empirical perspective the great sensitivity of parameter estimates to choice-set specification in the context of shopping destination choice.

these prediction errors may translate into the failure of a new brand of cereal or the underestimation of the demand and production capacity for a new car model, for instance, and ultimately, into tangible sums of money. With spatial choices, the connection between choice-set mis-specification and market share over- or under-estimation is more intricate due to the greater complexity of spatial decision-making and is often overlooked by practitioners. It should be no surprise therefore that the study of choice sets has attracted considerable attention in the field of marketing (Roberts and Nedungadi, 1995) and far less in spatial choice analysis.

In this paper we propose an extension of random-utility destination choice models that explicitly represents the formation of choice sets. The premise of the approach is that travel behaviour is inherently constrained: it is not only shaped by individual preferences but also by a host of limiting factors and circumstances that keep all choice options from being evaluated. Time is assumed to be the limiting resource that restricts the evaluation of alternatives and the selection to a sub-set of the universal choice set. This is consistent with the main thrust of Hägerstrand's (1970) time geography theory and with abundant empirical research on time constraints by Lenntorp (1976), Goodwin (1981), Kitamura and Kermanshah (1984), van der Hoorn (1983), and others who have established the link between the time budget available to individuals and households and destination choices. Within a given time budget, the amount of time that can be allocated for travel may result from a trade-off with other components of one's daily activity pattern, such as the time spent on the activity at the destination and the time spent on home activities. While we recognize that matters of scheduling of activities and allocation of time budgets between travel and non-travel activities are important, our goal is not to model daily activity patterns. Our approach, in contrast, aims at individual trips. We seek to obtain tractable models of destination choice at the trip level where choice is represent as being constrained. We also seek to understand whether such models are better than models that do not explicitly account for constraints.

The remainder of the paper is organized as follows. Section 10.2 summarizes a variety of approaches used to specify or characterize choice sets in destination choice models. The theoretical framework of our model is discussed in section 10.3, while some important estimation issues are covered in section 10.4. Section 10.5 describes the data used in the empirical tests of the constrained choice model against a matching conventional choice model and presents the results. The improvement in goodness-of-fit brought by the constrained-choice approach leads to conclusions and implications discussed in the final section of the paper.

10.2 Previous Research on Choice Set Modelling

A great many different approaches have been proposed to define or characterize choice sets for inclusion in discrete choice models. Two comprehensive reviews of these approaches are provided by Shocker *et al.* (1991) in the context of product or brand choices, and by Thill (1992) in the context of travel destination choices. It is common in random utility modelling to have choice sets defined by a few deterministic, *a priori*, rules-of-thumb. For instance, in Miller and O'Kelly (1983), the choice set of a certain individual is composed of all the destinations actually chosen by individuals from the same geographic area. In other cases, all individuals share an identical choice set made of all the destinations in the urban area of study (for instance, Ghosh, 1984; Thill, 1995). Black's (1984) approach involves drawing a circumference around each destination and place this destination in the choice set of all individuals enclosed by the perimeter. The radius of the circumference is derived heuristically to balance relative gains and losses in predictive power of the model in case of exclusion of destinations at the margins of the circumference. The limitations of these simple approaches to choice set definition are discussed in Thill (1992).

Early refinements of the delineation of the choice set focused on that portion of the geographic space --a prism-- that is deemed feasible because it fits within the constraints imposed by the limited time budget available to decision-makers. Possible trajectories in space-time are compiled for each individual by simulation of daily activity patterns, which typically involve combinatorial algorithms of exhaustive enumeration and evaluation. Lenntorp's (1976) PESASP model is the prototypical simulation study of movement possibilities. Recent work along this line has strived for a better treatment of information processing and of cognitive dimensions of spatial behaviour (Recker *et al.*, 1986a, b; Ettema *et al.* 1993; Golledge *et al.* 1994). The result of such classification scheme is a deterministic choice set that can be incorporated into a random utility model. The absence of statistical or mathematical modelling of choice set delineation precludes inference about the processes of choice set formation in a population of decision-makers. In a prediction mode, it also prevents the transferability of the results of the simulation over space and through time.

Some authors have viewed choice set generation as a search process by which information on choice options in the universal set is gathered and updated. Meyer (1980) presents a complex learning model that is coupled with random utility theory. The choice set is updated by addition of new options that succeed in satisfying noninferiority conditions thanks to new information collected, or by removal of options whose earlier selection turned out to fail to meet the expectations of the decision-maker. Other related work incorporating search processes is found in Richardson (1982), Hauser and Wernerfelt (1990), Horowitz (1991).

Fotheringham (1986, 1988) argued that decision-makers faced with a large set of spatial alternatives use a hierarchical strategy to process the information employed to reach a decision. In first instance, choice alternatives are perceived in clusters according to their similarities. The likelihood that an alternative is in the choice set is estimated by its similarity to all others alternatives. Similarity in the attribute space can be measured by a n-dimensional Manhattan distance (Borgers and Timmermans 1987), or a Pearson correlation coefficient (Meyer and Eagle 1982). A gravitational accessibility index usually serves to measure similarity in the geographic space (Fotheringham 1983).

While the idea that the choice process can be decomposed into a choice set generation stage and a choice generation stage can be traced back to the early 1960s, Manski (1977) is to be credited for formalizing it in the context of random-utility choice modelling. Choice probabilities are decomposed into probabilities for obtaining given choice sets and choice probabilities conditional upon choice sets. A probabilistic model describes choice set generation to reflect the incomplete information available to the analyst on the factors of generation. Constraints operating on individuals facing the choice of a transportation mode have been operationalized with some success within the framework proposed by Manski (1977). Gaudry and Dagenais' (1979) model is the simplest of all. It postulates that an individual's choice set either the universal choice set or a single, known, alternative. While the principle of "captivity" is tenable in transportation mode choice (Kitamura and Lam 1984, Swait and Ben-Akiva 1987b), it is not useful for destination choice modelling due to the typically large size of the universal choice set and the mobility that most people can enjoy to reach multiple destinations.

The independent availability model of Swait and Ben-Akiva (1987a), also derived from Manski's (1977) approach, formalizes choice set generation by means of stochastic membership functions that act as constraints on the availability of certain alternatives to an individual. Constraints work as exclusion criteria of the choice set. An extension by incorporating attitudinal data on choice alternatives is proposed by Ben-Akiva and Boccara (1995) to account for aspects of individual behaviour that cannot be inferred from observed choices. In their empirical application of the independent availability model and of its extension to travel mode choice in Baltimore, Maryland, they concluded that both models fit much better than a conventional multinomial logit model with identical specification of the systematic utility. For destination choice modelling, this approach has the drawback that it assumes that each alternative is included in the choice set or excluded from it independently of all the others. Thus, the set of all possible choice sets is the power set of all possible alternatives, which in destination choice is unmanageably large. In this paper, we propose an approach that is related to the independent availability model but avoids the need to deal with such a large number of choice sets.

10.3 Theoretical Framework

Random utility theory assumes that preferences among available alternatives of the choice set are described by a utility function expressed as the sum of two components: a systematic component that accounts for the effects of observable factors that influence choice, and a random component, that accounts for unobserved factors. Individuals choose the alternative with the highest utility. Since utilities depend on unobserved random variables, the choice of a certain alternative can only be predicted probabilistically, conditional on the observed characteristics. The probability P(j|C) that a randomly selected individual with choice set C chooses alternative j, conditional on the observed attributes of the individual and of the alternatives is given by

$$P(j|C) = P(U_j > U_k, \forall k \neq j, k \in C). \tag{10.1}$$

The multinomial logit model is obtained if the random components of the utility function are independently and identically distributed with the type I extreme value distribution. Other families of random utility models can be derived from other specifications of random components.

The choice set is that part of the universe of alternatives over which utility is maximized. Let us consider an example in which travel time is the only factor determining the choice set. Consistently with the time-geography theory, we further suppose that an individual's behaviour is controlled by his/her time budget. More specifically, the only constraint --or elimination criterion-- operating on the choice set is travel time. That is, an alternative is deemed feasible if, and only if, the travel time to reach it does not exceed that allowed by the individual's budget. Let T be the maximum travel time that delineates the choice set. Then choice set C_T consists of all alternatives j for which the travel time t_j satisfies $t_j \leq T$. If, conditional on C_T, choice is described by a logit model, the probability that alternative j is chosen is

$$(P(j|C_T) = \begin{cases} \exp(V_j)/\sum_{k \in C_T} \exp(V_k) & \text{if } t_j \leq T \\ 0 & \text{otherwise,} \end{cases} \tag{10.2}$$

where V_k is the systematic component of utility of alternative k. For the sake of simplification, the dependence of V_k on attributes of the individual is not represented in the notation of (10.2).

In revealed-preference data, the travel-time threshold T is not observable and is likely to differ among individuals. That is, the choice set cannot be known with certainty. Therefore equation (10.2) is not operational and the model given by (10.1) needs to be generalized to allow for a random component in the choice set. The choice set can be considered as a random element C of the set G of all conceivable choice sets (i.e., the power set of all possible alternatives). Let P(C) denote the probability that the choice set is C. Then the choice problem can be modeled as the determination of the choice set and of the choice conditional on the choice set. Manski (1977) suggested the following expression for the probability that alternative j is chosen:

$$P(j) = \sum_{C \in G} P(j|C)\, P(C), \tag{10.3}$$

where P(j|C) is the probability of choosing j given that alternatives in set C are considered. This formulation provides a natural framework for investigating the effects of constraints on choice. If an individual's time budget constrains him or her from choosing an alternative, then that alternative is in the universal choice set but not in C. Conversely, if an alternative is in C, the individual is not kept from choosing it.

Let us turn to the problem of incorporating travel-time constraints in the framework of equation (10.3). The travel-time threshold T that confines a certain individual's choice set can be represented as a random variable. Let $P_T(t;\theta)$ denote the cumulative distribution function of T, where θ is a vector of parameters on which P_T depends. Parametrization allows for a systematic dependence of P_T on observable attributes of individuals (say, income, family size, or gender). It follows from (3) that with a random T, the probability that alternative j is chosen is

$$P(j) = \int_{t=0}^{\infty} P(j|C_t)\; dP_T(t;\theta), \tag{10.4}$$

where $P(j|C_t)$ is choice probability conditional on choice set C_t given by (10.2).

Since destinations in the universal choice set are intrinsically discrete and mutually exclusive alternatives, equation (10.4) can be further simplified to a summation over all the nested sets of alternatives defined by incremental travel-time thresholds, once alternatives are sorted according to travel time. Therefore, an equivalent form of equation (10.4) is

$$P(j) = \sum_{k=1}^{J} P(j|C_k)\; p_T(k;\theta), \tag{10.5}$$

where $p_T(k;\theta)$ is the probability that travel-time threshold is between the travel times to destinations k and k+1:

$$p_T(k;\theta) = P_T(t_{k+1};\theta) - P_T(t_k;\theta). \tag{10.6}$$

Travel time to destinations k=1 to *J* provides the support for probability distribution function $p_T(k;\theta)$. This model, which we denominate the Nested Choice-Set Destination Choice (NCS-DC) model, considers as many possible choice sets as there are destinations in the universal choice set. It exploits the following attractive property of travel-time constraints, which is made possible by the unequivocal and non-random sorting of destinations according to their travel time from home: all destinations closer than any destination that meets the inclusion criterion set by the constraint are also in the choice set, and *vice versa*, all destinations less accessible than a destination that fails to meet the inclusion criterion are out of the choice set. Because not all subset combinations of the universal choice set need to be evaluated,

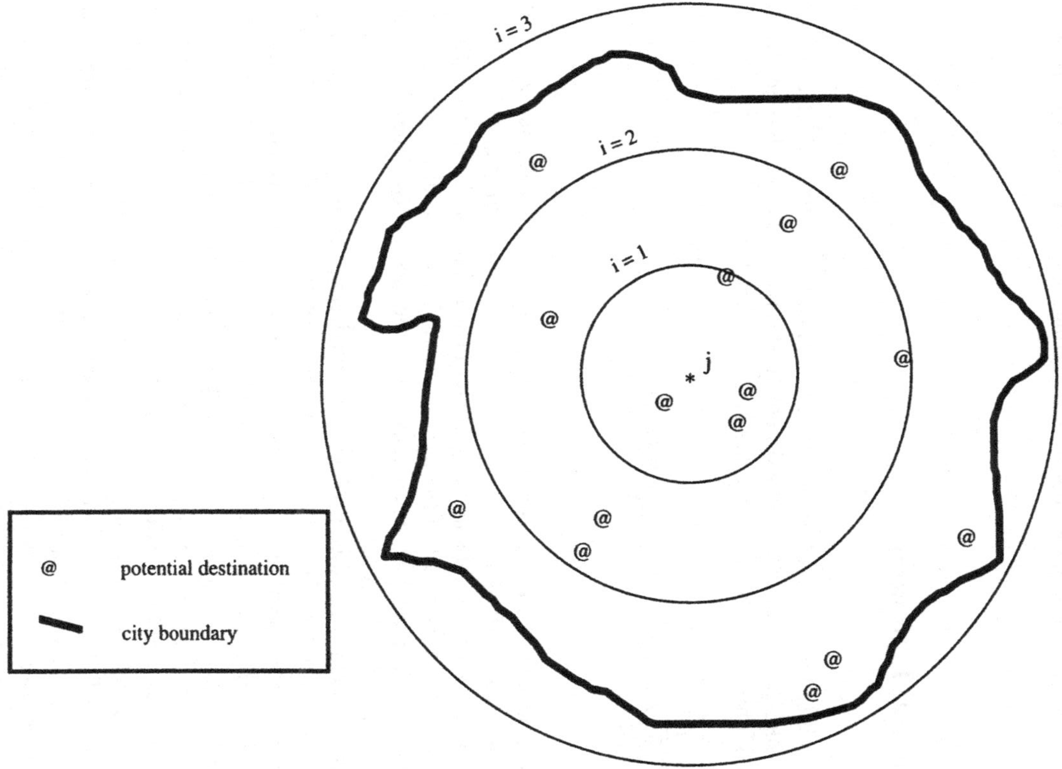

Fig. 10.1. Delineation of Possible Choice Sets in the Time-space of Individual j by means of Three Travel-time Threshold Points. Constraint T_3 indexes the universal choice set: the associated circumference encloses the whole urban area.

the proposed model is computationally less demanding than the independent availability logit model.

10.4 Statistical Estimation

Given an explicit specification of the utility function and of P_T, the NCS-DC model can, in principle, be estimated by maximum likelihood. The log likelihood of observations of choices by N randomly sampled individuals is

$$\log L = \sum_{n=1}^{N} \sum_{j=1}^{J} z_{nj} \log \sum_{k=1}^{J} P_n(j|C_{nk})\, p_T(k;n,\theta), \tag{10.7}$$

where J is the total number of alternatives in the *universal* choice set and $z_{nj} = 1$ if individual n chooses alternative j and 0 otherwise. The computational difficulty of the NSC-DC model remains substantial however because, as can be seen in (10.7), the log-likelihood depends on all of the alternatives in the universal choice set. For a large metropolitan area in which destinations are defined as traffic zones or census tracts, this can exceed 1000 destinations.

We propose to deal with this problem by restricting the support of P_T to a manageable set of discrete points. That is, the entire series of travel-time thresholds is approximated by assuming that they take only a few distinct values. Figure 1 illustrates the delineation of possible choice sets by travel-time thresholds in the time-space around a certain individual with such approximation. An indication of the accuracy of the approximation can be obtained empirically by comparing estimation results with different numbers of threshold points.

The approximation by few travel-time threshold points improves the tractability of the maximization of (10.7). Also, it makes it unnecessary to impose a parametric structure on the variation of P_T with T. To form the nonparametric representation of this variation, denote by T_i (I=1 to I) one of the I possible travel-time thresholds. The probability that the threshold is T_i is denoted by π_i. Then, if C_{ni}* is the set of all destinations k such that $t_{nk} \leq T_i$, the discrete-threshold approximation specifies that the probability that destination j is chosen is

$$P_n(j) = \sum_{i=1}^{I} P(j|C_{ni}^{*})\, \pi_i, \text{ with } 0 \leq \pi_i \leq 1, \forall i, \text{ and } \sum_{i=0}^{I} \pi_i = 1. \tag{10.8}$$

With the above non-parametric representation, parameters π_i may be parametrized to depend on observed attributes of individuals. This operational version of the NCS-DC model is name the Approximate Nested Choice-Set Destination Choice (ANCS-DC) model.

The values of the π_i and other parameters of this model can be estimated by maximum likelihood with the following log-likelihood function:

$$\log L = \sum_{n=1}^{N}\sum_{j=1}^{J} z_{nj} \log \sum_{i=1}^{I} P_n(j|C^*_{ni})\, \pi_i, \text{ with } 0 \le \pi_i \le 1, \forall i, \text{ and } \sum_{i=1}^{I} \pi_i = 1, \quad (10.9)$$

with $P_n(j|C^*_{ni})$ given by equation (10.2). Unlike the multinomial logit model, the log-likelihood of the ANCS-DC model is not globally concave. The search for an optimum is computationally more complex since a search of first order conditions may lead to a local extremum which is not the global maximum. It is important therefore to ascertain that the search algorithm converges to the same point when started at different initial values. [3]

10.5 Empirical Analysis

This section presents and discusses the estimation of an ANCS-DC model with data on the choice of non-work destinations in Minneapolis-St. Paul, Minnesota during 1990. It is compared to a multinomial logit (MNL) model of destination choice. The MNL model is first specified and estimated on the data so as to secure a good specification that minimizes the chances of spurious rejection of the MNL in favor of the ANCS-DC model. By doing this, we minimize the likelihood that an unconstrained (MNL) model is rejected because of misspecification arising from factors other than travel-time constraints. An ANCS-DC model is subsequently estimated with the same specification of the systematic utility. The MNL and ANCS-DC models are finally compared by means of a likelihood ratio test in which the MNL is the maintained model (Horowitz, 1988).

10.5.1 The Data

The Minneapolis-St. Paul, MN, Metropolitan Council conducted a large-scale travel behaviour inventory (TBI) in 1990 (Metropolitan Council, 1990). The main part of this effort was a home interview survey by mail and telephone to compile travel diaries during a 24-hour period. Detailed information on more than 100,000 trips over one block in length was recorded. Collectively, these trips provide a comprehensive depiction of the travel activities of all members aged five and over in 9746 households selected randomly from the metropolitan population according to a county stratification.

[3] Simulated annealing or another global optimization method could be used to guarantee convergence to the global optimum.

On the basis of previous research on travel time constraints (e.g., Goodwin 1981, Landau *et al.* 1981, Wigan and Morris 1981), we anticipate that travel-time constraints have different effects on choice sets for different classes of urban nonwork trips. Therefore, it is desirable to investigate the predictive power of the ANCS-DC model separately for a variety of different types of trips. For this study, we selected home-based, single-purpose shopping trips made by car within the metropolitan area. A total of 684 trips meeting these conditions were extracted from the entire database.

Respondents are geo-referenced by the traffic analysis zone (TAZ) in which their residence is located. For each respondent, the location of the chosen destination is known by traffic analysis zone. The Minneapolis-St. Paul metropolitan area is composed on 1165 internal traffic analysis zones, whose definition was revised in

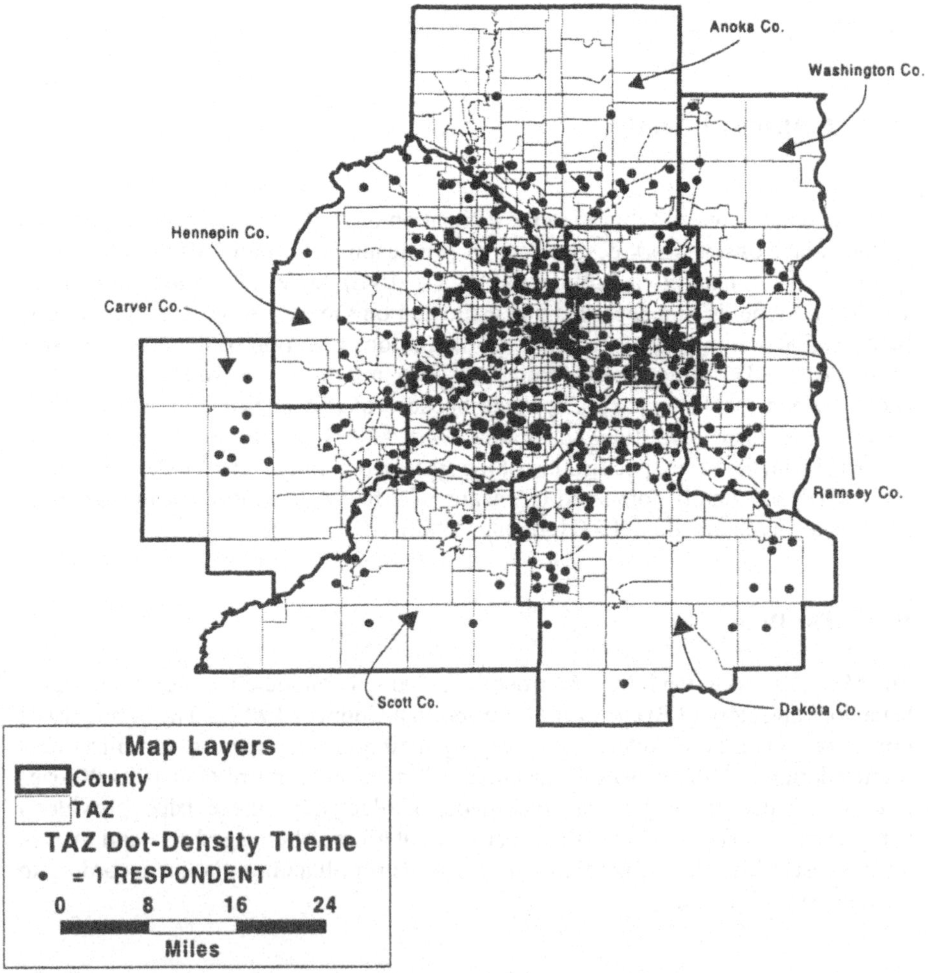

Fig. 10.2. Location of the Respondent Homes in Traffic Analysis Zones of the Minneapolis-St. Paul Metroploitan Area.

1990 in conjunction with the conduct of the travel behaviour inventory (Department of Transportation, n.d.). All 1165 TAZs form the universal choice set for the shopping destination choice problem considered here. The location of the respondent homes in the traffic analysis zones of the metropolitan area is mapped in Figure 10.2, while Figure 10.3 renders the location of the chosen shopping destinations.

Three sets of variables are used to predict the choice of a shopping destination, namely spatial separation between the individual's home location and the location of

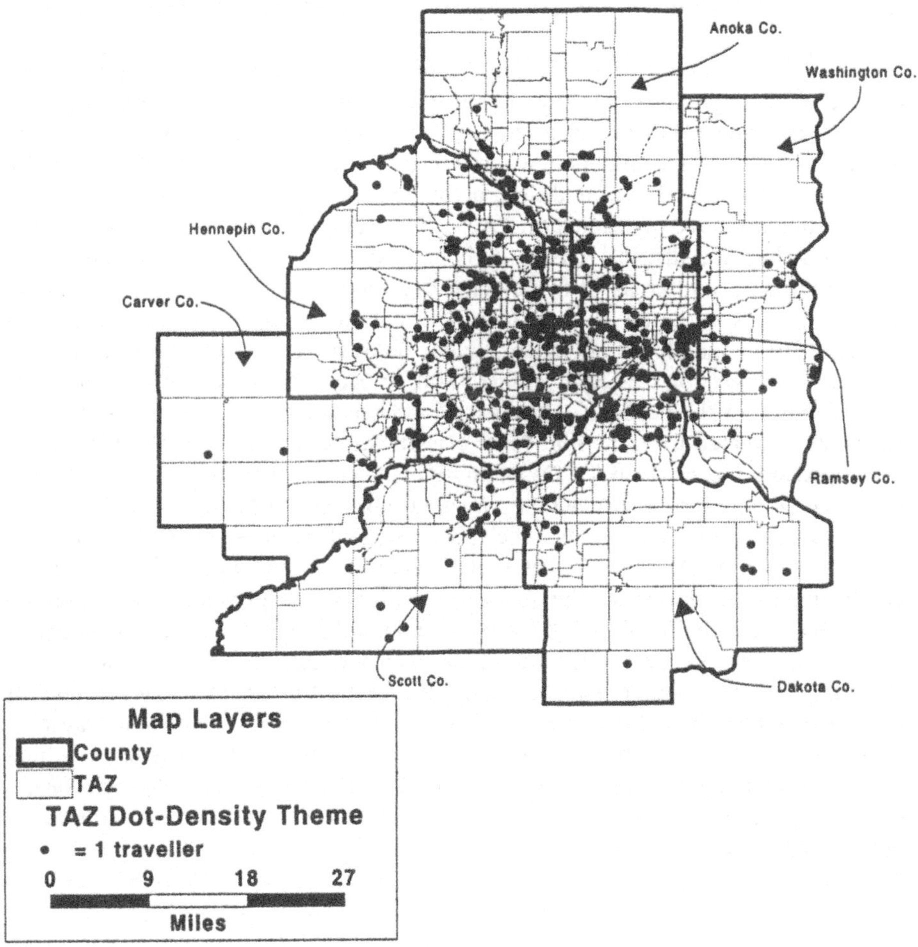

Fig. 10.3. Location of Shopping Destinations in Traffic Analysis Zones of the Minneapolis-St. Paul Metropolitan Area.

destinations, characteristics of the potential destinations, and attributes of the individual. Two related measures of spatial separation are used: distance (DIST) and travel time (TIME). DIST is the shortest distance measured on the highway network build by the Minnesota Department of Transportation (DOT). This network is formed of all roads in the metropolitan area with an annualized average daily traffic over 1,000 augmented by TAZ centroid connectors (Department of Transportation, n.d.). The TIME variable is the lowest travel time derived from network distance and a link speed matrix devised by the Minnesota DOT. Typical speed is assigned by the DOT to links in the highway network according to a two-way classification along two criteria: the facility type (e.g., divided arterial) and the area type, or geographic setting within the metropolitan area (e.g., central city, rural). Link directionality is accounted for in the computation of DIST and TIME. All these variables are part of an integrated Geographic Information System (GIS) made available by the Minnesota DOT.

Destination characteristics that may be relevant to shopping destination choice include TAZ population and household counts in 1990 (Census Transportation Planning Package (CTPP) released by the Bureau of the Census), TAZ employment in retail businesses and in personal services (CTPP) and the presence/absence of a regional shopping mall. In addition, a categorical variable describes the form of urbanization observed in destination TAZs. Four area types are distinguished: central city/central business districts (CBD), outlying business districts, developed areas, and developing and rural areas. It is expected that the geographic context of shopping destinations, as captured by the area type classification, influences destination choices. The geographic context may include objective and subjective elements such as traffic congestion, safety, architectural appeal and modernity, community atmosphere. For the purpose of inclusion in the choice model, a series of indicator variables is created from the four-level classification. Binary variables identifying developed areas, central city/CBDs and outlying business districts are used explicitly in the model, while rural and developing areas constitute the reference group.

The TBI database includes several characteristics of travelers believed to influence the formation of preferences for alternative shopping destinations. These attributes are the age of the individual, the gender (0 for females, 1 for males), the household size, annual household income (0 for income under \$35,000, 1 for income over this level), presence (1)/absence (0) of children under the age of 5, and car ownership (1 if 2 cars or more, 0 otherwise). Finally, the form of urbanization in the TAZ of residence of the individual is operationalized by the 'area type' classification variable mentioned above. Three binary variables are used to identify trip origins according to their location in a developed area, the central city/CBDs, and an outlying business district, respectively. Once again, rural and developing locations are implicitly defined in this coding scheme.

10.5.2 Estimation Results: Unconstrained Model

A large number of utility-function specifications can be constructed as linear combinations of the variables presented in the previous section. We use here a stepwise approach to explore a sub-set of these specifications. We start by estimating a plausible model of choice. Alternative specifications are subsequently compared to this initial maintained model. A specification that is significantly better than the current maintained model becomes the new maintained model. Since non-nested hypotheses are involved, a simple likelihood ratio test can be used for this purpose. This process stops when no reasonable utility specification can be found to reject the current maintained model. See Thill and Horowitz (1991) for another application of this approach to building choice models.

The aforementioned model selection method is used to obtain a good specification of the utility function of a multinomial logit model of shopping destination choice when all the TAZs in the metropolitan area are included in the choice set. The variables and parameter estimates of this model are given in Table 10.1. Only one parameter estimate is not significantly different from zero at the 0.05 level. As indicated by the ρ^2 and adjusted ρ^2 statistics, the multinomial logit model fits the destination choice data very well.

Coefficients have signs consistent with intuition and other studies of shopping destination choice. Travel time has the expected negative effect on destination utility. This effect is compounded by the age of the individual, but at a marginally decreasing rate. Other individual attributes (including gender, household income and the presence of infants in the household) are not found to predict as well as age the interpersonal variability of the intensity of travel time deterrence on the choice of a shopping destination.

Serving as a proxy for the count of elemental shopping alternatives in each TAZ, retail employment has a statistically significant effect on choice. This effect is positive, proportional to the square root of the number of retail workers and more pronounced among affluent travelers. In the same idea, regional shopping malls exert a unique attractiveness on shoppers. Other things being equal, resident population enhances the utility of a shopping destination, and this all the more if the shopper is affluent. Finally, the decision to shop in one of the four urban area types is influenced by the age of the traveler, the size of their household and the type of urban area where they live. More specifically, preference for shopping in a developed area or in an outlying business district rather than in a rural or developing area increases with the square root of age, while the preference for central city locations decreases with the square root of age. The likelihood of shopping in the central city or in a developed area instead of a developing or rural area is inversely related to the number of members in the shopper's household and directly related to the household's income. Affluent shoppers are also more attracted by shopping destinations in developing or rural areas than in those located in the outlying business districts. Residents of the central city and of developed areas have a statistically significant preference for destinations other than in rural or developing areas. Residents of outlying business districts prefer shopping in developing or rural areas than in

Table 10.1. Estimation results and summary statistics for the MNL and ANCS-DC models

Var. #	Variable Name	Unconstrained (MNL) Model		Constrained (ANCS-DC) Model	
		Coefficient Estimate	t statistic	Coefficient Estimate	t statistic
1	(Retail Employment) $^{1/2}$* Household Income	0.05067	12.8021	0.05241	12.8408
2	(Total Population) $^{1/2}$* Household Income	0.005952	2.5023	0.004718	2.0124
3	Travel Time * (Age) $^{1/2}$	-0.04284	-36.8992	-0.03397	-23.8352
4	Presence of Mall	1.1052	57.7907	1.0612	47.8385
	<u>Specific to Developed Area Destinations</u>:				
5	(Age)$^{1/2}$	0.1343	8.5304	0.1265	12.8613
6	Household Income	-0.1834	-9.6780	-0.0896	-4.7422
7	(Household Size)$^{1/2}$	-0.1549	-8.0326	-0.09967	-2.6738
8	Developed Area Origin	0.4990	27.6672	0.3763	13.6040
9	City Center/CBD Origin	0.08299	4.5962	0.1648	12.8666
10	Outlying BD Origin	-0.4111	-22.7185	-0.5376	-20.0334

Table 10.1. (Continued)

	Specific to City Center/CBD Destinations				
	$(\text{Age})^{1/2}$	-0.03410	-1.9741	-0.06810	-4.1866
11	Household Income	-0.6078	-33.6879	-0.6259	-18.4443
12	$(\text{Household Size})^{1/2}$	-0.2166	-12.0191	-0.1667	-5.7612
13	Developed Area Origin	0.8849	43.8612	1.2188	15.2443
14	City Center/CBD Origin	1.3003	71.9258	1.5094	19.6133
15	Outlying BD Origin	0.6712	37.1480	1.2621	51.4944
16					
	Specific to Outlying BD Destinations				
17	$(\text{Age})^{1/2}$	0.1749	9.1250	0.1513	6.9493
18	Household Income	-02909	-15.7430	-0.2555	-8.8547
19	$(\text{Household Size})^{1/2}$	-0.0001413	-0.007666	0.1098	2.7604
20	Developed Area Origin	0.3605	19.9758	0.2837	16.9399
21	City Center/CBD Origin	0.09270	5.1375	0.3022	4.6388
22	Outlying BD Origin	-0.9719	-53.8702	-1.1517	-51.2032
	Summary Statistics				
	Number of Observations	667		667	
	$L(0)$	-4709.338		-4709.338	
	$L(\beta)$	-3072.274		-3048.372	
	ρ^2	0.3476		0.3527	
	Adjusted ρ^2	0.3423		0.3463	

developed areas or in outlying business districts. On the other hand, they prefer the central city or the CBD over rural or developing areas of the metropolitan area.

10.5.3 Estimation Results: ANCS-DC Model

Estimation of the ANCS-DC model involves the constrained optimization of the log-likelihood function given by equation (9). The Davidon-Fletcher-Powell search algorithm in GAUSS is used for this purpose. For the sake of comparison with the estimation results of the multinomial logit model, the systematic utility of the ANCS-DC model follows the specification found to be the most powerful with the former model. We have experimented with various sets of discrete travel-time thresholds to identify a set offering a reasonable compromise between accuracy and computational tractability. The following scheme with nine threshold points results from this preliminary investigation: 5, 10, 15, 20, 30, 40, 50, 60 minutes, and a travel time set to an arbitrary large value to include all 1165 destinations in the universal choice set. To reflect that the marginal disutility of the time spent traveling is decreasing, threshold points are 10 minutes apart instead of 5 above 20 minutes of travel time. Estimation results under alternative schemes of discrete points indicate that this does not result in any loss of predictive power.

Estimates of the parameters of the systematic utility are reported in Table 1, along with summary statistics. The fit of the estimated ANCS-DC model to the choice data is very good. All twenty-two utility coefficients are statistically significantly different from zero at the 0.05 level. In most respects, signs are unchanged from those obtained with a multinomial logit model. The only exception is the parameter of the square root of household income for destinations in outlying business districts. While not statistically significant from zero at the 0.05 level in the MNL model, it reveals here a significantly higher responsiveness to outlying business district locations that to rural or developing locations, the more affluent the shopper is.

Let us now test whether the modelling of the choice set incorporated in the formulation of the ANCS-DC model (alternative model) enhances the representation of observed shopping destination choices in Minneapolis-St. Paul provided by the conventional MNL model (maintained model). The MNL model is a parametric special case of the ANCS-DC model with the same specification of the systematic utility. Hence our test involves nested specifications and a likelihood ratio test is well suited. The χ^2 statistic for this test is -2 (3072.274 - 3048.372) = 47.804 which is larger that the critical value of 20.090 with 8 degrees of freedom at the 0.01 significance level. This establishes the relevance of travel-time constraints in modelling our shopping destination choice problem and the statistical superiority of the ANCS-DC modelling approach over the conventional MNL model which does not distinguish between the two distinct impacts of spatial impedance in destination choice processes.

In the MNL model, the coefficient of travel time accounts for the sensitivity of destination choice probabilities to exclusion from the choice set due to limited time budget and to the marginal disutility of travel time. On the other hand, in the ANCS-DC model, this coefficient embodies exclusively the latter effects, while the fitting

Table 10.2. Tests of the different of coefficients between the MNL model and the ANCS-DC model

Var. #	Variable Name	T Statistic
1	(Retail Employment) $^{1/2}$ * Household Income	0.3052
2	(Total Population)$^{1/2}$ * Household Income	0.3685
3 **	Travel Time * (Age)$^{1/2}$	4.8255
4	Presence of Mall	1.4960
	<u>Specific to Developed Area Destinations</u>:	
5	(Age)$^{1/2}$	0.4202
6 **	Household Income	3.5083
7	(Household Size)$^{1/2}$	1.3166
8 **	Developed Area Origin	3.7141
9	City Center/CBD Origin	1.3579
10 **	Outlying BD Origin	3.9072
	<u>Specific to City Center/CBD Destinations</u>	
	(Age)$^{1/2}$	1.4329
11	Household Income	0.4703
12	(Household Size)$^{1/2}$	1.4639
13	Developed Area Origin	4.0729
14 **	City Center/CBD Origin	2.6448
15 **	Outlying BD Origin	10.4074
16 **		
	<u>Specific to Outlying BD Destinations</u>	
17	(Age)$^{1/2}$	0.8124
18	Household Income	1.0346
19	(Household Size)$^{1/2}$	0.7753
20	Developed Area Origin	1.7175
21 **	City Center/CBD Origin	3.0994
22 **	Outlying BD Origin	18.7209

** indicates that a difference is significant at the 0.05 level

Table 10.3. Estimates of travel-time thresholds

Coefficient	Estimate	T Statistic
Probability that T = 5 min.	0.1816	10.1257
Probability that T = 10 min.	0.1933	10.6761
Probability that T = 15 min.	0.0000	0.0000
Probability that T = 20 min.	0.0000	0.0000
Probability that T = 30 min.	0.0000	0.0000
Probability that T = 40 min.	0.0000	0.0000
Probability that T = 50 min.	0.2806	15.1494
Probability that T = 60 min.	0.1723	9.3030

of the choice set to operating time constraints is captured by the series of travel-time threshold probabilities. It is not a surprise, therefore, that the travel time parameter in the systematic utilities is lower in the ANCS-DC model than in the MNL model. In fact, the two parameters are significantly different at the 0.05 level. It can be seen from Table 2 that most other coefficients of geographic situation (numbers 8, 10, 14, 14, 16, 21, and 22) are also significantly different in the two models. The direction (positive or negative) of the adjustment of the latter coefficients from the MNL model to the ANCS-DC model is difficult to interpret dues to the complex geographic configuration of an urban area such as Minneapolis-St. Paul and to some collinearity between travel time and the binary area type variables. It is clear however that, as for the travel time coefficient, the existence of significant differences stems from the proper identification of the effect of travel time on choice set formation and on utilities. This is also confirmed by observing that all but one of the remaining coefficients, which are associated with non-situation variables, are not significantly different between the MNL model and the ANCS-DC model (Table 10.2).

The estimates of travel-time threshold probabilities (choice-set probabilities) are presented in Table 10.3. The probability that the time constraint operating on the size and composition of the choice set stands at 15, 20, 30, or 40 minutes is zero. The other four probabilities are significant at the 0.05 level. The implied probability that the time constraint is larger than 60 minutes, i.e, that the choice set is in fact the universal choice set, is 1 - (0.1816 + 0.1933 + 0.2806 + 0.1723) = 17.22 percent. This result is important in that it quantifies the magnitude of the problem of choice set misspecification that accompanies the MNL model in our empirical study. The chance that the choice set extends beyond the 60-minute range is in fact smaller than any of the four statistically significant choice-set probabilities. The MNL model estimated on the universal choice set turns out to be the correct model in fairly limited instances since there is an 82.78 percent chance that one's choice set is reduced to destinations within a one-hour drive from home.

Choice-set probabilities estimated with the ANCS-DC model have a bimodal distribution over time. The behavioural interpretation of the bimodality of choice-set probabilities is complicated by the fact that the data set contains whole sorts of shopping trips, from trips to the convenience store, to trips to a home improvement center, and finally to travel to the Mall of the Americas. If, as one would expect, the marginal disutility of travel time varies with the trip purpose, the model is misspecified and this effect may be picked up by the choice-set probabilities. Hence the bimodality of the probability distribution would be spurious: seemingly tight constraints standing for the high travel impedance of travel for convenience goods and services and looser constraints standing for the lower impedance of travel on which higher order goods and services are sought. Alternatively hypotheses can be offered. One plausible explanation for the bimodality that the data are sampled from a mixture of two populations, one having a lot of time and one having less. Choice set probabilities would then vary in a systematic fashion with some socio-economic, demographic, or locational characteristics of individuals, and that low and high constraints are respectively associated with distinct combinations of these characteristics. More empirical analysis is needed to establish whether such dependence exists.

10.6 Conclusions

In random utility models of destination choice, correct prediction of choices is conditional on correct information about the choice set. This paper has presented a new destination choice model consistent with Manski's (1977) two-stage decision process. The ANCS-DC model assumes that the choice set is contained within a travel perimeter defined by the individual's time budget. The nesting of choice sets and the approximation of the support of travel-time threshold probabilities by a manageable set of discrete points make the estimation of the ANCS-DC model computationally tractable with a maximum likelihood method.

The empirical analysis of shopping destination choice in Minneapolis-St. Paul tested the ANCS-DC model against a conventional MNL model and demonstrated a superior fit of the former over the latter. The results provide empirical evidence in support of the hypotheses that travel-time constraints matter in destination choice modelling. The impact of travel-time constraints on choice is not merely subsumed by the negative marginal utility of the time spent traveling. At a more conceptual level, the ANCS-DC also offers the advantage over the MNL model to disentangle the effect of travel impedance on choice formation and its effect on utilities. It formalizes the operation of external constraints on choice decisions.

Implications of the ANCS-DC model on planning and policy developments are twofold. First, the ANCS-DC model may be a better predictive model that a conventional MNL model. More empirical research is warranted to corroborate the superiority of the ANCS-DC model for other trip purposes and in other urban areas. Furthermore, the constraint-oriented approach of the ANCS-DC model offers a finer understanding of destination choice processes. It introduces the concepts of availability and feasibility of choice alternatives into plans to design and develop new retail outlets, new shopping centers, or initiate new marketing tools. It may help answer specific questions such as the following: can the market penetration of a store in a large, affluent neighborhood be increased by tailoring the merchandise mix to affluent tastes so as to overcome the deterrence of the long drive to the store, or is it a lost cause because the remoteness of the store excludes it from the choice set of many residents of the neighborhood?

If the findings reported in this paper answer some important questions, they also raise several issues that call for further research. Because different socio-economic and demographic groups organize daily activities differently and face different demands on their time budgets, choice-set probabilities can be hypothesized to vary among groups. In the same idea, time pressure is known to manifest itself differently in different choice contexts (different time of the day, different trip purposes, etc). If such dependence of the choice-set probabilities is confirmed by empirical analysis, the predictive power of the ANCS-DC model will be enhanced further. The quantification afforded by the ANCS-DC model will provide new insights in the interplay between individual free will and social and environmental contingencies.

Another important direction for future research on the ANCS-DC model concerns the extension of its framework to constraints other than those produced by limiting time budgets. It is often argued that awareness of and familiarity with choice alternatives are central considerations in the identification of alternatives effectively considered for selection. Many studies have revealed that the information individuals have on possible choice alternatives tends to decrease as travel time or distance from their home location increases. However, if a deterministic relationship between familiarity and travel time (or some other destination attribute) cannot be established, a deterministic ordering of destinations according to familiarity cannot be devised and the reduction in the number of possible choice sets needed for computational tractability cannot be achieved.

Acknowledgements. The authors gratefully acknowledge the financial support of NSF grant SBR-9308394. The first author is also grateful to the National Science Foundation for their support of the NCGIA through grant SES-8810917. We are particularly indebted to Rick Gelbmann, Robert Paddock, and Mark Philippi at the Metropolitan Council, Minneapolis-St. Paul, and William Barrett at the Office of Transportation Data Analysis, Minnesota Department of Transportation, for their tireless availability to answer our data-related questions. The empirical analysis of this project would not have been successfully completed without the dedicated research assistance of Aaron Wheeler.

References

Ben-Akiva, M., and B. Boccara. 1995. Discrete Choice Models with Latent Choice Sets. *International Journal of Research in Marketing*, **12**:9-24.

Black, W.C. 1984. Choice-Set Definition in Patronage Modeling. *Journal of Retailing*, **60**:63-85.

Borgers, A.W.J., and H.J.P. Timmermans. 1987. Choice Model Specification, Substitution and Spatial Structure Effects: A Simulation Experiment. *Regional Science and Urban Economics*, **17**:29-47.

Burnett, K.P. 1980. Spatial Constraints-Oriented Modelling as an Alternative Approach to Movement. Microeconomic Theory and Urban Policy. *Urban Geography*, **1**:151-66.

Burnett, K.P., and S. Hanson. 1979. Rationale for an Alternative Mathematical Approach to Movement as Complex Human Behavior. *Transportation Research Record*, **723**:11-24.

Burnett, K.P., and S. Hanson. 1982. The Analysis of Travel as an Example of Complex Human Behavior in Spatially-Constrained Situations: Definitions and Measurement Issues.

Transportation Research A, **16**:87-102.

Crompton, John L. 1992. Structure of Vacation Destination Choice Sets. *Journal of Tourism Research*, **19**:420-434.

Crompton, John L., and Paul K. Ankomah. 1993. Choice Set Propositions in Destination Decisions. *Annals of Tourism Research*, **20**:461-476.

Department of Transportation. *1990 Highway Network and TAZ Documentation*. N.d. Minnesota Department of Transportation, Office of Transportation Data Analysis, St. Paul, MN.

Desbarats, J. 1983. Spatial Choice and Constraints on Behavior. *Annals of the Association of American Geographers*, **73**:340-357.

Ettema, D., A. Borgers, and H.J.P. Timmermans. 1993. A Simulation Model of Activity Scheduling Behavior. *Transportation Research Record*, **1413**:1-11.

Fotheringham, A.S. 1983. A New Set of Spatial Interaction Models: The Theory of Competing Destinations. *Environment and Planning A*, **15**:15-36.

Fotheringham, A.S. 1986. Modelling Hierachical Destination Choice. *Environment and Planning A*, **18**:401-418.

Fotheringham, A.S. 1988. Consumer Store Choice and Choice Set Definition. *Marketing Science*, **7**:299-310.

Gaudry, M.J.I., and M.G. Dagenais. 1979. The Dogit Model. *Transportation Research B*, **13**:105-111.

Ghosh, A. 1984. Parameter Nonstability in Retail Choice Models. *Journal of Business Research*, **12**:425-436.

Golledge, R.G., M.-P. Kwan, and T. Garling. 1994. Computational Process Modeling of Household Travel Decision Using A Geographic Information System. *Papers in Regional Science*, **73**:99-117.

Goodwin, P.B. 1981. The Usefulness of Travel Budgets. *Transportation Research A*, **15**:97-106.

Hägerstrand, T. 1970. What about People in Regional Science? *Papers of the Regional Science Association*, **24**:7-21.

Hauser, J.R., and B. Wernerfelt. 1990. An Evaluation Cost Model of Consideration Sets. *Journal of Consumer Research*, **16**:393-408.

Horowitz, J.L. 1988. Specification Tests for Probabilistic Discrete Choice Models of Consumer Behaviour. In Golledge, R.G., and H.J.P. Timmermans, Ed., *Behavioural Modelling in Geography and Planning*, London: Croom Helm, pp. 124-137

Horowitz, J.L. 1991. Modeling the Choice of Choice Set in Discrete-Choice Random-Utility Models. *Environment and Planning A*, **23**:1237-1246.

Horowitz, J.L., and J. Louviere. 1995. What is the Role of Consideration Sets in Choice Modeling? *International Journal of Research in Marketing*, **12**:39-54.

Howard, J.A.1961. *Marketing Management Analysis and Planning*. Homewood, IL: Irwin.

Howard, J.A., and J.N. Seth. 1969. *The Theory of Buying Behavior*. New York: John Wiley.

Kitamura, R., and M. Kermanshah.1984. Sequential Model of Interdependent Activity and Destination Choices. *Transportation Research Record*, **987**:81-89.

Kitamura, R., and T.N. Lam. 1984. A Model of Constrained Binary Choice. In Volmuller, J., and R. Hamerslag (Editors) *Proceedings of the 9th international Symposium on Transportation and Traffic Theory*, Utrecht: VNU Science Press, pp. 493-512

Landau, U., J.N. Prashker, and M. Hirsh. 1981. The Effect of Temporal Constraints on Household Travel Behavior. *Environment and Planning A*, **13**:435-448.

Lenntorp, B. 1976. Paths in Space-Time Environments: A Time Geographic Study of Movement Possibilities of Individuals. *Lund Studies in Geography*, Series B, 44

Manski, C.F. 1977. The Structure of Random Utility Models. *Theory and Decision*, **8**:229-254.

McFadden, D. 1978. Modelling the Choice of Residential Location. In Karlquist, A., Lundqvist, L., Snickars, F., and J.W. Weibull (Editors) *Spatial Interaction Theory and Planning Models*, Amsterdam: North Holland, pp. 75-96

Metropolitan Council.1992. *Home Interview Survey. Methodology and Results*. Publication No. 550-92-061, Metropolitan Council, St. Paul, MN.

Meyer, R. 1980. Theory of Destination Choice-Set Formation under Informational Constraints. *Transportation Research Record*, **750**:6-12.

Meyer, R.J., and T.C. Eagle. 1982. Context-Induced Parameter Instability in a Disaggregate-Stochastic Model of Store Choice. *Journal of Marketing Research*, 19:62-71.

Miller, E.J., and M.E. O'Kelly. 1983. Estimating Shopping Destination Models from Travel Diary Data. *Professional Geographer*, **35**:440-449.

Pellegrini, P.A., A.S. Fotheringham, G. Lin 1997. Parameter Sensitivity to Choice Set Definition in Shopping Destination Choice Models. *Papers in Regional Science*, forthcoming.

Recker, W.W., M.G. McNally, and G.S. Root. 1986a. A Model of Complex Travel Behavior: Part I: Theoretical Development. *Transportation Research A*, **20**:307-318.

Recker, W.W., M.G. McNally, and G.S. Root. 1986b. A Model of Complex Travel Behavior: Part II: An Operational Model. *Transportation Research A*, **20**:319-330.

Richardson, A. 1982. Search Models and Choice Set Generation. *Transportation Research A*, **16**:403-419.

Roberts, J., and P. Nedungadi. 1995. Studying Consideration in the Consumer Decision Process: Progress and Challenges. *International Journal of Research in Marketing*, **12**:3-7.

Shocker, A.D., M. Ben-Akiva, B. Boccara, P. Nedungadi. 1991. Consideration Set Influences on Consumer Decision-Making and Choice: Issues, Models, and Suggestions. *Marketing Letters*, **2**(3):181-197.

Sheppard, E.S. 1980. The Ideology of Spatial Choice. *Papers of the Regional Science Association*, **45**:197-213.

Sheppard, E.S. 1984. The Distance-Decay Gravity Model Debate. In G.L. Gaile, and C.J. Wilmott (Editors). *Spatial Statistics and Models*, Dordrecht: Reidel, pp. 367-388

Swait, J.D., and M.E. Ben-Akiva. 1987a.Incorporating Random Constraints in Discrete Models of Choice Set Generation. *Transportation Research B*, **21**:91-102.

Swait, J.D., and M.E. Ben-Akiva. 1987b. Empirical Test of a Constrained Choice Discrete Model: Mode Choice in Sao Paulo, Brazil. *Transportation Research B*, **21**:103-115.

Stetzer, F.C, and A.G. Phipps. 1977. Spatial Choice Theory and Spatial Indifference: A Comment. *Geographical Analysis*, **9**:400-403.

Thill, J.-C. 1992. Choice Set Formation for Destination Choice Modelling. *Progress in Human Geography*, **16**:361-382.

Thill, J.-C. 1995. Modeling Store Choices with Cross-Sectional and Pooled Cross-Sectional Data: A Comparison. *Environment and Planning A*, **27**:130-1315.

Thill, J.-C., and J.L. Horowitz. 1991. Estimating a Destination-Choice Model from a Choice-Based Sample with Limited Information. *Geographical Analysis*, **23**:298-315.

Thill, J.-C., and J.L. Horowitz. 1997. Travel-Time Constraints on Destination Choice Sets. *Geographical Analysis*, forthcoming

van der Hoorn, T. 1983. Experiments with an Activity-Based Travel Model. *Transportation*, **12**:61-77.

Wigan, M.R., and J.M. Morris. 1981. The Transport Implications of Activity and Time Budget Constraints. *Transportation Research A*, **15**:63-86.

Williams, H.C.W.L., and J.D. Ortuzar. 1982. Behavioural Theories of Dispersion and the Misspecification of Travel Demand Models. *Transportation Research B*, **16**:167-219.

11 Space-Time Consumer Modelling, Store Wars, and Retail Trading Hour Policy in Australia

Robert G. V. Baker

Department of Geography and Planning, University of New England, Armidale, 2351, Australia.

11.1 Introduction

"Baker's comments on the extension or deregulation of trading hours are no more than his personal opinions. They are not substantiated by any empirical evidence. On the contrary, market research suggests that community fabric is built around modern shopping centres, large and small, while the traditional strip shopping areas become productively used by newer types of retailers and other small business." Keil and Haberkern (1994)

Space-time modelling of aggregate consumer shopping behaviour has been put on the political agenda across Australia with the predictions of significant changes of 'when' and 'where' consumers, on aggregate, undertake their shopping trips in a deregulated trading hour environment. Originally conceived in Baker (1985), the retail aggregate space-time trip (RASTT) model is constructed around a differential equation of spatial and temporal operators, where space is partially differentiated once and time twice. The gravity model of trip distance and a periodic shopping-time function appear in the solutions of this equation. For the boundary conditions, the trading period was assumed constant, which was the case in the regulated pre-1992 Sydney shopping environment. Baker (1994a) showed that the gravity coefficient, β, and average visits per week k to a planned centre, were theoretically and empirically linked (as shown in the regression analysis of fifteen samples of five such centres in Sydney during 1980/81 and 1988/89). Furthermore, centre scale produced non-linearities in both the trip distance and frequency constructs, which allowed for the definition of a number of characteristics for aggregate consumption at smaller community planned centres (or 'small' centre behaviour) and larger regional planned centres (or 'large' centre behaviour).

The RASTT model was tested extensively using a Sydney 1988/89 data set (Baker, 1994a; 1996a). The localities selected, in Morning (M) and Afternoon (A) samples, ranged from two community centres [Ashfield Mall (AM) and Market Place Leichhardt (ML)]; sub-regional centres [Westfield Burwood (WB) and Westfield Chatswood WC)]; and a regional centre [Bankstown Square (BS)]. These five planned suburban shopping centres (PSSCs) are located within the inner city, western and northern regions of Sydney and the data set consisted of 1,522 interviews.

When the NSW Government deregulated shopping hours in 1992, the obvious research question arose concerning how the space-time distributions would be affected when trading period (T) per week was increased substantially for retail centres. What became immediately apparent for 'small' centre behaviour (of which the supermarkets are a fundamental construct) was that the $1/T^2$ factor would considerably lower the gravity coefficient and hence, substantially expand the market areas of the major supermarket anchors. This result was presented in Baker (1993) and, after some publicity, Coles-Myer (Australia's largest retailer and major supermarket chain) mounted a public campaign against this result. They used the report they commissioned from Keil and Haberkern (1994), in the media and subsequent trading hour inquiries.

Therefore, space-time modelling has now become intimately bound up across Australia with what Wrigley (1994) termed 'store wars'. He used the term to describe the corporate strategies of food retailers in Britain over the 1980's. In the UK, new store development programs were the engine for corporate growth. Yet in Australia, it is trading hour policy that is generating short term growth in turnover, and is currently a major platform in the corporate strategies of the major supermarkets as they try to deregulate trading hours in the other states of Australia. One reason for this variation may be the market concentration in Australia, where 71% of the packaged grocery market in 1994 was controlled by three major supermarket chains (Woolworths, Coles and Franklins).

Therefore, this paper summarises the evolution of the RASTT and SASTV models, and is at the stage now where a corresponding estimate of trading hours per week can be made, based on the average number of shops visited and the amount of time spent shopping at a retail centre per week. Trading hours are shown to vary with the socio-economics of the primary trade area and as such, shopping-time policy should be determined on a local or regional rather than a state or national level.

11.2 Space-Time Modelling of the Interlocational Trip Frequency

Thill and Thomas (1987) state that it should be a major aim of theoretical research to conceive a framework for trip behaviour that combines both spatial and temporal aspects of travel choices. Traditional studies have focused on the effect of distance in the construction of the gravity model of spatial interaction, yet Baker (1994a) has established a strong empirical association between the gravity coefficient of shopping centre patronage and mean trip frequency to that centre. Space and time constructs appear to have a conceptual translation. Retail space-time modelling suggests that a substantial extension of hours, changes on aggregate, "when" people shop, and as such, there should be a corresponding shift in "where" they shop over a seven day-a-week cycle. This is at the heart of the relationship between this type of modelling and the broader context of 'store wars' in Australia. Large retailers have mounted a public campaign against this connection, arguing that there is no relationship between the extension of hours and 'where' consumers undertake their shopping (Keil and Haberkern, 1994). This relationship is therefore critical to explore in the context of the politics of retail policy in Australia.

The key trip equation in space-time modelling was developed in Baker (1985) where, contrary to the usual form of classical diffusion, time was partially differentiated twice and space once to yield an equation defined as:

$$\frac{\partial^2 \phi}{\partial t^2} = M \frac{\partial \phi}{\partial x} \qquad (11.1)$$

where M is defined as a measure of the aggregate household mobility. The solution takes the form

$$\phi = A \exp(-\beta D)\sin(k\ t) \qquad (11.2)$$

where β is centre loyalty (or the gravity coefficient), D is the trip distance and the separation constant k is defined to be the mean interlocation trip frequency (ITF). The gravity coefficient is not just the spatial impedance, but rather defines the spatial influence that the centre has on the surrounding trade area (see Baker, 1994a). The ITF frequency is the average number of trips undertaken per week, by the sampled population from an origin (usually a residence) to the shopping centre. Patronage from the planned centre follows a gravity distribution that is periodically reconstructed over time. The interlocational trip frequency defines the average number of trips per week to a particular centre and is related to the gravity coefficient by:

$$\beta = \frac{k^2}{M} \qquad (11.3)$$

This equation has been confirmed by Baker (1994a) with an R-squared of .53 from fifteen samples taken at planned shopping centres in Sydney, Australia from 1980/81 and 1988/89. There, the gravity coefficient is inherently a function of time through the frequency of centre visitation. This is why the traditional interpretation of β as the spatial impedance is not appropriate. A high number of trips corresponds to large β coefficients (with the converse for a low number of trips) and this fusion of space-time constructs is a significant result. The gravity coefficient is more an indication of influence than impedance and defines aggregate consumer 'loyalty' to patronise their nearest centre. There is now a fundamental relationship in the 'when' and 'where' consumers travel to do their shopping and as such β must be reinterpreted in the context of Equation 11.3.

The frequency of purchasing a range of commodities has been developed as a construct of central place theory by Losch (1954). Curry (1962) and Nystuen (1967) describe consumer demands of shopping centres in space as part of the satisfaction by commodities of different purchasing frequencies. Bacon (1971) and Eaton and Lipsey (1982) consider potential shopping activities in space and time as individual demand functions, and purchasing frequencies as exogenous variables. Papageorgiou and Brummel (1975) reviewed the relative trip patterns, travel frequencies and purchase frequencies within a hierarchy at Toronto and made inferences according to the corresponding generalised density gradients. Their data suggested that second order centres were by-passed within the prevailing trip patterns but they viewed, with suspicion, their partial estimates of alternative purchase frequencies. However, their work is important, since it identified the relevance of the multipurpose trip and the quadratic form of linking frequencies and distance. Therefore, Lange (1978), in assessing the constraints acting on shopping frequencies and commodity mixes, concludes that a dynamic theory of regional structure is still very much incomplete. Yet Equation 11.2 defines the relationship between spatial character (through the β coefficient) and shopping time allocation (through the number of trips) and may provide a key to understanding the dynamics of intraurban central places. Indeed, Warnes and Daniels (1979), in their review of the spatial aspects of an intrametropolitan central place hierarchy, state that:

"There is evidence of an ordered spatial structure (in the intraurban distribution of shops): this not only reflects the very strong influence of the friction of distance on shopping travel, but also the effect of other principles of personal movement, such as the temporal ordering of travel and the relationship between journey distance and the mode of the frequency of trips." (p.402)

Equation 11.3 summarises this temporal ordering of travel under the constraint of distance minimisation. The result should empirically highlight some form of ordered spatial structure and the differences should be apparent in the empirical testing of Equation 11.3 over a range of planned shopping centres.

This equation appears empirically relevant for community and sub-regional PSSC samples in the Sydney 1980/82 and 1988/89 data sets in Baker (1994a), but there is a quadratic relationship for β and k with centre size, where both gradients become positive for regional centres with over 150 retail destinations or 45,000 sq. metres of floor space (Fotheringham, 1982; Fotheringham, 1985; Baker, 1994a). Such an agglomerative effect destroys the linearity of the space-time correspondence in shopping trip behaviour and there are distinct types of behaviour dependent on the time of day and year as well as the size of the centre. The differences in the relationships are summarised in Table 11.1. Equation 11.3 seems more appropriate to 'small centre' shopping behaviour where consumers periodically visit the centre, via the grocery trip, and the trip frequency distribution follows a normal distribution centred on the once-a-week shopping. Conversely, large centre behaviour is characterised by a more complex interaction of a substantial underestimation from a

Table 11.1 Characteristics of 'Small' and 'Large' Centre Behaviour

Small Centre Behaviour

1. There are high R-squared values from the gravity hypothesis.

2. The trip frequency distribution follows the normal distribution of the Fourier hypothesis and therefore shopping behaviour is simple harmonic.

3. Patronage is concentrated in the local region and the centre exerts a strong spatial influence upon regular shopping behaviour.

4. It may be characterised by grocery trips and may be patronised by less mobile households (particularly in the morning).

5. There is no change in the gravity distributions from morning to afternoon samples.

6. It is defined by an aggregate household mobility $M < 2.8$.

7. The gradient of the gravity coefficient to centre scale is negative.

Large Centre Behaviour

1. It occurs for shopping centres with the number of outlets $N>150$.

2. The gravity hypothesis is still relevant, but with lower R-squared.

3. The trip frequency distribution does not follow the normal distribution of the Fourier hypothesis, where there is substantial underestimation of once-a-week and low frequency visitation and as such, shopping behaviour is not simple harmonic.

4. Patronage is dispersed over a wider area throughout the day and consumers are less 'bonded' to that particular centre.

5. The centre is patronised by more mobile consumers in the afternoon.

6. The gradient of the gravity coefficient to centre scale is positive.

7. It is defined by an aggregate household mobility of $M > 2.8$.

normal distribution (from the Fourier transform of the spatial distribution), for once-a-week and low frequency visits. Therefore, it was concluded that aggregate shopping trips to a regional PSSC does not follow a simple periodic shopping trip hypothesis (Baker,1994a).

These fundamental differences with the size of the shopping centre, suggest that the concept of a centre hierarchy is relevant and counters the claims by Beavon (1977, quoted in Warnes and Daniels; 1979) that hierarchies have been more readily recognised than have been justified by the data and that intrametropolitan size distribution of central places is a continuum. However, in this analysis, lower order community centres and higher order regional shopping centres have distinctive and discrete characteristics. Distance minimisation is no longer the appropriate behavioural construct at a regional centre and, as Christaller (1966, quoted in Warnes and Daniels; 1979) recognises:

"In practice, heterogeneous influences generate more complex landscapes than those produced by a single locating principle (p.385)."

Baker (1996a) has provided a socio-economic basis for these types of behaviour by empirically connecting, in the Sydney 1988/89 data set, the standardised number of multi-purpose shoppers (MPS) or (ω) and high disposable income consumers (HDI) or (I) with the mean interlocational trip frequency (k) in the equations:

$$\varphi = hk \tag{11.4}$$

$$\varphi = \pm mI\sqrt{\beta M} \tag{11.5}$$

for h, m constants. This research supported the West (1993) hypothesis that trips to supermarkets are the basis for MPS and identifies it as a relevant shopping strategy for both high disposable income households and low household mobility households at planned centres with the number of outlets, N > 50. HDI consumers construct MPS around the grocery trip at both small and large planned centres as a time minimisation strategy. Conversely, low-mobility consumers use MPS around the supermarket visit as a distance minimisation mechanism aiming to minimise the total effort of shopping to obtain the desired set of goods and services (Hanson,1980). The important conclusions are that there is a strong linear relationship between the standardised number of MPS consumers and the interlocational trip frequency, and that this type of shopping is the basis for the development and maintenance of a shopping-centre hierarchy. The association between the combinations of the purchases of different commodity types and trip frequency suggests that the multi-purpose trip is a further construct of a retail central place hierarchy.

11.3 Structural Change in 'Small' and 'Large' Centre Behaviour from a Major Extension of Trading Hours

The interlocational trip frequency defines travel to and from a shopping centre. Yet, consumers travel to different destinations *within the centre* as part of this trip and as such, there is assumed to be a second frequency construct, termed the intracentre shopping frequency (ISF) that is equally important to the understanding of the origin-destination(s) of shopping journeys. The centre destination of the trip encompasses a subset of many intracentre destinations, (n), that may be used to define the characteristics of shopping at that centre. The key to this concept is understanding how shoppers, on aggregate, distribute their shopping time to the different destinations in the centre. This is the basis for the shopping aggregate space-time visitation (SASTV) model. For the Sydney 1988/89 data set there was a very strong relationship between the mean number of shops visited in the shopping centre (*n*) and the mean time spent shopping (*t*) with R^2 = .96 and the Durbin-Watson statistic showed no autocorrelation of d = 2.07, d_v = 1.32 < d < 4- d_v at the .05 level of significance (see Figure 11.1). This distribution of shopping time then appears to be related to centre scale and hence, there should be some equivalence to an (ISF) definition of n/T with t/T per trading week (T). From the (SASTV) model, it becomes apparent that the number of shop visitations within the shopping centre per week is governed by the trading boundaries (as well as other variables, such as centre scale). The initial trading hour condition can be defined for a population density (ϕ) as ϕ = 0 at t = 0 and ϕ = 0 at t = T. In other words, there would be no consumers on t week (T). From the (SASTV) model, it becomes apparent that the number of shop visitations within the shopping centre per week is governed by the trading boundaries (as well as other variables, such as centre scale). The initial trading hour condition can be defined for a population density (ϕ) as ϕ = 0 at t = 0 and ϕ = 0 at t = T. In other words, there would be no consumers on the floor-space at the beginning or end of the trading week. For simple periodic shopping trips, with an intracentre shopping frequency f, assume that sin ft = 0, and for defined trading boundaries, f = n π/ T, where n is assumed to be the mean aggregate number of destination visited per trading he floor-space at the beginning or end of the trading week. For simple periodic shopping trips, with an intracentre shopping frequency f, assume that sin ft = 0, and for defined trading boundaries, f = n π/ T, where n is assumed to be the mean aggregate number of destination visited per trading
week. It is further postulated that there is some relationship between (ITF) and (ISF) of the order k= af, where a is an unknown function assumed, at this stage, to be unity. The result is that the (ISF) has been redefined as:

$$f = \frac{n\pi}{T} \quad \text{or} \quad f_o = \frac{n}{4T} \quad \text{for} \quad f = 2\pi f_o \tag{11.6}$$

where f_o is the fundamental intracentre shopping frequency (FISF) for the positive frequency domain 1/4T (remembering that there is also a negative frequency domain

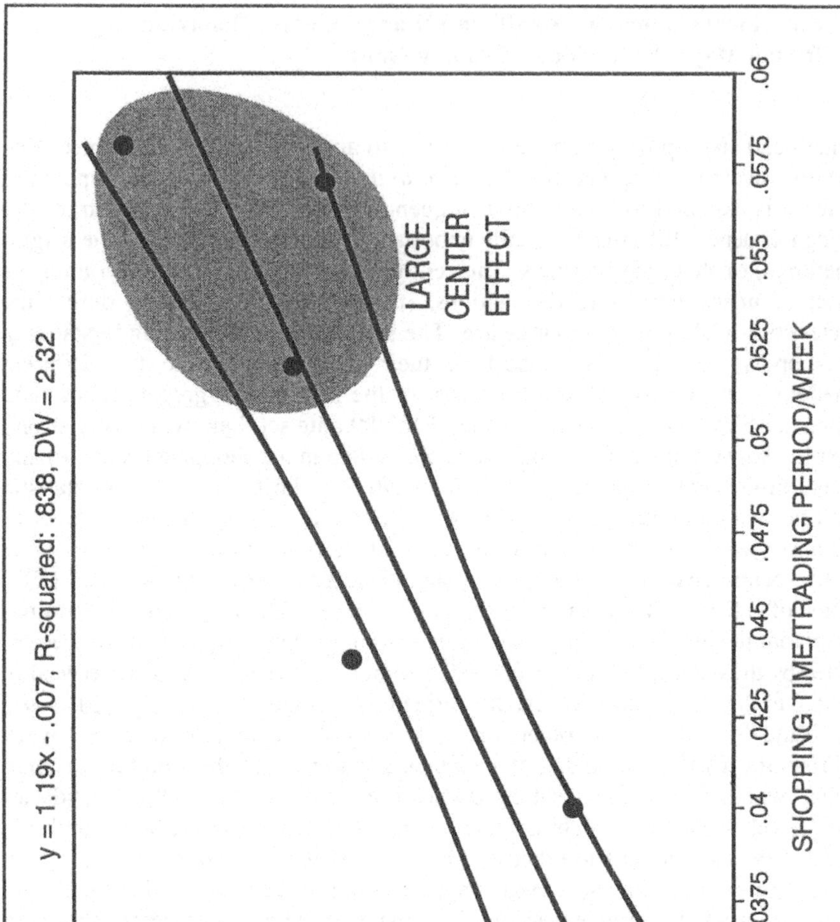

Fig. 11.1 The Relationship between Mean Number of Shops Visited and Mean Shopping Time in the Sydney Data Set, 1988/89

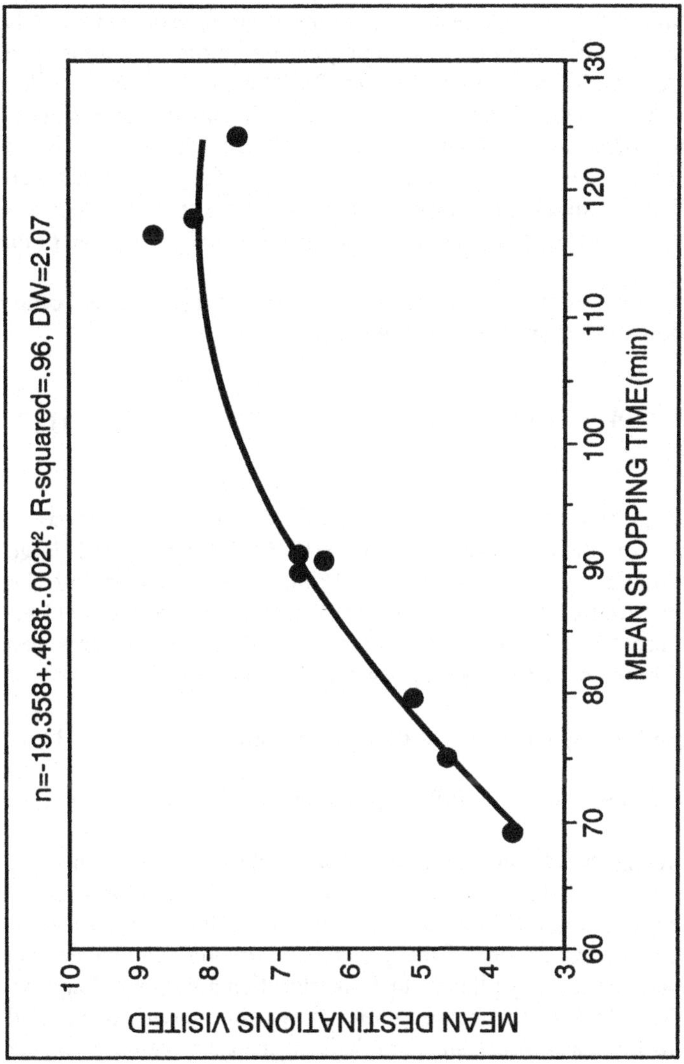

Fig. 11.2. The Relationship between n/4T and t/T in the Sydney Data Set, 1988/89

with a 1/4T contribution that is not meaningful to the shopping trip model). The (FISF) can be represented alternatively, as being equivalent to the mean time spent shopping divided by the trading period per week (t / T). This association is checked empirically in the Sydney 1988/89 data set, in Figure 11.2, showing an R-squared value of 0.84 and the Durbin-Watson statistic with no autocorrelation ($d = 2.32$, $d_v = 1.32 < d < 4- d_v$ at the .05 level of significance). It is important to note that there is a centre scale effect on the (FISF), which is not surprising since shoppers, on aggregate, spend more time shopping and visit more destinations as the size of the centre increases.

By using Equation 11.6, the space-time shopping patronage for a planned centre can now be rewritten, using the (**n** π/T) form of the (ISF), as:

$$\phi = A \exp(-\beta x) \sin(\frac{n\pi}{T} t) \qquad (11.7)$$

This equation can be plotted for values of A, β, n and T and examples of 'small' centre (Market Place Leichhardt surveyed on the 8/12/88 Afternoon) and 'large' centre (Bankstown Square surveyed on 23/3/89 Morning) behaviour from Baker (1994a) are shown in (Figures 11.3 and 11.4). The equations (11.8a) and (11.8b) define the regulated pre-1992 distributions for these samples, where centres were allowed to trade 49.5 hours over six days per week (144 hours/week).

$$14.8 \exp(-.84\ x) \sin(.352\ t) \text{ for } n = 5.55 \text{ destinations /week} \qquad (11.8a)$$

$$2.2 \exp(-.47\ x) \sin(.689\ t) \text{ for } n = 10.87 \text{ destinations /week} \qquad (11.8b)$$

The Marketown Leichhardt sample for small centre behaviour shows periodic behaviour for the six days-a-week, but the gravity distribution changes for a number of days and there is a mid-week lull in trading for the centre. This has ramifications for sampling design, since the nature of the gravity distribution can change, as a function of how consumers regularly allocate their interlocational shopping frequency over a six day cycle. The Bankstown Square sample shows, for the regulated period, a lower frequency behaviour with much smaller relative densities spread over a wider market area. This periodicity would be masked by the greater proportion of low frequency and random patronage that are not recorded in this graph. Yet, for two of the six days, some periodicity may be evident in the space-time distributions. Once again, the day of the survey may produce variations that may be just part of the dynamics of that particular shopping centre.

The impact of the extension of trading hours on these 'small' and 'large' shopping centres may be simulated for 70 and 100 hours per week under deregulation. The centres now operate seven days-a-week over a period of 168 hours. The β and n values are assumed constant for the purpose of this simulation. For Marketown Leichhardt, the impact of Sunday trading sees the distribution skewed towards the end of the week, with patronage on Mondays and Tuesdays now substantially reduced, whilst for Bankstown Square, there is now only one regular shopping period towards the end of the week.

For 100 hours trading per week, Market Place Leichhardt remains periodic, but this is reduced from seven to five days a week. The linear k = af assumption which states that a reduction in the (ITF) would result in a corresponding decline in the (ISF), seems justified. For Bankstown Square, there is a substantial increase in the number of periodic shopping days. This is a surprising result since, at this level of trading hours, the periodicity of small and large centre behaviour seems to converge. Regional planned shopping centres appear to mirror a form of 'small' centre behaviour. What does this mean? With the decrease in intracentre shopping frequency, there is an increase in the interlocational frequency and as such, the linearity assumption for large centres does not seem to be appropriate. This is not unexpected, since non-linearity plague 'large' centre shopping behaviour. Yet the change to a degree of aggregate regular shopping at regional planned centres under extended hours is intriguing. Indeed, there has been a major collapse in the retail property market in traditional shopping centres adjoining planned regional centres in Sydney since the trading hour deregulation in 1992. Liverpool, Parramatta and Penrith unplanned shopping centres have had a 45%, 24% and 21% reduction, respectively, in shop rentals from 1991-95 (see Figure 11.5). A shop in Liverpool that was appraised at $A750,000 in 1991, is now only valued at $A 375,000 (NSW Valuer General, 1995). Does this shift to a large centre periodicity under deregulation explain this collapse? Remember this result comes from a simulation and the centres would have to be resurveyed to check if, in fact, this has happened. If this was the case, then a space-time modelling explanation would be that an appearance of periodic shopping under a major trading hour extension at regional planned centres, would lead to a corresponding rise in the interlocational trip frequency and multipurpose shopping (from Equation 11.4) to a point where, if all the retail functions of the adjoining traditional shopping centre are duplicated on the floor space of the regional centre, there would be little need for consumers to shop outside the planned centre. Hence, there is a very good explanation for the dramatic drop in passing trade, retail rentals and property values in traditional regional centres, since trading hour deregulation in 1992.

For 'small' centre behaviour, there appears to be an equivalence between the (ITF) and (ISF) and as such, a mathematical statement can be made to explain the effect that the trading period has on 'where' consumers, on aggregate, undertake their shopping. Therefore, for 'small centre' behaviour, the effect of trading hours on the gravity coefficient can be restated from substituting Equation (11.6) into (11.3) to produce:

$$\beta = \frac{a^2 n^2}{16 M T^2} \quad . \tag{11.9}$$

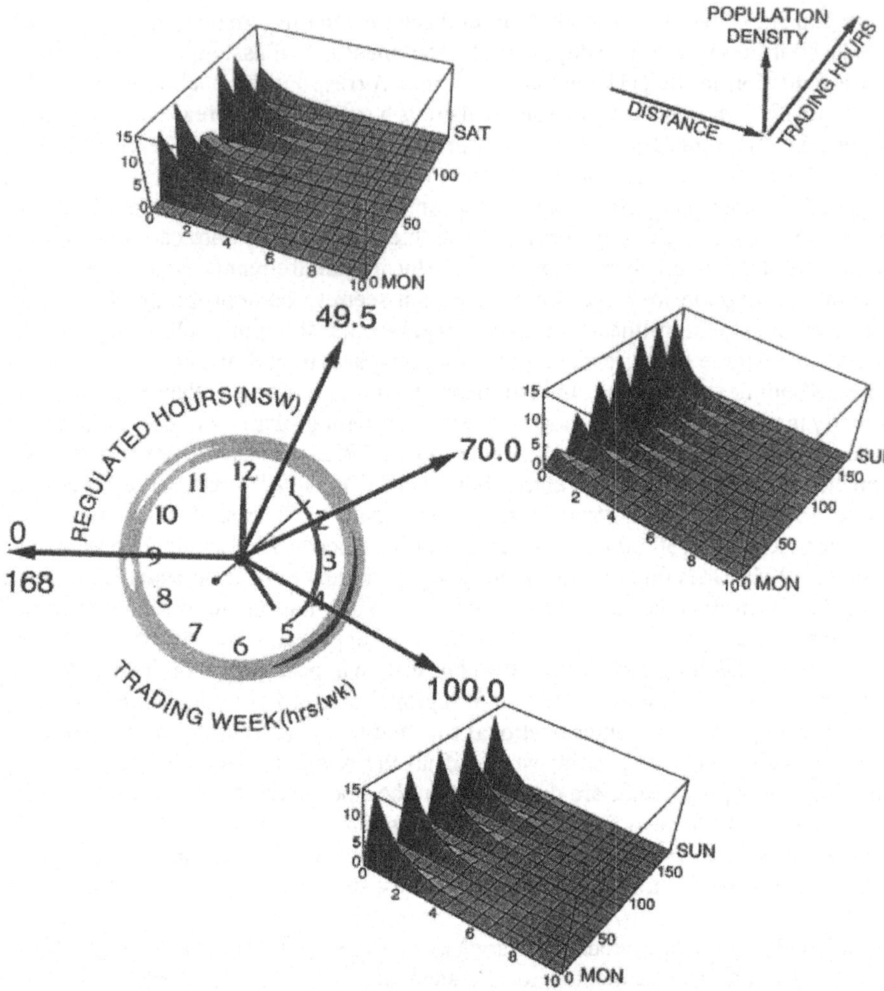

Fig. 11.3. Simulation of the Space-Time Shopping Distributions for 49.5, 70 and 100 Trading Hours per week for 'Small Centre' Behaviour

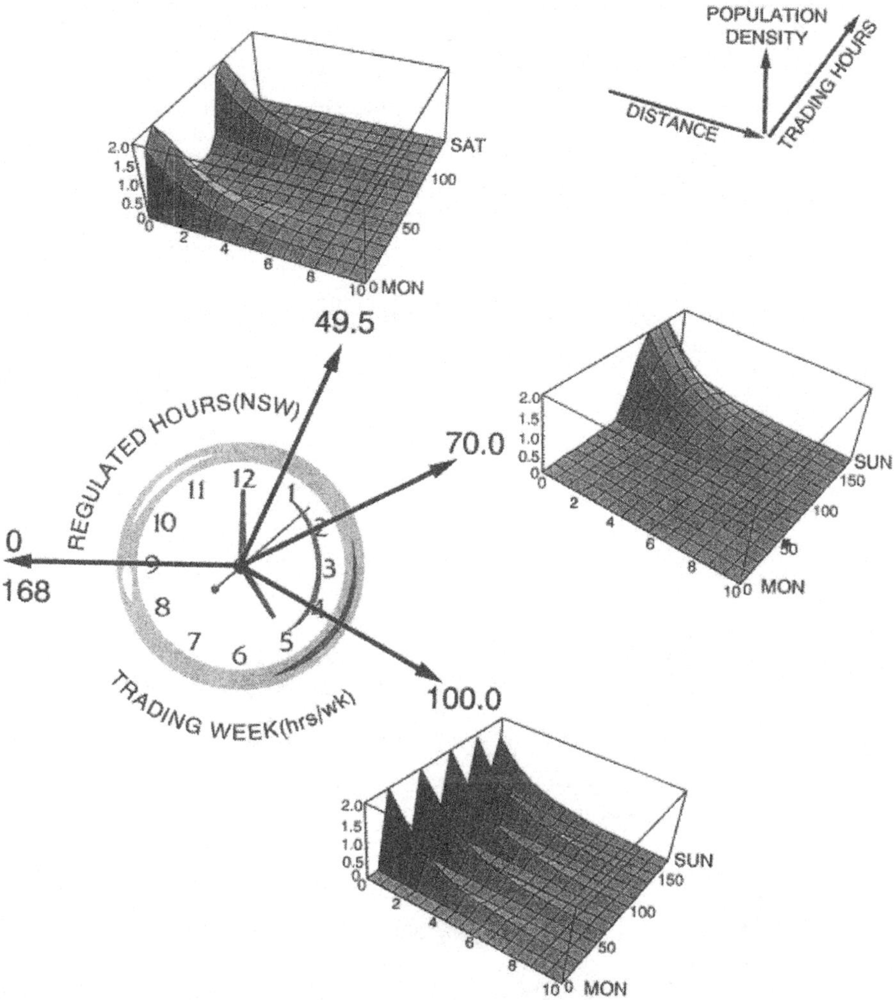

Fig. 11.4. Simulation of the Space-Time Shopping Distributions for 49.5, 70 and 100 Trading Hours per week for 'Large Centre' Behaviour

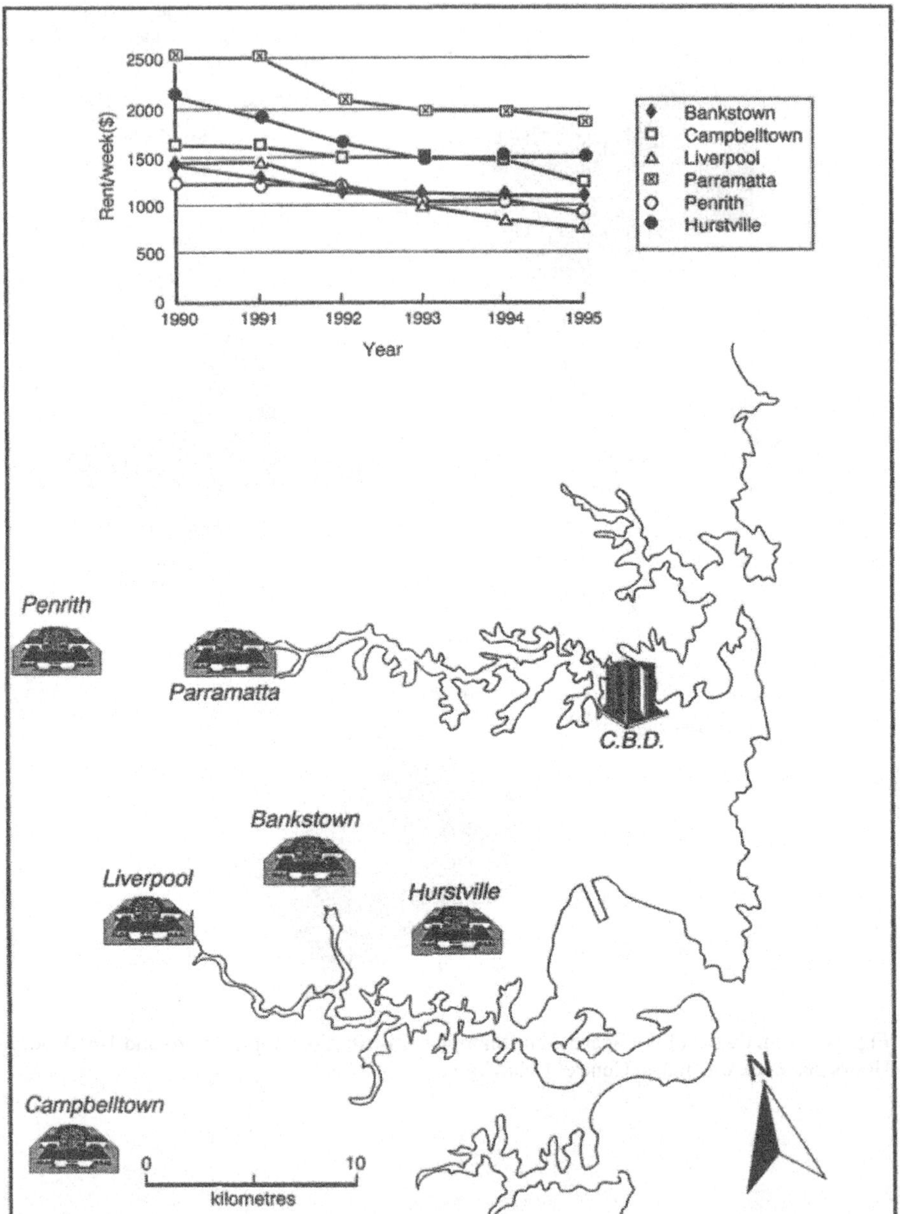

Fig. 11.5. The Decline of Traditional Regional Retail Property Markets in Sydney between 1991-1995.

The result suggests that there would be a substantial drop in the gravity coefficient ($\Delta\beta$) from the ($1/\Delta T^2$) factor from the extension of trading hours. There would also be a substantial depreciation in centre loyalty (that is, the aggregate propensity for shoppers to travel to their nearest centre). Shopping would become less regular at centres with $N < 150$ outlets and there would be a propensity for less multipurpose shopping and more random single purpose trips. The fundamental construct of 'small' centre behaviour are supermarkets, who would benefit from this shift from the expanded market area (from lower β values) and more single purpose trips (from less multipurpose trips).

Conversely, as indicated in the simulation of Equation 11.8a, a substantial extension of hours would increase the periodicity of shopping at large centres, and by space-time complementarity, increase the corresponding β values for regional PSSCs. This rise in centre loyalty would indicate an increase in aggregate preference to travel more regularly to their nearest regional shopping centre. This scenario is summarised in Figure 11.6. The extension of trading hours would substantially change the characteristics of 'small' and 'large' centre behaviour. Therefore, Baker (1993; 1994b) argued that the major beneficiaries of the deregulation of retail trading hours would appear to be supermarkets and large regional shopping centres.

Publication of this result in the national media (for example, *The Age*, 8/11/93, p.7 and p.27; *Australian Financial Review*, 17/11/93, p.29; *West Australian*, 6/11/93, p.30; and *Sydney Morning Herald*, 24/11/93, p.28) provoked an immediate reaction from the large retailer, Coles-Myer, who employed Keil and Haberkern (1994) to mount a public campaign against the 'where' result in Equation (11.9). They argued that the idea of 'wave nature' for regular shopping was too abstract for retailing and that there was no complementarity between shopping time and place. Their study used the Australian Bureau of Statistics (ABS) on the number of small shops in the retail industry from 1948 to 1992. They concluded that the number of small businesses was on the rise, despite the increases in retail trading hours and their number, in relation to population, has been steady at approximately ten shops per thousand over the last two decades. Baker (1994b) responded arguing that the use of the number of small business was a poor measure of the impact of the deregulation of trading hours. Their study had no basis to project the implications beyond 1992 of deregulated trading hours. Furthermore, it ignored the effect of the 1989-1992 recession, where self-employment was the only net positive growth area from people beginning businesses from redundancy payouts. Their study would also include the rapid rise in home-based small business and, on aggregate, it would be very difficult to filter the impact of deregulated hours from this scenario.

Yet the lines were drawn in 'store wars' with the Keil and Haberkern (1994) report being presented at every subsequent inquiry into trading hours in Australia. Since trading hour deregulation in NSW and extensions in Victoria, Queensland and Tasmania, supermarket turnover has increased 40.4% ($A7.806 to $A10.959 billion) for Woolworths and 18.0% ($A6.202 to $A7.340 billion) for Coles from 1991/92 to 1994/95. Furthermore, there has been a rapid decline of traditional retail property markets adjoining regional planned shopping centres in suburbs of Sydney. The space-time models suggest that there is an association, but to show, conclusively, these connections is a difficult exercise in the retail politics of data acquisition.

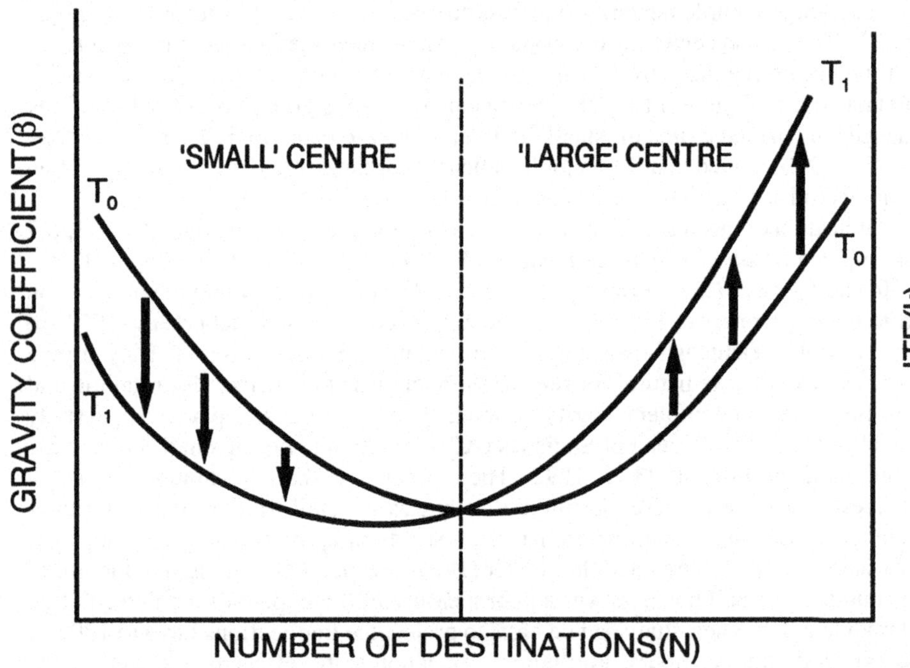

Fig. 11.6. Structural Changes in Predicted Small and Large Centre Behaviour at Planned Shopping Centres from a Major Extension of Trading Hours

11.4 Space-Time Modelling of the Intracentre Shopping Frequency

The development of Equations 11.1 to 11.3 is based on the separation of the space-time operators by the interlocational frequency, k, and the definition of 'small' and

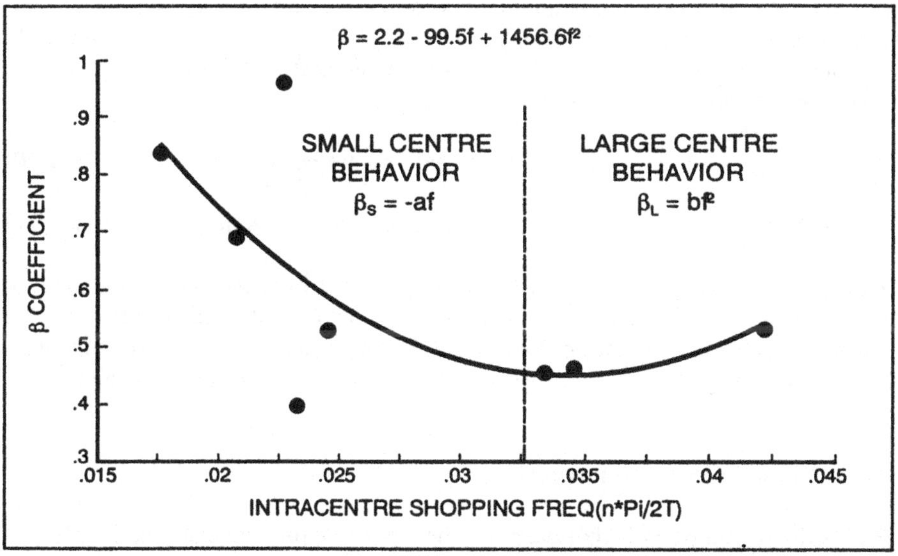

Fig. 11.7. The Distribution of the Fundamental Shopping Frequency with the Gravity Coefficient for the Sydney Data Set, 1988/89

'large'centre behaviour comes from the lack of correspondence of 'large' centre behaviour to 'simple harmonic' shopping. The socio-economic reason for this is that the once-a-week grocery trip is a fundamental construct of 'small' centre behaviour and is less important in 'large' centre behaviour under regulated hours. The steps in the modelling involve the postulation of Equation 11.1, solving it to yield the solution in 11.2 and determining the relationship between the gravity coefficient, β and the interlocational trip frequency, k, in 11.3. The empirical relevance of steps two and three provide the justification that the differential equation in 11.1was the appropriate guess.

Yet, the interlocational frequency provides no mechanism to deal with the non-linearities in the various space-time distributions for 'large' centre behaviour. Furthermore, it is the intracentre shopping frequency that provides the information on the relationship between trading period and aggregate space-time consumer behaviour.

As such, the paper will now focus on the (ISF) distribution with centre scale in the 1988/89 Sydney data set, and endeavour to seek a solution to the problem of adequately describing 'large' centre shopping behaviour.

For the intracentre shopping frequency, the analysis will follow a rearranged sequence of steps to the evolution of the (ITF) relationships, since we will begin with an empirical association between the intracentre shopping frequency, f, and the gravity coefficient, β, then derive the appropriate differential equation(s) and finally, the appropriate solution(s) for the trading period T.

The empirical relationship between f and β for the Sydney 1988/89 data set (excluding the leverage of the anomalous Ashfild Mall 23/3/89 Afternoon sample) is shown in Figure 11.7 and is quadratic with an R-squared of .45 and the Durbin-Watson statistic with no autocorrelation ($d = 1.97$, $dv = 1.32 < d < 4- dv$ at the .05 level of significance). The non-linear relationship suggests the discrete range of both 'small' and 'large' centre behaviour, such as:

$$\beta_S = - af \qquad \text{for 'small' centre behaviour} \qquad (11.10)$$

$$\beta_L = bf^2 \qquad \text{for 'large' centre behaviour} \qquad (11.11)$$

To deal with the non-linearity, we can break up the distributions into Equations 11.10 and 11.11 and review them separately, both mathematically and empirically. The results are dealt with in detail in Baker (1996b), but a summary of the key outcomes of this shopping aggregate space-time visitation (SASTV) model are presented below.

11.4.1 'Small' Centre Behaviour

The starting assumption is that the consumers are shopping regularly at the planned shopping centre in the form of a complex space-time periodic function:

$$\phi = exp\ i(f\,t - \beta_S\, x) \qquad (11.12)$$

In this form, the assumption is still that regular shopping interacts with the gravity form of spatial interaction.

Differentiating space and time partially twice from Equation 11.12, and substituting into the square of Equation 11.10 yields the equation:

$$\frac{1}{\varepsilon^2}\frac{\partial^2\phi}{\partial x^2} = \frac{\partial^2\phi}{\partial t^2} \qquad (11.13)$$

This is a form of the wave equation, which should be of no surprise, since one of the characteristics of 'small' centre behaviour is that aggregate trip frequency distribution follows the normal distribution of the Fourier hypothesis, and therefore shopping behaviour is 'simple harmonic'. The use of such a physical term simply means that there is an aggregate dominance of the once-a-week grocery trip. The ϵ^2 term is not the classical velocity (which is usually represented by a $1/v^2$ term on the right-hand side of the equation). This term may be visualised to a measure of accessibility defined by the equation:

$$\varepsilon = \frac{\beta\mu}{k} \tag{11.14}$$

where μ is a convenience factor, which is calculated from k/f and has an empirical distribution in the Sydney 1988/89 data set (see Figure 11.8).

A solution of Equation 11.13 can be written in the standard form for the initial trading hour condition for the population density (ϕ) as ϕ = 0 at t = 0 and ϕ = 0 at t = T, namely:

$$\phi = \begin{Bmatrix} \sin\dfrac{n\pi t}{T} & \sin\dfrac{n\pi\varepsilon x}{T} \\ \sin\dfrac{n\pi t}{T} & \cos\dfrac{n\pi\varepsilon x}{T} \end{Bmatrix} \tag{11.15}$$

The determination that $(1+\eta)\pi/2T = 2\pi f_o$ for regular shopping is in the same form

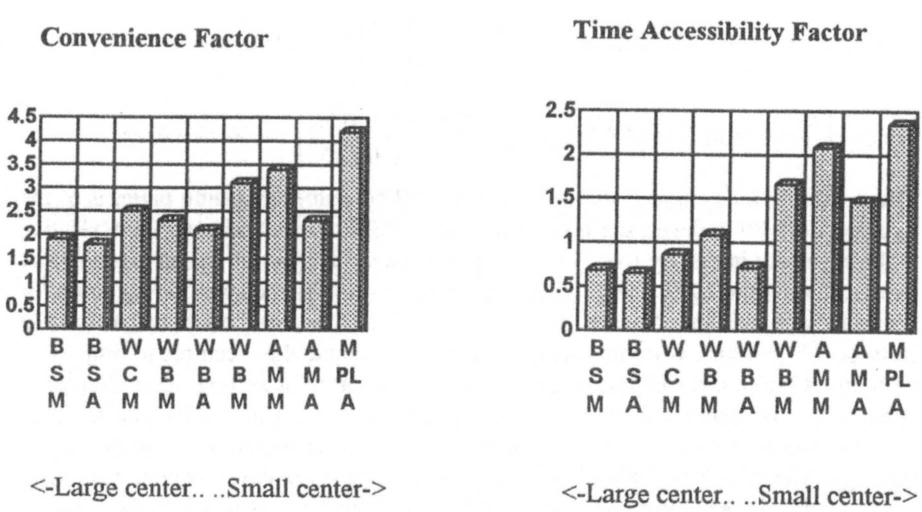

Fig. 11.8. The Distribution of the Convenience and Accessibility Factors in the Sydney Data Set, 1988/89

as Equation 11.6, except that now there is a $1 + \eta$ term introduced instead of just the n term. The reason is that the first destination of a shopping trip is assumed to be a parking position and as such, the $1 + \eta$ summarises the first parking stop (1) plus the subsequent mean number of shops visited (η) per sample. Therefore, the average trading period for 'small' centre shopping behaviour is redefined for the positive trading period as:

$$T = \left(\frac{1 + \eta}{4 f_o} \right) \tag{11.16}$$

There is a shopping contribution of $(\eta/4f_o)$ and parking addition of $(1/4f_o)$ to the trading hour calculation. If $\eta = 0$, there is still a period available for browsing around the shopping centre. However, the question arises in retail policy on whether this component should be included in the estimation of the shopping hours for a centre. The aggregate quantities of n and f_o and the parking contribution, if appropriate, may be converted to per week values by multiplying the sample quantities by the (ITF). It must be stressed that the mean trading hour estimation is made relative to the present assignment of trading hours, since f_o is determined from the average time spent shopping *per trading week.*

For the three periodic samples of planned centres (Market Place Leichhardt, 8/12/88 Afternoon; Ashfield Mall, 23/3/89 Morning and Westfield Burwood 23/3/89 Morning) in Baker (1994a), the relative mean trading hours per week can be calculated from Equation 11.16:

Market Place Leichhardt, 8/12/88 Afternoon $= \left(\frac{1.5 + 5.55}{4 X 0.035} \right) = 50.4 hr/week$

Ashfield Mall, 23/3/89 Morning $= \left(\frac{1.57 + 7.22}{4 X 0.040} \right) = 54.9 hr/week$

Westfield Burwood, 23/3/89 Morning $= \left(\frac{1.29 + 6.58}{4 X 0.035} \right) = 56.2 hr/week$

These calculations suggest that on the pre-1992 consumer shopping patterns of the Sydney 1988/89 data set, the hours for 'small' PSSCs were undersupplied, relative to the regulated hours, up to eight hours per week on average. It was interesting that two years after the 1992 deregulation (in December, 1994), the core trading hours for Market Place Leichhardt were 78.5 hours, Ashfield Mall, 90.5 hours and Westfield Burwood 71.5 hours, with the supermarket anchors within these centres trading 78.5, 91.5 and 87 hours, respectively. Supermarkets appear to be setting the trading hour agenda for the centres well beyond the eight hours extra per week that the pre-deregulation shopping patterns indicated. The previous simulation suggests that, by trading in the vicinity of 80-90 hours per week over seven days-a-week, there is an effective shift in demand to Saturday and Sunday from Monday and Tuesday trading. This redistribution may just be part of the socio-economic adjustment in Australian households, but part of this growth in demand comes from shifts in preference (from lower β values) to travel further to planned centres and large supermarkets from traditional shopping centres in the surrounding market areas. The space-time model suggests that there would be a substantial short term change to the gravity coefficient ($\Delta\beta$) from the ($1/\Delta T^2$) factor under deregulation. These impacts are still being assessed, but a critical determinant in evaluating the validity of this prediction, is whether there has been a significant rise in retail vacancies in traditional shopping centres since 1992.

11.4.2 'Large' Centre Behaviour

Under the previous simulation of Bankstown Square for 50 to 70 hours per week, it is observed that there is some periodicity underlying the distributions, yet with a significant extension of trading hours from 70 to 100 hours per week, day-specific regularity returned to aggregate shopping distributions at the centre. Therefore, the (ISF) has the advantage over the (ITF) in that this underlying periodicity may allow for a calculation of trading hours that previously could not be undertaken.

The starting assumption is that consumers are once again shopping regularly in the form of a complex space-time periodic function:

$$\phi = \exp I(ft - \beta_L x) \tag{11.17}$$

This is constrained by $\beta_L = b f^2$, where $b = c / N$ for the number of retail outlets, N, for $N > 150$.

Differentiating partially space once and time twice from Equation 11.17, and substituting into Equation 11.11 yields the equation:

$$i\frac{\partial\phi}{\partial x} = -\frac{c}{N}\frac{\partial^2\phi}{\partial t^2} \tag{11.18}$$

Baker (1996b) shows that part of the solution of this equation for a positive trading period, T, is the same trading hour definition as that for 'small' centre behaviour, namely:

$$T = \left(\frac{1 + \eta}{4f_o}\right) \tag{11.19}$$

For two 'large centre' samples (Bankstown Square, 23/3/89 Morning and Bankstown Square, 23/3/89 Afternoon) in Baker (1994a), the mean relative trading hours from the sampled shopping patterns can be calculated from Equation 11.19 as:

$$\text{Bankstown Square, 23/3/89 Morning} = \left(\frac{1.39 + 10.87}{4 \, X \, 0.052}\right) = \mathbf{58.9}\, \mathbf{\textit{hr/week}}$$

$$\text{Bankstown Square, 23/3/89 Afternoon} = \left(\frac{1.49 + 13.26}{4 \; X \; 0.058}\right) = 63.6 \, hr/week$$

The aggregate shopping patterns suggest that this large regional centre was under-supplied in trading hours of up to 14 hours per week on average. The reason for the longer estimated shopping hours here could be that there is a greater proportion of comparison shopping undertaken within the higher order centre .

In December 1994, Bankstown Square assumed the core trading period of 72 hours/week under deregulation, although the supermarkets there were trading 84 and 94 hours per week. The decline in the rentals in the adjoining traditional shopping centre at Bankstown from 1991 to 1995 was 13.7% (NSW Valuer General, 1995). For similarly-sized Penrith Plaza, the core trading period at the same time was 94 hours. There, the supermarket anchors were trading 102 and 91.5 hours per week and in surrounding retail rentals there has been a more substantial slump of 21% from 1991 to 1995. The relationship between the trading hours of regional planned centres, their supermarket anchors and the reductions in passing trade and rentals in adjoining traditional regional shopping centres needs to be empirically assessed. Yet, the simulation of the space-time model predicts a rise in the interlocational trip frequencies and corresponding multi-purpose shopping with a substantial extension of trading hours. Whether there is any significant empirical connection awaits further research.

11.5 The Socio-Economic Basis of Trading Hour Estimation

The development of the socio-economic basis for the space-time model has come from Baker (1996a) establishing the empirical relevance between the standardised number of multipurpose shoppers (ω) and high income consumers (I) with the gravity coefficient (β), the interlocation trip frequency (k) and the aggregate mobility (M) from Equations 11.4 and 11.5. This allows for the estimation of a socio-economic basis for trading hours for 'small' and 'large' centre behaviour which is summarised below from Baker (1996b).

11.5.1 'Small' Centre Behaviour

The socio-economic base for the fundamental intracentre shopping frequency, f_o, for 'small' centre behaviour can be defined from Equations 11.4 and 11.5 by,

$$f_o = \frac{\varphi^2}{AI^2M} \tag{11.20}$$

and as such, the trading hour period estimation (without the parking component) can be restated from Equation 11.16 as,

$$T = \sqrt{\frac{A_1^2 n^2 I^4 M^2}{16\varphi^4}} \tag{11.21}$$

The principal contribution for the extension of trading hours comes from the number of high disposable income (HDI) households in the primary trade area of the centre. A substantially higher than average proportion of double income households represents more time-constrained household shopping opportunities over the week and as such, there would be greater demand for longer shopping hours at this centre. Other positive contributions for an extension of hours are the number of shops visited per week and aggregate household mobility.

According to the space-time model, the major force operating against an extension of trading hours, for 'small' centre behaviour, is the number of consumers multipurpose shopping on the planned centre floor space. Multipurpose shopping (MPS) is a time-efficient method of combining shopping opportunities in the one trip. The model suggests that this could be a way that high disposable income households undertook their shopping in the regulated environment of pre-1992 in Sydney. It would be the mechanism to combat limited shopping time budgets and, Baker (1996a) argues, that this time minimisation strategy uses the supermarket visit as the fundamental construct at both 'small' and 'large' planned shopping centres. This may be the reason why consumers in NSW were not active in seeking an extension of trading hours before 1992 in Sydney, because they were using MPS effectively to satisfy their shopping requirements. Therefore, the expectation is that an extension of trading hours by the NSW Government, would attract 'time poor' HDI consumers to shop at night or on the weekends and as such, there would be a corresponding reduction in the number of multipurpose trips from such households. With a major extension of trading hours , it may no longer be necessary for double income households to employ the MPS trip as a mechanism to combat a limited trading hour regime. This hypothesis is currently being tested.

However, (Baker, 1996a) has shown that multipurpose shopping is not solely used by HDI households but also by low-mobility consumers who use MPS exclusively around the grocery trip. Such groups combine shopping activities periodically at their nearest centre and MPS is just an extension of a distance minimisation strategy. Therefore, a higher than average number of lower mobility households with their propensity for MPS would lower the M values and increase the δ values to create an environment for a smaller trading period per week. A community shopping centre's trading hours in a deregulated environment should then reflect the socio-economic base of the surrounding primary trade area.

11.5.2 'Large' Centre Behaviour

The socio-economic base for 'large' centre shopping can be defined by the fundamental intracentre shopping frequency as,

$$f_o = \frac{\varphi^2 N}{AI^2 cM} \tag{11.22}$$

and as such, the trading hour period estimation (without the parking component) can be restated from Equation 11.16 as,

$$T = \sqrt{\frac{A_3^2 \pi^2 c^{3/2} n^2 I \, M^{1/2}}{4\varphi \, N^{3/2}}} \tag{11.23}$$

for c, A_2 and A_3 as unspecified constants. Once again, the positive forces behind the extension of trading hours appears to be the number of shops visited at the centre per week, the standardised number of HDI households and aggregate mobility, whilst those against are MPS and the size of the centre. Contrary to 'small' centre behaviour, the number of shops visited is the major relative contributor to the trading hour definition, whilst the size of the centre, rather than the propensity for MPS, is the principal force dampening an extension of hours. This result is interesting, since it suggests that the 100 hour trading period described previously in the simulation for a large regional centre, is only a short term scenario, since the size of the centre and propensity for MPS would reduce such an oversupply to a sustainable shopping period. However, this may not occur if there were substantial shop visits from comparison shopping, a significantly higher than average disposable income or aggregate household mobility in the primary trade area, that would negate the centre scale effect.

Therefore, the socio-economics of the primary trade area should be the major determinant for the allocation of trading hours for shopping centres. Trading hour policy should not be set by governments on national or large areal assignments or from large supermarkets with market growth strategies. The space-time model suggests that hours should be set more on local area or regional allocations. Yet, under 'store wars' in Sydney, some large regional shopping centres are currently overtrading in retail hours, which may be having a substantial effect on the turnover and capital markets of adjoining traditional shopping centres. Plunging retail rentals and escalating vacancy rates may be a characteristic of this time competition. Over the long term, there may be a new equilibrium established of less trading hours, but this is dependent on the socio-economics of the trade area.

11.6 Concluding Remarks

The analysis of the implications of an extension of trading hours has been central to the recent retail political agenda in Australia. The RASTT and SASTV models allow for some appreciation of the structural change implicit in a substantial extension of the trading period per week. There is going to be a radical shift in 'when' and 'where' some consumer sub-populations are going to undertake their shopping trips. This is very much determined by the socio-economics of households within the primary trade areas of the retail centres. The RASTT model is determined from aggregate trip behaviour to PSSCs and is structured around the mean interlocational trip frequency, whilst the SASTV model, through the intracentre shopping frequency, determines the allotment of shop visitations relative to the time allocated to shopping per week.

The RASTT model suggests, for 'small' centre behaviour, that there would be a substantial drop in the gravity coefficient ($\Delta\beta$) from the ($1/\Delta T^2$) factor from the extension of trading hours. Since large supermarkets are a fundamental construct of 'small' centre behaviour, the implication is that they are going to substantially expand their primary trade areas, at the expense of competing small businesses in traditional central places. Furthermore, this problem is compounded by the current 'store war' policy of duplicating high turnover traditional small business stock within the floorspace of the supermarket.

The SASTV model allows for the calculation of mean trading hours and suggests that the hours for Bankstown Square should have only been increased, on average, by 15 hours per week. The simulation of this regional centre shows that doubling the hours would create a day-specific periodicity in the space-time distributions, which may be the cause of problems for the adjoining traditional shopping centres. The socio-economic solution of the 'large' centre SASTV model implies that the size of the centre would be a factor operating against an oversupply of trading hours yet, in the interim, substantial damage may be done to the retail amenity of adjoining shopping centres, before the long term equilibrium is attained. This is the problem of allowing market forces to set the boundaries, since the large supermarkets are using the readjustment period to eliminate small business competition by oversupplying the market with trading hours. This is, then, one scenario of 'store wars' within the current retail context in NSW, Australia.

The socio-economic basis of trading hour determination has important consequences for retail planning. Firstly, it suggests that trading hour policy should be determined on a local or regional basis rather than a 'blanket' state or national policy and secondly, that the weightings should vary according to the type of space-time behaviour observed at the shopping centre. For 'small' centre behaviour, the standardised number of HDI households and aggregate household mobility are major forces for extending the trading period, whilst the number of multi-purpose shoppers on the centre floor space depreciates the trading period. For 'large' centre behaviour, the same forces are present, but these factors have less influence on the estimation of hours. Rather, it is the size of the centre that negates an oversupply of the trading period. These conclusions should form the basis of future hypothesis testing, but the important result is that there is a socio-economic basis underlying the physical constructs and differences within both space-time models.

Finally, this research raises another interesting issue that needs to be carefully considered. The RASTT and SASTV models suggest that, in retail geography, time constructs are an integral part of spatial analysis. There should be a fundamental link between market area analysis and the trading hours of retail centres. Yet this correspondence, between the spatial and temporal planes, does not mean that there is a direct transformation of a time 'mass' into a traditional central place hierarchy. There are economic opportunities to exploit locational shifts from a manipulation of the temporal plane through out-of-town shopping centres and stand-alone supermarkets. Time convenience in a highly mobile society does not necessarily translate itself into a central place hierarchy. Indeed, there could be a central time hierarchy quite different to traditional locations of shopping places that are open for commercial exploitation by large PSSCs and supermarkets. Therefore, this lack of correspondence may necessitate stronger landuse planning as well as shopping-time planning. This, of course, would further raise the ire of the larger players in the market but, as is the case in the UK, with revised Planning Guideline No.6, the government there has moved to actively discourage any further out-of-town planned centres in a deregulated trading hour environment. The relationship between landuse and trading hour planning should be a further corollary of this research and as such, the RASTT and SASTV models have much to say, in this regard, towards retail policy formulation in Australia and overseas.

References

Australian Financial Review, Poor future for corner store: academic, 17/11/93, 29.

Bacon, R. 1971. An approach to the theory of consumer shopping behaviour, *Urban Studies,* **5**:55-64.

Baker, R.G.V. 1985. A dynamic model of spatial behaviour to a planned suburban shopping centre, *Geographical Analysis*, **17**:331-338.

Baker, R.G.V. 1993. The regionalising of consumer behaviour: Sydney and beyond. *Paper presented at the Australian and New Zealand Regional Science Association Conference*, Armidale, December, 1993.

Baker, R.G.V. 1994a. An assessment of the space-time differential model for aggregate trip behaviour to planned suburban shopping centres, Geographical *Analysis,* **26**:341-362.

Baker, R.G.V. 1994b. The impact of trading hour deregulation on the retail sector and the Australian community, *Urban Policy and Research*, **12**:104-117.

Baker, R.G.V. 1996a. Multi-purpose shopping behaviour at planned suburban shopping centres: a space-time analysis, *Environment and Planning A*, **28**:611-630.

Baker, R.G.V. 1996b. On the development of the SASTV model in estimating retail trading hours and its application to market area analysis, Unpublished paper, available from the author.

Beavon, K.S.O. 1977. *Central Place Theory: A Reinterpretation* Longman, London.

Christaller, W. 1966. *Central Places in Southern Germany*. Translated by C.W. Baskin Prentice Hall, Englewood Cliffs, NJ.

Curry, L. 1962. The geography of service centres within towns: the elements of an operational approach. *Lund Studies in Geography*, Series B, **24**:31-53.

Eaton, B. and Lipsey, R. 1982. An economic theory of central places, *Economic Journal*, **92**:56-72.

Fotheringham, A.S. 1982. Distance decay parameters: a reply, *Annals Association of American Geographers*, **72**:552-554.

Fotheringham, A.S. 1985. Spatial competition and agglomeration in urban modelling, *Environment and Planning A*, **17**:213-230.

Hanson. S. 1980. Spatial diversification and multi-purpose travel, *Geographical Analysis*, **12**:245-257.

Keil, G. and Haberkern, G.P. 1994. A review of the methodology and implication of the work of Dr R.G.V. Baker on retail trading hours, *Report Commissioned by Coles-Myer*, Marketshare, Brisbane.

Lange, S. 1978. the role of consumer behaviour in the distribution of shopping centres, in Funck, R. and Parr, J.B. (Editors), *The Analysis of Regional Structure: Essays in Honour of August Losch, Karlsruhw Papers in Regional Science*, No.2, Pion, London., 62-73.

Losch, A. 1954. *The Economics of Location* Yale University Press, New Haven.

NSW Valuer General Blue Book 1995.

Nystuen, J. 1967. A theory and simulation of intraurban travel, in Garrison, W., and Marble, D. (Editors), *Quantitative Geography Pt.1: Economic and Cultural Topics*, Northwestern University Press, Evanston..

Papageorgiou,Y.Y. and Brummell, A.C. 1975. Crude inferences on spatial behaviour, *Annals Association of American Geographers*, **65**:1-12.

Sydney Morning Herald, Deregulation a threat to small shops, 24/11/93, 28.

The Age, Study urges stop to deregulating hours, 8/11/93, 7 and 27.

Thill, J.-C. and Thomas, I. 1987. Towards conceptualising trip-chaining behavior, *Geographical Analysis*, **19**:1-17.

Warnes, A.M. and Daniels, P.W. 1979. Spatial aspects of an intrametropolitan central place hierarchy, *Progress in Human Geography*, **3**:384-406.

West Australian, Academic warns on trading hours, 6/11/93, 30.

Wrigley, N. 1994. After the store wars. Towards a new era of competition in UK food retailing, *Retailing and Consumer Services*, **1**:1-17.

12 A Multi-Objective Model for Developing Retail Location Strategies in a DSS Environment

Theo A. Arentze, Aloys W.J. Borgers and Harry J.P. Timmermans

Eindhoven University of Technology, Urban Planning Group, P.O. Box 513, 5600 MB Eindhoven, The Netherlands

12.1 Introduction

In developing retail location policies, planners face major uncertainties regarding the questions how the consumer population will develop, where to locate facilities and how consumers and producers will react to new developments. The central aim of spatial decision support system (DSS) is to improve the effectiveness of locational decisions by making data and (analytic) models accessible to decision makers (Densham and Rushton, 1988; Armstrong and Densham, 1990; Densham, 1991). It has been argued on several occasions that recent advances in spatial modelling, information technology and data availability have favoured the cost-benefit ratio of these systems (Bertuglia *et al.* 1994, Birkin *et al.* 1996).

Several applications of DSS or customised GIS in retail planning have been described in the literature. Most systems focus on the feasibility and impact assessment of location plans using spatial interaction or discrete choice models (Roy and Anderson 1988; Borgers and Timmermans 1991; Kohsaka 1993; Birkin *et al.* 1994, 1996; Clarke and Clarke 1995; Arentze *et al.* 1996a). Generating retail plans in the earlier stage of the decision making process has received less attention. Generally, local and regional planners are concerned with developing a vision how the retail system under study should develop to attain a set of planning objectives. Such a vision or plan can serve as a basis for the formulation of location policies.

There exists a vast body of literature on the multiple facility location problem that is potentially relevant for retail planning. The methods developed involve the representation of a location problem in the form of an algebraic location-allocation model and the use of a standard algorithm for solving the model. For an overview of methods and applications see Ghosh and Rushton (1987) and Drezner (1995). Examples of applications in spatial DSS or customised GIS are reported in Armstrong *et al.* (1991), Kohsaka (1993), Densham (1994), Birkin *et al.* (1996) and Arentze *et al.* (1996a). In a former study, we have investigated the use of a knowledge-based system as an alternative approach to supporting retail plan formulation (Arentze *et al.*, 1996b).

The purpose of the present study is to develop a model based on the multiple facility problem literature that should be useful in a DSS for retail planning. To be useful in a DSS the model should meet the following requirements:

a. required data for using the model should be readily available in a standard DSS-database;

b. the model supports an interactive use so that decision makers can participate in the spatial search process;

c. the model can handle the multiple objectives involved in location planning;

d. the model is flexible in the sense that it does not depend on a specific formulation of model components (e.g., the shopping model).

The second and third requirement are particularly important for the acceptance of the system. The importance of an active involvement of decision makers in the spatial search process has been stressed by Malczewski and Ogryczak (1990), Densham (1991) and Armstrong *et al.* (1991). The ability to handle multiple objectives is important for the relevance of model results for real-world problems that are often ill-structured. To develop a model that meets these requirements, this paper is structured as follows. First, section 2 reviews current modelling approaches and clarifies the assumptions and objectives of our approach. The sections that follow describe the specification of the model and a case-study illustrating numerical properties of the model and its application in a DSS. Finally, the last section discusses the major conclusions and possible ways of future research.

12.2 Modelling Approaches

12.2.1 Current Approaches

Since the early sixties, location-allocation models have been widely used to solve multiple facility location problems. These models simultaneously optimise the location of facilities and the allocation of consumers to those facilities. A variety of generic problems has been formulated dependent on assumptions regarding the allocation and location rule. For example, the so-called *p*-median model allocates consumers to the nearest facility and selects the facility locations that minimise aggregate travel.

With regard to the location rule used, it is useful to make a well-known distinction between competitive facility problems and central facility problems. The first problem type is typical for private sector planning. It involves locating a network of facilities in a competitive market environment. In retailing, various location rules have been used for maximising different aspects of network performance. These include market share, profits, accessibility and demand in the catchment area (for a review see Ghosh and McLafferty 1987b; Kohsaka, 1989; Ghosh *et al.*, 1995). The central facility location problem, on the other hand, is relevant for planners who are concerned with finding a balance between consumer's benefit and the cost of supplying facilities. Beaumont (1987) gives an overview of models that can operationalise central place concepts. Leonardi (1981a, b) describes a unifying framework for public facility location models.

In the present context, we focus on the latter type of models. The majority of these models use the nearest-centre rule for allocating users to facilities. These models do not provide an adequate representation of retail location problems. In retail systems, consumers determine which facilities to patronise and, hereby, they make a trade-off between the attractiveness of facilities and travel distance. Two lines of research are potentially relevant for the present study. The first tradition has focused on making basic models more realistic by replacing the nearest-centre rule by a spatial interaction or choice model that can account for these trade-offs. Examples of this approach can be found in Hodgson (1978) and Beaumont (1987). Ghosh and McLafferty (1987a) used an allocation rule that accounts for multipurpose shopping. This rule assumes that consumers minimise a cost function and may, therefore, be criticised for not taking into account facility attractiveness.

Another tradition has focused on extending spatial interaction models to consider facility locations simultaneously. A number of approaches deserve attention. The models developed by Coelho and Wilson (1976) and Leonardi (1978) simultaneously optimise the location and size of facility centres based on maximising consumer surplus and a measure of accessibility (log-accessibility), respectively. Both models are formulated in the form of a mathematical program such that they can be solved using available non-linear programming methods. Furthermore, it is shown that in the optimum consumers are allocated to facility centres in accordance to a production constrained interaction model.

A closely related family of models is concerned with the dynamics of retail systems given assumptions of consumer and producer behaviour. Harris and Wilson (1978) introduced a retail equilibrium model that has invoked a large number of follow-up studies. The model is derived by solving the centre attractiveness term in a production constrained interaction model for which the costs of supplying facilities balances revenues. The costs-revenues balancing condition reflects equilibrium if producers maximise profits and there is enough competition between producers. The model describes a non-linear system in which producers and consumers react on each others actions. Follow-up studies have focused on numerical properties and extensions of this model (e.g., Beaumont *et al.* 1981; Clarke and Wilson 1983; Lombardo and Rabino 1989; Wilson 1990).

The above mentioned interaction-based location models appear to be equivalent. Leonardi (1978) shows that maximisation of consumer surplus, maximisation of log-accessibility and balancing of costs and revenues give the same solutions. Roy and Johansson (1984) interpret these solutions as Nash equilibria in a two-player game involving producers and consumers both pursuing their own interests. They suggest an extension to a three-player game by including the planning authority as an additional player. In their model, the role of the planner consists of formulating macro-location policies for allocating floor space across locations. The behaviour of producers and consumers is represented by entropy maximisation. The aim of the planner is to ensure overall efficiency and equity of consumer' and producer' benefits in the retail system. The efficiency criterion is a compromise between the interests of the three groups involved. The criterion is given by a weighted sum of factors related to transaction profits (retailers), travel costs (consumers) and operating and capital costs of the required public infrastructure (public interests). The equity criterion, on the other hand, is given by multiple objective functions related to equity in profits (retailers) and accessibility (consumers). Roy and Johansson suggest a satisficing approach to solve the multiple objective function problem.

12.2.2. Assumptions of our approach

Our approach combines elements of the approaches reviewed above motivated by the criteria a DSS-model should meet outlined in the introduction section. Following Roy and Johansson, we consider consumers, producers and planners as the three groups involved. We suggest that they have the following roles:

1. consumers decide on the allocation of expenditure across centres and pay for the services and travelling;
2. producers (developers and retailers) take facility investment decisions and pay for the investment and operating costs of supplying facilities;
3. planners decide on the potential location and available land for retail facilities and pay for the required public infrastructure costs (e.g., the road network and parking facilities).

We emphasise that the assumed role of the planner is somewhat more modest than previous approaches have assumed (e.g., Roy and Johansson's model). We assume that planners can create conditions for establishing facilities by designating shopping locations and providing public infrastructure. Whether or not opportunities offered are utilised eventually depends on producers' decisions. Implied by this assumption is that centre size is the outcome of interaction between consumers and producers rather than the decision of planners. To make effective location decisions, planners must, however, anticipate on centre size. The practical consequence of this assumption is that the location model should be based on realistic assumptions on consumer and producer behaviour. We further assume that the planning objective is to balance the interests of all the groups involved. In global terms, these include opportunities of consumers, opportunities of producers and public investments costs and externalities (Van der Heijden, 1986; Oppewal 1995). It follows that planning decision support requires a multiple objective analysis.

12.3 Model Specification

The proposed location model consists of two components. The first component describes the equilibrium state of the retail system with respect to the behaviour of consumers and producers. This component differs from the equilibrium model developed by Harris and Wilson (1978) on minor points related to the specification of the shopping model. However, in contrast to the work of Harris and Wilson, the model is a component of a broader model representing the location problem from the planner's point of view. The second component defines this location problem in terms of a set of objective functions. This section describes these two components in turn and, next, considers procedures for solving the location problem.

12.3.1 The Equilibrium model

The retail system is disaggregated into G branch sectors, such as for example convenience goods, semi-durable goods and durable goods. To reduce the complexity of the problem, location and size decisions are considered one branch sector at a time. In the following, we consider a certain branch sector, g, of the system, but for simplicity of presentation we will leave out the g-subscripts of the model variables.

The study area is subdivided into a set of residential zones $i \in \boldsymbol{I}$ with retail demand $\{ E_i \}$. Retail facilities are supplied at a set of shopping locations $j \in \boldsymbol{J}$ with centre size $\{ W_j \}$. Consumer choice behaviour is represented by an array $\{ p_{ij} \}$ indicating the probability of a consumer in zone i selecting shopping destination j. A discrete choice model or production constrained spatial interaction model (or any other allocation model) is used to predict these probabilities based on centre size W_j, optionally one or more centre attributes X_{sj} and travel time or distance C_{ij}. For example, a discrete choice model of the multinomial logit (MNL) type can be written in general form as:

$$p_{ij} = \begin{cases} \dfrac{\exp(V_{ij})}{\sum_K \exp(V_{ij})}, & \text{if } W_j > 0 \quad j,k \in \boldsymbol{J}_i \\ 0 & , \text{otherwise} \end{cases} \tag{12.1}$$

where:

$$V_{ij} = \alpha W_j + \Sigma_s \beta_s X_{sj} + \theta C_{ij} \tag{12.2}$$

p_{ij} probability of a consumer in i selecting centre j;
$J_i \subseteq J$ a location-specific choice set;
X_{sj} value of the j-th centre on the s-th attribute;
α weight of centre size (referred to as the scale parameter);
β_s weight of the s-th attribute;
θ distance decay parameter;
C_{ij} travel time or distance from the i-th zone to the j-th centre.

The size variable W_j is a fixed component of the model and can be taken as a measure of the choice range of products offered by the centre. The additional attributes, s, are optional and should be chosen in such a way that W_j can be manipulated independent of these attributes. The choice set $\boldsymbol{J}_i$ is location-specific and is usually defined as the set of centres reachable within a given travel time or distance (Borgers and Timmermans, 1991).

An estimate of centre revenues (sales) is then simply given by:

$$D_j = \sum_i p_{ij} E_i \quad \forall j \in \boldsymbol{J} \tag{12.3}$$

where:

D_j estimated sales (revenues) of centre j;
E_i amount of retail expenditure in zone i.

It should be noted that by using a constant for the expenditure term, E_i, this equation does not account for demand elasticity with respect to available supply. In an useful extension of the model, E_i is modelled as a function of the W_j's, possibly in the way described by Ghosh and Mclafferty (1987b). Furthermore, this equation assumes that the study area is closed in the sense that incoming expenditure and outgoing expenditure flows are zero. In reality, however, these flows are nonzero and depend on the available supply in the study area W_j.

Following much work in dynamic retail models, we assume that the costs of supplying facilities is given by $p_j W_j$, where p_j is a location-specific constant denoting the average cost per unit size. These costs include rent price, average wage and overhead for stores at location j. In particular, due to rent price these costs will vary across locations. Transaction profits are a proportion of revenues given by aD_j, whereby $0 < a < 1$. Now, we define $k_j = p_j/a$ to indicate the breakeven revenue per unit size.

Centre size W_j is the outcome of investment decisions of individual producers. However, the possible range of W_j is restricted by a given minimum W^{min} and a given location-specific maximum W_j^{max}. The minimum reflects the minimum scale of a centre required for an economically feasible exploitation. The maximum, on the other hand, is defined by the available land for retail activities designated by the planner. The value $W_j = 0$ indicates the absence of retail facilities at the possible shopping location j.

Following Harris and Wilson's equilibrium model, we assume that the collective behaviour of producers can be described as minimising the imbalance between costs and revenues. That is, investments continue as long as revenues exceeds costs and available land (W_j^{max}) allows further growth. Vice versa, de-investments (closure of stores) continue as long as costs exceeds revenues. If the minimum size W^{min} is reached, then facilities are not feasible at the location under concern and W_j drops to zero. Therefore, in equilibrium the following conditions hold:

$$W_j = \begin{cases} \dfrac{D_j}{k_j} \; \mathit{if} \; k_{jW}^{\min} \leq D_j \leq k_j W_j^{\max} \\ W_j^{\max} \; \mathit{if} \; D_j > k_j W_j^{\max} \quad \forall j \in \boldsymbol{J} \\ 0 \; \mathit{if} \; D_j > k_j W_j^{\min} \end{cases} \tag{12.4}$$

where:

k_j breakeven revenue per size unit;
W^{min} minimum scale required for a viable centre;

W_j^{max} metrage of the area available for retail facilities at location j.

Hence, for any centre: costs and revenues are in balance (12.4a), centre size has reached the maximum (12.4b) or facilities are not feasible (12.4c). If 'normal' profits are included in k_j, these conditions imply that retail activities do not generate more than normal profits unless the maximum size constrains further growth (4b).

The system defined by equations 12.1-12.4 is non-linear with respect to the W_j's. A required change in W_j to bring size in balance with revenues D_j affects the relative attractiveness of j and therefore causes a change not only in D_j but also in all D_k's, $k \neq j$. Equilibrium values of W_j's are therefore interdependent. The non-linear behaviour of a costs-revenues balancing condition in a spatial interaction model of the Huff-type has been extensively studied in the context of Harris and Wilson's model. A review of the findings can be found in Lombardo and Rabino (1989). The finding that is presently relevant indicates that within a reasonable range of the scale parameter of the interaction model ($0 < \alpha \leq 1$) there exists a positive ($W_j > 0$), unique and globally stable equilibrium solution (Rijk and Vorst, 1983). This finding cannot be readily applied to the present model. The minimum size constraint may prohibit nonzero solutions particularly when a relatively high minimum level is used. Furthermore, a system based on a MNL-formulation of the shopping model may have different properties in this respect. At present it is important to establish that a solution for W_j in equations 12.1-12.4 is unique and stable for reasonable values of the scale parameter.

An iterative procedure is commonly used for computing equilibrium points in systems like the one defined by equation 12.1-12.4. Bertuglia and Leonardi (1980) describe heuristic algorithms for solving this type of problems. The computed array of equilibrium values $\{ W_j \}$ is taken to represent the equilibrium state of the retail system with respect to consumer and producer behaviour. Obviously, the validity of this model depends on the validity of the shopping model and the balancing mechanism. The first assumes utility (or entropy) maximising shopping behaviour. The latter assumes profit-maximising behaviour of producers and a competitive retail market structure. With respect to the latter, Roy (1995) suggests that the competitive market assumption may be reasonable for low-order good retail markets (e.g., convenience goods), but that an oligopolistic model is needed to describe market equilibrium in higher-order good sectors. Therefore, a useful extension of the equilibrium model allows different market equilibrium conditions dependent on the retail sector under concern. A more general discussion on market structures and urban modelling can be found in Anas (1990).

12.3.2 Planning Objectives

Given the uniqueness of the solution of equations 12.1-12.4 for W_j, planners cannot exercise any influence on the equilibrium retail structure $\{ W_j \}$, once shopping location J and size $\{ W_j^{max} \}$ decisions have been made. Or, if planners do have the power to decide on $\{ W_j \}$, these decisions are completely restricted by the consumer and producer imposed conditions 1-4. In both perspectives, only the set $\boldsymbol{J}$ and array $\{ W_j^{max} \}$ are subject to planning. We assume that the available land $\{ W_j^{max} \}$ is given by higher level plans. Therefore, the only decision variable left for retail planners is the set of possible shopping locations $\boldsymbol{J}$. We emphasize that for any $\boldsymbol{J}$ the equilibrium model makes sure that demand matches supply for all centres included. Given the minimum size constraint W^{min}, the matching constraint implies that in equilibrium not necessarily all locations in a given $\boldsymbol{J}$ locate a facility (i.e., possibly $W_j = 0$).

We assume as given the set $\boldsymbol{K}$ containing all locations, $k \in \boldsymbol{K}$, considered as optional for inclusion in $\boldsymbol{J}$. In many cases, these will include existing shopping locations and new locations where centres can be built. We suggest that planners try to balance the interests of consumers, producers and the community in general (public). Then, the planning problem can be formulated as finding $\boldsymbol{J} \subseteq \boldsymbol{K}$ optimising or satisficing multiple objective functions $Z_q(\{ W_j \mid j \in \boldsymbol{J} \})$ related to interests of the different groups. The following objective functions may represent the planning problem.

Related to consumers:

•maximisation of aggregate utility for the consumer population:

$$\sum_{i \in I} \left(P_i \ln \sum_{j \in Ji} \exp(V_{ij}) \right) \tag{12.5}$$

•maximisation of minimum utility for the consumer population:

$$\min_{i \in I} \left(\ln \sum_{j \in Ji} \exp(V_{ij}) \right) \tag{12.6}$$

Related to producers:

•maximisation of total amount of viable floor space:

$$\sum_{j \in Ji} W_j \tag{12.7}$$

Related to the community in general:

•minimisation of aggregate shopping travel:

$$\sum_{i \in I} \left(P_i \ln \sum_{j \in Ji} (p_{ij} C_{ij}) \right) \tag{12.8}$$

•minimisation of retail infrastructure costs:

$$\sum_{j \in Ji} z_j \tag{12.9}$$

where:

P_i consumer population size in zone i;
Z_j costs of public infrastructure for retail activities at j.

and other elements are defined as above. Criterion 12.5 is a measure of the utility consumers derive from available facilities. Ben Akiva (1978) developed this measure based on assumptions of multinomial logit choice models. Maximisation of this measure tends to result in a few large centres near high population concentrations. In contrast, the next objective 12.6 tends to result in a larger set of (smaller) centres which are more evenly spread across the area. By maximizing the minimum utility across zones this objective reduces differences in shopping opportunities between locations. Criterion 12.7 favours locations with relatively low capital and operating costs (i.e., k_j) and favourable growth possibilities (large W_j^{max} compared to attracted demand D_j). Criterion 12.8 favours solutions that reduce aggregate travel so that these solutions have positive value for improving environmental quality and reducing traffic congestion. The last criterion 12.9 gives preference to locations where public infrastructure costs are low, for example, because infrastructure can be shared with other activities.

Objectives 12.5-12.9 are constrained only by the requirement that $\boldsymbol{J} \subseteq \boldsymbol{K}$. Generally, it is useful to impose as an additional constraint that some of the elements of $\boldsymbol{J}$ are fixed, i.e. must recur in any solution. Typically, planners may want to fix those existing centres that are not considered candidates for closure (or relocation). If $\boldsymbol{F} \subseteq \boldsymbol{K}$ is the set of fixed locations, then the more restrictive constraint can be written as $\boldsymbol{F} \subseteq \boldsymbol{J} \subseteq \boldsymbol{K}$. Finally, we emphasise that a solution for $\boldsymbol{J}$ indicates *potential* locations for retail activities. The equilibrium model determines which of these locations will be effective and will contribute to the performance of $\boldsymbol{J}$ in terms of the objective functions. In this sense, the equilibrium model constrains the extend to which solutions are effective.

12.3.3 An Iterative Multi-Objective Analysis

Malczewski and Ogryczak (1990, 1995, 1996) formulate a useful interactive approach for handling multiple objectives in central facility location problems. Their approach is based on a reference-point method, which, in contrast to goal-programming techniques, guarantees efficient (Pareto-optimal) solutions and does not require the specification of weights. In this approach, a solution to the problem emerges through interaction between the decision maker and the model (i.e., computer system). In global terms, the decision maker sets an aspiration level and a reservation level for each objective and the model generates efficient (Pareto-optimal) solutions. In an iterative process, the decision maker adjusts aspiration and reservation levels until a satisficing solution is obtained (for details, see the above mentioned studies).

12.3.4 A Possible Solution Algorithm

The reference-point method discussed above can be used to solve the multiple objective problem, provided that there is a solution procedure for solving each of the objective functions 12.5-12.9 separately. The objective functions incorporate the equilibrium model 12.1-12.4, in the sense that to evaluate a solution for $\boldsymbol{J}$, the equilibrium values of the W_j's must be known. A straightforward method of solving each single-objective function involves embedding an algorithm for calculating equilibrium values of W_j in an algorithm for optimising the number and location of centres. Informally, a procedure for solving an objective function, q, can be described in terms of the following steps:

a. set the number of centres N to a reasonable minimum
b. initialise N elements in a solution for $\boldsymbol{J}$ and make sure that $\boldsymbol{F} \subseteq \boldsymbol{J} \subseteq \boldsymbol{K}$
c. define starting values $W_j^{(0)}$ such that:

$$W_j^{(0)} = W_j, \qquad \text{if } j \in F$$
$$W_j^{(0)} = \tfrac{1}{2}\,(W^{min} + W_j^{max}), \qquad \text{else}$$

d. solve the equilibrium model (eq. 12.1-12.4) given starting values $\{ W_j^{(0)} \}$
e. calculate the objective function value of the equilibrium: $Z_q\,(\{ W_j^{(t)} \})$
f. set $\boldsymbol{J}$ to the next solution satisfyng $\boldsymbol{F} \subseteq \boldsymbol{J} \subseteq \boldsymbol{K}$ and repeat from step 3
g. increase N
h. repeat from step 2 until N exceeds a reasonable maximum
i. return the best solution

This algorithm optimises locations for each possible number of centres. To reduce required computer time, the number of centres is varied within a reasonable minimum and maximum defined by the planner. For generating new solutions for a given number of centres (step 6) a complete enumeration or a heuristic method can be used. The heuristic vertex substitution method developed by Teitz and Bart (1968) searches solution spaces in an efficient way and frequently converges to the optimum solution (Rosing *et al.*, 1979). To solve the equilibrium model for each solution (step 4) the iterative method mentioned earlier can be used, provided that suitable starting values $W_j^{(0)}$ are chosen (step 3). Starting values must be larger than zero, because the balancing mechanism operates only on centres with positive size. To facilitate rapid convergence and to ensure suitable starting positions, the size of new centres is set halfway between the minimum and maximum value and the size of existing centres is set to the existing size.

12.4 Illustration

The central component of the location model described in the former section is the equilibrium system based on the costs-revenues balancing mechanism. The behaviour of such systems has been extensively studied in the past, as noted before. However, no attention has been paid to the dynamic behaviour of MNL-based shopping models. This section describes a case-study conducted to investigate some numerical properties of this type of model. Furthermore, the case serves to demonstrate the application of the equilibrium model in a DSS-environment. To this end, we have implemented the equilibrium model in the retail planning DSS that we developed in earlier studies (Arentze *et al.* 1996a, 1996c, 1996d).

The case concerns convenience good facilities in Maastricht - a middle-sized city in the Netherlands with approximately 117.000 inhabitants. Figure 12.1 shows the spatial pattern of facilities in the situation that existed in the early nineties. Typically, convenience good facilities are supplied in a relatively dense set of small neighbourhood centres. In addition, convenience good facilities are present two larger centres with an above-local service area. The most centrally located one is the major shopping centre. As in many other larger cities in the Netherlands, the increase in scale of retail facilities that occurred over the last decade has threatened the viability of the small neighbourhood centres. To test its face validity and use in a DSS, we used the
model to find a new equilibrium state of the retail system when a larger minimum scale (W^{min}) is imposed on the existing centres.

The study area was subdivided into ten residential zones, for which estimates of retail demand, E_i, were available. With respect to the supply side, table 12.1 shows

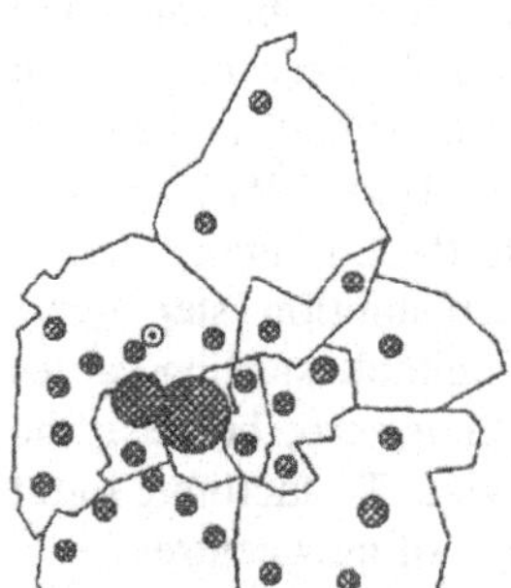

centre size

- 10 000 m^2

- $<3\,500$ m^2

Fig. 12.1. Spatial Pattern of Convenience Good Facilities in the Present State

Table 12.1. Data and Solutions of a MNL-based Equilibrium Model in the Maastricht case. Minimum centre size is 750 sq.m. and the scale parameter (α) is 0.00022 (solution 1), 0.00027 (solution 2) and 0.00032 (solution 3).

Present State		Solution (1)		Solution (2)		Solution (3)	
W_j (sq.m.)	D_j/k_jW_j	Δwj (sq.m.)	D_j/k_jW_j	ΔW_j (sq.m.)	D_j/k_jW_j	ΔW_j (sq.m.)	D_j/k_jW_j
12398	0.91	-1076	1.00	2602	1.09	2602	1.13
1962	0.83	-372	1.01	-412	1.01	-994	1.03
3136	0.63	-1093	1.00	-1258	1.01	-2051	1.01
229	1.25	-229	-	-229	-	-229	-
1905	0.51	-498	1.00	-909	1.00	-1905	-
351	1.08	-351	-	-351	-	-351	-
575	0.55	-575	-	-575	-	-575	-
1685	0.50	-540	1.00	-888	1.01	-1685	-
8710	1.08	-5889	1.00	-6232	1.01	6290	1.22
3755	0.56	-908	1.00	-1348	1.00	-3755	-
1061	1.33	1581	1.00	1308	1.01	-1061	-
932	1.44	1461	1.01	1192	1.00	-932	-
1084	1.27	1316	1.01	1092	1.01	-1084	-
1961	0.62	-332	1.01	-447	1.02	-1961	-
716	1.63	1447	1.01	1241	1.01	-716	-
1671	0.57	-424	1.00	-642	1.01	-1671	-
638	0.82	112	1.00	-638	-	-638	-
931	1.34	581	0.99	355	1.01	-113	0.99
1088	0.86	-59	0.97	-215	0.98	-1088	-
4490	0.80	-1259	0.99	-1454	1.01	-2855	1.00
3051	1.27	854	1.00	677	1.00	-1819	1.01
1500	1.19	455	0.99	258	0.99	-448	0.99
725	1.56	653	1.00	635	1.01	176	1.01
532	1.31	428	1.02	-532	-	-532	-
306	1.75	-306	-	-306	-	-306	-
2142	0.55	-938	0.99	-998	1.01	-1392	1.01
742	1.94	922	0.99	602	0.98	-742	-
1987	1.57	1887	0.99	1400	1.00	-1987	-
5074	1.24	1118	1.00	1590	1.00	9926	1.11

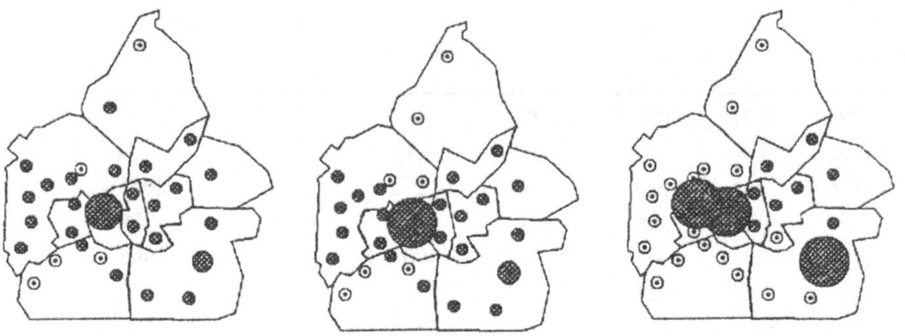

solution (1) *small scale effect* — solution (2) *medium scale effect* — solution (3) *large scale effect*

Fig.12.2. Graphical Display of Model Solutions 1-3.

the centre size, W_j, data that were used. The breakeven revenue per size unit, k_j, was set to a relatively low value (6 500 Dutch guilders) for the neighbourhood centres and a relatively high value (7 500 Dutch guilders) for the two larger above-local centres, to reflect differences in rent price. Travel time data, C_{ij}, were based on travel times reported by respondents in a consumer survey. Consumer choice behaviour was described by a MNL-model that included as explanatory variables floor space size in the convenience good sector (W_j), floor space size in the comparison good sector (say X_{1j}) and travel time (C_{ij}). Model parameters were estimated based on a survey of consumers residing in the area.

Equilibrium points were calculated for various values of the scale parameter, α, of the shopping model, using an iterative method. The minimum size W^{min} was (arbitrarily) set to 750 square meter. All maximum values, W_j^{max}, were set to the same arbitrarily large value (15 000 sq.m.), so that in effect the growth of all centres was unconstrained. In all cases convergence was reached within 8-10 iterations. The results are shown numerically in table 12.1 and graphically in figure 12.2. The solutions 1-3 are ordered based on increasing values of the scale parameter. Not surprisingly, this series is characterised by an increasing degree of concentration of retail facilities. Solution three represents the upper extreme case. In this solution three centres have reached the maximum size (15 000 square meter) at the expense of the smaller centres. In general, the solutions 1-3 seem to indicate that the floor space distribution pattern is relatively insensitive for variation in the scale parameter up to a certain point.

Beyond that point (solution 3), centres with a large starting size value tend to grow strongly at the expense of smaller competing centres, giving rise to highly unequal distributions of floor space across locations. More systematic computer experiments are needed to verify these statements and investigate other ones.

Given appropriate values of model parameters, the equilibrium points can provide useful information for plan decision making. A model solution shows the equilibrium state of the retail system under given starting conditions. Users can vary these starting conditions to evaluate the implications for the equilibrium. In this way, the impact of various assumptions regarding population and economic (per capita expenditure) developments can be investigated. Of particular interest are those starting conditions that can be controlled by planners in real-world situations. For example, the quality of the retail environment influences the competitive strength of centres and may be used by planners as an instrument to support weak centres. If the model is sensitive to such qualitative aspects of centres, then planners can use the model interactively to determine whether such supportive actions are likely to be effective for a particular centre.

12.5 Conclusions and Discussion

This paper described a multi-objective optimisation model for use in a DSS for retail planning. As a distinguishing characteristic of our approach, a retail equilibrium model is embedded in objective functions for evaluating retail plans. A retail plan specifies the possible locations and available space for developing retail activities. The equilibrium model then generates the equilibrium state of the retail system with respect to consumer and producer behaviour. Thus, the equilibrium model makes sure that the solutions generated are consistent not only with the goals of planners (as in most location-allocation models), but also with the behaviour of consumers (as in interaction-based location models) and the behaviour of producers (as in dynamic retail models). In that sense, the model is consistent with micro-economic theory. A case-study demonstrated some of the numerical properties of a MNL-based equilibrium model and the use of such models in a DSS.

This approach seems to have advantages particularly in an interactive DSS-environment. There are, however, also some potential problems that ask for further research. First, the computing time needed to solve the optimisation model may prohibit a highly interactive use. Efficient solution algorithms need to be developed. Second, it should be noted that the predictive validity of the shopping model is critical, as prediction errors may propagate in unknown ways across iterations. Therefore, in applications it is important to assess the sensitivity of the model for reasonable variations in parameter values. Furthermore, the shopping model should be extended to relax unrealistic assumptions, such as demand inelasticity and single-purpose shopping trips. Third, the balancing mechanism used in the present model may produce unrealistic equilibrium points in the case of market structures characterised by oligopolistic or monopolistic competition. This property is a potential limitation only if the equilibrium model is used for prediction. Alternatively, model outcomes can be taken to indicate the state of the retail system that planners can attain through negotiation with producers. In the latter case, the balancing condition is appropriate as it complies with conditions for optimal service provision (in terms of consumer satisfaction).

Besides the latter normative use, the retail equilibrium model is potentially useful for analysing impacts of decision scenario's for longer time horizons than would be possible based on a shopping model alone. A promising line of future research focuses on a model type in which equilibrium conditions are replaced by empirically estimated models of retailer's reactive behaviour. A framework for such models is developed in Van der Heijden (1986) and elaborated in Oppewal (1995). Then, successive steps in model solutions can be interpreted as describing retail development trajectories in time. Within limited time horizons the trajectories may give reliable information that planners can use to formulate actions in a time frame to guide developments.

Acknowledgements
This research is supported by the Technology Foundation (STW).

References

Anas, A. 1990. General Economic Principles for Building Comprehensive Urban Models, in C.S. Bertuglia, G. Leonardi and A.G. Wilson (eds.) *Urban Dynamics: Designing an Integrated Model*, Routledge, London, 7-44.

Arentze, T.A., A.W.J. Borgers, and H.J.P. Timmermans 1996a. A Generic Spatial Decision Support System for Planning Retail Facilities, in M. Craglia and H. Couclelis (eds.) *Geographic Information Research: Bridging the Atlantic*, Forthcoming.

Arentze, T.A., A.W.J. Borgers, and H.J.P. Timmermans 1996b. A Knowledge-Based Model for Developing Location Strategies in a DSS for Retail Planning, Proceedings of the 3rd Design & Decision Support Systems in Architecture & Urban Planning Conference, Spa, Belgium, 17-38.

Arentze, T.A., A.W.J. Borgers, and H.J.P. Timmermans 1996c. Design of a View-based DSS for Location Planning, *International Journal of Geographical Information Systems*, **10**:219-236.

Arentze, T.A., A.W.J. Borgers, and H.J.P. Timmermans 1996d. The Integration of Expert Knowledge in Decision Support Systems for Facility Location Planning, *Computers, Environment and Urban Systems*, **19**:227-247.

Armstrong, M.P. and P.J. Densham 1990. Database Organization Strategies for Spatial Decision Support Systems, *International Journal of Geographical Information Systems* **4**:3-20.

Armstrong, M.P., G. Rushton, R. Hoeney, B.T. Dalziel, P. Lolonis, S. De, P.J. Densham 1991. Decision Support for Regionalization: A Spatial Decision Support System for Regionalizing Service Delivery Systems, *Computers, Environment and Urban Systems*, **15**:37-53.

Beaumont, J.R. 1987. Location-Allocation Models and Central Place Theory, in Ghosh A., Rushton G. (eds.) *Spatial Analysis and Location-Allocation Models*, Von Nostrand Reinhold Company, New York, 21-53.

Beaumont, J.R. 1981. The Dynamics of Urban Retail Structure: Some Exploratory Results Using Difference Equations and Bifurcation Theory, *Environment and Planning A*, **13**:1473-1483.

Ben-Akiva, M. and S.R. Lerman 1978. Disaggregate Travel and Mobility-Choice Models and measures of Accessibility, in A. Hensher and P.R. Stopher (eds.) *Behavioural Travel Modelling*, Croom Helm, London, 654-679.

Bertuglia, C.S., G.P.Clarke and A.G. Wilson 1994. Models and Performance Indicators in Urban Planning: The Changing Policy Context, in C.S. Bertuglia, G.P.Clarke and A.G. Wilson (eds.) *Modelling the City: Performance, Policy and Planning*, Routledge, London, 20-36.

Bertuglia, C.S. and G. Leonardi 1980. Heuristic Algorithms for the Normative Location of Retail Activities Systems, *Papers of the Regional Science Foundation*, **44**:149-159.

Birkin, M., G.P. Clarke, M. Clarke and A.G. Wilson 1994. Applications of Performance Indicators in Urban Modelling: Subsystems Framework, in C.S. Bertuglia, G.P. Clarke and A.G. Wilson eds.. *Modelling the City: Performance, Policy and Planning*, Routledge, London, 121-150.

Birkin, M., G. Clarke, M. Clarke, A. Wilson 1996. *Intelligent GIS: Location Decisions and Strategic Planning*, Geoinformation International, Cambridge.

Borgers, A.W.J. and H.J.P. Timmermans 1991. A Decision Support and Expert System for Retail Planning, *Computers, Environment and Urban Systems*, **15**:179-188.

Clarke C. and M. Clarke 1995. The Development and Benefits of Customized Spatial Decision Support Systems, in P. Longley and G. Clarke (eds.) *GIS for Business and Service Planning*, Geoinformation International, Cambridge, 227-254.

Clarke, M. and A.G. Wilson 1983. The Dynamics of Urban Spatial Structure: Progress and Problems, *Journal of Regional Science*, **23**:1-18.

Coelho, J.D., and A.G. Wilson 1976. The Optimum Location and Size of Shopping Centres, *Regional studies*, **10**:413-421.

Densham, P.J. 1991. Spatial Decision Support Systems, in D.J. Maguire, M.F.Goodchild and D.W. Rhind (eds.), *Geographical Information Systems: Principles*, John Wiley & Sons, New York, 403-412.

Densham, P.J., and G. Rushton 1988. Decision Support Systems for Locational Planning, in R. Golledge and H. Timmermans (eds.) *Behavioural Modelling in Geography and Planning*, Croom-helm, London, 65-90.

Densham, P.J. 1994. Integrating GIS and spatial modelling: Visual interactive modelling and location selection, *Geographical Systems*, **1**:203 - 221.

Drezner, Z. (ed.) 1995. *Facility Location: A Survey of Applications and Methods*, Springer, New York.

Ghosh, A. and S.L. McLafferty 1987a. Optimal Location and Allocation with Multipurpose shopping, in Ghosh A., Rushton G. (eds.) *Spatial Analysis and Location-Allocation Models*, Von Nostrand Reinhold Company, New York, 55-75.

Ghosh, A. and S.L. McLafferty 1987b. *Location Strategies for Retail and Service Firms*, Lexington Books, Lexington.

Ghosh, A. and G. Rushton (eds.) 1987. *Spatial Analysis and Location-Allocation Models*, Von Nostrand Reinhold Company, New York.

Ghosh, A., S. McLafferty and C.S. Craig 1995. Multifacility Retail Networks, in Z. Drezner (ed.) *Facility Location: A Survey of Applications and Methods*, Springer, New York, 301-330.

Harris, B., and A.G. Wilson 1978. Equilibrium Values and Dynamics of Attractiveness Terms in Production-Constrained Spatial-Interaction Models, *Environment and Planning A*, **10**:371-388.

Heijden van der R.E.C.M. 1986. *A Decision Support System for the Planning of Retail Facilities: Theory, Methodology and Application*, Ph.D-Dissertation, Eindhoven University of Technology, Eindhoven.

Hodgson, M.J. 1978. Toward More Realistic Allocation in Location-Allocation Models: An Interaction Approach, *Environment and planning A*, **10**:1273-1285.

Khosaka, H. 1989. A Spatial Search-Location Model of Retail Centers, *Geographical Analysis*, **21**:338-349.

Khosaka, H. 1993. A Monitoring and Locational Decision Support System for Retail Activity, *Environment and planning A*, **25**:197-211.

Leonardi, G. 1978. Optimum Facility Location by Accessibility Maximizing, *Environment and Planning A*, **10**:1287-1305.

Leonardi, G. 1981a. A Unifying Framework for Public Facility Location Problems-Part 1: A

Critical Overview and Some Unsolved Problems, *Environment and Planning A*, **13**:1001-1028.

Leonardi, G. 1981b. A Unifying Framework for Public Facility Location Problems-Part 2: Some New Models and Extensions, *Environment and Planning A*, **13**:1085-1108.

Lombardo, S.T., and G.A. Rabino 1989. Urban Structures, Dynamic Modelling and Clustering, in J. Hauer, H. Timmermans and N. Wrigley (eds.) *Urban Dynamics and Spatial Choice Behaviour*, Kluwer Academic Publishers, Dordrecht, 203-217.

Malczewski, J. and Ogryczak 1990. An Interactive Approach to the Central Facility Location Problem: Locating Pediatric Hospitals in Warsaw, *Geographical Analysis*, **22**:244-258.

Malczewski, J. and W. Ogryczak 1995. The Multiple Criteria Location Problem: 1. A Generalized Network Model and the Set of Efficient Solutions, *Environment and Planning A*, **28**:1931-1960.

Malczewski, J. and W. Ogryczak 1996. The Multiple Criteria Location Problem: 2. Preference-Based Techniques and Interactive Decision Support, *Environment and Planning A*, **28**:69-98.

Oppewal, H. 1995. *Conjoint Experiments and Retail Planning*, Ph.D-Dissertation, Eindhoven University of Technology, Eindhoven.

Rijk, F.J.A. and A.C.F. Vorst 1983. On the Uniqueness and Existence of Equilibrium Points in an Urban Retail Model, *Environment and Planning A*, **15**:475-482.

Rosing, K.E., E.L. Hillsman, H. Rosing-Vogelaar 1979. The Robustness of Two Common Heuristics for the p-Median Problem, *Environment and Planning A*, **11**:373-380.

Roy, J.R. 1995. The Use of Spatial Interaction Theory in Spatial Economic Modeling, in R. Wyatt and H. Hossain (eds.) *Proceedings off the 4th International Conference on Computers in Urban Planning and Urban Management*, Melbourne, Australia, 139-150.

Roy, J.R., and M. Anderson 1988. Assessing Impacts of Retail Development and Redevelopment, in P.W. Newton, M.A.P. Taylor and R. Sharpe (eds.) *Desktop Planning: Microcomputer Applications for Infrastructure and Services Planning and Management*, Hargreen, Melbourne, 172-179.

Roy, J.R., and B. Johansson 1984. On Planning and Forecasting the Location of Retail and Service Activity, *Regional Science and Urban Economics*, **14**:433-452.

Teitz, M.B. and P. Bart 1968. Heuristic Methods for Estimating the Generalised Vertex Median of a Weighted Graph, *Operations Research*, **16**:955-961.

Wilson, A.G. 1990. Services 1: A Spatial Interaction Approach, in C.S. Bertuglia, G. Leonardi and A.G. Wilson (eds.) *Urban Dynamics: Designing an Integrated Model*, Routledge, London, 251-273.

13 Recent Developments in the Modelling of Strategy Reformulation

James O. Huff[1] and Anne S. Huff[2]

[1]Department of Geography, University of Colorado, Boulder, CO 80309, USA
[2]College of Business Administration, University of Colorado, Boulder, CO 80309, USA; and Cranfield School of Management, Cranfield, UK

13.1 Introduction

In economic geography there is a persistent tendency to assume that decision making processes internal to the firm are outside the realm of inquiry either because they are inherently idiosyncratic or because they are irrelevant when gauged against structural conditions shaping the overall direction of change within the industry. An unfortunate consequence of this limiting mind set is the pervasive implicit or explicit assumption that firms within the same industry make locational decisions based on a strategy that is essentially the same for all firms in the industry and that this strategy changes very slowly (and uniformly across the firms in the industry) relative to the speed and frequency of locational decisions taken by each firm in the industry. This is certainly an attractive simplifying assumption; unfortunately, there are very good theoretical and empirical reasons to challenge this assumption which is precisely what this paper aims to do.

The analysis in this paper emphasizes the advantages of framing questions of why and when firms will change strategy in terms of the interdependence between organization stress and inertia. The central concerns of the strategy field have always included the questions of when and why firms change strategic direction. The answer often involves the suggestion that one or more stressors -- in the form of new technology, new or changed competition, altered regulation, new leadership within the firm, and so on -- triggers strategic realignment efforts (e.g. Learned, Christensen, Andrews, and Guth, 1965; Andrews, 1971; Porter, 1980; Rumelt, 1984). Recently, more attention has been given to the inertial factors that balance this basic evolutionary process (Hannon and Freeman, 1984; Tushman and Romanelli, 1985; Schwenk and Tang, 1989; Gersick, 1991). An important contribution of this stream of work is to remind theorists that forces for and against strategic change are simultaneously at work within the organization. Recognizing inertia, in the form of prior investments, commitments to customers, and organization routines, makes answering the questions of when and why firms change strategies more problematic, but helps account for the fact that even under very stressful situations firms vary in the speed and scope of their realignment efforts (Barr, 1991).

Particular attention must be paid to the *history* of accumulating stress and inertia. Firms that have had the same general strategy for some time are likely to view the prospect of strategic change very differently than firms that have recently changed strategy. Firms "see" the same set of potentially destabilizing events through different thus triggering qualitatively different strategic responses. The firm draws upon past experience as well as the borrowed knowledge of the successes and failures of its competitors to craft a strategy that capitalizes on the firm's resources, embodied in its current strategic position, while responding to emerging competitive opportunities (Huff, 1982).

When thinking about strategic change, it also is important to distinguish between *first order change* involving incremental adjustments to current strategy that occur as part of day-to-day operations of the firm and *second order change* associated with major shifts in strategy that precipitate fundamental changes within the structure and function of the organization.

A key premise in this paper is that first and second order change arise out of fundamentally different decision making processes. For most organizations, most of the time, strategic renewal involving second order change is *not* a topic of sustained consideration consuming significant amounts of the organization's resources. Instead, incremental adjustments to strategy are accomplished within an ongoing, problem solving,"business as usual", mode of operation. The shift to an active mode of strategic evaluation and the consideration of strategic alternatives signals a shift into a second order change decisioning process triggered by the accumulation of unresolved stress surrounding the current strategy and supplanting arguments in favor of continuing with business as usual.

The paper begins with an examination of the strategy reformulation process from an explicitly behavioural perspective. The observed strategic changes made by the sample of pharmaceuticals firms are then used as the empirical basis for evaluating several key premises in the behavioural explanation of the strategic change process. The first is that a firm's competitive position and subsequent changes in strategy are mappable within the strategic space defined in terms of a limited set of strategic dimensions critical to firms in the pharmaceuticals industry. In support of the earlier theoretical arguments, second order changes in strategy are shown to be associated with high stress / low inertia conditions.

The paper concludes with a behavioural model of strategic change that highlights the interdependencies between stress and inertia. The firm's ability to make frame sustaining, homeostatic adjustments to strategy which effectively reduce stress depends upon the level of commitment / inertia surrounding the current strategy. On the inertia side, the model suggests that increasing stress can erode commitment / inertia; and when stress exceeds inertia, a crisis of confidence concerning the viability of the current strategy can develop quite rapidly. A Markov model of the transition between high and low stress and inertia states provides a strong empirical basis for making a distinction between "movers" and "stayers" within the set of pharmaceutical firms in our study.

13.2 Stress, Inertia and Second Order Change Among Firms in the Pharmaceuticals Industry

The discussion of strategic positioning and strategic change begins with an assumption that firms within the same industry are operating within the same "strategic space". Their strategies are similar in that the strategies are formulated in terms of the same set of key strategic dimensions (Fiegenbaum, 1987; Reger, 1988). The range of choices made by individual firms along these critical dimensions are assumed to define the industry. This initial assertion is based on the idea that external circumstances require surviving firms to attend to key issues; it also implies that the content of a newly constructed strategy necessarily builds upon the resources and the knowledge embodied in previous strategies identified by both the firm in question and others responding to similar conditions.

13.2.1 Defining Second Order Strategic Change

For the pharmaceutical industry, positioning within the industry and change in strategy are defined in terms of two strategic variables identified in past research (Cool, 1985; Barr, 1991) as critical to firms in the pharmaceutical industry: *research and development* effort as measured by annual expenditures on R&D as a percentage of sales (RADS) and attention to *advertising* as measured by annual expenditures on advertising as a percentage of sales (ADS). The data on these variables as well as other independent variables in subsequent analysis are from COMPUSTAT/PC. In schematic terms, the structure of the industry is described by the competitive positions of a set of firms with similar strategies (see Figure 13.1).

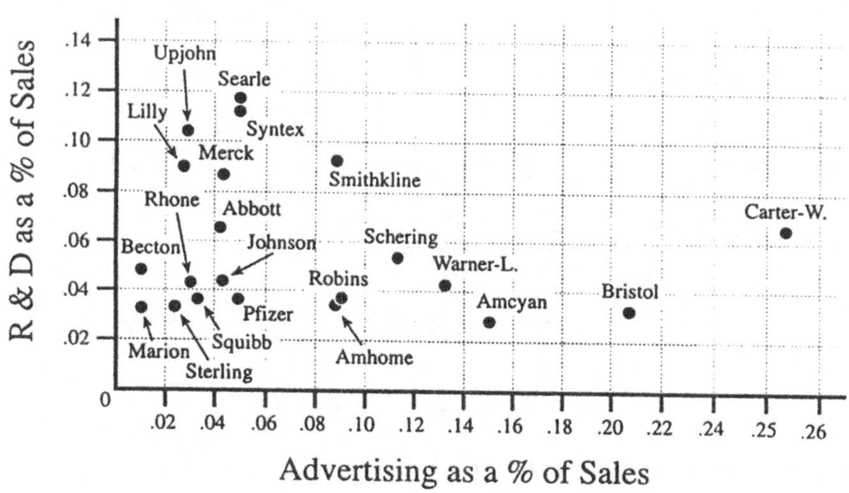

Fig. 13.1. Strategic Positions 1970

The second half of the operational definition of strategic change involves the selection of unit time interval over which change is defined and the specification of a minimum amount of change along each strategic dimension that is assumed to constitute a major or second order change in strategy. Four year time intervals serve as the units of observation in the analysis. Relatively long time intervals were chosen in an effort to capture longer term changes in strategy and to minimize the effects of short term fluctuations in strategic variables. Furthermore, the time intervals closely parallel the stable strategic time periods identified by Cool and Dierickx (1993) in their analysis of the pharmaceutical industry. All changes in strategy are defined in terms of observed differences in R&D expenditures as a percentage of sales and Advertising expenditures as a percentage of sales at the beginning and end of each four year time period.

Given a graphic means of "positioning" a firm within the constellation of strategies defining the industry, it is a simple matter to represent the magnitude and direction of strategic change for individual firms as well as industry evolution arising from the associated reconfiguration of firms in strategic space . The map of strategic change shown in Figure 2, for example, is based on the observed behaviours of a sample of twenty-nine firms in the pharmaceutical industry, 1970-1974.

Subsequent empirical analysis is concerned specifically with the prediction of major (second order) changes in strategy as represented by the longer arrows on the maps in Figure 2; changes that tend to reach more deeply into the character and structure of the organization as distinguished from first order, homeostatic, changes designed to maintain the essential character of the current strategy. (Argyris, 1977; Meyer Brooks and Goes, 1989; Watzlawick et.al., 1974).

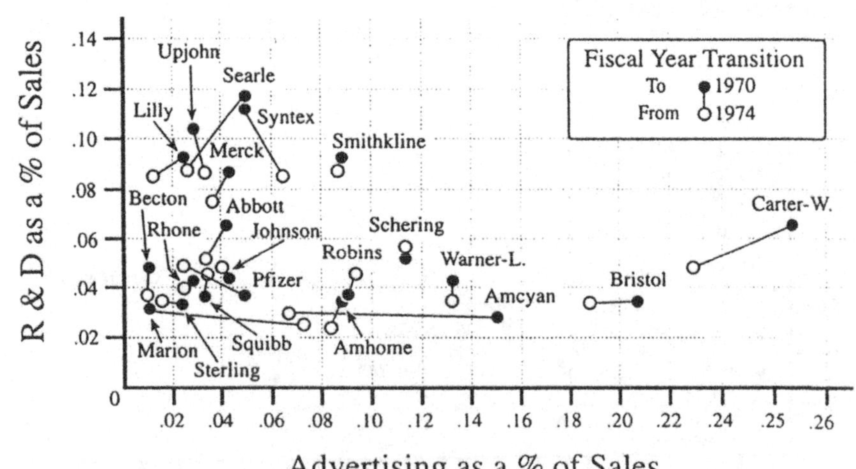

Fig. 13.2. Strategic Position Transition 1970-1974

The empirical problem of defining the threshold between incremental changes in strategy and second order changes in strategy is difficult because there are no clearly discernible breakpoints in the distribution of observed changes in R&D expenditures as a percentage of sales. In the present analysis, the threshold values are defined as 15% of the average ranges for RADS and ADS. We have experimented with other threshold values in this general range and the results are not significantly different from those presented below. Given this definition, there are fifty-two second order changes in strategy and eighty instances in which firms retain the same strategy over the four year period in question. Approximately 40% of the firms move during a given four year period.

13.2.2 Evidence of Strategic Change under High Stress, Low Inertia Conditions

Before proceeding to more sophisticated analysis of the strategic change process, it is instructive to demonstrate that pharmaceutical firms experiencing high levels of stress and low levels of inertia are indeed the firms that tend to make major changes in strategy as reflected in significant reallocation of resources for R&D and Advertising.

As noted earlier, organization stress arises from a mismatch between the changing demands placed upon the organization and the capacity of the current strategy to respond to those changes (Huff, Huff & Thomas, 1992; Huff & Huff, 1995). In the present analysis, organization stress and the associated pressure for second order change in strategy is measured in terms of a firm's performance relative to industry norms. Performance levels near the industry average are assumed to be associated with low levels of organization stress and correspondingly low propensities to make second order changes in strategy. Consequently, firms experiencing moderate performance levels are expected to engage in incremental modifications to the firm's current strategy, strategy sustaining mechanisms to keep organization stress within manageable bounds. On the other hand, firms following strategies associated with *either* high performance *or* low performance relative to industry norms are assumed to be experiencing high organization stress and are expected to exhibit relatively high propensities to make second order changes in strategy.

The hypothesized "U-shaped" relationship between stress and performance differs from standard practice in the strategy field wherein stress is generally assumed to be associated only with poor performance. However high performance conditions are likely to be stressful because they tend to raise expectations beyond the firm's capacity to consistently perform at that level, thereby pushing the firm into a search for yet higher return strategies. A further source of stress is that higher return strategies are likely to be higher risk strategies which are inherently unstable.

The argument is illustrated graphically in Figure 13.3. The estimated performance surface summarizes the observed performance data for the pharmaceutical firms in the sample. High stress firms with high propensities to make second order changes in strategy are expected to be positioned near the lowest and the highest points on the performance surface.

Table 13.1 provides direct empirical support for our contention that both high and low performance is likely to be a high stress condition that can precipitate a second order

Table 13.1. Relationship Between Stress and Strategic Change

	High Stress Firms		Low Stress Firms
	High Performance	Low Performance	Moderate Performance
Change Strategy	20	21	11
Retain Strategy	11	10	59

change in strategy. The sample was split into quartiles based on each firm's reported rate of sales growth, SALG, (during each of the five time intervals). The "High Stress" group consists of firms in the top quartile, "high performance" firms, and firms in the bottom quartile,"low performance" firms. The "Low Stress" group consisted of the firms in the second and third quartiles.

As expected, High Stress firms (high performance and low performance firms combined) were much more likely to change strategy than were Low Stress firms (X^2 = 35.10, 1 df); and there is no significant difference in the propensity to change strategy between high and low performance firms (X^2 = .07, 1 df).

A stress/inertia perspective on strategic change recognizes that organizations are very resistant to change and that significant changes in strategy are not without cost and are

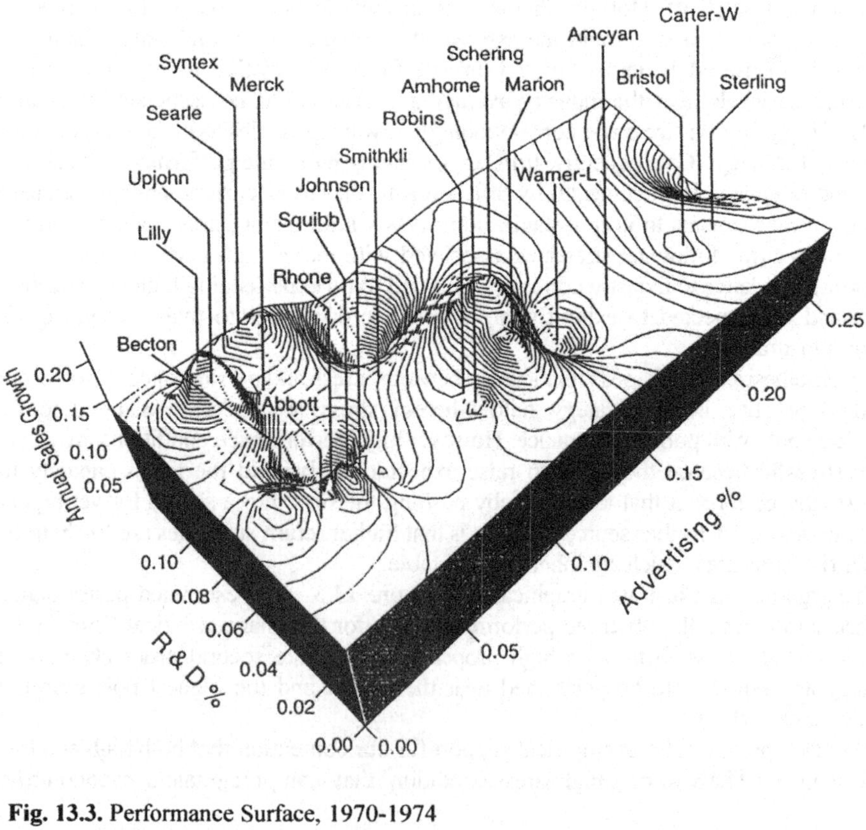

Fig. 13.3. Performance Surface, 1970-1974

not undertaken lightly even under high stress conditions. Furthermore, organization inertia associated with a particular strategy tends to increase over time as the strategy becomes institutionalized.

The distinction between firms experiencing high inertia and firms experiencing low inertia is defined in terms of the time since the firm's last major change in strategy. If the firm changed strategy during the previous four year time period, the firm is assigned to a low inertia state. If no major changes occurred last time period, the firm is assigned to a high inertia state.

Table 13.2. Relationship Between Inertia and Strategic Change

	Low Inertia	High Inertia
Change Strategy	26	26
Retain Strategy	15	65

The importance of strategic inertia in the strategic change decision is illustrated in Table 13.2. The data from the sample of pharmaceutical firms indicate a clear relationship between the decision to retain the current strategy and the retention of the strategy in the previous time interval (X^2 = 14.37, 1 df).

Before moving into a discussion of more sophisticated strategic change models, it is instructive to summarize the joint effects of stress and inertia on the strategic change decision as shown in Table 13.3. When firms are partitioned on the basis of both stress and inertia, it is evident that the probability of changing strategy is quite sensitive to the combined effects of stress and inertia (X^2 = 39.8, 3 df).

The overall message from the foregoing analysis is that firms experiencing high levels of stress (as indicated by performance levels significantly above as well as significantly below industry norms) are more likely to make major changes in strategy than are low stress firms. Our findings also support the cumulative inertia hypothesis outlined in the previous chapter. When taken together, organization stress and organization inertia clearly can be used to differentiate between perspective movers and stayers within the pharmaceutical industry as shown in Table 13.3.

13.3 A Behavioural Model of Strategic Change

This section of the paper outlines a behavioural model of the strategy reformulation process which is consistent with earlier empirical findings. The model is framed in explicitly behavioural terms while continuing to build on the main premise of this paper which is that the key to understanding the strategic change process is to be found in an analysis of the interaction between organization stress and organization inertia. In this model, the stress and inertia experienced within an organization at a given point in time are endogenous to the strategy reformulation process and are functionally interdependent. Although every firm's strategic response is informed by a unique set of prior experiences and capabilities, it is argued that the process involves a recognizable and predictable set of transitions between organization states associated with different aspects of the reformulation process (Huff, Huff, and Thomas, 1994; Huff & Huff, 1995). The thinking behind this model can be traced directly back to earlier work on stress-inertia models of the decision to move (Huff & Clark, 1978). The ideas in this section also draw upon significant theoretical and modelling advances pertaining to the dynamics of strategic change as illustrated in the work of Hannon and Freeman (1984) and Nelson and Winter (1982) on the theoretical side and Zajac and Kraatz (1993), Gersick (1991), Ambergey and Miner (1992), and Barnett and Carroll (1995) on the modelling side.

Table 13.3. The Joint Effects of Stress and Inertia on the Strategic Change Decision

	High Stress/ Low Inertia	Low Stress/ Low Inertia	High Stress/ High Inertia	Low Stress/ High Inertia
Change Strategy	22	4	19	7
Retain Strategy	6	9	15	50

Stress, inertia, and strategic change are as defined in Tables 13.1 and 13.2

13.3.1 Elements of the Strategy Reformulation Process.

Organization inertia in the form of escalating commitment to the current strategy at the organization level can be thought of as a process of innovation adoption (Rogers, 1962; Lave & March, 1975; Abrahamson, 1991) with interesting parallels to the process Kuhn (1970) describes for the adoption of a new scientific paradigm (Huff, 1982; Pitt & Johnson, 1987; Rumelt, 1984; Sheldon, 1980).

As time passes, commitment to the current strategy is likely to increase if basic task demands are met. As policies and procedures are developed that produce the outputs anticipated from a new strategy, initial commitment will increase. Operating efficiencies due to experience curve effects will further reinforce commitment over time. Capital expenditures for buildings, equipment and training that are uniquely tied to current strategy also cause commitment to grow; less tangible, but often equally important, are accumulating good will assets with suppliers, buyers and others that can not be completely transferred to any other strategy (Williamson 1979). A "framework" of assumptions about strategy that no longer need to be examined grows under these conditions and channels managerial perception such that the question of changing strategy is less likely to arise (Minsky, 1977; Gioia and Sims, 1986; Johnson, 1989).

At the individual level, escalating commitment can be thought of as the adoption of new strategic ideas by an increasing number of actors within the organization. If the strategy satisfactorily meets current demands, managers are unlikely to look for greater perfection (Simon, 1945). As a new strategy is more strongly supported and more completely implemented, even individuals who are not completely convinced are motivated by self interest to find ways to accommodate themselves to the confines of current strategy, and thus have at least some stake in the status quo.

While the forces of inertia in general gain strength, often binding organizations to one strategy for long periods of time, the grounds for changing strategy are always present as well. Various events (poor performance, new technology, changes in the number and activity of competitors, demographic and social shifts, internal reorganization and new leadership) make past commitments less appropriate (e.g. Learned, Christensen, Andrews, and Guth, 1965; Andrews, 1971; Rumelt, 1984). Organization stress summarizes ways in which current strategy is not satisfactory; it reflects doubts and dissatisfactions of individual actors and imperfects in the fit between the organization and its environment.

Stress is always present because no strategy is perfect. This stress will increase if implementation falls short of expectations, a fairly frequent occurrence because abstract plans almost always incorporate inconsistencies that only become obvious with experience. Stress also increases because the organization setting is itself dynamic. As opportunities develop, as new technologies and new ideas become available, the inadequacies of current strategy are underscored. New strategies of other firms, including new entrants to the industry, also are likely to increase organizational stress. If these competitors achieve results that the focal organization does not, the problems of fit are further highlighted.

Organizations are designed as active problem solving entities which means that many dissatisfactions can be quickly addressed by small changes in operations, personnel reassignment, product improvements and the like. In addition to internal adjustments in current practice, stress also tends to dissipate as attention to and memory of specific stressful events fades. Fortuitous changes in circumstance can also reduce stress. These adjustment processes are all aspects of *organizational homeostasis*: "the tendency of a system ... to maintain internal stability owing to the coordinated response of its parts to any situation tending to disturb its normal condition or function" (Stein, 1966, 679). However, not all problems can be satisfactorily resolved within one strategic framework, and thus it is unlikely, over time, that homeostatic changes can completely counteract change and dissatisfaction.

When inertia exceeds stress, the most likely state of organization activity involves incremental homeostatic renewal processes within the framework of current strategy. Only if and when stress exceeds inertia will the organization begin to actively consider other strategic alternatives. If the homeostatic capability of the organization inadequately reduces new stress, serious questioning of day to day decision making activity becomes more likely. In the face of increasing but unresolved stress, individuals within the organization begin to ask whether the current situation could be better dealt with in some other way. The key characteristic of this organization state is that important actors within the organization are forced by unresolved stress to consider the pluses and minuses of current strategy in an abstract and holistic way that is quite different from the unquestioning problem solving characteristic of day to day organizational activity.

More dramatic, synoptic renewal efforts begin with the overt recognition of tension between voices for change and other conservative voices that typically argue for a renewed commitment to find adaptive solutions within the framework of current strategy. Discussion among individuals with different orientations toward a possible change have the interesting effect of surfacing issues that have been submerged under "business as usual". New issues in turn tend to require the collection of new information, or new interpretations of existing information. With discussion comes the critical examination of the schemata under which the individuals within the organization have been operating.

The effectiveness of further homeostatic renewal efforts increasingly come to depend upon the level of organization commitment to the current strategy. The overt evaluation of current strategy signals the *possibility* of a major change in strategy. If stress levels continue to increase, current stress and the possibility of change retard continued growth in commitment to the current strategy, making a major renewal effort more and more likely.

13.3.2 A Formal Statement of the Strategic Change Model

A simple model of stress / inertia dynamics will serve to illustrate the main features of the strategic change process outlined thus far. The organization is assumed to be in one of two states: State I – Business as Usual or State II-- Strategic Evaluation. At any point in time **t**, the organization is in State I when inertia , **I(t)**, exceeds stress, **S(t)**; or the organization is in State II if stress exceeds inertia. Furthermore, the probability that the organization makes a second order change in strategy during time **t**, **P(t)**, is positive only if the organization is in State II and is proportional to the positive difference between stress and inertia such that

$$\mathbf{P(t)} \quad = \mathbf{k}\,[\,\mathbf{S(t)} - \mathbf{I(t)}\,],\ \text{if}\ [\mathbf{S(t)} - \mathbf{I(t)}] > 0 \\ = 0,\ \text{otherwise.} \qquad (13.1)$$

Organization stress at time t+1, **S(t+1)**, is assumed to be governed by: the unresolved stress from the previous time period, **S(t)**, the homeostatic reduction in S(t) during the time interval (t, t+1), **H(t)**, and the additional stress arising during (t,t+1), **z(t)**, such that

$$\mathbf{S(t+1)} = \mathbf{S(t)} - \mathbf{H(t)} + \mathbf{z(t)}. \qquad (13.2)$$

Although it is recognized that new stress, **z(t)**, is a random variable best modeled as a Poisson process, for example see Huff, Huff, and Thomas, 1992, the focus in this discussion is on the non-linearities in the system and so the model will be simplified by assuming that new stress, **z(t)**, is a constant such that

$$\mathbf{z(t)} = \mathbf{z}\,. \qquad (13.3)$$

The effectiveness of the homeostatic response to stress is assumed to be affected by the level of inertia, **I(t)**, reflecting the level of commitment to the current strategy such that

$$\mathbf{H(t)} = \mathbf{aI(t)S(t)}. \qquad (13.4)$$

Inertia associated with the current strategy at time t+1, **I(t+1)**, is assumed to increase with increasing commitment to the current strategy in the previous time period, **I(t)** and decrease with increasing unresolved stress associated with the current strategy, **S(t)**, such that

$$\mathbf{I(t+1)} = \mathbf{I(t)} + \mathbf{b}[\mathbf{I(t)} - \mathbf{S(t)}][1 - \mathbf{I(t)}]. \qquad (13.5)$$

In the absence of the negative effects of stress, the above expression simply indicates that inertia tends to describe an "S - shaped" growth path. However, increasing stress levels tend to undercut commitment and inertia actually declines when stress exceeds inertia.

To complete the system, it is assumed that: any major change in strategy is accompanied by a return to some baseline level of inertia, $\underline{\mathbf{I}}$, reflecting organization commitment to the new strategy; whereas organization stress is assumed to experience a proportionate reduction reflecting the actual and/or expected improvement in performance under the new strategy such that stress, **S*(t+1)** and inertia, **I*(t+1)** at the beginning of the next strategic cycle are

$$\begin{aligned} \mathbf{S^*(t+1)} &= \mathbf{S(t) - aS(t)} \\ \mathbf{I^*(t+1)} &= \underline{\mathbf{I}}\,. \end{aligned} \tag{13.6}$$

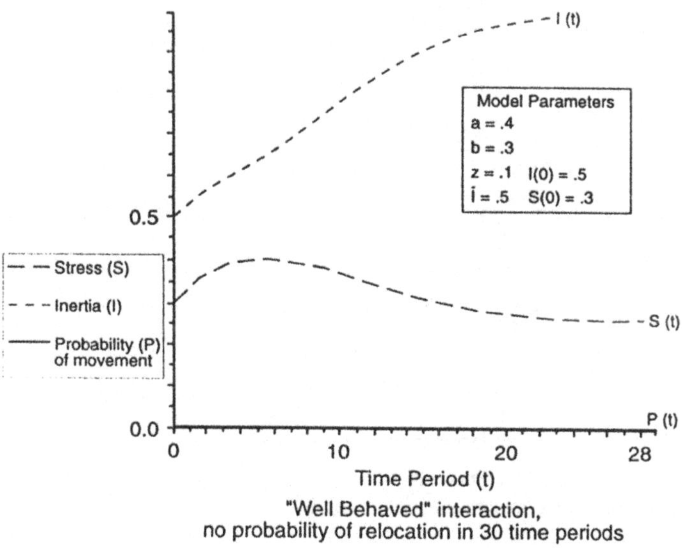

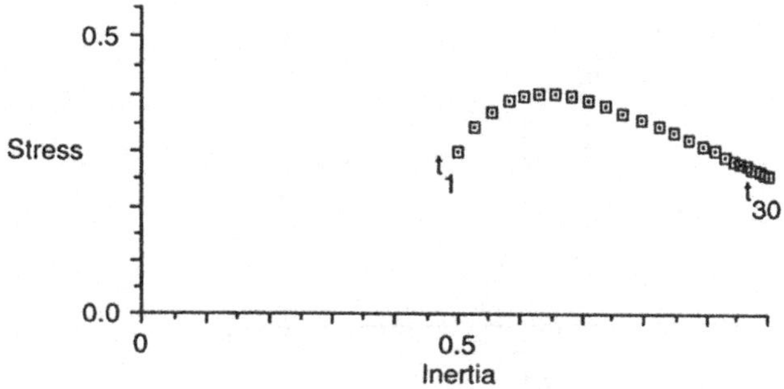

Fig. 13.4. Strategic Change Simulation #1

As expected, when the model is run as a simulation, the resulting histories of strategic change are quite sensitive to the values of the four parameters in the model. The probability of a significant change in strategy decreases with: a) reductions in the magnitude of new stress, **z**, impinging on the organization; b) increases in **a**, as a measure of the efficiency of the organization's homeostatic response to rising stress levels; c) increases in **b**, as a measure of the rate at which the organization creates or sheds commitment to the current strategy; and d) increases in **I̲**, the basic level of inertia or resistence to change following any significant change in strategy.

In addition to the general relationships between the parameter values and the probability of making a second order change in strategy, the above equations describe a very simple but quite interesting non-linear system. Figures 13.4 and 13.5 describe the orbits of firms under identical values for the parameters, **a,b,z,** and **I̲** and but with differing initial conditions, **S(0)** and **I(0)**.

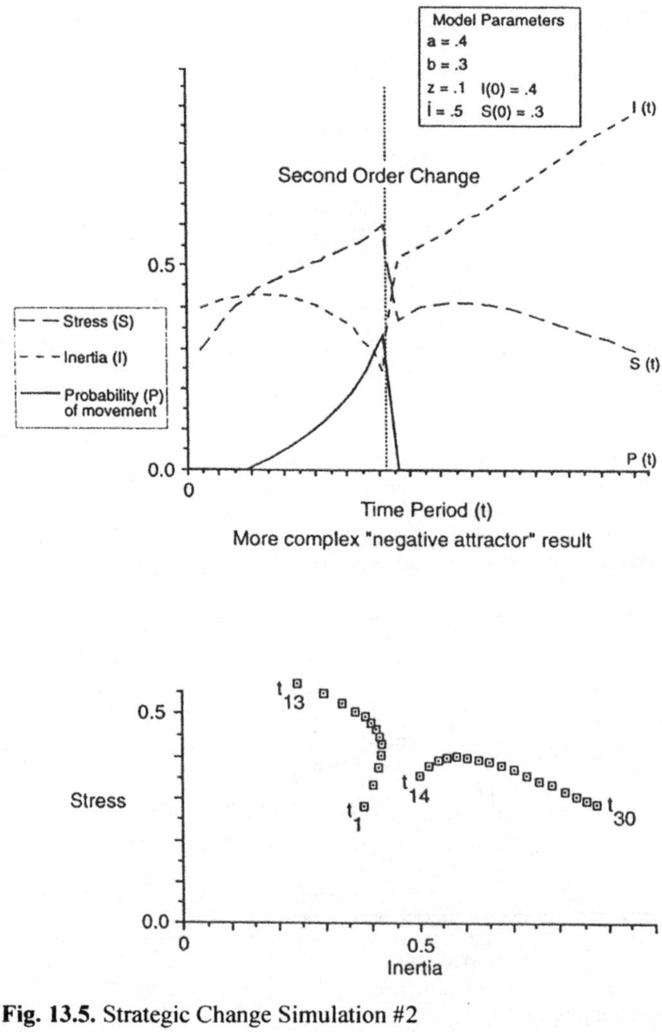

Fig. 13.5. Strategic Change Simulation #2

Under the assumptions of this model, organizations tend to evolve toward either a high inertia / low to moderate stress condition that is associated with "stayer" firms in State I or they enter State II and tend toward a high stress / low inertia condition which precipitates a second order change in strategy. The change in strategy reduces stress but also reduces inertia. Low stress / low inertia nor high stress / high inertia conditions tend to be quite unstable conditions in this model; and there is a marked tendency to orbit back into a high stress / low inertia, "mover", state or to settle into a low stress / high inertia, "stayer", state. As suggested in the simulations shown in Figures 13.4 and 13.5, small changes in initial stress and inertia levels clearly can lead to very different patterns of behaviour. In fact the configuration of the orbits illustrates the existence of a negative attractor in the neighborhood of $S(t) = I(t) = \sqrt{(z / a)}$.

13.3.3 Movers and Stayers

An important hypothesis suggested by the foregoing analysis is that firms achieving a high inertia state will tend to remain in that state and firms experiencing high stress will tend to remain in a high stress condition over time. Firms tend to move along certain predictable and distinctive orbits suggesting a typology of "movers" and "stayers" -- that is, once firms begin making strategic changes they tend to continue to do so; and once they have been at rest they are more likely to continue following the same strategy. This distinction between movers and stayers and the associated differences in organization response under similar environmental conditions has been a topic of considerable interest in the organization change literature (Ambergey, Kelly, and Barnett, 1993; Gersick, 1991; and Bennett and Carroll, 1995); however, empirical evidence in support of the theoretical models is rare. Consequently, the following results are of interest beyond their connection to the specific theoretical model presented in the previous section.

A four state Markov chain will be employed to explore these strategic orbits -- the four states being the four stress/inertia categories used in the previous analysis. In this instance, however, the focus will be on the time paths (through these four states) followed by individual firms over the twenty year period, 1970 - 1990. Two relatively distinct patterns emerge. Firms either tend to orbit around the high stress/low inertia state, "movers", or to orbit around the low stress/high inertia state, "stayers" with the low stress/low inertia state and the high stress/high inertia state representing transient or unstable strategic conditions. The existence of these two strategic orbits is best illustrated by a transition matrix, **T**, summarizing the conditional probabilities of shifting from state i to state j during a given time interval. Table 13.4 presents the transition matrix summarizing the experiences of the pharmaceutical firms in our sample. The transition probabilities are estimated from the 132 observed transitions between the four stress/inertia states made by the sample of firms over four time periods (each period is five years in duration).

This table shows, for example, that a total of 34 firms (in one of five time periods) originate in a High Stress/High Inertia state. Given that a firm begins a four year period in a High Stress/ High Inertia State, there is a .382 probability that the firm will end the

Table 13.4. Transition Probabilities Between Stress / Inertia States

FROM/TO	HS/LI	LS/LI	HS/HI	LS/HI
HS/LI (N=28)	.643	.143	.107	.107
LS/LI (N=13)	.231	.077	.462	.231
HS/HI (N=34)	.382	.176	.205	.235
LS/HI (N=57)	.035	.088	.228	.649

four year period in a High Stress/ Low Inertia State and a .235 probability that the firm will end the period in a low Stress/High Inertia State.

From the matrix **T** of transition probabilities shown in Table 13.4, it is evident that firms tend to follow a rather limited set of evolutionary paths. The critical distinction is between "movers" and "stayers." A firm with high stress and low inertia not only is very likely to change strategy (move) but also is likely to remain in that same high mobility state in the next time period (t_{11} = .643); whereas firms in the low stress/high inertia state are likely to retain the current strategy and to remain in that low stress/high inertia state from one time period to the next (t_{44} = .649). The low stress/low inertia state is clearly unstable (t_{22} = .077) with a marked tendency to shift into a high stress state. The high stress/high inertia state isn't stable either. Only twenty percent of the firms remain in that state from one time period to the next, they ultimately spin off along a "stayer" orbit (t_{34} = .235) or they change strategy ($t_{31} + t_{32}$ = .558) and enter the "mover" orbit.

The structure of the transition matrix thus suggests a pattern of strategic change in which a limited set of firms do most of the changing while the majority of the firms in the industry retain the same strategy for relatively long periods of time. Some volatile firms exist. They can and apparently do reduce stress through changes in strategy, thereby moving the firm toward a more stable strategic position; and once stabilized, the firm tends to remain so. On the other hand, rising stress levels may precipitate a change in a long established strategy which, in reducing inertia, may stimulate further change thereby transforming a "stayer" into a "mover".

13.3.4 Strategic Change and Performance Improvement

To analyze performance levels before and after a major change in strategy, as compared to changes in performance levels for firms making no major change in strategy, annual return on sales measures were used. The performance categories were: high -- top quartile in sales growth; medium -- the two middle quartiles; and low -- bottom quartile. A change in strategy was assumed to be successful if high performers remained high, medium performers became high performers, or low performers became medium or high performers.

The results of the analysis are shown in Tables 13.5 and 13.6. Fifty-three percent of the firms making a change in strategy were successful in relative performance terms. Of the firms retaining the current strategy, only twenty percent maintained a high performance level or improved their performance from one period to the next.

A more detailed analysis of performance changes following a significant change in strategy suggests that high performance firms have the most to gain from major changes

Table 13.5. Changes in Performance for Movers

	AFTER		
	High	Medium	Low
BEFORE			
High (N =18)	.667	.167	.167
Medium (N=11)	.364	.545	.091
Low (N=20)	.150	.350	.500

Table 13.6. Change in Performance for Stayers

	AFTER		
	High	Medium	Low
BEFORE			
High (N=11)	.182	.727	.091
Medium (N=58)	.121	.690	,190
Low (N=13)	.154	.385	.462

in strategy. For high performers, there is a .667 chance of remaining in the high performance category after a major change in strategy whereas high performers that do not make significant changes in strategy have only a .182 chance of retaining their high performance status suggesting that high performers can't follow a "stayer" strategy for very long and hope to maintain their high performance position within the industry. Even though the large majority of middle performance firms are "stayers" and visa versa, a comparison of performance changes for movers and stayers for this group suggests that movers have a greater chance of improving their competitive position than do stayers since middle performance movers become high performance firms 36.4% of the time whereas middle performance stayers achieve high performance status 12.1% of the time. Low performance firms have about the same chance of improving their performance position irrespective of their decision to "move" or "stay".

13.4 Conclusion

The central question driving this investigation is deceptively simple: why do firms within the same industry and operating within the same competitive environment adopt qualitatively different strategic responses to changing environmental conditions? This general question encompasses a series of more specific questions of concern to theorists and practitioners alike. Why do some firms within an industry retain the same basic strategy over long periods of time whereas other firms in the same competitive environment make numerous changes in strategy? Is it possible to predict which firms are likely to make major changes in strategy given prevailing environmental conditions? How do the past experiences of the firm affect present and future strategic decisions?

How does a better understanding of the strategic change process contribute to work on firm behaviour in economic geography?

A good deal has been written about why firms change strategies -- to respond to new regulation, new technology and other threats and opportunities in the environment, to carry out the vision of a new leader, to erect or solidify barriers to entry or otherwise gain competitive advantage. While these and other descriptions have theoretic appeal, several problems make them difficult to apply to actual case histories. First, firms almost always feel buffeted by their environments, and thus almost always have reason to make a strategic move, even in eras that in retrospect seem minimally stressful in comparison with current pressures. Conversely, not all firms change strategy, even when simultaneously faced with several factors that might be expected to induce a significant strategic response. Finally, it is not always easy to transform vague and contradictory signals from the firms environments into messages about threats, opportunities, rents, barriers, or competitive advantages.

Attention to the forces that bind firms to one strategy has done a great deal to summarize the micro and macro factors that reduce the likelihood that a given firm will be able to respond (or quickly respond) to indicators that strategic change might be desirable. While this general approach appears promising, more needs to be done to specify the interplay between stress and inertia, and how it might affect the timing of strategic change. The analysis in this paper speaks directly to this issue.

A formal model of strategic change highlights the interdependencies between stress and inertia. The firm's ability to make frame sustaining, homeostatic adjustments to strategy which effectively reduce stress depends upon the level of commitment/inertia surrounding the current strategy. On the inertia side, the model suggests that increasing stress can erode commitment/inertia; and when stress exceeds inertia, a crisis of confidence concerning the viability of the current strategy can develop quite rapidly. In so far as this model does capture important aspects of the strategic change dynamic, it suggests that much closer attention should be paid to stress and inertia conditions surrounding a given strategy in the initial implementation stages because relatively small differences in these initial conditions can result in radically different strategic responses over time. Too often, a study of strategic change ends if and when a change occurs which is precisely when closest attention should be paid to what happens next as the organization more or less successfully adjusts to the new strategy.

As expected, empirical data from the pharmaceutical industry show that firms following the same strategy for a period of time were much less likely to change strategy in the future. One distinctive feature of the analysis is that moderate performance levels are associated with low levels of organization stress and relatively low propensities to make second order changes in strategy whereas firms following strategies associated with *either* high performance *or* low performance relative to industry norms are likely to experience high organization stress and relatively high propensities to make significant changes in strategy.

A Markov model of the transitions between high and low stress and inertia states provides a strong empirical basis for making a distinction between "movers" and "stayers" within the set of pharmaceutical firms in our study. Two thirds of the cases fall into either the high stress/low inertia category or the low stress/high inertia category; and once in this mover or stayer state, firms have a high probability of remaining in that state. Of particular interest, the analysis shows that firms which change strategy in the pharmaceutical industry tend to show subsequent performance improvements as the result of changing strategy. Results from the empirical analysis of the evolving stress, inertia conditions and the resulting histories of strategic change closely parallel the predicted changes arising out of the behavioural model.

The foregoing analysis is certainly relevant to those interested in questions of strategic change; but how does an analysis of the strategic change process contribute to behavioural approaches in economic geography? The first implication from the study is that geographic studies of firm behaviour should not ignore the firm's strategic position as a potentially important variable in the analysis. It is evident that a wide range of strategies are potentially viable within what appears to be a relatively homogeneous industry and that different strategies have very different spatial/ geographic implications. In the case of the pharmaceuticals industry for example, there is likely to be a close mapping between the firm's strategic position in terms of R&D activity and Advertising activity and the relative importance of locational considerations pertaining to the availability of a skilled labor force essential for R&D development versus the locations of major markets which focus advertising efforts.

The second implication is that even within a relatively stable and homogeneous industry, as in the present analysis, firms exhibit widely divergent levels of profitability, and differ widely in the propensity to make significant changes in strategy along specific dimensions. Firms differ widely in the timing and the magnitude of their strategic and locational responses to organization stress arising from shifts in environmental conditions. What the foregoing analysis shows is that these differences in response are predictable and need to be explained with reference to the firm's prior history of strategic change. Significant changes in strategy are likely to precipitate significant locational consequences that in turn affect specific sectors of local labor markets as these firms redirect investment so as to achieve outcomes consistent with the revised strategy.

References

Abrahamson, E. 1991. Managerial fads and fashions: the diffusion and rejection of innovations. *Academy of Management Review*, **16**:586-612.

Ambergey, T. L., Kelly, D., Barnett, W. P. 1993. Resetting the clock: the dynamics of organizational change and failure. *Administrative Science Quarterly*, **38**:51-73.

Ambergey, T.L., and Miner, A.S. 1992. Strategic momentum: the effects of repetitive, positional, and contextual momentum on merger activity. *Strategic Management Journal*, **13**:335-48.

Andrews, K. R. 1971. *The concept of corporate strategy*. Homewood, IL: Dow-Jones Irwin.

Argyris, C. 1977. Double loop learning in organizations. Harvard Business Review.**55**:115-25.

Barnett, W.P. & Carroll, G.R. 1995. Modeling internal organizational change. *Annual Review of Sociology*, **21**:217-36.

Barr, P. 1991. *Organization stress and mental models*. Unpublished doctoral dissertation, University of Illinois.

Bogner, W.C., Pandian. J.R., Thomas, H. 1994. Modeling Strategic Group Movements. In H. Daems and H. Thomas (eds.), *Strategic groups, strategic moves and performance*. Tarrytown, New York: Elsevier Science.

Cool, K.O. & Dierickx, I. 1993. Rivalry, strategic groups and firm profitability. *Strategic Management Journal*, **14**:47-59.

Fiegenbaum, A. 1987. *Dynamic aspects of strategic groups and competitive strategy: Concepts and empirical examination in the insurance industry*. Unpublished doctoral dissertation, University of Illinois, Urbana-Champaign.

Gersick, C.J.G. 1991. Revolutionary change theories: a multilevel exploration of the punctuated equilibrium paradigm. *Academy of Management Review*, **16**:10-36.

Gioia, D. & Sims, H., Jr. (Eds.) 1986. *The thinking organization*. San Francisco: Jossey-Bass.

Hannon, M.T. & Freeman, J. 1984. Structural inertia and organizational change. *American Sociological Review*, **49**:149-64.

Huff, A. S. 1982. Industry influences on strategy reformulation. *Strategic Management Journal*, **3**:119-131.

Huff, J. O. & Clark, W. A. V. 1978. Cumulative stress and cumulative inertia: A behavioral model of the decision to move. *Environment & Planning A*, **10**:1101-19.

Huff, J.O. & Huff, A.S. 1995. Stress, inertia, opportunity, and competitive position: A SIOP model of strategic change in the pharmaceuticals industry. *Best Papers Proceedings*,. Vancouver: Academy of Management.

Huff, J. O. , Huff , A. S.& Thomas, H. 1994. In H. Daems and H. Thomas (eds.), *Strategic groups, strategic moves and performance*. Tarrytown, New York: Elsevier Science.

Johnson, G. 1988. Rethinking incrementalism. Strategic Management Journal, **9**:75-91.

Kuhn, T.S. 1970. *The structure of scientific revolutions*. Chicago: The University of Chicago Press.

Lave, C. A. & March, J. G. 1975. *An introduction to models in the social sciences*. New York: Harper & Row.

Learned, E. P., Christensen, C. R., Andrews, K. R. & Guth, W. D. 1965. *Business Policy*. Homewood, IL: Irwin.

Meyer, A. D., Brooks, G. R. & Goes, J. B. 1989. *Environmental jolts and industry revolutions: Organizational responses to discontinuous change*. Working Paper, Graduate School of Management, University of Oregon.

Minsky, M. 1977. Frame-system theory. In P. N. Johnson-Laird & P. C. Wason, *Thinking*. Cambridge: Cambridge University Press.

Nelson R. R. & Winter, S. G. 1982. *An evolutionary theory of economic change*. Cambridge: Cambridge University Press.

Pitt, M. & Johnson, G. 1987. Managing strategic change. In G. Johnson (Ed.). *Business strategy and retailing*. Chichester: Wiley.

Porter, M. E. 1980. *Competitive strategy*. New York: Free Press.

Reger, R. K. 1988. *Competitive positioning in the Chicago banking market: Mapping the mind of the strategist*. Unpublished doctoral dissertation, University of Illinois, Urbana-Champaign.

Rogers, E. M. 1962. *Diffusion of innovations*. New York: Free Press.

Rumelt, R. P. 1984. Towards a strategic theory of the firm. In R. Lamb, (Ed.), *Competitive strategic management*. Englewood Cliffs, NJ: Prentice-Hall.

Schwenk, C. & Tang, M. 1989. Persistence in questionable strategies: explanations from the economic and psychological perspectives. *OMEGA: International Journal of Management Science*, **17**:559-570.

Sheldon, A. 1980. Organizational paradigms. *Organization Dynamics*, **8**:61-71.

Simon, H. A. 1945. *Administrative behavior*. New York: MacMillan.

Stein, J. (Ed.) 1966. *The Random House dictionary of the English language*. New York: Random House.

Tushman, M. L. & Romanelli, E. 1985. Organizational evolution. *Research in Organization Behavior*, **7**:171-222.

Watzlawick, P., Weakland, J. H. & Fisch, R. 1974. *Change*. New York: Norton.

Williamson, O. E. 1979. Transaction-cost economics: The governance of contractual relations.

Journal of Law and Economics, **22**:233-260.

Zajac, E.J. & Kraatz, M.S. 1993. A diametric forces model of strategic change: assessing the antecedents and consequences of restructuring in the higher education industry. *Strategic Management Journal*, **14**:83-102.

14 Some Implications of Behaviour in Agricultural Markets

Gordon F. Mulligan

Department of Geography and Regional Development, University of Arizona, Tucson, AZ 85721, USA

14.1 Introduction

Modern agricultural location theory stems from the seminal contributions of David Ricardo (1772-1834) and Johann Heinrich von Thünen (1783-1850) nearly two hundred years ago. Today, in the capitalist economies, much of the spatial variation in agricultural land use depends on but two factors: first, differences in physical features, where soil fertility, climate, and topography affect productivity; and second, differences in relative location, where farms face different transportation costs in delivering their produce to either markets or points of export. Unfortunately, even after the suggestion of McCarty and Lindberg (1966) some thirty years ago, few serious attempts have been made to synthesize these two complementary approaches to agricultural location theory.

This paper focuses largely on von Thünen's model, although the last section addresses environmental variability. As Dunn (1955) pointed out, the traditional interpretation of von Thünen is a partial equilibrium model where all land-use decisions are based on exogenous market prices. While Dunn acknowledged that market demand should be considered for each agricultural activity, he never did demonstrate how to operationalise closure of the model. Consequently, the traditional von Thünen model for agriculture, with its sole emphasis on production or supply, must be considered seriously deficient as a predictive tool.

The main intention of the paper is to outline various extensions of the traditional model, based on new research directions formulated by other analysts during recent years. In order to simplify the mathematics, all analysis of the paper focuses on two activities competing for land in a one-dimensional economy. However, all results are generalisable to many activities in a two-dimensional economy. Section 14.2 reviews the traditional von Thünen model in some detail. Then Section 14.3 closes the model for demand, and provides partial equilibrium solutions for agricultural prices, land areas, and outputs. Comparative static predictions are next given for both the traditional and the closed model. Section 14.4 briefly considers how the size of the economy's market town affects agricultural prices and land uses in two separate ways: first, through the actual physical extent of the town; and second, through the demand exerted by the residents of the town. Section 14.5 addresses the somewhat neglected issue of climatic variability by introducing two possible environmental states, termed wet years and dry years, which influence only the annual yields of the two activities. Landowners are then allowed to pursue two competing objectives in light of this climatic uncertainty. First, they can maximise their long-run average return by choosing that one activity that maximises their expected rent. Or, they can opt for an annual maximum guaranteed return by mixing their two activities so that, at least at those locations near the market town, rent is identical in either environmental state. Land uses are shown to vary significantly with the strategic behaviour adopted by landowners.

The analysis employs several well-known and acceptable assumptions. The regional economy is assumed to be isotropic where environmental conditions and transportation rates are the same everywhere. All agricultural produce is shipped to a market town located in the centre of the economy. Mobility of all factors is assumed to be perfect and all production units (points in space) exhibit constant returns to scale. The market demand for both activities is also assumed to be linear in price. These last two assumptions, while somewhat restrictive, allow most of the solutions to be derived through matrix algebra. Finally, it should be stressed that the paper provides only partial equilibrium and not general equilibrium solutions for activity prices, land areas, and outputs.

14.2 The Traditional Model

Denote the yield, price at the market town, non-land production cost, and transportation rate for activity i by E_i, p_i, a_i, and f_i, respectively. Then the rent $R_i(k)$ earned at distance k from the market is

$$R_i(k) = E_i(p_i-a_i) - E_if_ik = R_i(0) - T_ik \tag{14.1}$$

where $R_i(0)$ is rent at the market town and T_i, the transportation cost per unit of land, is the slope of the rent gradient. Note that the rent earned by activity i just falls to zero at distance

$$k_{i,max} = (p_i-a_i)/f_i = R_i(0)/T_i \tag{14.2}$$

which indicates activity i's extensive margin (spatial margin, economic limit).

In order to simplify matters assume that only i=2 rent-earning activities compete for the land surrounding the market town. Assume that activity 1 dominates the land closest to the town, while activity 2 dominates the land further out. Then, adapting an argument from Dunn (1955), necessary and sufficient conditions for this two-zone domination pattern are $R_1(0)>R_2(0)>0$ and $k_{2,max}>k_{1,max}>0$. These conditions imply that $T_1>T_2>0$, indicating that activity 1 has the steeper rent gradient. The point of indifference k_{12} can be calculated as

$$\begin{aligned} k_{12} &= [E_1(p_1-a_1)-E_2(p_2-a_2)]/(E_1f_1-E_2f_2) \\ &= [R_1(0)-R_2(0)]/(T_1-T_2) \end{aligned} \tag{14.3}$$

where $k_{12}<k_{2,max}$ from the two conditions stated above. This means that activity 1 dominates from k=0 to $k=k_{12}$ and activity 2 dominates from $k=k_{12}$ to $k=k_{2,max}$. In the one-

dimensional economy the areas dominated by the two activities are of the following size:

$$A_1 = 2k_{12} \quad (14.4a)$$
$$A_2 = 2(k_{2,max}-k_{12}) \quad (14.4b)$$

and the outputs (quantities supplied) of the two activities are as follows:

$$Q_1^S = E_1A_1 \quad (14.5a)$$
$$Q_2^S = E_2A_2 \quad (14.5b)$$

To illustrate these properties, consider the numerical values (in appropriate units) given for the various parameters in Table 14.1. The rent earned at the market town is $R_1(0)=55$ for activity 1 and $R_2(0)=20$ for activity 2, the slopes of the rent gradients are $T_1=4$ and $T_2=0.5$, the extensive margins are $k_{1,max}=13.75$ and $k_{2,max}=40$, and the point of indifference is $k_{12}=10$. Activity 1 dominates on both sides of the market town from k=0 to k=10 while activity 2 dominates from k=10 to k=40. Consequently, the areas dominated by the two activities are $A_1=20$ and $A_2=60$, and the two output levels are $O_1=400$ and $O_2=600$.

As pointed out earlier, this traditional model focuses solely on the production or supply side of each agricultural activity. Nevertheless, predictions can be made regarding land-use changes by examining the comparative statics of this model. An incremental shift in price or non-land production cost induces a parallel shift in the appropriate rent gradient while an incremental shift in yield or transportation rate induces a rotational shift in the rent gradient. Comparative static analysis, as performed by Peet (1969), Foust and DeSouza (1978), Jones (1991), and others, generally shows how changes in the two zones of domination depend upon shifts in only two locations: the point of indifference k_{12} and the economic limit $k_{2,max}$.

In the two-activity case the eight parameters in Table 14.1 are assumed to be independent of one another and, therefore, eight different predictions regarding land-use changes can be made. However, the traditional model is obviously underspecified, since causal relationships between the different parameters are not properly articulated, and this can lead to erroneous predictions. Besides it makes very little sense to undertake a comparative static prediction where price is the exogenous variable. At the very least it must be recognised that the two prices are not independent of the other parameters, since, from equations 14.2 and 14.3:

$$p_1 = a_1 + (E_1f_1-E_2f_2)k_{12}/E_1 + E_2(p_2-a_2)/E_1 \quad (14.6a)$$
$$p_2 = a_2 + f_2k_{2,max} \quad (14.6b)$$

Equation 14.6a indicates that p_1 is directly influenced by the seven other parameters (holding k_{12} constant), including p_2, while equation 14.6b indicates that p_2 is directly influenced by two other parameters (holding $k_{2,max}$ constant).

Table 14.1. Input data

	Activity 1	Activity 2
Yield	E_1=20	E_2=10
Price	p_1= 6	p_2= 4
Cost	a_1= 3.25	a_2= 2
Rate	f_1= 0.20	f_2= 0.05

14.3 The Closed Model

As Jones et al (1978), Samuelson (1983), Nerlove and Sadka (1991), and others have argued, the traditional production-only model should be closed for demand. For the case of i=2 activities, introduce two linear demand curves as follows:

$$Q_1^D = \alpha_1 - \beta_{11}p_1 - \beta_{12}p_2 \tag{14.7a}$$
$$Q_2^D = \alpha_2 - \beta_{21}p_1 - \beta_{22}p_2 \tag{14.7b}$$

where the intercept terms $\alpha_1,\alpha_2>0$, the own-price slope terms $\beta_{11},\beta_{22}>0$, and the cross-price slope terms $\beta_{12},\beta_{21}\geq 0$. As only two activities are considered these agricultural goods are necessarily substitutes for one another. Note that the two demand relationships can also be expressed as:

$$p_1 = \delta_1 - \lambda_{11}Q_1^D - \lambda_{12}Q_2^D \tag{14.8a}$$
$$p_2 = \delta_2 - \lambda_{21}Q_1^D - \lambda_{22}Q_2^D \tag{14.8b}$$

where the same ordering conditions hold on the intercept, own-demand, and cross-demand terms. Equations 14.7a and 14.8a and equations 14.7b and 14.8b are simply linear transformations of one another. The versions given in equations 14.7a and 14.7b are easier to use and are adopted for the analysis of this paper.

Next consider two equations that directly relate distances to the market town and quantities supplied, as standardised by yields:

$$k_{12} = Q_1^S/2E_1 \tag{14.9a}$$
$$k_{2,max} = Q_1^S/2E_1 + Q_2^S/2E_2 \tag{14.9b}$$

Rearrange equations 14.6a and 14.6b, consider equations 14.9a and 14.9b, and then introduce equations 14.7a and 14.7b by setting $Q_1^S=Q_1^D$ and $Q_2^S=Q_2^D$ in order to clear output for both activities. It follows that:

$$\mathbf{X}\mathbf{p} = \mathbf{Y}\mathbf{a} + \mathbf{Z} \tag{14.10}$$

where **X** is a 2 by 2 coefficient matrix, **p** is the 2 by 1 price vector, **Y** is a 2 by 2 coefficient matrix, **a** is the 2 by 1 non-land production cost vector, and **Z** is a 2 by 1 coefficient vector. The elements of the two coefficient matrices and the coefficient vector are as follows:

$$\begin{aligned}
x_{11} &= E_1 + (E_1f_1\text{-}E_2f_2)\beta_{11}/2E_1 \\
x_{12} &= \text{-}E_2 + (E_1f_1\text{-}E_2f_2)\beta_{12}/2E_1 \\
x_{21} &= f_2(\beta_{11}/2E_1 + \beta_{21}/2E_2) \\
x_{22} &= 1 + f_2(\beta_{12}/2E_1 + \beta_{22}/2E_2) \\
y_{11} &= E_1 \\
y_{12} &= \text{-}E_2 \\
y_{21} &= 0 \\
y_{22} &= 1 \\
z_1 &= (E_1f_1\text{-}E_2f_2)\alpha_1/2E_1 \\
z_2 &= f_2(\alpha_1/2E_1 + \alpha_2/2E_2)
\end{aligned}$$

From equation 14.10 it follows that equilibrium prices can be endogenously solved as

$$\mathbf{p} = \mathbf{X}^{-1}\mathbf{Y}\mathbf{a} + \mathbf{X}^{-1}\mathbf{Z} \tag{14.11}$$

Now, for illustrative purposes, consider the very simplest demand situation where the two cross-effect terms are both set equal to zero:

$$Q_1^D = 460 - 10p_1 \tag{14.12a}$$
$$Q_2^D = 680 - 20p_2 \tag{14.12b}$$

Adopting the input data from Table 14.1, it follows that the two equilibrium prices are p_1=6 and p_2=4. Therefore, all of the other numerical properties of the traditional model given earlier in the paper, pursuant to the data of that table, also hold for the closed model with the two particular demand relationships shown in equations 14.12a and 14.12b.

Comparative static analysis should always be based on the algebraic properties of the model, as this allows assertions to be formulated in a general and (hopefully) unambiguous manner. However, even in this simple model the algebra is messy and derivatives are difficult to sign. As a reasonable substitute, then, numerical simulations for different parametric shifts can be considered.

14.3.1 Non-land Production Costs

First consider a shift in the non-land production cost for activity 2 from a_2=2 to a_2=1.95. In the traditional model, a parallel shift outward occurs in the rent gradient for activity 2, thereby decreasing the point of indifference to k_{12}=9.857 and extending the economic limit of activity 2 out to $k_{2,max}$=41. Thus activity 2 expands its zone of domination in two directions, both taking land away from activity 1 near the prior point of indifference and adding land past the prior extensive margin. Table 14.2 shows the expected changes in land use and output that result from this small decrease in non-land production cost. In the closed model, using the demand parameters given above, the price of activity 1 is hardly affected by the shift in non-land production cost a_2, but the price of activity 2 is reduced to p_2=3.952. The extensive margin of activity 2 is slightly increased to $k_{2,max}$=40.047 while the point of indifference k_{12} remains the same. As Table 14.2 clearly shows, the traditional model severely overpredicts both the decrease in land area for activity 1 and the increase in land area for activity 2 that would accompany this cost shift.

Table 14.2. Comparative statics for two cost shifts

	a_2: 2.00→1.95		f_1: 0.20→0.25	
	Traditional	Closed	Traditional	Closed
p_1	6.000	6.000	6.000	6.471
p_2	4.000	3.952	4.000	3.994
k_{12}	9.857	10.000	7.778	9.882
$k_{2,max}$	41.000	40.047	40.000	39.889
A_1	19.714	20.000	15.556	19.764
A_2	62.286	60.094	64.444	60.013
Q_1	394.280	399.988	311.120	395.289
Q_2	622.860	600.952	644.440	600.128

The interrelationship apparent between equilibrium prices in the closed model, even when the cross-price terms are set to zero, is especially noteworthy. Of course, the anticipated shifts for the closed model depend intimately upon the parametric values actually chosen for the two demand curves.

14.3.2 Transportation Rates

Now consider a shift upward in the transportation rate for activity 1 from f_1=0.20 to f_1=0.25. In the traditional model, a rotational shift occurs in the rent gradient as the extensive margin of activity 1 is reduced from $k_{1,max}$=13.75 to $k_{1,max}$=11. This moves the point of indifference inward to k_{12}=7.778, so that activity 1 loses land and activity 2 gains land in the neighbourhood of the prior point of indifference k_{12}=10. Table 14.2 shows the expected changes in land use and output that result from this small increase in the transportation rate. In the closed model, the shift upward in f_1 drives the price of activity 1 up to p_1=6.471 while the price of activity 2 remains nearly the same, thereby only slightly reducing the point of indifference to k_{12}=9.882. The traditional model overpredicts the amount of land taken out of activity 1 and placed into activity 2 because that model cannot compensate for the rate increase with an accompanying price increase. It is also worth noting that when transportation rates fall for both activities, the closed model predicts that prices will fall for both activities.

14.3.3 Demand Properties

Now consider the comparative statics of market demand. Table 14.3 considers small changes in the intercept term α_1 of equation 14.12a and examines how the equilibrium solutions for prices, land areas, and outputs respond to these changes. As the demand curve for activity 1 shifts outward, there is both an upward shift in p_1 and an upward shift in p_2, and both an outward shift in the point of indifference k_{12} and the extensive margin $k_{2,max}$. However, the output of activity 1 monotonically increases while the output of activity 2 monotonically decreases as more land is put into the former use and less land into the latter use. The important result to note is that a parametric shift in the intercept term for one demand curve has implications for the amount of land placed into either use.

Similar effects can be discerned for a parametric shift in either the own-price or cross-price terms of equations 14.7a and 14.7b.

14.4 The Market Town and the Closed Model

The closed von Thünen model can be easily adapted to account separately for both the areal size of the market town and the demand exerted by the residents of that town. Return to the closed model of the previous section and assume that the market town is no longer dimensionless.

Suppose, instead, that the boundary (extensive margin) of the town is k_0 (on both sides). Recall that landowners must still absorb all transportation costs in hauling their produce from the farm to the point k=0 in the centre of the town. Recognising this, introduce the shift parameter k_0 in equations 14.9a and 14.9b as follows:

$$k_{12} = Q_1^S/2E_1 + k_0 \tag{14.13a}$$
$$k_{2,max} = Q_1^S/2E_1 + Q_2^S/2E_2 + k_0 \tag{14.13b}$$

Note that k_0 represents the urban-rural boundary where landowners are indifferent between urban and agricultural land uses. Consequently, the coefficients in matrix **Z** must also be adjusted before applying equation 14.11 to solve for the equilibrium properties of the model; note that

$$z_1 = (E_1f_1\text{-}E_2f_2)[(\alpha_1/2E_1) + k_0]$$
$$z_2 = f_2(\alpha_1/2E_1 + \alpha_2/2E_2 + k_0)$$

Assume, for illustrative purposes, that the population size of the market town is N=100,000 and that all residents have the same demand for each good. Then the demand equations 14.7a and 14.7b can be standardised into per capita form by simply dividing

Table 14.3 Effects of a shift in the demand curve for activity 1

α_1	p_1	k_{12}	Q_1	p_2	$k_{2,max}$	Q_2
440	5.905	9.524	380.952	3.977	39.546	600.452
460	6.000	10.000	400.000	4.000	40.000	600.000
480	6.095	10.476	419.057	4.023	40.454	599.545
500	6.189	10.953	438.110	4.045	40.907	599.091

Table 14.4. Two effects of the market town

N	k_0	p_1	p_2	k_{12}	$k_{2,max}$
100,000	0	6.000	4.000	10.000	40.000
100,000	5	6.947	4.226	14.763	44.536
100,000	10	7.894	4.454	19.527	49.073
200,000	10	10.219	6.177	27.891	83.537

each of the coefficients in the two equations by population N; the per capita versions of equations 14.12a and 14.12b are:

$$Q_1^D = 0.00460 - 0.00010p_1 \quad (14.14a)$$
$$Q_2^D = 0.00680 - 0.00020p_2 \quad (14.14b)$$

The above equations can be used to isolate two different growth effects of the market town. On the one hand, the town might expand areally, extending its boundary outward, while holding its population N constant; this would simply reflect a reduction in population density. On the other hand, the town might experience population growth while holding its area constant; this would constitute an increase in population density. In this case there would be an outward shift in aggregate demand for each of the two goods.

Note the four cases shown in Table 14.4. Case 1 refers to the usual situation of a dimensionless town as examined earlier in the paper. Case 2 shows the equilibrium solutions for a town with boundary k_0=5. Obviously the prices of both goods are driven up as both outputs must now be transported further to the market than in case 1. In case 3 the density of the town is halved as the urban boundary is extended even further out to k_0=10. Again, both prices are driven upward. Finally, case 4 shows equilibrium solutions when the population size is doubled to N=200,000. Here p_1, p_2, k_{12}, and $k_{2,max}$ are all substantially increased over their counterpart values in case 3.

Urban growth has two interdependent effects on agricultural land uses. First, areal growth alone drives prices upward as producing units become more distant from markets. However, with this effect there is very little change in the land areas devoted to the two activities. Only the locations of these land uses change. Second, holding per capita demand constant, population growth shifts the aggregate demand curve of each activity outward. Agricultural land uses move even further outward to accommodate the shift in demand for the two goods. The changes between cases 2 and 4 in Table 14.4 roughly depict what happens in the real world when both areal and population growth simultaneously occur.

14.5 Varying Environmental States

Since Davenport (1960), Gould (1963), and Wolpert (1964) geographers have shown some interest in understanding how decision makers adopt agricultural strategies in the face of environmental uncertainty. These contributions, while stimulating, have never been fully adapted to a spatial or locational framework, although the contributions of

Smith (1977), Cromley (1982), Jones (1983), and Cromley and Hanink (1989) are all important and worthy of note.

For this paper consider the simplest possible situation where there are j=2 environmental states, which are simply called wet years and dry years. Assume that yield E_{ij} for activity i varies with the environmental state but that the non-land production cost a_i and the transportation rate f_i do not. As price p_i in the traditional model does not depend upon the environmental state, the analysis of this section is largely concerned with the more interesting case of the closed model.

Now the rent $R_{ij}(k)$ earned at distance k from the (dimensionless) market town depends upon the state of the environment; that is

$$R_{ij}(k) = E_{ij}(p_i - a_i) - E_{ij}f_i k \qquad (14.15)$$

which is the counterpart to equation 14.1 for environmental state j. To facilitate the analysis assume that:

(a) $E_{11}>E_{12}$ and $E_{22}>E_{21}$, meaning that activity 1 has its highest yield in environmental state 1 (wet year) and activity 2 has its highest yield in environmental state 2 (dry year);
(b) $R_{11}(0)>R_{21}(0)$, meaning that activity 1 dominates near the market town in environmental state 1;
(c) $k_{2,max}>k_{1,max}$, meaning that activity 2 always has a greater extensive margin than activity 1; and
(d) environmental state 1 occurs with probability q_1 and environmental state 2 with probability q_2, where $q_1+q_2=1$.

Assumption (a) ensures that the analysis is meaningful while assumptions (b) and (c) are the usual ordering principles of Thünian analysis. With the above assumptions it is an easy matter to estimate the expected yield E_1^* and E_2^* for each of the two activities in any given year:

$$E_1^* = q_1E_{11} + q_2E_{12} \qquad (14.16a)$$
$$E_2^* = q_1E_{21} + q_2E_{22} \qquad (14.16b)$$

Consider the numerical values (in appropriate units) for yields given in Table 14.5. Finally, assume that wet years occur with probability q_1=0.60 and dry years with probability q_2=0.40. The values chosen for these yields and probabilities ensure that activity 1 always dominates the land closest to the market town; that is, per assumption (b), no reversal of the rent gradients occurs where activity 2 outbids activity 1 for land near the market town. These particular parametric values mean that the two expected yields in the average year are E_1^*=16 and E_2^*=14, respectively.

Table 14.5. Yield data for two environmental states

	Activity 1	Activity 2
State j=1 (wet year)	20	10
State j=2 (dry year)	10	20

14.5.1 Maximum Expected Return

Cromley (1982) has pointed out that the landowner maximises the long-run return on land by estimating the highest expected rent at each distance k from the market town. The expected return $R_i*(k)$ depends upon the magnitudes of the two state probabilities as well as the associated yield for each state. In the case of the traditional model the two expected-value rent gradients can be computed directly as follows:

$$R_1*(k) = q_1R_{11}(k) + q_2R_{12}(k) = 44 - 3.20k \quad (14.17a)$$
$$R_2*(k) = q_1R_{21}(k) + q_2R_{22}(k) = 28 - 0.70k \quad (14.17b)$$

where the point of indifference is k_{12}=6.4 and the extensive margin for activity 2 is $k_{2,max}$=40. In other words, landowners maximise their long-run rent by placing land between k=0 and k=6.4 into activity 1 and land between k=6.4 and k=40 into activity 2. It is worth stressing that landowners cannot guarantee this average return to land in any one given year; instead the rent gradients now represent the return to land they can expect to receive, on the average, over a period of many years.

Equilibrium solutions for the closed model can be specified by adjusting the coefficients of the matrices **X**, **Y**, and **Z** stated earlier in equation 14.10. Simply substitute E_1^* for E_1 and E_2* for E_2 and then calculate the price vector **p** in the usual way per equation 14.11. This operation leads to solutions for k_{12} and $k_{2,max}$ as before. However, now the solution space is more constrained than earlier and two different demand curves are introduced:

$$Q_1^D = 225 - 10p_1 \quad (14.18a)$$
$$Q_2^D = 1100 - 30p_2 \quad (14.18b)$$

The data in the first column of Table 14.6 show the equilibrium solutions for the expected-value model using the cost data from Table 14.1 and the yield data from Table 14.5. The long-run annual expectations for prices are p_1=5.823 and p_2=4.010. At k=0, the expected or average rent in any given year is R(0)=41.168. The prices p_1 and p_2 should be charged each and every year, irrespective of whether environmental state 1 or state 2 actually occurs in a given year. When wet years occur, with probability q_1, there is an annual surplus from activity 1 and an annual deficit from activity 2; when dry years occur, with probability q_2, there is a deficit from activity 1 and a surplus from activity 2. However, if landowners can freely store their surpluses, in the long run they will maximise their return on land by choosing the prices shown in Table 14.6.

Table 14.6. Alternative models for environmental uncertainty

	Maximum Expected Return	Guaranteed Return
p_1	5.823	6.008
p_2	4.010	4.007
k_{12}	5.211	10.027
$k_{2,max}$	40.201	40.148
A_1	10.423	10.307
A_2	69.978	69.988
O_1	166.768	164.914
O_2	979.698	979.811

Of course, these solutions not only depend upon the adopted input data but also the values of the two probabilities q_1, q_2. If environmental state 1 were to become more certain, for example, the results of the expected-value model would increasingly resemble those of the environmentally-certain model discussed in Section 14.3 earlier. Future research should clarify how equilibrium prices, land areas, and outputs move in response to different values for the two environmental probabilites.

14.5.2 Maximum Guaranteed Return

Cromley (1982) also pointed out that Gould's (1963) approach to subsistence farming behaviour leads to a maximum guaranteed return for landowners, which is independent of the probabilities of the two environmental states. This approach satisfies Wald's (1950) criterion that decision makers might assume the very worst will happen and then adjust their behaviour accordingly.

The most important observation, though, is that this guaranteed return can be identified continuously away from the market town, right up to the point of indifference. At each distance k ($0 \leq k \leq k_{12}$) a monotonically increasing function d (k), denoting the proportion of land that is placed into activity 1, can be solved as follows:

$$d_1(k) = [R_{22}(k) - R_{21}(k)] / \{[R_{11}(k) - R_{12}(k)] + [R_{22}(k) - R_{21}(k)]\} \tag{14.19a}$$

where

$$d_2(k) = 1 - d_1(k) \tag{14.19b}$$

For example, using the data from Table 14.1 for the traditional model, the proportion of land placed into activity 1 is

$$d_1(k) = (20 - 0.5k)/(47.5 - 2.5k) \tag{14.20}$$

which means that landowners at the market town place 42.11 percent of their land in activity 1 and 57.89 percent of their land in activity 2, thereby guaranteeing an annual return of R(0)=34.74 irrespective of the ensuing environmental state. As k becomes larger, more land is placed into activity 1 and less into activity 2 until the point of indifference for environmental state 1 is reached; at k_{12}=10, 66.67 percent of the land is in activity 1 and 33.33 percent is in activity 2, and the guaranteed annual return is R(10)=15. From k_{12} to $k_{2,max}$, all land is placed in activity 2, where the landowner assumes that environmental state 1 (wet year), the worst of the two possibilities, will prevail each year.

In order to substantiate the claim that the point of indifference k_{12} for state 1 represents the outer limit for mixed uses, first note that the guaranteed return at distance k near to the market town is:

$$GR_{12}(k) = A/B \quad (14.21)$$

where

$$A = (E_{11}E_{22}-E_{12}E_{21})(p_1-a_1-f_1k)(p_2-a_2-f_2k)$$
$$B = (E_{11}-E_{12})(p_1-a_1-f_1k) + (E_{22}-E_{21})(p_2-a_2-f_2k)$$

Next set

$$GR_{12}(k) = E_{21}(p_2-a_2-f_2k) \quad (14.22)$$

where the right-hand side shows rent for activity 2 under the worst possible condition (state 1). Then it can be shown that the point of indifference k_{12} for a wet year is

$$k_{12} = [E_{11}(p_1-a_1) - E_{21}(p_2-a_2)]/(E_{11}f_1-E_{21}f_2) \quad (14.23)$$

Further derivations show that at k_{12} the proportion of land in activity 1 must always be

$$d_1(k_{12}) = [E_{11}(E_{22}-E_{21})]/(E_{11}E_{22}-E_{12}E_{21}) \quad (14.24)$$

where the proportion in activity 2 is $d_2(k_{12})=1-d_1(k_{12})$. Finally, recall from equation 14.2 that the extensive margin for activity 1 in environmental state 1 is

$$k_{2,max} = (p_2-a_2)/f_2$$

In order to calculate the land area devoted to activity 1 it is necessary to integrate the land-use function $d_1(k)$ between the market town and the outer limit of mixed uses (on both sides of the town); that is

$$A_1 = 2\int_0^{k_{12}} d_1(k)\, dk \quad (14.25)$$

where the area devoted to activity 2 is

$$A_2 = 2k_{2,max} - A_1 \quad (14.26)$$

Individual behaviour, then, is based on an assessment of extreme environmental variability. Landowners, in perceiving uncertainty about their economic return, decide to mix the two activities together, at least up to the point of indifference. Expected yields E_1^* and E_2^* are used as before to determine the expected outputs for the average year, per equations 14.5a and 14.5b.

The right-hand data of Table 14.6 show solutions for the guaranteed return model.

Note that land out to k_{12}=10.027 is mixed between both activities while all land beyond that point is placed only in activity 2. Equilibrium prices, land areas, and outputs are only marginally different from those of the expected-value model but the land-use pattern is very different. As before, landowners can have surpluses or deficits for either of the two activities in any given year, and the solutions depict average behaviour over the long run. At k=0, 42.13 percent of land is placed in activity 1 and 57.87 percent is placed in activity 2; here the return to land, irrespective of the environmental state, is always R(0)=34.850.

Based on these numerical results it is difficult to accept that landowners would ever choose the guaranteed-return strategy in a capitalist economy; on the other hand, in Gould's (1963) case, where exchange does not take place, the strategy is perfectly rational to adopt as long as the annual guaranteed (physical) return exceeds the subsistence level. As the results of Table 14.6 do indicate, location-specific rents are always higher when landowners adopt the maximum expected-return strategy as opposed to the guaranteed-return strategy, and this considerable rent differential perists throughout the entire von Thünen economy.

So why are real world agricultural landscapes often a mosaic of different land uses, even in the capitalist societies? One possible answer rests in a key assumption of the two models, that related to the free storing of all nonperishable surpluses. If landowners could not hoard their surplus output during good years, and all surplus output was disposed of on the market, then prices would drop dramatically during those good years. Likewise prices would rise steeply during years of shortage. As in a portfolio model (Cromley and Hanink 1989), landowners might very well be willing to trade off some expected rent for less variability in annual rent and, possibly, this might persuade some to choose the second strategy over the first. (In fact these annual prices can be solved by applying the alternative demand relationships shown in Section 14.3.) Besides, in the real world landowners might have to meet financial commitments that vary from year to year, a factor that is not adequately addressed in this paper where constant non-land production costs are assumed. Finally, it is well known that information of various types is neither perfectly diffused nor perfectly used (Wolpert 1964): richer corporate farmers simply might make better use of long-run climatic information than poorer family farmers. More likely, though, corporate farmers have better information regarding future prices, another source of variation that is not addressed in the present paper.

This entire issue certainly requires more attention, especially when it is recalled that the results of Table 14.6 are based on specific parametric values adopted for yields, demand levels, and costs. In any case, given the aforementioned rent differentials, strict adoption of the guaranteed-return strategy seems unlikely in capitalist economies; in all likelihood the mosiac of agricultural land uses is due to the adoption of other strategies by landowners.

14.6 Conclusions

This paper has examined some implications of behaviour in agricultural markets. The analysis has dealt with a one-dimensional von Thünen model that has been closed for market demand. While the analysis has only dealt with i=2 activities, the analysis is fully generalisable to many activities. The traditional, production-oriented version of the von Thünen model was shown to be deficient for comparative static predictions, especially since that version cannot capture price interdependencies.

The paper is preliminary in certain respects and it has various shortcomings. However, it does suggest several avenues of further inquiry. First, more attention should be paid to the geographic sources of demand. Intraregional and interregional components for demand could be identified and the effects of shifts in either source of demand could be examined. Besides, localized rent surfaces around secondary market towns could be examined. Second, more study could be devoted to the interdependency between urban growth and agricultural land-use change. The demand relationships for residents of the city should be changed to allow specification of trade-offs between the consumption of urban land, urban transportation, and agricultural goods. Also, the issue of environmental uncertainty requires much more thought. Doubtless the portfolio approach can be adapted to the closed von Thünen model. Finally, yields might be considered more satisfactorily in a production-function model as this would allow scale economies to be introduced.

References

Cromley, R.G. 1982. The von Thünen model and environmental uncertainty. Annals of the Association of American Geographers, **72**:404-410.

Cromley, R.G. and D.M. Hanink 1989. A financial-economic von Thünen model. Environment and Planning A, **21**:951-960.

Davenport, W. 1960. Jamaican fishing: a game theory analysis. Yale University Publications in Anthropology, **59**:3-11.

Dunn, Jr., E.S. 1955. The equilibrium of land-use patterns in agriculture. Southern Economic Journal, **21**:173-187.

Foust, J.B. and A.R. de Souza 1978. The Economic Landscape. Charles Merrill, Columbus, Ohio.

Gould, P.R. 1963. Man against his environment: a game theoretic framework. Annals of the Association of American Geographers, **53**:290-297.

Jones, A., W. McGuire and A. Witte 1978. A reexamination of some aspects of von Thünen's model of spatial location. Journal of Regional Science, **18**:1-15.

Jones, D.W. 1983. Location, agricultural risk, and farm income diversification. Geographical Analysis, **15**:231-246.

Jones, D.W. 1991. An introduction to the Thünen location and land use model. Research in Marketing, **5**:35-70.

McCarty, H.H. and J.B. Lindberg 1966. A Preface to Economic Geography. Prentice-Hall, Englewood Cliffs, N.J.

Nerlove, M.L. and E. Sadka 1991. Von Thünen's model of the dual economy. Journal of Economics, **54**:97-123.

Peet, R. 1969. The spatial expansion of commercial agriculture in the nineteenth century: a von Thünen interpretation. Economic Geography, **45**:283-301.

Samuelson, P.A. 1983. Thünen at two hundred. Journal of Economic Literature, **21**:1468-1488.

Smith, T.R. 1977. Uncertainty, diversification, and mental maps in spatial choice problems. Geographical Analysis, **10**:120-140.

Wald, A. 1950. Statistical Decision Functions. John Wiley, New York.

Wolpert, J. 1964. The decision process in a spatial context. Annals of the Association of American Geographers, **54**:537-558.

Part C

CI-based Spatial Analysis

15 Neurocomputing for Earth Observation - Recent Developments and Future Challenges

Graeme G. Wilkinson

Joint Research Centre, European Commission, 21020 Ispra, Varese, Italy.

15.1 Introduction

Earth observation, or satellite remote sensing, is now well-established as one of the principle methods of obtaining spatial information concerning the terrestrial environment. Its primary advantages over ground-based survey approaches are timeliness, spatial coverage, and low cost, though these advantages are often bought at the expense of descriptive precision. However, the enormous wealth of timely and low cost information, coupled with the need to generate high precision spatial products descriptive of landscapes poses some of the most challenging computational problems in geographical science -both in terms of data throughput and complexity of analysis. Neurocomputing is one tool that has recently made a significant and growing contribution to this discipline.

Above all, there is a requirement in remote sensing for tools which can easily extract useful products from complex data sets which result from the fusion of multispectral, multitemporal and multi-sensor observations. There has been widespread criticism in recent years that the efforts devoted to the development of satellites and sensor systems have not been met with equivalent efforts to exploit the data which they generate. This is certainly a valid criticism, though one which is slowly being acknowledged resulting in new integrated research, development and market stimulation activities, such as the European "Centre for Earth Observation Programme" (CEO, 1995), to rectify the problem. Underlying the whole field, however, is the basic need for algorithms which optimize the spatial products extracted from remotely-sensed data sets without the need for excessive human intervention. The need to develop "intelligent" methods to transform raw data into usable products is strong. The integration of numerical, symbolic and connectionist processing together with the use of advanced high speed computer architectures is likely to feature highly in the long term development in this field. This chapter will concentrate primarily on the examination of the state of the art with regard to connectionist or neural computation in remote sensing and on future expectations from such approaches. Neurocomputing has been recognized as a potentially valuable technique in geographical data modelling and problem solving in the last few years (Fischer and Gopal, 1993) and the field of remote sensing is one example of an area in which the benefits are beginning to be exploited.

15.2 Neural Networks in Remote Sensing To Date

15.2.1 The Growth of Neurocomputing in Remote Sensing

The first results demonstrating the use of neural networks in the analysis of Earth observation data appeared in the year 1988. Since that time, there has been a rapid growth in the amount of research devoted to neurocomputing in remote sensing as indicated by the increase in the number of papers appearing in the principal journals and major international conferences (figure 15.1).

The range of uses of neural networks in remote sensing over the past few years is impressive (table 15.1), covering many different operations on remote sensing data and a variety of thematic applications. By far the largest body of research has been devoted to the classification problem, however.

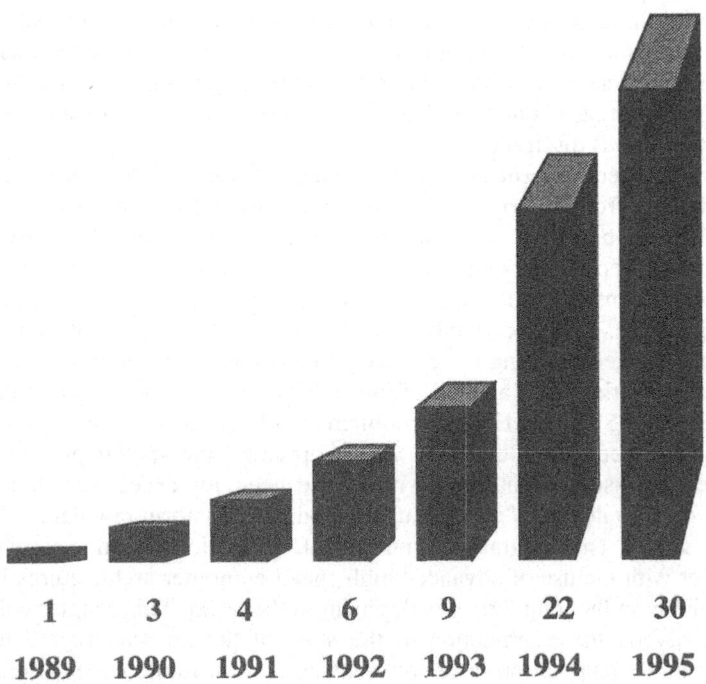

Fig. 15.1. Growth in Number of Research Papers Devoted to Neurocomputing in Remote Sensing Found by the Author in the Principal Remote Sensing Journals and Major International Conferences for the Period 1989-1995.

Table 15.1. Examples of uses of neural networks in data preprocessing and /or generic analysis in remote sensing

Use of Neural Networks	Example Relevant References
Data Pre-Processing / Generic Operations	
Multi-sensor data fusion / classification	Kanellopoulos et al. 1994, Key et al. 1989, Wilkinson et al. 1995b
Multi-date image classification	Kanellopoulos et al. 1991
Geometrical rectification	Liu and Wilkinson, 1992
Texture classification	Kanellopoulos et al. 1994, Shang and Brown 1994
Linear object detection	Chao and Dhawan 1994, Hellwich 1994
Multi-view angle image analysis	Abuelgasim and Gopal 1994
Stereo image matching	Loung and Tan 1992
Smoothing SAR imagery	Ellis et al. 1994
Multi-scale data analysis	Moody et al. 1994
Integration of prior knowledge	Foody 1995
Contextual classification	Wan and Fraser 1994
Very high dimensional data analysis	Benediktsson et al. 1993
Unsupervised data clustering	Wilkinson et al. 1993
Environmental Thematic Applications	
Land cover mapping	Civco 1993, Dreyer 1993 Kanellopoulos et al. 1992
Forest mapping	Wilkinson et al. 1995b
Crop recognition from SAR imagery	Foody et al. 1994
Rainfall estimation	Zhang and Scofield 1994
Precipitation cell top estimation	Spina et al. 1994
Cloud classification	Lee et al. 1990, Wilkinson et al. 1993
Sahelian land cover change mapping	Gopal et al. 1993
Sea ice mapping	Kwok et al. 1991, Maslanik et al. 1990
Vegetation canopy scattering inversion	Pierce et al. 1994
Ship wake detection	Fitch et al. 1991
Shoreline feature mapping	Ryan et al. 1991

15.2.2 Neural Network Classification Experience in Earth Observation

The classification process is one of the most fundamental in the operational use of satellite remote sensing for mapping. Historically, although the idea of building pattern classification algorithms around simple interconnected processing elements (e.g. the perceptron) emerged in the 1960's, theoretical limitations to their capabilities -as identified by Minsky and Papert (1969) reduced the pace of development of connectionist approaches for many years. The enthusiasm was not renewed until the 1980's when new connectionist models emerged, such as the "self-organizing map" (Kohonen, 1984) and the "adaptive resonance theory" (Carpenter and Grossberg, 1987). A further land-mark event was the development of the "backpropagation" algorithm for training multi-layer perceptron systems (Werbos, 1974; Rumelhart et al., 1986). The relative simplicity of the multi-layer perceptron model and the backpropagation algorithm led to an explosion of trial applications from the late 1980's -often with interesting and unexpected results in remote sensing. This explosion of trials also encompassed other types of neural networks and various hybrid models.

The detailed characteristics of the various neural network models here will not be examined here as there are numerous texts covering such models (e.g. Simpson, 1990; Kung, 1993). It is nevertheless important to briefly consider the main features and common elements of typical neural network systems. Essentially, all neural networks are based around the idea that a set of primitive processing elements ("neurons"), when linked together by connections which carry variable strengths or "weighting" factors, are able to encode relatively complex mathematical transformations. These transformations appear to be suitable for a wide class of mathematical problems -especially those which concern supervised or unsupervised pattern recognition. Connectionist systems can also be viewed as "adaptive" systems, and even "learning" systems, since the strengths or weights of the connections are normally learned from training data. A connectionist or neural network system can therefore be seen as a scalable trainable transformer of data from an input space to an output space. The precise nature of the mathematical input and output spaces will depend on the application and the form of the training data. One of the important consequences of the rather general behaviour of most neural network models, is that they potentially have many different uses. This partly explains the current explosion in neural network research -both on theory and applications- worldwide. A second possible reason for the explosion or rapid growth in the use of connectionist approaches is that since they consist of large groups of interconnected processing elements, they can be potentially mapped (or even hard-wired) into parallel computer hardware. However, it is notable that few good parallel learning algorithms yet exist for most of the current neural network models. The appeal of connectionist algorithms therefore lies in their achievement of complexity from simple building blocks and their apparent suitability for parallel computing. These are both highly desirable characteristics for application domains which involve huge quantities of data requiring relatively complex processing such as Earth observation.

To date most of the research on the application of neural networks for classification in remote sensing has been based on the multi-layer perceptron network (though see chapter 17 for an interesting exception to this). Much of this work has been reviewed by Paola and Schowengerdt (1995). Numerous comparisons have been made between the multi-layer perceptron networks and the more traditional statistical classification approaches such as the maximum-likelihood classifier (e.g. Benediktsson et al., 1990). Overall the results of such comparisons are mixed, though the neural networks appear to win (in the sense of giving better overall accuracy in spatial products) more often than the statistical classifiers. However this finding depends significantly on the network architectures used (i.e. numbers of layers and neurons per layer) and on the fact that it is very difficult to establish a network which is equivalent to a given statistical model in terms of its ability to distinguish classes by virtue of the very different underlying principles of operation. Whether or not neural networks are regarded as being better classifiers than statistical methods, it has been argued that they can produce better results in the hands of inexperienced users than statistical methods (Furby, 1996). In general, the success of neural network classification has been found to depend significantly on data regularization and training procedures.

15.2.3 Data Regularization and Network Training Issues

The regularization of remotely sensed data is an essential process in their classification on account of the way the data are recorded. Most spectral measurements recorded by satellite sensors are discretized and coded as integer numbers in the range 0-255 (though there are some exceptions to this). A multi-layer perceptron neural network, however, applies an activation function at the first layer nodes to weighted sums of the inputs which come from the satellite measurements. This activation function (usually a sigmoid function or hyperbolic tangent) saturates within the approximate bounds [-1.7,+1.7] and is most sensitive in the middle of this range. In order to ensure that a network has the possibility to learn how to classify it is necessary to scale input values to the range of the activation function in order to avoid saturation. This is a straightforward operation, though one which is essential otherwise a network will not descend in the error space to achieve a high quality classification. Indeed the achievement of a good classification is effectively a global minimization problem involving the search for minimum error in a high-dimensional weight space. The attainment of the minimum error can not be guaranteed in less than infinite time which is a fundamental objection of many practitioners. Usually a network can be trained to descend to a level of error which is acceptable for the mapping purpose of interest. However the fact that this can not be guaranteed, coupled with the fact that it depends on an iterative training process which starts off with random weights, often makes the neural method less acceptable in remote sensing than traditional statistical approaches.

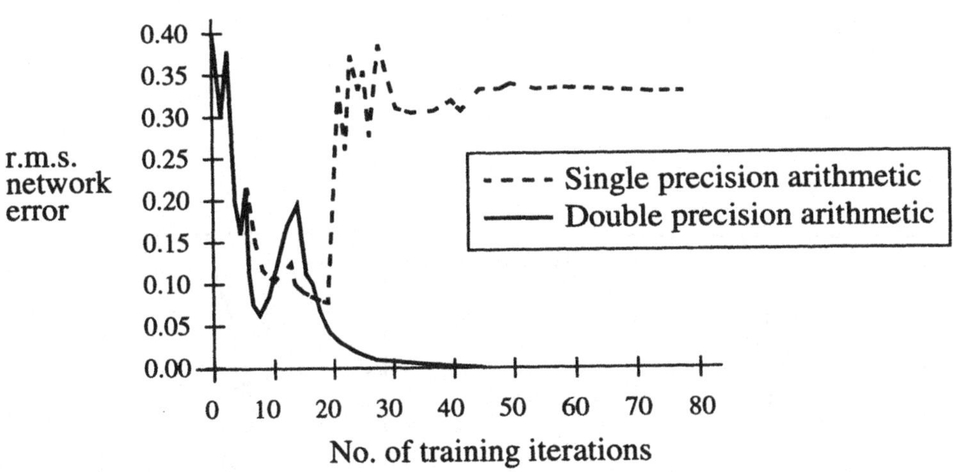

Fig. 15.2. Manifestation of Chaos in Multi-Layer Perceptron Training. Network performance as indicated by the r.m.s. error is over-sensitive to arithmetical rounding errors. (Training commenced with identical network architectures and weight sets in both single and double precision experiments).

A further problem encountered with neural networks in certain circumstances is that since they are based on non-linear activation functions they are susceptible to falling into "chaotic" regimes (Van der Maas et al., 1990) in which results are highly unpredictable and very sensitive to starting conditions and even to arithmetical rounding errors (figure 15.2).

15.2.4 The Generalization and Over-fitting Problem

One of the critical factors in determining the success of a neural network classifier is its performance in generalization -i.e. how well it classifies unseen data samples (i.e. pixels) after it has been trained. Ideally a neural network's classification performance must be tested on "unseen" data (i.e. data not used in training) at regular intervals during the training process and the accuracy of classifying such data should improve. However, networks may have a tendency to over-fit the training data and to match their discriminant surfaces too closely to the training data as illustrated in figure 15.3. This tends to happen with large networks in particular. One of the problems is assessing whether over-fitting is occurring is that it is extremely difficult to visualize the behaviour of neural classifiers in feature space -especially in remote sensing applications. In remote sensing, it is usually necessary to classify more than three spectral or temporal "channels" at a time. This creates a mathematical feature space which is extremely difficult to visualize. Interest is now developing in the use of tools to reduce dimensionality such as projection pursuit (Huber, 1985; Jimenez and Landgrebe, 1994) and to use advanced visualization hardware -e.g. using "virtual reality" devices which give three-dimensional viewing and the possibility to interact with, and even manually "sculpt", class discrimination surfaces, though this work is currently in its infancy (Fierens et al., 1994).

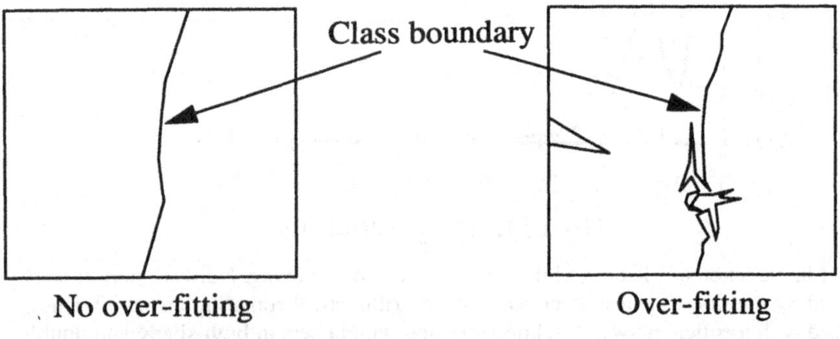

Fig. 15.3. Effect of Over-Fitting by a Multi-Layer Perceptron in a Two-Dimensional Feature Space Derived from Landsat Thematic Mapper Bands. The class discrimination boundary on the right has matched far too closely to the training data and is unlikely to give a good general separation of two broad land cover classes.

15.2.5 The Development and Potential of Hybrid Classifiers and Alternative Network Models

Interestingly, recent work on studying the feature space behaviour of neural classifiers has demonstrated that there can be significant differences between the way in which neural networks and statistical classifiers separate data classes. These differences come from the underlying mathematical models and explain, for example, why a statistical classifier can be found to perform well with a certain group of data classes whilst a neural network trained on the same data will perform less well on that particular group of classes but much better on another group. This has led to the development of simple strategies to integrate statistical and neural classification. A typical approach is to use statistical and neural network classifiers in parallel and to pass pixels on which the two types of classifier disagree to a third classifier (neural network) which has been specifically trained to deal with "difficult" samples (Wilkinson et al., 1995a). Adoption of such strategies has been found to give significant improvement in total classification accuracy which therefore results in thematic spatial products of higher quality and utility. In most cases of such strategies, however, the underlying neural model is the multi-layer perceptron. Recent work by Gopal et al. (1993) and Fischer and Gopal (1996) has indicated, however, that very high classification performance can be reached in remote sensing by use of alternative -and much rarer- network models i.e. the Fuzzy ARTMAP (see chapter 17 in this volume) without the need for hybrid classifier development. It is clear that hybrid techniques and some of the newer neural network models deserve more attention in remote sensing in the future and could form the basis of a new generation of classifier tools.

15.3 The Hard Versus Soft Classification Dilemma

Our discussion so far has been based on the underlying assumption that the aim of classification -neural or non-neural- is to transform spectral measurements made by satellites into fixed thematic categories which can describe a landscape -i.e. as a thematic map. This is certainly a gross simplification of reality. Landscapes are notoriously difficult to describe by fixed class labels even though we have a natural cartographic tendency to do so. Cartography originated in the very human desire to simplify, make sense of, and label the natural world in a way that would facilitate easy communication of information and navigation. Yet recent experience with land cover mapping has demonstrated the quasi-impossibility of finding an adequate set of simple labels that can describe all situations found in nature. Recent efforts to create land cover maps at global or continental scale have often been plagued by difficulties concerning the definition of appropriate nomenclatures. Interestingly, even in cases where nomenclatures have been carefully defined, considerable subjectivity is found to exist in maps and it is quite common for different map producers working on adjacent map sheets to label boundary-crossing parcels with different categories. This has been a typical experience, for example, with the European Commission's continent-wide CORINE Land Cover project. Modern geographical information systems now permit us to store much more sophisticated information about landscape parcels than single fixed labels and so we can begin to move away from the concept of single fixed category labels -i.e. away from "hard" classification to "soft" classification which permits the assignment of multiple labels -with some attached "abundance" measure. In theory this can allow for natural intergrades to exist between categories in a numerically defined way -e.g. on the basis of fuzzy class membership functions. This transformation from hard to soft classification may actually help remote sensing to generate products which are more acceptable to end users and decision makers. Unfortunately most evidence to date points to a rather spectacular failure of the hard classification approach. For typical experiments involving hard classification of whole scenes taken from the Landsat Thematic Mapper instrument, for example, average total classification accuracies on "unseen" test sets rarely exceed 90% with figures more commonly of the order of 80%. The trend over the last few years is not encouraging even though there has been a rapidly rising use of neural network methods, notwithstanding the earlier comments on the potential of hybrid systems and new network models. Table 15.2 shows some statistics on total classification accuracies as reported in peer-reviewed papers in one journal over a recent three year period. Total classification accuracies of the order of 80% are inadequate for a significant majority of applications of Earth observation on land. For example in measuring the area of land covered by forest in the European Union, an error rate of 20% would be approximately equivalent to omitting three times the forest area in the whole of France. This level of error is insufficient for most policy or decision-making purposes in the environmental field.

In order to produce "soft" classification from remote sensing via neural networks, several approaches are possible. One simple method is to define a "fixed set of mixed classes" and to carry out a normal hard training and classification process -i.e. the classification is a hard classification into a fixed set of classes, but some class labels actually represent pure categories and others some mixed categories. This approach was used for example by Wilkinson et al. (1995b) who classified forested areas in Portugal from multisource imagery. In that experiment,it was found that the existence

Table 15.2. Average total classification accuracies reported in papers in the journal Photgrammetric Engineering and Remote Sensing 1993-1995

Year	No. of papers examined	No. classification experiments	Average Correct	Standard Deviation
1993	9	11	82.35%	3.75%
1994	12	37	77.2%	9.38%
1995	9	23	80.4%	13.36%

Note: in some cases more meaningful classification accuracy measures such as the Kappa statistic were given by authors. however the total Percentage Correctly Classified (PCC) value for image pixels is the one which is quoted most frequently and which thus allows the widest possible sample and intercomparison.

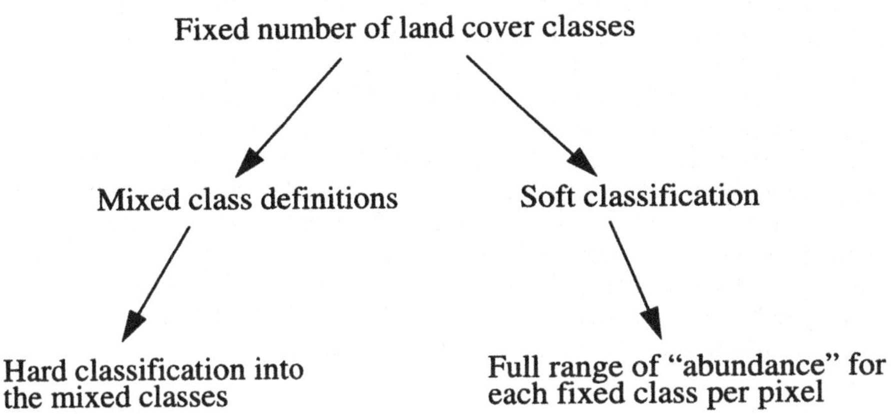

Fig. 15.4. Different Approaches to "Soft" Classification in Earth Observation and Land Cover Mapping.

of mixed-species forest stands required such an approach. A multilayer perceptron network was specifically trained with examples of mixed forest areas from ground truth information and then used to hard classify the image into a fixed set of classes of which some were mixed.

An alternative approach, is to use a totally soft classification process in the sense that a fixed set of class labels are used, with one neural network output node per class, but the strength of each output is interpreted as a measure of the "abundance" or fuzzy membership value of that class for the pixel of interest. Theoretically a multi-layer perceptron network's output, for example, should approximate the a posteriori Bayesian probabilities of each class for the pixel (Wan, 1990; Kanaya and Miyake, 1991), which we can assume are closely related to the percentage coverages of each particular category present in the "mixed" pixel. These alternative approaches to soft classification are illustrated in figure 15.4.

If a "totally soft classification" process is used, there is then a choice of strategy for training a neural network. For example, should a network be trained only with examples of "100% pure classes" or should it be trained with pure and mixed classes? In either case the network output can still be given a soft interpretation. In recent research at the Joint Research Centre it has been found that training with mixed classes as well as pure is best (figure 15.5). The network training set included "2-component" mixtures (i.e. pixels which were known to be mainly mixtures of two classes -a dominant and subsidiary- in known proportions). Clearly in such cases, the network output target vector has to be defined in a way which relates to the class proportions -different methods for doing this were used in the work described based either on linearly scaling the proportions within the range [-1,+1] or by binning the proportions (Bernard and Wilkinson, 1995).

Leaving soft training issues aside, overall there have been several recent applications of soft or fuzzy classification in Earth observation with good results (e.g. Foody 1996a, 1996b).

Although the recent trend towards soft classification is to be welcomed, it does have one significant drawback. Ultimately complex land cover descriptions coming from soft classification of Earth observation data have enormous potential and can be stored in geographical information systems in full. The percentage cover of each category within each pixel or spatial unit can be given a non-integer real value if necessary permitting an infinite number of continuous mixed classes. Although this can be extremely flexible, the generation of "maps" (in their traditional sense) from such continuously varying landscape descriptors is not easy. Conventional maps require a fixed easy-to-understand set of thematic categories which can be easily printed on a coloured map and understood by an end-user. It therefore becomes necessary to "harden" or discretize the soft landscape information again to render it into a form which is usable by human decision makers.

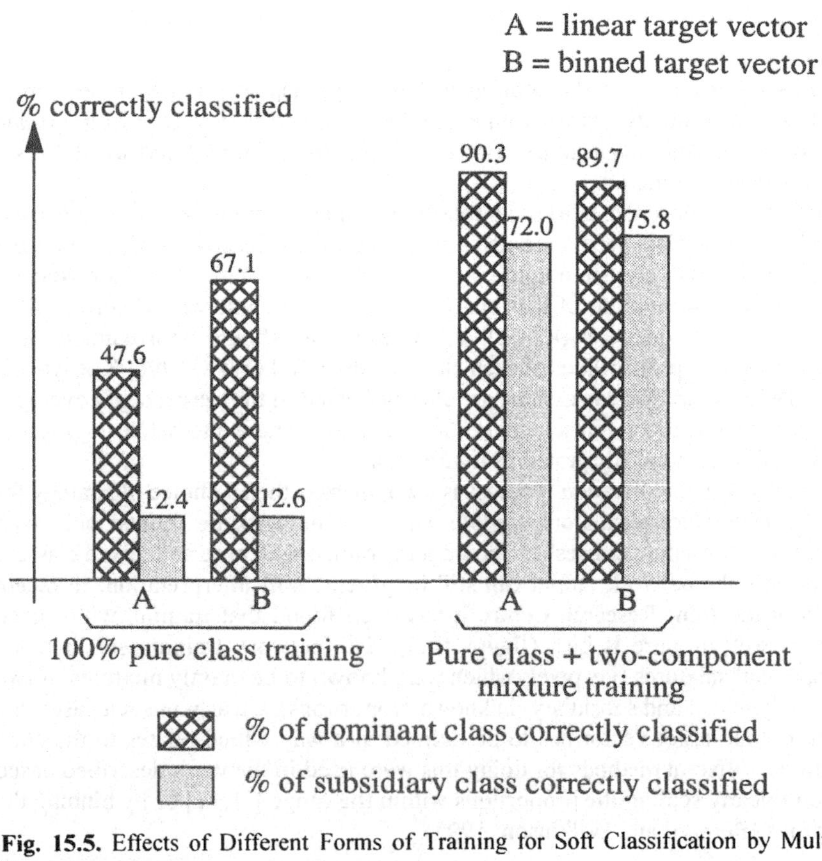

Fig. 15.5. Effects of Different Forms of Training for Soft Classification by Multi-Layer Perceptron.

15.4 Future Challenges

15.4.1 Operational Large Classification Problems and Classifier Portability

One of the main advantages of remote sensing over other more conventional territorial survey methods is that in principle large areas can be mapped very quickly and on a regular basis. This relative advantage increases with the size of the area mapped. For operational purposes, it is useful to be able to map areas as large as whole countries or continents. Indeed many international authorities are now interested in global coverage of land cover for example. The classification of image mosaics covering large areas (i.e. national -- multi-national coverage) is extremely difficult, mainly because examples of the same land cover type can change significantly from one region to another e.g. through differences in soil background, geology, climate, vegetation phenology, construction techniques and materials. Indeed the challenge of large area mapping is arguably the single most important challenge in remote sensing to date. It is extremely difficult to train classifiers so that they generalize well to an area of more than 100 x 100 km. Recently experiments have been conducted to evaluate the performance of networks trained with data from one part of Europe and used to classify land cover in another. Table 15.3 shows the results of attempting to classify land cover in the Département Ardèche in southern France with a neural network trained to recognize the same classes in the Lisbon area in Portugal and vice versa. The table shows that when networks are trained in one geographical area and used for classification in a foreign test area, the overall classification accuracy is extremely low. By continuing the training with some local ground truth the "foreign" classifier can be made to adapt to "local" conditions though it can not be made to perform as well as a classifier which has been trained only with local ground truth data.

15.4.2 Automating the Neural Process

A critical step in the development of neural systems in Earth observation or in any other field of spatial data analysis must ultimately be the creation of fully automatic systems which require zero human intervention. Most geographers do not want to actually have to think about issues such as which neural network architecture to use, how to set up training data, how to organize training and evaluate performance. Systems are needed which are essentially "switch-on, learn, and classify systems" (SOLACS) -without human interaction. The user should not even need to be aware which kind of algorithm has been employed. In such systems all aspects of architecture selection, parameter setting, evaluating results and, if necessary, re-training should be handled automatically. There is no reason why this can not take place, however it requires a new emphasis on neural software engineering with end-user requirements in mind. Neural network systems need to move from the research arena to the operational arena as parts of a geographical data analysis tool-box -which ideally should be totally user-friendly requiring no deep understanding of what is going on inside the "box" in just the same way as a user of

Table 15.3. Average total classification accuracies from an experimental test of classifier portability across Europe.

	Nature of Experiment	PCC
LOCAL	Classification of Lisbon area with network trained with local ground truth	71.1%
LOCAL	Classification of Ardèche area with network trained with local ground truth	63.9%
FOREIGN	Classification of Lisbon area with network trained with Ardèche area ground data	14.7%
FOREIGN	Classification of Ardèche area with network trained using Lisbon area ground data	12.7%
FOREIGN + RE-TRAIN	Classification of Lisbon area with network trained using Ardèche area ground data with partial re-training with local ground data	67.1%
FOREIGN + RE-TRAIN	Classification of Ardèche area with network trained using Lisbon area ground data with partial re-training with local ground data	51.2%

The table shows the results of classifying 9 classes of land cover in the Ardèche area of France and the Lisbon area in Portugal (both representative of Mediterranean landscapes) from 3 spectral + 3 textural image features. The table shows the comparative performance of: networks trained locally and used to classify locally (LOCAL); networks trained on one site and used to classify the other (FOREIGN); and networks trained on one site and used to classify the other after some additional re-training of the net with local ground truth (FOREIGN + RE-TRAIN). Multi-layer perceptron networks were used in each case. [Adapted from DIBE (1995)]

a word processing system has no need to know how the system goes about complex tasks such as text justification.

An important requirement in the automatic use of neural systems is the possibility for training to be controlled automatically so that good performance and generalization capability are achieved. Since it is not possible to give reliable heuristics for specifying the ideal network architecture or learning parameters for a one-off training session, it is necessary to allow an automatic system to experiment with different configurations in a short time frame -i.e. one which is realistic from the point of view of a waiting user.

A key requirement is for computer hardware which offers fast network prototyping -i.e. by rapidly configuring, training and testing a variety of different networks and comparing ultimate performance levels against user-defined or system-set goals. This is only likely to become feasible by exploiting special purpose hardware which is adapted specifically for fast neural network operations. A typical example of such a system is the SYNAPSE-1 computer built by the Siemens-Nixdorf Advanced

Technologies company. This machine has special hardware memories for storing network weights plus special VLSI microchips for fast network node operations (Ramacher et al. 1994) -figure 15.6. This machine has recently been used in trials on remote sensing problems. The SYNAPSE-1 also has software tools which make neural systems development relatively easy --these include a high-level "neural algorithms programming language" (nAPL) together with an even higher-level visual programming environment "ECANSE" -Environment for Computer-Assisted Neural Systems Engineering. Although such hardware and programming tools are intended for the neural network software engineer, ultimately it should be possible to deliver turn-key "SOLAC" systems based on this technology which are directly under the control of non-expert end users.

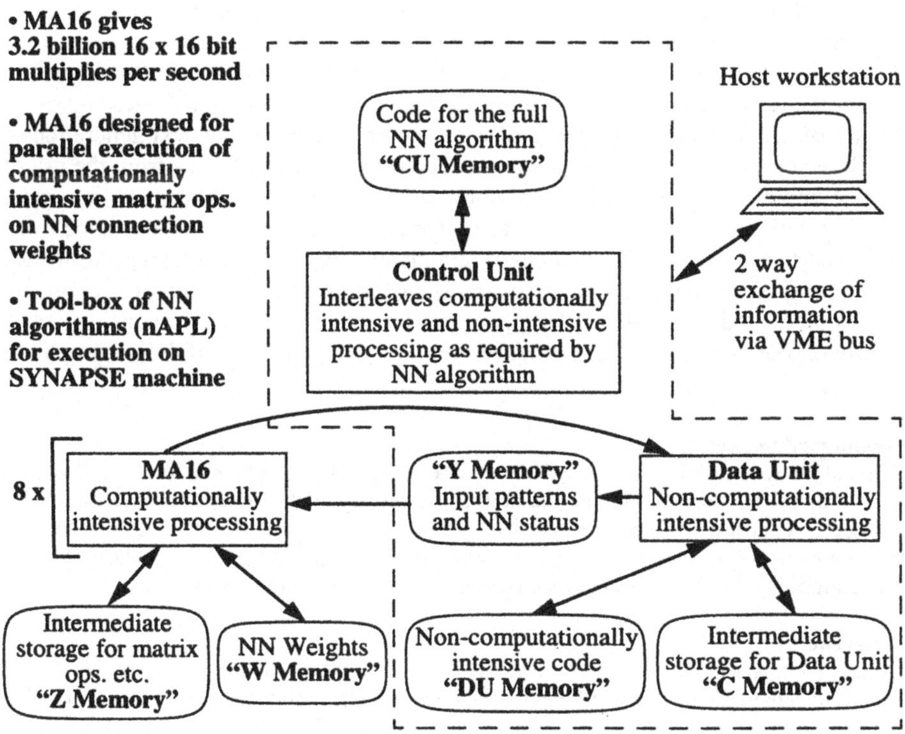

Fig. 15.6. An Overview of the Siemens-Nixdorf SYNAPSE-1 Neurocomputer (SYnthesis of Neural Algorithms on a Parallel Systolic Engine).

15.4.3 New Applications of Neurocomputing within Earth Observation

So far this chapter has been concerned primarily with the use of neural networks in data classification, which has been by far their greatest use in remote sensing to date. However, it is worth reflecting on the fact that neural networks can be trained to encode any arbitrary data transformation and could therefore be used within Earth observation in a wider range of applications such as geometrical matching of imagery and ground data, data inversion, stereo matching, atmospheric correction etc. To date there have been limited experiences with the first three of these (e.g. Liu and Wilkinson, 1992; Pierce et al., 1994; Loung and Tan, 1992)

15.5 Conclusions

Neurocomputing has recently been making a significant and worthwhile contribution to the science of Earth observation --principally in the extraction and interpretation of valuable spatial information from space imagery. Neural network techniques have now reached a mature level of application in remote sensing and a vast body of evidence has been accumulated about optimal architectures, training approaches etc. The main challenges for the future on the algorithmic front are to extend existing approaches to cope with more complex problems -e.g. larger data volumes and much higher dimensional feature spaces offered by the sensors on the next generation of Earth observing systems. Also we can expect to see major benefits from the use of neurocomputing for a few additional complex data processing procedures in remote sensing which currently rely on more conventional approaches. At the same time, however, there are also new challenges associated with delivering high performance computing on special purpose hardware to enable rapid prototyping of networks to take place, and with the delivery of zero-user-intervention systems to make neurocomputing just another element in the every-day tool box of the geographer.

Acknowledgments
The author would like to thank the many colleagues and co-workers at the Joint Research Centre of the European Commission, and also external collaborators, who have directly or indirectly contributed over the last few years to the generation of the results and the development of the views which are expressed in this chapter, particularly Ioannis Kanellopoulos, Freddy Fierens, Alice Bernard, Charles Day and Aristide Varfis.

References

Abuelgasim, A. and Gopal, S. 1994. Classification of multiangle and multispectral ASAS data using a hybrid neural network model. *Proceedings International Geoscience and Remote Sensing Symposium (IGARSS 94)*, held at Pasadena, 8-12 August. IEEE Press, Piscataway, NJ. Vol. 3, 1670-1672.

Benediktsson, J. A., Swain, P. H., and Ersoy, O. K. 1990. Neural network approaches versus statistical methods in classification of multisource remote sensing data. *IEEE Transactions on Geoscience and Remote Sensing*, **28**(4):540-552.

Benediktsson. J. A., Swain. P. H., and Ersoy. O. K. 1993. Conjugate gradient neural networks in classification of multisource and very-high dimensional remote sensing data. *International Journal of Remote Sensing*. **14**(15):2883-2903.

Bernard. A. C. and Wilkinson. G. G. 1995. Neural network classification of mixed pixels. *Proc. International Workshop on "Soft Computing in Remote Sensing Data Analysis"*. Milano. Italy. 4-5 December 1995. Organized by Consiglio Nazionale Delle Ricerche. Italy.

Carpenter. G. A. and Grossberg. S. 1987. ART2 - self-organization of stable category recognition codes for analog input patterns. *Applied Optics*. **26**:4919-4930.

CEO. 1995. *Centre for Earth Observation (CEO) Concept Document*. European Commission. Joint Research Centre. Document Reference CEO/160/1995.

Chao, C.-H. and Dhawan, A.P. 1994. Edge detection using a Hopfield neural network. *Optical Engineering* **33**(11):3739-3747.

Civco. D. L. 1993. Artificial neural networks for land-cover classification and mapping. *International Journal of Geographical Information Systems*. 7(2):173-186.

DIBE. 1995. Portability of neural classifiers for large area land cover mapping by remote sensing. final report on Contract 10420-94-08 ED ISP I. to Joint Research Centre. European Commission. by Department of Biophysical and Electronic Engineering (DIBE). University of Genova. Italy.

Dreyer. P. 1993. Classification of land cover using optimized neural nets on SPOT data, *Photogrammetric Engineering and Remote Sensing*. **59**(5):617-621.

Ellis, J., Warner, M. and White. R. G. 1994. Smoothing SAR images with neural networks. *Proc. International Geoscience and Remote Sensing Symposium (IGARSS 94)*, held at Pasadena, 8-12 August, IEEE Press. Piscataway. NJ. 4, 1883-1885.

Fierens. F., Wilkinson. G. G. and Kanellopoulos. I. 1994. Studying the behaviour of neural and statistical classifiers by interaction in feature space. *SPIE Proceedings*. **2315**:483-493.

Fischer. M. M. and Gopal. S. 1993. Neurocomputing and spatial information processing -from general considerations to a low dimensional real world application, *Proc. Eurostat/DOSES Workshop on New Tools for Spatial Data Analysis*. Lisbon. November. EUROSTAT. Luxembourg. Statistical Document in Theme 3. Series D. 55-68.

Fischer. M. M. and Gopal, S. 1996. Spectral pattern recognition and Fuzzy ARTMAP classification: design features. system dynamics. and real world simulations. chapter 16. this volume.

Fitch. J. P. et al. 1991. Ship wake detection procedure using conjugate grained trained artificial neural networks. *IEEE Transactions on Geoscience and Remote Sensing*, GE-**29**(5):718-726.

Foody, G. M. 1996a. Approaches for the production and evaluation of fuzzy land cover classifications from remotely-sensed data. *International Journal of Remote Sensing*. **17**(7):1317-1340.

Foody. G. M. 1996b. Relating the land-cover composition of mixed pixels to artificial neural network classification output. *Photogrammetric Engineering and Remote Sensing*, **62**(5):491-499.

Foody. G. M. 1995. Using prior knowledge in artificial neural network classification with a minimal training set. *International Journal of Remote Sensing*. **16**(2):301-312.

Foody. G. M., McCulloch. M. B. and Yates. W. B. 1994. Crop classification from C-band polarimetric radar data. *International Journal of Remote Sensing*. **15**(14):2871-2885.

Furby. S. 1996. A comparison of neural network and maximum likelihood classification. Technical Note No. I.96.11. Joint Research Centre. European Commission. Ispra. Italy.

Gopal. S., Sklarew. D. M., and Lambin. E. 1993. Fuzzy neural networks in multitemporal classification of landcover change in the Sahel. *Proc. Eurostat/DOSES Workshop on New Tools for Spatial Data Analysis*. Lisbon. November. EUROSTAT. Luxembourg.

Statistical Document in Theme 3, Series D, 69-81.

Hellwich, O. 1994. Detection of linear objects in ERS-1 SAR images using neural network technology. *Proc. International Geoscience and Remote Sensing Symposium (IGARSS 94)*, held at Pasadena, California, 8-12 August, IEEE Press, Piscataway, NJ, **4**, 1886-1888.

Huber, P. J. 1985. Projection pursuit. *Annals of Statistics*, **13**(2):435-475.

Jimenez, L. and Landgrebe, D. 1994. High dimensional feature reduction via projection pursuit. *Proc. International Geoscience and Remote Sensing Symposium (IGARSS '94)*, held at Pasadena, California, **2**:1145-1147, IEEE Press, Piscataway, NJ.

Kanaya, F. and Miyake, S. 1991. Bayes statistical classifier and valid generalization of pattern classifying neural networks. *IEEE Transactions on Neural Networks*, **2**(4):471-475.

Kanellopoulos, I., Varfis, A., Wilkinson, G. G. and Mégier, J. 1991. Neural network classification of multi-date satellite imagery. *Proc. International Geoscience and Remote Sensing Symposium (IGARSS'91)*, 3-6 June, Espoo, Finland, IEEE, Piscataway, NJ., Vol IV, 2215-2218.

Kanellopoulos, I., Varfis, A., Wilkinson, G. G., and Mégier, J. 1992. Land cover discrimination in SPOT imagery by artificial neural network -a twenty class experiment. *International Journal of Remote Sensing*, **13**(5):917-924.

Kanellopoulos, I., Wilkinson, G. G. and Chiuderi, A. 1994. Land cover mapping using combined Landsat TM imagery and textural features from ERS-1 Synthetic Aperture Radar imagery, *Proc. European Symposium on Satellite Remote Sensing*, held in Rome, 26-30 September 1994, European Optical Society / International Society for Optical Engineering.

Key, J., Maslanik, A., and Schweiger, A. J. 1989. Classification of merged AVHRR and SMMR arctic data with neural networks. *Photogrammetric Engineering and Remote Sensing*, **55**(9):1331-1338.

Kohonen, T. 1984. *Self Organization and Associative Memory*, Springer Series in Information Sciences Vol. 8, Springer-Verlag, Berlin.

Kung, S. Y. 1993. *Digital Neural Networks*, Prentice Hall, Englewood Cliffs, NJ.

Kwok, R. et al. 1991. Application of neural networks to sea ice classification using polarimetric SAR images. *Proc. International Geoscience and Remote Sensing Symposium (IGARSS 91)*, held at Espoo, Finland, 3-6 June, IEEE Press, Piscataway, NJ, **1**, 85-88.

Lee, J., Weger, R. C., Sengupta, S. K., and Welch, R. M. 1990. A neural network approach to cloud classification, *IEEE Transactions on Geoscience and Remote Sensing*, **28**(5):846-855.

Liu, Z. K. and Wilkinson, G. G. 1992. A neural network approach to geometrical rectification of remotely-sensed satellite imagery, Technical Note No. 1.92.118, Joint Research Centre, Commission of the European Communities, Ispra, Italy.

Loung, G. and Tan, Z. 1992. Stereo matching using artificial neural networks. International Archives of Photogrammetry and Remote Sensing, Vol. 29, B3, 417-421.

Maslanik, J., Key, J., and Schweiger, A. 1990. Neural network identification of sea-ice seasons in passive microwave data. *Proc. International Geoscience and Remote Sensing Symposium (IGARSS '90)*, held in Maryland, USA, IEEE Press, Piscataway, NJ, 1281-1284.

Minsky, M. L. and Papert, S. A. 1969. Perceptrons. MIT Press, Cambridge, Mass.

Moody, A., Gopal, S., Strahler, A. H., Borak, J. and Fisher, P. 1994. A combination of temporal thresholding and neural network methods for classifying multiscale remotely-sensed image data, *Proc. International Geoscience and Remote Sensing Symposium (IGARSS 94)*, held at Pasadena, California, 8-12 August, IEEE Press, Piscataway, NJ, Vol. 4, 1877-1879.

Paola, J. D. and Schowengerdt, R. A. 1995. A review and analysis of backpropagation neural networks for classification of remotely-sensed multi-spectral imagery. *International Journal of Remote Sensing*, **16**(16):3033-3058.

Pierce, L.E., Sarabandi, K., and Ulaby, F.T. 1994. Application of an artificial neural network in canopy scattering inversion. *International Journal of Remote Sensing*, **15**(16):3263-3270.

Ramacher, U., Raab, W., Anlauf, J., Hachmann, U. and Wesseling, M. 1994. *SYNAPSE-1 -a general purpose neurocomputer*, Technical Report, Corporate Research and Development Division, Siemens-Nixdorf.

Ryan, T. W., Sementelli, P., Yuen, P. and Hunt, B. R. 1991. Extraction of shoreline features by neural nets and image processing. *Photogrammetric Engineering and Remote Sensing*, **57**(7):947-955.

Shang, C. and Brown, K. 1994. Principal features-based texture classification with neural networks, *Pattern Recognition*, **27**(5):675-687.

Simpson, P. K. 1990. *Artificial Neural Systems: Foundations, Paradigms, Applications and Implementations*, Pergamon Press, New York.

Spina, M.S., Schwartz, M.J., Staelin, D.H., and Gasiewski, A.J. 1994. Application of multilayer feedforward neural networks to precipitation cell-top altitude estimation. *Proc. International Geoscience and Remote Sensing Symposium (IGARSS '94)*, held at Pasadena, California, 8-12 August, IEEE Press, Piscataway, NJ., Vol. 4, 1870-1872.

Van der Maas, H. L. J., Vershure, P. F. M. J., and Molenaar, P. C. M. 1990. A note on chaotic behaviour in simple neural networks, *Neural Networks*, **3**:119-122.

Wan, E. A. 1990. Neural network classification: a Bayesian interpretation, *IEEE Transactions on Neural Networks*, **1**(4):303-305.

Wan, W. and Fraser, D. 1994. A self-organizing map model for spatial and temporal contextual classification, *Proc. International Geoscience and Remote Sensing Symposium (IGARSS 94)*, held at Pasadena, California, 8-12 August, IEEE Press, Piscataway, NJ, Vol. 4, 1867-1869.

Werbos, P. 1974. *Beyond regression: new tools for prediction and analysis in the behavioural sciences*, Unpublished PhD thesis, Harvard University, Cambridge, Mass.

Wilkinson, G. G., Kontoes, C., and Murray, C. N. 1993. Recognition and inventory of oceanic clouds from satellite data using an artificial neural network technique. *Proc. International Symposium on "Dimethylsulphide, Oceans, Atmosphere and Climate"*, held at Belgirate, Italy, 13-15 October 1992, G. Restelli and G. Angeletti (editors), Kluwer Academic Publishers, Dordrecht, 393-399.

Wilkinson, G. G., Fierens, F. and Kanellopoulos, I. 1995a. Integration of neural and statistical approaches in spatial data classification, *Geographical Systems*, **2**:1-20.

Wilkinson, G. G., Folving, S., Kanellopoulos, I., McCormick, N., Fullerton, K. and J. Mégier. 1995b. Forest mapping from multi-source satellite data using neural network classifiers -an experiment in Portugal, *Remote Sensing Reviews*, **12**:83-106.

Zhang, M. and Scofield, R.A. 1994. Artificial neural network technique for estimating heavy convective rainfall and recognizing cloud mergers from satellite data, *International Journal of Remote Sensing* **15**(16):3241-3261.

16 Fuzzy ARTMAP - A Neural Classifier for Multispectral Image Classification

Sucharita Gopal[1] and Manfred M. Fischer[2]

[1]Department of Geography, Boston University, Boston, MA USA
[2]Department of Economic & Social Geography, Vienna University of Economics and Business Administration, A-1090 Vienna, Augasse 2-6, Austria and the Institute for Urban and Regional Research, Austrian Academy of Sciences, A-1010 Vienna, Postgasse 4

16.1 Introduction

Spectral pattern recognition deals with classifications that utilize pixel-by-pixel spectral information from satellite imagery. The literature on neural network applications in this area is relatively new, dating back only about six to seven years. The first studies established the feasibility of error-based learning systems such as backpropagation (see Key et al., 1989, McClellan et al., 1989, Benediktsson et al., 1990, Hepner et al., 1990). Subsequent studies analysed backpropagation networks in more detail and compared them to standard statistical classifiers such as the Gaussian maximum likelihood (see Bischof et al., 1992, Kanellopoulos et al., 1993, Fischer et al., 1994).

In this chapter we analyse the capability and applicability of a different class of neural networks, called fuzzy ARTMAP, to multispectral image classification. Fuzzy ARTMAP synthesizes fuzzy logic and Adaptive Resonance Theory (ART) models by describing the dynamics of ART category choice, search and learning in terms of analog fuzzy set-theoretic rather than binary set-theoretic operations. The chapter describes design features, system dynamics and simulation algorithms of this learning system, which is trained and tested for classification of a multispectral image of a Landsat-5 Thematic Mapper (TM) scene (270x360 pixels) from the City of Vienna on a pixel-by-pixel basis. Fuzzy ARTMAP performance is compared with that of a backpropagation system and the Gaussian maximum likelihood classifier on the same database.

The chapter is organized as follows. Section 16.2 gives a brief mathematical description of the unsupervised learning system, called ART 1, which is a prerequisite to understanding the ARTMAP system. Section 16.3 shows how two ART 1 modules are linked together to form the ARTMAP supervised learning system for binary pattern recognition problems. Section 16.4 leads to one generalization of ARTMAP, called fuzzy ARTMAP, that learns to classify continuous valued rather than binary patterns, and to a simplified version of the general fuzzy ARTMAP learning system, which will be used as general purpose remote sensing classifier in this study. Section 16.5 describes the remote sensing classification problem which is used to test the

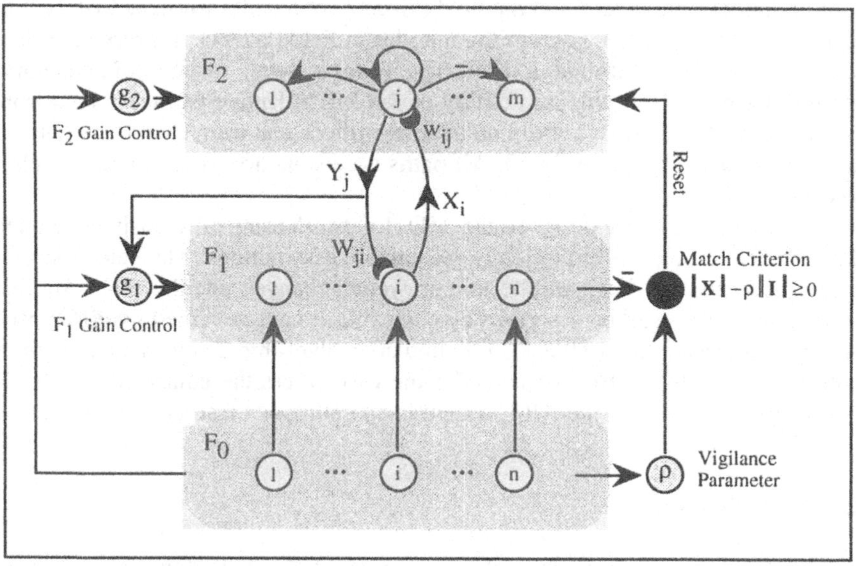

Fig. 16.1. The ART 1 Architecture

classifier's capabilities. The simulation results are given in 16.6 and compared with those obtained by the backpropagation network and the conventional maximum likelihood classifier. The final section contains a some conclusions.

16.2 Adaptive Resonance Theory and ART 1

The basic principles of adaptive resonance theory (ART) were introduced by Stephen Grossberg in 1976 as a theory of human cognitive information processing (Grossberg, 1976 a, b). Since that time the cognitive theory has led to a series of ART neural network models for category learning and pattern recognition. Such models may be characterized by a system of ordinary differential equations (Carpenter and Grossberg, 1985, 1987a) and have been implemented in practice using analytical solutions or approximations to these differential equations.

ART models come in several varieties, most of which are unsupervised, and the simplest are ART 1 designed for binary input patterns (Carpenter and Grossberg, 1987a) and ART 2 for continuous valued (or binary) inputs (Carpenter and Grossberg, 1987b). This section describes the ART 1 model which is a prerequisite to understanding the learning system fuzzy ARTMAP. The main components of an ART 1 system are shown in Fig. 16.1. Ovals represent fields (layers) of nodes, semicircles adaptive filter pathways and arrows paths which are not adaptive. Circles denote nodes (processors), shadowed nodes the vigilance parameter, the match criterion and gain control nuclei that sum input signals. The F_1 nodes are indexed by i and the F_2 nodes by j [categories, prototypes]. The binary vector $\mathbf{I}=(I_1,...,I_n)$ forms the bottom-up input (input layer F_0) to the field (layer) F_1 of n nodes whose activity vector is denoted by $\mathbf{X}=(X_1,...,X_n)$. Each of the n nodes in field (layer) F_2 represents a class or category of inputs around a prototype (cluster seed, recognition category) generated during self-organizing activity of ART 1. Adaptive pathways lead from each F_1 node to all F_2 nodes (bottom up adaptive filter), and from each F_2 node to all F_1 nodes (top down adaptive filter). All paths are excitatory unless marked with a minus sign.

Carpenter and Grossberg designed the ART 1 network using previously developed building blocks based on biologically reasonable assumptions. The selection of a winner F_2 node, the top down and bottom up weight changes, and the enable/disable (reset) mechanism can all be described by realizable circuits governed by differential equations. The description of the ART 1 simulation algorithm below is adapted from Carpenter et al. (1991a, b). We consider the case where the competitive layer F_2 makes a choice and where the ART system is operating in a fast learning mode.

16.2.1. F_1-Activation

Each F_1 node can receive input from three sources: the $F_0 \rightarrow F_1$ bottom-up input; non-specific gain control signals; and top-down signals from the m nodes (winner-take-all units) of F_2, via an $F_2 \rightarrow F_1$ adaptive filter. A node is said to be *active* if it generates an output signal equal to 1. Output from inactive nodes equals 0. In ART 1 a F_1 node is active if at least two of the three input signals are large. This *rule for F_1 activation* is called the *2/3 Rule* and realized in its simplest form as follows: The ith F_1 node is active if its net input exceeds a fixed threshold:

$$X_i = \begin{cases} 1 & \text{if } I_i + g_1 + \sum_{j=1}^{n} Y_j W_{ji} > 1 + \bar{W} \\ 0 & \text{otherwise} \end{cases} \tag{16.1}$$

where I_i is the binary F_0-F_1 input, g_1 the binary non-specific F_1 gain control signal, and term $\sum Y_j W_{ji}$ the sum of F_2-F_1 signals Y_j via pathways with adaptive weights W_{ji}, and $\bar{W}(0<\bar{W}<1)$ is a constant. Hereby the F_1 gain control g_1 is defined as

$$g_i = \begin{cases} 1 & \text{if } F_0 \text{ is active and } F_2 \text{ is inactive} \\ 0 & \text{otherwise} \end{cases} \tag{16.2}$$

It is important to note that F_2 activity inhibits g_1, as shown in Fig. 16.1. These laws for F_1 activation imply that, if F_2 is inactive, then

$$X_i = \begin{cases} 1 & \text{if } I_i = 1 \\ 0 & \text{otherwise} \end{cases} \tag{16.3}$$

If exactly one F_2 node J is active, the sum $\sum X_i W_{ji}$ in equation (16.1) reduces to the single term W_{Ji}, so that

$$X_i = \begin{cases} 1 & \text{if } I_i = 1 \text{ and } W_{Ji} > \bar{W} \\ 0 & \text{otherwise} \end{cases} \tag{16.4}$$

16.2.2. Rules for Category Choice |F_2 choice|

F_2 nodes interact with each other by lateral inhibition. The result is a competitive winner-take all response. The set of committed F_2 nodes (prototypes) is defined as follows. Let T_j denote the total input from F_1 to the jth F_2 processor, given by

$$T_j = \Sigma_{i=1}^{n} X_i \, w_{ij} \,, \tag{16.5}$$

where w_{ij} represent the $F_1 \rightarrow F_2$ (i.e. bottom-up or forward) adaptive weights. If some $T_j > 0$, define the F_2 choice index J by

$$T_J = \max_{j=1,\dots,m} T_j \,. \tag{16.6}$$

Characteristically, J is uniquely defined. Then the components of the F_2 output vector $\mathbf{Y} = (Y_1,\dots,Y_m)$ are

$$Y_i = \begin{cases} 1 & \text{if } j = J \\ 0 & \text{if } j \neq J \end{cases} \tag{16.7}$$

If two or more indices j share maximal input, then one of these is chosen at random.

16.2.3. Learning Laws: Top Down and Bottom Up Learning

The *learning laws* as well as the rules for choice and search, may be described, using the following notation. Let $\mathbf{A} = (A_1,\dots,A_m)$ be a binary m-dimensional vector, then the norm of $\mathbf{A}$ is defined by

$$\|\mathbf{A}\| = \sum_{i=1}^{m} |A_i| \tag{16.8}$$

Let $\mathbf{A}$ and $\mathbf{B}$ be binary m-dimensional vectors, then a third binary m-dimensional vector $\mathbf{A} \cap \mathbf{B}$ may be defined by

$$(\mathbf{A} \cap \mathbf{B}) = 1 \text{ if and only if } A_i = 1 \text{ and } B_i = 1. \tag{16.9}$$

All *ART 1 learning* is gated by F2 activity. That is, the bottom up (forward) and the top down (backward or feedback) adaptive weights w_{iJ} and W_{Ji} can change only when the Jth F2 node is active. Both types of weights are functions of the F1 vector **X**. Stated as a differential equation, the *top-down* or *feedback learning rule* is

$$\frac{d}{dt}W_{ji} = Y_j(X_i - W_{ji}), \tag{16.10}$$

where learning by W_{ji} is gated by Y_j. When the Y_j gate opens (i.e., when $Y_j>0$), then learning begins and W_{ji} is attracted to X_i:

$$W_{ji} \rightarrow X_i. \tag{16.11}$$

In vector terms: if $Y_j > 0$, then $\mathbf{W}_j = (W_{j1},...,W_{jn})$ approaches $\mathbf{X} = (X_1,...,X_n)$. Such a learning rule is termed *outstar learning rule* (Grossberg 1969). Initially, all W_{ji} are maximal, i.e.

$$W_{ji}(0) = 1. \tag{16.12}$$

Thus (with fast learning where the adaptive weights fully converge to equilibrium values in response to each input pattern) the top-down (feedback) weight vector $\mathbf{W}_J$ is a binary vector at the start and end of each input presentation. By (16.3), (16.4), (16.9), (16.11) and (16.12), the *binary* F1 *activity* (output) vector is given by

$$\mathbf{X} = \begin{cases} \mathbf{I} & \text{if } F_2 \text{ is inactive} \\ \mathbf{I} \cap \mathbf{W}_J & \text{if the jth } F_2 \text{ node is active} \end{cases} \tag{16.13}$$

When F_2 node J is active, by (16.4) and (16.10) learning causes

$$\mathbf{W}_J \rightarrow \mathbf{I} \cap \mathbf{W}_J(\text{old}) \tag{16.14}$$

In this *learning update rule* $\mathbf{W}_J$ (old) denotes $\mathbf{W}_J$ at the start of the current input presentation. By (16.11) and (16.13), **X** remains constant during learning, even though $|\mathbf{W}_J|$ may decrease. The first time an F2 node J becomes active it is said to be *uncommitted.* Then, by (16.12) - (16.14)

$$\mathbf{W}_J \rightarrow \mathbf{I}. \tag{16.15}$$

he bottom up or forward weights have a slightly more complicated learning rule which leads to a similar, but normalized result. The combination with F2 nodes which undergo cooperative and competitive interactions is called *competitive learning.* Initially all F2 nodes are uncommitted. Forward weights w_{ij} in F1 → F2 paths initially satisfy

$$w_{ij}(0) = \alpha_j, \tag{16.16}$$

where the parameters α_j are ordered according to $\alpha_1>\alpha_2>...>\alpha_n$ for any admissible $F_0 \to F_1$ input **I**.

Like the top-down weight vector $\mathbf{W}_J$, the bottom-up $F_1 \to F_2$ weight vector $\mathbf{w}_J=(w_{1J},...,w_{iJ},...,w_{nJ})$ also becomes proportional to the F_1 output vector **X** when the F_2 node J is active. But in addition the forward weights are scaled inversely to $\| \mathbf{X} \|$, so that

$$w_{iJ} \to \frac{X_i}{\beta + \|\mathbf{X}\|} \tag{16.17}$$

with $\beta > 0$ (the small number β is included to break ties). This $F_1 \to F_2$ learning law (Carpenter and Grossberg, 1987a) realizes a type of competition among the weights $\mathbf{w}_J$ adjacent to a given F_2 node J.

By (16.13), (16.14) and (16.17), during learning

$$\mathbf{w}_J \to \frac{\mathbf{I} \cap \mathbf{W}_J(\mathrm{old})}{\beta + \|\mathbf{I} \cap \mathbf{W}_J(\mathrm{old})\|} \, . \tag{16.18}$$

(16.18) establishes the update rule for forward weights. The w_{ij} initial values are required to be sufficiently small so that an input **I** which perfectly matches a previously learned vector $\mathbf{w}_J$ will select the F_2 node J rather than an uncommitted node. This is accomplished by assuming that

$$0 < \alpha_j = w_{ij}(0) < \frac{1}{\beta + \|\mathbf{I}\|} \tag{16.19}$$

for all $F_1 \to F_2$ inputs **I**. When **I** is first presented, $\mathbf{X} = \mathbf{I}$, so by (16.5), (16.14), (16.16), and (16.18), the $F_1 \to F_2$ input vector $T = (T_1,...,T_m)$ obeys

$$T_j(\mathbf{I}) = \sum_{i=1}^{n} I_i \, w_{ij} = \begin{cases} \|\mathbf{I}\|\alpha_j & \text{if j is an uncommitted node} \\ \dfrac{\|\mathbf{I} \cap \mathbf{W}\text{-}j\|}{\beta + \|\mathbf{W}\text{-}j\|} & \text{if j is a committed node.} \end{cases} \tag{16.20}$$

(16.20) is termed the choice function in ART 1, where β is the choice parameter and $\beta \neq 0$. The limit $\beta \to 0$ is called *conservative limit*, because small β-values tend to minimize recoding during learning. If β is taken so small then - among committed F_2 nodes - T_j is determined by the size $\| \mathbf{I} \cap \mathbf{W}_j \|$ relative to $\| \mathbf{W}_j \|$. Additionally, α_j values are taken to be so small that an uncommitted F_2 node will generate the maximum T_j value in (16.20) only if $\| \mathbf{I} \cap \mathbf{W}_j \| = 0$ for all committed nodes. Larger values of α_j and β bias the system toward earlier selection of uncommitted nodes when only poor matches are to be found among the committed nodes (for a more detailed discussion see Carpenter and Grossberg, 1987a).

16.2.4. Rules for Search

It is important to note that ART 1 overcomes the stability - plasticity dilemma by accepting and adapting the prototype of a category (class) stored in F_2 only when the input pattern is *"sufficiently similar"* to it. In this case, the input pattern and the stored prototype are said to *resonate* (hence the term *resonance* theory). When an input pattern fails to match any existing prototype (node) in F_2, a new category is formed (as in Hartigan's, 1975, leader algorithm), with the input pattern as the prototype, using a previously uncommitted F_2 unit. If there are no such uncommitted nodes left, then a novel input pattern gives no response (see Hertz et al., 1991).

A dimensionless parameter ρ with $0<\rho\leq 1$ which is termed *vigilance parameter* establishes a matching (similarity) criterion for deciding whether the similarity is good enough for the input pattern to be accepted as an example of the chosen prototype. The degree of match (similarity) between bottom-up input **I** and top-down expectation $\mathbf{W}_j$ is evaluated at the orienting subsystem of ART 1 (see Fig. 16.1) which measures whether prototype J adequately represents input pattern **I**. A *reset* occurs when the match fails to meet the criterion established by the parameter ρ.

In fast-learning ART 1 with choice at F_2, the *search process* may be characterized by the following steps:

Step 1: Select one F_2 node J that maximizes T_J in (16.20), and read-out its top-down (feedback) weight vector $\mathbf{W}_J$.

Step 2: With J active, compare the F_1 output vector $\mathbf{X} = \mathbf{I} \cap \mathbf{W}_J$ with the $F_0 \rightarrow F_1$ input vector **I** at the orienting subsystem (see Fig. 16.1).

Step 3A: Suppose that $\mathbf{I} \cap \mathbf{W}_J$ fails to match **I** at the level required by the ρ-criterion, i.e. that

$$\| \mathbf{X} \| = \| \mathbf{I} \cap \mathbf{W}_J \| < \rho \, \| \mathbf{I} \|. \tag{16.21}$$

This mismatch causes the system to reset and inhibits the winning node J for the duration of the input interval during which **I** remains on. The index of the chosen prototype [F_2 node] is reset to the value corresponding to the next highest $F_1 \rightarrow F_2$ input

T_J. With the new node active, steps 2 and 3A are repeated until the chosen prototype satisfies the similarity [resonance] criterion (16.21).

Step 3B: Suppose that $\mathbf{I} \cap \mathbf{W}_J$ meets the similarity (match function) criterion, i.e.

$$\| \mathbf{X} \| = \| \mathbf{I} \cap \mathbf{W}_J \| \geq \rho \| \mathbf{I} \|, \tag{16.22}$$

then ART 1 search ends and the last chosen F_2 node J remains active until input **I** shuts off (or until ρ increases).

In this state, called *resonance*, both the feedforward ($F_1 \rightarrow F_2$) and the feedback ($F_2 \rightarrow F_1$) adaptive weights are updated if $\mathbf{I} \cap \mathbf{W}_J(\text{old}) \neq \mathbf{W}_J(\text{old})$. If ρ is chosen to be large (i.e. close to 1), the similarity condition becomes very stringent so that many finely divided categories (classes) are formed. A ρ-value close to zero gives a coarse categorization. The vigilance level can be changed during learning.

Finally, it is worth noting that ART 1 is exposed to discrete presentation intervals during which an input is constant and after which F_1 and F_2 activities are set to zero. Discrete presentation intervals are implemented by means of the F_1 and F_2 gain control signals (g_1, g_2). Gain signal g_2 is assumed (like g_1 in (16.2)) to be 0 if F_0 is inactive. When F_0 becomes active, g_2 and F_2 signal thresholds are assumed to lie in a range where the F_2 node which receives the largest input signal can become active.

16.3 The ARTMAP Neural Network Architecture

ARTMAP is a neural network architecture designed to solve supervised pattern recognition problems. The architecture is called ARTMAP because it maps input vectors in $\Re^n$ (such as feature vectors denoting spectral values of a pixel) to output vectors in $\Re^m$ (with m<n), representing predictions such as land use categories, where mapping is learned by example from pairs $\{\mathbf{A}^{(p)}, \mathbf{B}^{(p)}\}$ of sequentially presented input and output vectors p=1,2,3,... and $\mathbf{B}^{(p)}$ is the correct prediction given $\mathbf{A}^{(p)}$. Fig. 16.2 illustrates the main components of a binary ARTMAP system. The system incorporates two ART 1 modules, ART_a and ART_b. Indices a and b identify terms in the ART_a and ART_b modules, respectively. Thus, for example ρ_a and ρ_b denote the ART_a and ART_b vigilance (similarity) parameters, respectively.

During supervised learning ART_a and ART_b read vector inputs **A** and **B**. The ART_a complementing coding preprocessor transforms the vector $\mathbf{A}=(A_1,...,A_{na})$ into the vector $\mathbf{I}_a= (\mathbf{A}, \mathbf{A}^C)$ at the ART_a field F_0^a, where $\mathbf{A}^C$ denotes the complement of **A**. The complement coded input $\mathbf{I}_a$ to the recognition system is the 2na-dimensionable vector

$$\mathbf{I}_a = (\mathbf{A}, \mathbf{A}^C) = (A_1, \dots, A_{na}; A^C_1, \dots, A^C_{na};), \tag{16.23}$$

where

$$A^C_i = 1 - A_i \tag{16.24}$$

Complement coding achieves normalization while preserving amplitude information (see Carpenter et al., 1991a). $\mathbf{I}_a$ is the input to the ARTa field F_1^a. Similarly, the input to the ARTb field F_1^b is the vector $\mathbf{I}_b=(\mathbf{B}, \mathbf{B}^C)$.

If ARTa and ARTb were disconnected, each module would self-organize category groupings for the separate input sets $\{\mathbf{A}^{(p)}\}$and $\{\mathbf{B}^{(p)}\}$, respectively, as described in section 16.2. In an ARTMAP architecture design, however, ARTa and ARTb are connected by an inter-ART module, including a map field that controls the learning of an associative map from ARTa recognition categories (i.e. compressed representations of classes of examplars $\mathbf{A}^{(p)}$) to ARTb recognition categories (i.e. compressed representations of classes of examplars $\mathbf{B}^{(p)}$). Because the map field is the interface, where signals from F_2^a and F_2^b interact, it is denoted by F^{ab}. The nodes of F^{ab} have the same index j [j=1,...,m_b] as the nodes of F_2^b because there is a one-to-one correspondence between these sets of nodes.

ARTa and ARTb operate as outlined in section 16.2 with the following additions. First, the ARTa vigilance (similarity) parameter ρ_a can increase during inter-ART reset according to the match tracking rule. Second, the map field F^{ab} can prime ARTb. This means, if F^{ab} sends nonuniform input to F_2^b in the absence of an $F_0^b \rightarrow F_1^b$ input $\mathbf{B}$, then F_2^b remains inactive. But as soon as an input arrives, F_2^b selects the node J receiving the largest $F^{ab} \rightarrow F_2^b$ input. Node J, in turn, sends to F_1^b the top-down input weight vector W_J^b. Rules for the control strategy, called match tracking, are specified in the sequel (Carpenter et al., 1991a).

Let $\mathbf{X}^a = (X_1^a, \dots, X_{na}^a)$ denote the F_1^a output vector and $\mathbf{Y}^a= (Y_1^a, \dots, Y_{ma}^a)$ the F_2^a output vector. Similarly, let denote $\mathbf{X}^b = (X_1^b, \dots, X_{nb}^b)$ the F_1^b output vector and $\mathbf{Y}^b = (Y_1^b, \dots, Y_{mb}^b)$ the F_2^b output vector. The map field F^{ab} has m_b nodes and binary output vector $\mathbf{X}^{ab}$. Vectors $\mathbf{X}^a$, $\mathbf{Y}^a$, $\mathbf{X}^b$, $\mathbf{Y}^b$ and $\mathbf{X}^{ab}$ are set to the zero vector, $\mathbf{0}$, between input presentations.

The $F_2^a \rightarrow F^{ab}$ adaptive weights z_{kj} with k=1,...,ma and j=1,...,mb obey an outstar learning law similar to that governing the $F_2^b \rightarrow F_1^b$ weights, namely

$$\frac{d}{dt} z_{kj} = Y_k^a (X_k^{ab} - z_{kj}). \tag{16.25}$$

Each vector $(z_{k1}, \dots, z_{kmb})$ is denoted by $\mathbf{z}_k$. According to the learning rule established by (16.25), the $F_2^a \rightarrow F^{ab}$ weight vector $\mathbf{z}_k$ approaches the map field F^{ab} activity vector X^{ab} if the k-th F_2^a node is active. Otherwise $\mathbf{z}_k$ remains constant. If node k has not yet learned to make a prediction, all weights z_{kj} are set equal to 1, using an assumption, analogous to equation (16.12), i.e. $z_{kj}(0)=1$ for k=1,...,ma, and j=1,...,mb.

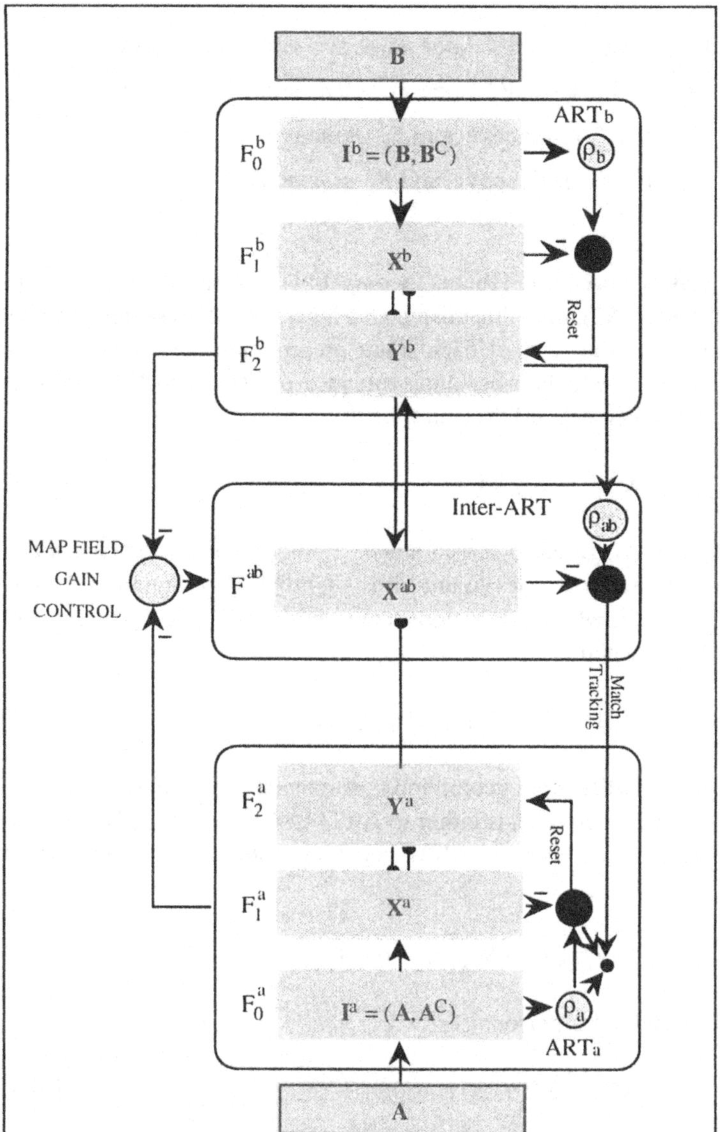

Fig. 16.2. Block Diagram of an ARTMAP System

During resonance with ARTa category k active, $\mathbf{z}_k \rightarrow \mathbf{X}^{ab}$. In fast learning, once k learns to predict the ARTb category J, that association is permanent (i.e. $z_{kJ} = 1$ and $z_{kj} = 0$ with $j \neq J$ for all time). The F^{ab} output vector $\mathbf{X}^{ab}$ obeys

$$X^{ab} = \begin{cases} Y^b \cap z_k & \text{if the k-th } F_2^a \text{ node is active and } F_2^b \text{ is active} \\ z_k & \text{if the k-th } F_2^a \text{ node is active and } F_2^b \text{ is inactive} \\ Y^b & \text{if } F_2^a \text{ is inactive and } F_2^b \text{ is active} \\ 0 & \text{if } F_2^a \text{ is inactive and } F_2^b \text{ is inactive} \end{cases} \tag{16.26}$$

When ARTa makes a prediction that is incompatible with the actual ARTb input (i.e. $\mathbf{z}_k$ is disconfirmed by $\mathbf{Y}^b$), then this mismatch triggers on ARTa search for a new category as follows. At the start of each input presentation the ARTa vigilance (similarity) parameter ρ_a equals a baseline vigilance $\bar{\rho}_a$. The map field vigilance parameter is ρ_{ab}. If a mismatch at F^{ab} occurs, i.e. if

$$\| \mathbf{X}^{ab} \| < \rho_{ab} \; \| \mathbf{I}^b \|, \tag{16.27}$$

then match tracking is triggered to search a new F_2^a node. Match tracking starts a cycle of ρ_a adjustment and increases ρ_a until it is slightly higher than the F_1^a match value $\| \mathbf{A} \cap \mathbf{W}_k^a \| \; \| \mathbf{I}^a \|^{-1}$, where $\mathbf{W}_k^a$ denotes the top-down $F_2^a \rightarrow F_1^a$ ARTa weight vector $(\mathbf{W}_1^a, \ldots, \mathbf{W}_{na}^a)$. Then

$$\| \mathbf{X}^a \| = \| \mathbf{I}^a \cap \mathbf{W}_k^a \| < \rho_a \| \mathbf{I}^a \| \tag{16.28}$$

where $\mathbf{I}^a$ is the current ARTa input vector and k is the index of the active F_2^a node. When this occurs, ARTa search leads either to ARTMAP resonance, where a newly chosen F_2^a node K satisfies both the ARTa matching criterion (see also equation (16.21)):

$$\| \mathbf{X}^a \| = \| \mathbf{I}^a \cap \mathbf{W}_K^a \| \geq \rho_a \| \mathbf{I}^a \| \tag{16.29}$$

and the map field matching criterion:

$$\| \mathbf{X}^{ab} \| = \| \mathbf{I}^b \cap \mathbf{z}_k \| \geq \rho_{ab} \| \mathbf{Y}^b \| \tag{16.30}$$

or, if no such node K exists, to the shut-down of F_2^a for the remainder of the input presentation (Carpenter et al., 1993).

	Binary ARTMAP	Fuzzy ARTMAP
ARTa Category Choice [β choice parameter]	$T_k(\mathbf{I}^a) = \dfrac{\lVert \mathbf{I}^a \cap \mathbf{W}_k^a \rVert}{\beta + \lVert \mathbf{W}_k^a \rVert}$	$T_k(\mathbf{I}^a) = \dfrac{\lVert \mathbf{I}^a \wedge \mathbf{W}_k^a \rVert}{\beta + \lVert \mathbf{W}_k^a \rVert}$
ARTb Category Choice [β choice parameter]	$T_j(\mathbf{I}^b) = \dfrac{\lVert \mathbf{I}^b \cap \mathbf{W}_j^b \rVert}{\beta + \lVert \mathbf{W}_j^b \rVert}$	$T_j(\mathbf{I}^b) = \dfrac{\lVert \mathbf{I}^b \wedge \mathbf{W}_j^b \rVert}{\beta + \lVert \mathbf{W}_j^b \rVert}$
ARTa Matching Criterion [ρ_a ARTa vigilance parameter]	$\lVert \mathbf{I}^a \cap \mathbf{W}_K^a \rVert \geq \rho_a \lVert \mathbf{A} \rVert$	$\lVert \mathbf{I}^a \wedge \mathbf{W}_K^a \rVert \geq \rho_a \lVert \mathbf{A} \rVert$
ARTb Matching Criterion [ρ_b ARTb vigilanc parameter]	$\lVert \mathbf{I}^b \cap \mathbf{W}_J^b \rVert \geq \rho_b \lVert \mathbf{B} \rVert$	$\lVert \mathbf{I}^a \wedge \mathbf{W}_J^b \rVert \geq \rho_b \lVert \mathbf{B} \rVert$
Map Field F^{ab} Matching Criterion [ρ_{ab} Map field vigilance parameter]	$\lVert \mathbf{X}^{ab} \rVert = \lVert \mathbf{Y}^b \cap \mathbf{z}_K \rVert \geq \rho_{ab} \lVert \mathbf{Y}^b \rVert$	$\lvert \mathbf{X}^{ab} \rVert = \lVert \mathbf{Y}^b \wedge \mathbf{z}_K \rVert \geq \rho_{ab} \lVert \mathbf{Y}^b \rVert$
ARTa $F_2^a \rightarrow F_1^a$ **Learning Weight Updates** [γ learning parameter]	$\mathbf{W}_K^a\,(\text{new}) = \gamma\,(\mathbf{A} \cap \mathbf{W}_K^a\,(\text{old})) + (1-\gamma)\,\mathbf{W}_K^a\,(\text{old})$	$\mathbf{W}_K^a\,(\text{new}) = \gamma\,(\mathbf{A} \wedge \mathbf{W}_K^a\,(\text{old})) + (1-\gamma)\,\mathbf{W}_K^a\,(\text{old})$
ARTb $F_2^b \rightarrow F_1^b$ **Learning Weight Updates** [γ learning parameter]	$\mathbf{W}_J^b\,(\text{new}) = \gamma\,(\mathbf{B} \cap \mathbf{W}_J^b\,(\text{old})) + (1-\gamma)\,\mathbf{W}_J^b\,(\text{old})$	$\mathbf{W}_J^b\,(\text{new}) = \gamma\,(\mathbf{B} \wedge \mathbf{W}_J^b\,(\text{old})) + (1-\gamma)\,\mathbf{W}_J^b\,(\text{old})$

Fig. 16.3. The Fuzzy ARTMAP Classifier: A Simplified ARTMAP Architecture

16.4 Generalization to Fuzzy ARTMAP

Fuzzy ARTMAP has been proposed by Carpenter et al. (1991b) as a direct generalization of ARTMAP for supervised learning of recognition categories and multidimensional maps in response to arbitrary sequences of continuous-valued (and binary) patterns not necessarily interpreted as fuzzy set of features. The generalization to learning continuous and binary input patterns is achieved by using fuzzy set operations rather than standard binary set theory operations (see Zadeh, 1965). Fig. 16.3 summarizes how the crisp logical ARTMAP operations of category choice, matching and learning translate into fuzzy ART operations when the crisp (non-fuzzy or hard) intersection operator ($\cap$) of ARTMAP is replaced by the fuzzy intersection or (component-wise) minimum operator (Λ). Due to the close formal homology between ARTMAP and fuzzy ARTMAP operations (as illustrated in Fig. 16.3), there is no need to describe fuzzy ARTMAP in detail here, but for a better understanding it is important to stress differences to the ARTMAP approach.

Fuzzy ARTMAP in its most general form inherits the architecture as outlined in Fig. 16.2 and employs two fuzzy ART modules as substitutes for the ART 1 subsystems. It is noteworthy that fuzzy ART reduces to ART 1 in response to binary input vectors (Carpenter et al., 1993). Associated with each F_2^a [F_2^b] node k=1,...,m_a [j=1,...,m_b] is a vector $\mathbf{W}_k^a$ [$\mathbf{W}_j^a$] of adaptive weights which subsumes both the bottom-up and top-down weight vectors of ART 1.

Fuzzy ARTMAP dynamics are determined by a choice parameter $\beta > 0$, a learning parameter $\gamma \in [0,1]$; and three vigilance (similarity) parameters: the ART$_a$ vigilance parameter ρ_a, the ART$_b$ vigilance parameter ρ_b and the map field vigilance parameter ρ_{ab} with $\rho_a, \rho_b, \rho_{ab} \in [0,1]$. The choice functions $T_k(\mathbf{A})$ and $T_j(\mathbf{B})$ are defined as in Fig. 16.3, where the fuzzy intersection (Λ) for any n-dimensional vectors $\mathbf{S}=(S_1,...,S_n)$ and $\mathbf{T}=(T_1,...,T_n)$ is defined by

$$(\mathbf{S}\Lambda\mathbf{T})_1 = \min(S_1, T_1) \tag{16.31}$$

The fuzzy choice functions $T_k(\mathbf{A})$ and $T_j(\mathbf{B})$ (see Fig. 16.3) can be interpreted as a fuzzy membership of the input $\mathbf{A}$ in the k-th category and the input $\mathbf{B}$ in the j-th category, respectively. In the conservative limit (i.e. $\beta \rightarrow 0$) the choice function $T_k(\mathbf{A})$ primarily reflects the degree to which the weight vector W^a_k is a fuzzy subset of the input vector $\mathbf{A}$. If

$$\frac{\| \mathbf{I}_1^a \, \Lambda \, \mathbf{W}_k^a \|}{\| \mathbf{W}_k^a \|} = 1 \, , \tag{16.32}$$

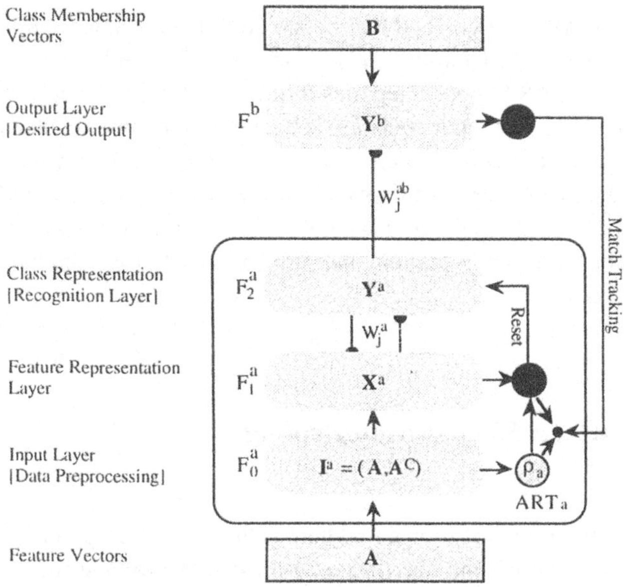

Fig. 16.4. The Fuzzy ARTMAP classifier: A Simplified ARTMAP Architecture

then W^a_k is a fuzzy subset of $\mathbf{I}^a$ and category k is said to be a fuzzy subset choice for input $\mathbf{I}^a$. When a fuzzy subset exists, it is always selected over other choices. The same holds true for $T_j(\mathbf{I}^b)$. (Carpenter et al., 1992). Resonance depends on the degree to which $\mathbf{I}^a[\mathbf{I}^b]$ is a fuzzy set of W^a_k [W^b_k], by the matching criteria (or functions) outlined in Fig. 16.3. The close linkage between fuzzy subsethood and ART choice, matching and learning forms the foundations of the computational features of fuzzy ARTMAP (Carpenter et al., 1992). Especially if category K is a fuzzy subset ARTa choice, then the ARTa match function value ρ_a is given by

$$\rho_a = \frac{\|\mathbf{I}^a \wedge \mathbf{W}_K^a\|}{\|\mathbf{I}^a\|} = \frac{\|\mathbf{W}_K^a\|}{\|\mathbf{I}^a\|} . \tag{16.33}$$

Once the search ends, the ARTa weight vector $\mathbf{W}_K^a$ is updated according to the equation

$$\mathbf{W}_K^a\,(\text{new}) = \gamma\,(\mathbf{A} \wedge \mathbf{W}_K^a\,(\text{old})) + (1 - \gamma)\,\mathbf{W}_K^a\,(\text{old}) \tag{16.34}$$

and similarly the ARTb weight vector $\mathbf{W}_J^b$:

$$\mathbf{W}_J^b\,(\text{new}) = \gamma\,(\mathbf{B} \wedge \mathbf{W}_J^b\,(\text{old})) + (1 - \gamma)\,\mathbf{W}_J^b\,(\text{old}) \tag{16.35}$$

where $\gamma = 1$ corresponds to fast learning as described in Fig. 16.3.

The aim of fuzzy ARTMAP is to correctly associate continuous valued ART_a inputs with continuous valued ART_b inputs. This is accomplished indirectly by associating categories formed in ART_a with categories formed in ART_b. For a pattern classification problem at hand, the desired association is between a continuous valued input vector and some categorical code which takes on a discrete set of values representing the a priori given classes. In this situation the ART_b network is not needed because the internal categorical representation which ART_b would learn already exists explicitly. Thus, the ART_b and the map field F^{ab} can be replaced by a single F^b as shown in Fig. 16.4.

16.5 The Spectral Pattern Recognition Problem

The spectral pattern recognition problem considered here is the supervised pixel-by-pixel classification problem in which the classifier is trained with examples of the classes (categories) to be recognized in the data set. This is achieved by using limited ground survey information which specifies where examples of specific categories are to be found in the imagery. Such ground truth information has been gathered on sites which are well representative of the much larger area analysed from space. The image data set consists of 2,460 pixels (resolution cells) selected from a Landsat-5 Thematic Mapper (TM) scene (270 x 360 pixels) from the city of Vienna and its northern surroundings (observation date: June 5, 1985; location of the center: 16°23' E, 48°14' N; TM Quarter Scene 190-026/4). The six Landsat TM spectral bands used are blue (SB1), green (SB2), red (SB3), near IR (SB4), mid IR (SB5) and mid IR (SB7), excluding the thermal band with only a 120 m ground resolution. Thus, each TM pixel represents a ground area of 30 x 30 m^2 and has six spectral band values ranging over 256 digital numbers (8 bits).

Table 16.1. Categories used for classification and number of training/testing pixels

Category	Description of the Category	Pixels	
		Training	Testing
C1	Mixed grass and arable farmland	167	83
C2	Vineyards and areas with low vegetation cover	285	142
C3	Asphalt and concrete surfaces	128	64
C4	Woodland and public gardens with trees	402	200
C5	Low density residential and industrial areas (suburban)	102	52
C6	Densely built up residential areas (urban)	296	148
C7	Water courses	153	77
C8	Stagnant water bodies	107	54
Total Number of Pixels for Training and Testing		1640	820

The purpose of the multispectral classification task at hand is to distinguish between the eight categories of urban land use listed in Table 16.1. The categories chosen are meaningful to photo-interpreters and land use managers, but are not necessarily spectrally homogeneous. This prediction problem, used to evaluate the performance of fuzzy ARTMAP in a real world context, is challenging. The pixel-based remotely sensed spectral band values are noisy and sometimes unreliable. The number of training sites is small relative to the number of land use categories (one-site training case). Some of the urban land use classes are sparsely distributed in the image. Conventional statistical classifiers such as the Gaussian maximum likelihood classifier have been reported to fail to discriminate spectrally inhomogeneous classes such as C6 (see, e.g., Hepner et al., 1990). Thus, there is evidently a need for new more powerful tools (Barnsley, 1993).

Ideally, the ground truth at every pixel of the scene should be known. Since this is impractical, one training site was chosen for each of the eight above mentioned land use categories. The training sites vary between 154 pixels (category: suburban) and 602 pixels (category: woodland and public gardens with trees). The above mentioned six TM bands provide the data set input for each pixel, with values scaled to the interval [0,1]. This approach resulted in a data base consisting of 2,460 pixels (about 2.5 percent of all the pixels in the scene) that are described by six-dimensional feature vectors, each tagged with its correct category membership. The set was divided into a training set (two thirds of the training site pixels) and a testing set by stratified random sampling, stratified in terms of the eight categories. Pixels from the testing set are not used during network training (parameter estimation) and serve only to evaluate out-of-sample test (prediction, generalization) performance accuracy when the trained classifier is presented with novel data. The goal is to predict the correct land use category for the test sample of pixels.

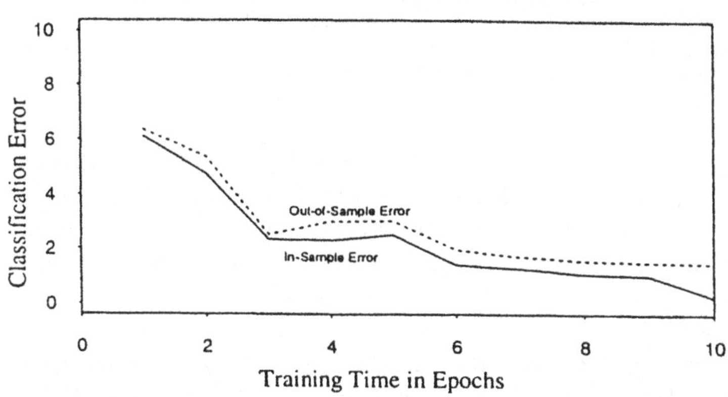

Fig. 16.5. In-Sample and Out-of Sample Classification Error During Training (β=0.001, γ=1.0, ρ_a=0.001)

A good classifier is one which after training with the training set of pixels is able to predict pixel assignments over much wider areas of territory from the remotely sensed data without the need for further ground survey (see Wilkinson et al., 1995). The performance of any classifier is, thus, dependent upon three factors: the adequacy of the training set of pixels and, therefore, the choice of the training sites; the in-sample performance of the classifier; and the out-of-sample or generalization performance of the trained classifier. Of these three factors the first is often outside the control of the data analyst, and thus outside of the scope of this chapter.

16.6 Fuzzy ARTMAP Simulations and Classification Results

In this real world setting, fuzzy ARTMAP performance is examined and compared with that of the multi-layer perceptron and that of the conventional maximum likelihood classifier. (In-sample and out-of sample) Performance is measured in terms of the fraction of the total number of correctly classified pixels (i. e. the sum of the elements along the main diagonal of the classification error matrix).

During training and testing, a given pixel provides an ART_a input $\mathbf{A}=(A_1, A_2, A_3, A_4, A_5, A_6)$ where A_1 is the blue, A_2 the green, A_3 is the red, A_4 the near infrared, A_5 and A_6 the mid infrared (1.55-1.75 μm and 2.08-2.35 μm, respectively) spectral band values measured at each pixel. The corresponding ART_b input vector **B** represents the correct land use category of the pixel's site:

$$\mathbf{B} = \begin{cases} (1,0,0,0,0,0,0,0) & \text{for mixed grass and arable farmland; category 1} \\ (0,1,0,0,0,0,0,0) & \text{for vineyards and areas with low vegetation cover; category 2} \\ (0,0,1,0,0,0,0,0) & \text{for asphalt and concrete surfaces; category 3} \\ (0,0,0,1,0,0,0,0) & \text{for woodland and public gardens with trees; category 4} \\ (0,0,0,0,1,0,0,0) & \text{for low density residential and industrial areas;category 5} \\ (0,0,0,0,0,1,0,0) & \text{for densely built up residential areas; category 6} \\ (0,0,0,0,0,0,1,0) & \text{for watercourses; category 7} \\ (0,0,0,0,0,0,0,1) & \text{for stagnant water bodies; category 8} \end{cases}$$

During training vector **B** informs the fuzzy ARTMAP classifier of the land use category to which the pixel belongs. This supervised learning process allows adaptive weights to encode the correct associations between **A** and **B**. The remote sensing problem described in section 16.5 requires a trained fuzzy ARTMAP network to predict the land use category of the test set pixels, given six spectral band values measured at each pixel.

Following a search, if necessary, the classifier selects an ART_a category by activating an F_a^2 node K for the chosen pixel, and learns to associate category K with the ART_b land use category of the pixel. With fast learning (γ =1), the class prediction of each ART_a category K is permanent. If some input **A** with a different

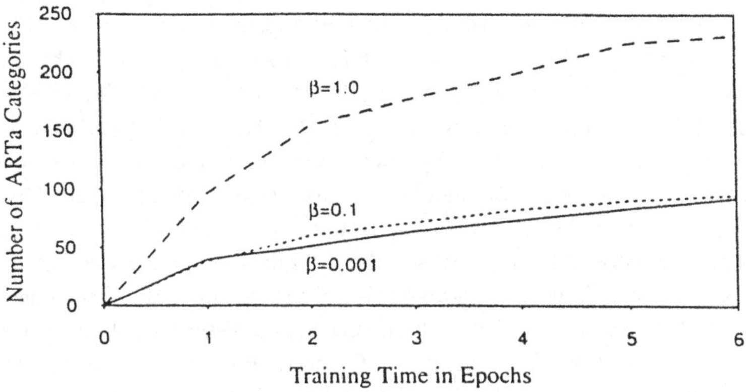

Fig. 16.6. Effect of Choice Parameter β on the Number (m_a) of ART_a Categories (γ=1.0, ρ_a=0.001)

class prediction later chooses this category, match tracking will raise vigilance ρ_a just enough to trigger a search for a different ART_a category. If the finite input set is presented repeatedly, then all training set inputs learn to predict with 100% classification accuracy, but start to fit noise present in the remotely sensed spectral band values.

Fuzzy ARTMAP is trained incrementally, with each spectral band vector **A** presented just once. Following a search, if necessary, the classifier selects an ART_a category by activating an F^a_2 node K for the chosen pixel, and learns to associate category K with the ART_b land use category of the pixel. With fast learning (γ =1), the class prediction of each ART_a category K is permanent. If some input **A** with a different class prediction later chooses this category, match tracking will raise vigilance ρ_a just enough to trigger a search for a different ART_a category. If the finite input set is presented repeatedly, then all training set inputs learn to predict with 100% classification accuracy, but start to fit noise present in the remotely sensed spectral band values.

All the simulations described below use the simplified fuzzy ARTMAP architecture outlined in Fig. 16.3, with three parameters only: a choice parameter β>0, the learning parameter γ=1 (fast learning), and an ART_a vigilance parameter $\rho_a \in [0,1]$. In each simulation, the training data set represents 1,640 pixels and the testing data set 820 pixels. Fuzzy ARTMAP was run with five different random orderings of the training and test sets, since input order may affect in-sample and out-of-sample performance. All simulations were carried out at the Department of Economic Geography (WU-Wien) on a SunSPARCserver 10-GS with 128 MB RAM.

Table 16.2 summarizes out-of-sample performance (measured in terms of classification accuracy) on 15 simulations, along with the number of ART_a categories generated and the number of epochs needed to reach any asymptotic training set performance (i. e. about 100% in-sample classification accuracy; Fig. 16.5 shows how in-sample and out-of-sample performance changes depending on the number of training epochs of fuzzy ARTMAP). Each run had a different, randomly chosen presentation order for the 1,640 training and the 820 testing vectors. The choice parameter β was set, first, near the conservative limit at value β=0.001, and then at the higher values of β=0.1 and β=1.0. These β-value inputs were repeatedly presented in a given random order until 100% training classification accuracy was reached. This required six to eight epochs in the cases of β=0.001 and β=0.1, while for β=1.0 eight to ten epochs were necessary. There seems to be a tendency that the number of epochs needed for 100% training set performance is increasing with higher βb-values. All simulations used fast learning (γ=1.0), which generates a distinct ARTa category structure for each input ordering. The number of F^a_2 nodes ranged from 116 to 148 in the case of β=0.001, 115 to 127 in the case of β=0.1, and 202 to 236 in the case of β=1.0. This tendency of increasing number of ARTa categories with increasing

Table 16.2. Fuzzy ARTMAP simulations of the remote sensing classification problem: The effect of variations in choice parameter β (ρ_a=0.0)

Choice Parameter β		Out-of-Sample Performance	Number of F^2_a Nodes	Number of Epochs
β = 0.001				
	Run 1	98.54	125	6
	Run 2	99.26	116	8
	Run 3	99.51	121	6
	Run 4	99.39	126	6
	Run 5	99.75	148	7
	Average	99.29	127	6.5
β = 0.1				
	Run 1	99.26	126	7
	Run 2	99.90	115	6
	Run 3	99.36	115	7
	Run 4	99.51	124	7
	Run 5	99.26	127	7
	Average	99.26	121	7
β = 1.0				
	Run 1	99.98	218	10
	Run 2	98.17	202	8
	Run 3	98.90	212	8
	Run 4	98.50	236	10
	Run 5	98.40	232	10
	Average	98.75	220	9

Table 16.3. Fuzzy ARTMAP Simulations of the remote sensing classification problem: The effect of variations in vigilance ρ_a (β = 0.001)

Vigilance (Similarity) Parameter ρ_a	In-Sample Performance	Out-of-Sample Performance	Number of F^a_2 Nodes
ρ = 0.95			
Run 1	97.01	96.20	285
Run 2	97.00	96.20	298
Run 3	96.15	95.60	276
Run 4	96.21	95.36	276
Run 5	95.06	94.39	286
Average	96.36	95.82	284
ρ = 0.75			
Run 1	93.00	92.00	52
Run 2	92.26	93.29	47
Run 3	91.82	90.00	42
Run 4	93.00	93.04	53
Run 5	90.31	91.83	53
Average	92.08	92.03	50
ρ = 0.50			
Run 1	92.20	91.40	43
Run 2	90.20	89.51	43
Run 3	94.45	94.76	44
Run 4	93.35	93.42	43
Run 5	92.98	93.90	45
Average	92.62	92.59	44
ρ= 0.0			
Run 1	90.70	90.60	35
Run 2	92.26	91.22	44
Run 3	90.97	90.30	34
Run 4	91.95	90.73	40
Run 5	92.56	92.44	32
Average	91.69	91.06	37

β-values and increasing training time is illustrated in Fig. 16.6. All simulations used ρ_a= 0.0 which tends to minimize the number of F^a_2 nodes compared with higher ρ_a-values not shown in Table 16.2. The best average result (averaged over five independent simulation runs) was obtained with β=0.01 and 6.5 epoch training (99.29% classification accuracy). All the 15 individual simulation runs reached an out-of-sample performance close to 100% (range: 98.40 to 99.90%).

Table 16.3 shows how in-sample and out-of-sample performance changes depending on the number of F^a_2 nodes with ρa=0.95, 0.75, 0.50 and 0.0. In these

Table 16.4. Fuzzy ARTMAP simulations of the remote sensing classification problem: The effect of variations in training size (ρa= 0.0, β= 0.001)

Number of Training Pixels	In-Sample Performance	Out-of-Sample Performance	Number of F^a_2 Nodes
164	83.2	80.1	19
1,640	93.0	92.0	33
16,400	99.3	99.2	135
164,000	99.3	99.2	225

simulations, learning is incremental, with each input presented only once (in ART terminology: one epoch training). The choice parameter is set to β=0.001. The best overall results, in terms of average in-sample and out-of-sample performance were obtained with an ARTa vigilance close to one (96.36% and 95.82%, respectively). For ρa=0.0 the in-sample and out-of-sample performances decline to 91.69% and 91.06%, respectively. But the runs with ρa= 0.0 use much fewer ARTa categories (32 to 44) compared to ρa= 0.95 (276 to 298 ARTa categories), and generate stable performance over the five runs. Increasing vigilance creates more ARTa categories. One final note to be made here is that most fuzzy ARTMAP learning occurs on the first epoch, with the test set performance on systems trained for one epoch typically over 92% that of systems exposed to inputs for six to eight epochs (compare Table 16.3 with Table 16.2).

Table 16.5. Performance of fuzzy ARTMAP simulations of the remote sensing classification problem: Comparison with the Multi-Layer Perceptron and the Gaussian Maximum Likelihood classifier

Classifier	Epochs[1]	Hidden Units/ ARTa Categories	Adaptive Weight Parameters	In-Sample Classification Accuracy	Out-of-Sample Classification Accuracy
Fuzzy ARTMAP	8	116	812	100.00	99.26
Multi-Layer Perceptron	92	14	196	92.13	89.76
Gaussian Maximum Likelihood	-	-	-	90.85	85.24

[1]one pass through the training data set

Fuzzy ARTMAP: $\beta = 00.1$, $\gamma = 1.0$, $\rho_a = 0.0$, asymptotic training

Multi-Layer Perceptron: logistichidden unit activation,softmax output unit activation, network pruning, epoch based stochastic version of back-propagation with epoch size of three, learning rate $\gamma = 0.8$

Table 16.4 summarizes the results of the third set of fuzzy ARTMAP simulations carried out, in terms of both in-sample and out-of-sample performance along with the number of F_2^a nodes. The choice parameter β was set near the conservative limit at value β=0.001 and ARTa vigilance at ρ_a= 0.0. Training lasted for one epoch only. As training size increases from 164 to 164,000 pixel vectors both in-sample and out-of-sample performances increase, but so does the number of ARTa category nodes. In-sample classification accuracy increases from 83.2% to 99.3%, and out-of-sample classification accuracy from 80.1% to 99.2%, while the number of ARTa category nodes increases from 19 to 225. Each category node K requires six learned weights $\mathbf{w}^a_K$ in ARTa. One epoch training on 164 training pixels creates 19 ARTa categories and so uses 72 ARTa adaptive weights to achieve 80.1% out-of-sample classification accuracy (820 test pixels), while one epoch training on 164,000 pixels requires 225 ARTa categories and, thus, 1,350 ARTa adaptive weights to arrive at an out-of-sample performance of 99.2%. Evidently, the fuzzy ARTMAP classifier becomes arbitrarily accurate provided the number of F^a_2 nodes increases as needed.

Finally, fuzzy ARTMAP performance is compared with that of a multi-layer perceptron classifier as developed and implemented in Fischer et al. (1994), using the same training and testing set data. Table 16.5 summarizes the results of the comparison of the two neural classifers in terms of the in-sample and out-of-sample classification accuracies along with the number of epochs (i. e. one pass through the training data set) and the number of hidden units/ARTa category nodes (a hidden unit is somewhat analogous to an ARTa category for purposes of comparison) to reach asymptotic convergence, and the number of adaptive weight parameters. The fuzzy ARTMAP classifier has been designed with the following specifications: choice parameter near the conservative limit at value β=0.001, learning parameter γ=1.0, constant ARTa vigilance ρ_a=0.0, repeated presentation of inputs in a given order until 100% training set performance was reached. Stability and match tracking allow fuzzy ARTMAP to construct automatically as many ARTa categories as are needed to learn any consistent training set to 100% classification accuracy. The multi-layer perceptron classifier is a pruned one hidden layer feedforward network with 14 logistic hidden units and eight softmax output units, using an epoch-based stochastic version of the backpropagation algorithm (epoch size: three training vectors, no momentum update, learning parameter γ=0.8).The Gaussian maximum likelihood classifier based on parametric density estimation by maximum likelihood was chosen because it represents a widely used standard for comparison that yields minimum total classification error for Gaussian class distributions.

The fuzzy ARTMAP classifier has an outstanding out-of-sample classification accuracy of 99.26% on the 820 pixels testing data set. Thus the error rate (0.74%) is less than 1/15 that of the multi-layer perceptron and 1/20 that of the Gaussian maximum likelihood classifier. A more careful inspection of the classification error (confusion) matrices (see appendix) shows that there is a some confusion between the urban (densely built-up residential) areas and water courses land use categories in the case of both the multi-layer perceptron and the Gaussian maximum likelihood classifiers, which is peculiar. The water body in this case is the river Danube that flows through the city and is surrounded by densely built up areas. The confusion could be caused by the 'boundary problem' where there are mixed pixels at the boundary. The fuzzy ARTMAP neural network approach evidently accommodates more easily a heterogeneous class label such as *densely built-up residential areas* to produce a visually and numerically correct map, even with smaller numbers of training pixels (see Fig. 16.7).

The primary computational difference between the fuzzy ARTMAP and the multi-layer perceptron algorithms is speed. The backpropagation approach to neural network training is extremely computation-intensive, taking about one order of magnitude more time than the time for fuzzy ARTMAP, when implemented on a serial workstation. Although this situation may be alleviated with other, more efficient training algorithms and parallel implementation, it remains one important drawback to the routine use of multi-layer perceptron classifiers. Finally, it should be mentioned that in terms of total number of pathways (i.e. the number of weight parameters) needed for the best performance, the multilayer perceptron classifier is superior to fuzzy ARTMAP, but at the above mentioned higher computation costs and the lower classification accuracies.

16.7 Summary and Conclusions

Classification of terrain cover from satellite imagery represents an area of considerable current interest and research. Satellite sensors record data in a variety of spectral channels and at a variety of ground resolutions. The analysis of remotely sensed data is usually achieved by machine-oriented pattern recognition techniques of which classification based on maximum likelihood, assuming Gaussian distribution of the data, is the most widely used one. We compared fuzzy ARTMAP performance with that of an error-based learning system based upon the multi-layer perceptron and the Gaussian maximum-likelihood classifier as conventional statistical benchmark on the same database. Both neural network classifiers outperform the conventional classifier in terms of map user's, map producer's and total classification accuracies. The fuzzy ARTMAP simulations did lead by far to the best out-of-sample classification accuracies, very close to maximum performance.

Evidently, the fuzzy ARTMAP classifier accommodates more easily a heterogenenous class label such as densely built-up residential areas to produce a visually and numerically correct urban land use map, even with smaller numbers of training pixels. In particular, the Gaussian maximum likelihood classifier tends to be sensitive to the purity of land use category signatures and performs poorly if they are not pure.

Fig. 16.7. The Fuzzy ARTMAP Classified Image
(Note: A coloured version of this figure can be found on-line under the URL
http://wigeomac6.wu-wien.ac/Forschung/Publikationen/Fig-16-3.html)

The study shows that the fuzzy ARTMAP classifier is a powerful tool for remotely sensed image classification. Even one epoch of fuzzy ARTMAP training yields close to maximum performance. The unique ART features such as speed and incremental learning may give the fuzzy ARTMAP multispectral image classifier the potential to become a standard tool in remote sensing especially when it comes to use data from future multichannel satellites such as the 224 channel Airborne Visible and Infrared Imaging Spectrometer (AVIRIS), and to classifiying multi-data and multi-temporal imagery or when extending the same classification to different images. To conclude the chapter, we would like to mention that the classfier leads to crisp rather than fuzzy classifications, and, thus, loses some attractiveness of *fuzzy* pattern recognition systems. This is certainly one direction for further improving the classifier.

Acknowledgements

The authors thank Professor Karl Kraus (Department of Photogrammetric Engineering and Remote Sensing, Vienna Technical University) for his assistance in supplying the remote sensing data used in this study. This work has been funded by the Austrian Ministry for Science, Traffic and Art (funding contract no. EZ 308.937/2-W/3/95). The authors also would like to thank Petra Staufer (Wirtschaftsuniversität Wien) for her help.

References

Barnsley M. 1993. Monitoring urban areas in the EC using satellite remote sensing, *GIS Europe*, **2**(8):42-4.

Benediktsson J.A., Swain P.H., Ersoy O.K. 1990. Neural network approaches versus statistical methods in classification of multisource remote sensing data, *IEEE Transactions on Geoscience and Remote Sensing*, **28**:540-52.

Bezdak J.C., Pal S.K. (eds.) 1992. *Fuzzy Models for Pattern Recognition*, IEEE, New York.

Bischof H., Schneider W., Pinz, A.J. 1992. Multispectral classification of Landsat-images using neural networks, *IEEE Transactions on Geoscience and Remote Sensing*, **30**(3): 482-90.

Carpenter G.A. 1989. Neural network models for pattern recognition and associative memory, *Neural Networks*, **2**:243-57.

Carpenter G.A., Grossberg S. 1985. Category learning and adaptive pattern recognition, a neural network model, *Proceedings of the Third Army Conference on Applied Mathematics and Computing*, ARO-Report 86-1, 37-56.

Carpenter G.A., Grossberg S. 1987a. A massively parallel architecture for a self-organizing neural pattern recognition machine, *Computer Vision, Graphics, and Image Processing*, **37**:54-115.

Carpenter G.A., Grossberg S. 1987b. ART 2 Stable self-organizing of pattern recognition codes for analog input patterns, *Applied Optics*, **26**:4919-30.

Carpenter G.A., Grossberg S. eds. 1991. *Pattern Recognition by Self-organizing Neural Networks*, MIT Press, Cambridge MA

Carpenter G.A., Grossberg S. 1995. Fuzzy ART, in Kosko B. ed. *Fuzzy Engineering*, Prentice Hall., Carmel.

Carpenter G.A., Grossberg S., Reynolds J.H. 1991a. ARTMAP Supervised real-time learning and classification of nonstationary data by a self-organizing neural network, *Neural Networks,* **4**:565-88.

Carpenter G.A., Grossberg S., Rosen, D.B. 1991b. Fuzzy ART Fast stable learning and categorization of analog patterns by an adaptive resonance system, *Neural Networks,* **4**:759-71.

Carpenter G.A., Grossberg S., Ross, W.D. 1993. ART-EMAP: A neural network architecture for object recognition by evidence accumulation, Proceedings of the World Congress on Neural Networks (WCNN-93), Lawrence Ertbaum Associates, III, Hillsdak N.J., 643-56.

Carpenter G.A., Grossberg S., Markuzon N., Reynolds J.H., Rosen D. B. 1992. Fuzzy ARTMAP A neural network architecture for incremental supervised learning of analog multidimensional maps, *IEEE Transactions on Neural Networks,* **3**:698-713.

Dawson M.S., Fung A.K., Manry M. T. 1993. Surface parameter retrieval using fast learning neural networks, *Remote Sensing Reviews,* **7**:1-18.

Fischer M.M., Gopal S., Staufer P., Steinnocher K. 1994. *Evaluation of neural pattern classifiers for a remote sensing application*, Paper presented at the 34th European Congress of the Regional Science Association, Groningen, August 1994.

Grossberg S. 1969. Some networks that can learn, remember, and reproduce any number of complicated space-time patterns, *Journal of Mathematics and Mechanics,* **19**:53-91.

Grossberg S. 1976a. Adaptive pattern classification and universal recoding, I: Parallel development and coding of neural feature detectors, *Biological Cybernetics*, **23**:121-34.

Grossberg S. 1976b. Adaptive pattern classification and universal recoding, II: Feedback, expectation, olfaction and illusion, *Biological Cybernetics*, **23**:187-202.

Grossberg S. 1988. Nonlinear neural networks Principles, mechanisms, and architectures, *Neural Networks,* **1**:17-61.

Hara Y., Atkins R.G., Yueh S.H., Shin R.T., Kong J. A. 1994. Application of neural networks to radar image classification, *IEEE Transactions on Geoscience and Remote Sensing,* **32**:100-9.

Hartigan J. 1975. *Clustering Algorithms,* Wiley, New York.

Hepner G.F., Logan T., Ritter, N., Bryant N. 1990. Artificial neural network classification using a minimal training set comparison of conventional supervised classification, *Photogrammetric Engineering and Remote Sensing,* **56**:469-73.

Hertz J., Krogh A., Palmer, R.G. 1991. *Introduction to the Theory of Neural Computation,* Addison-Wesley, Redwood City CA

Kanellopoulos I., Wilkinson G.G., Mégier J. 1993. Integration of neural network and statistical image classification for land cover mapping, *Proceedings of the International Geoscience and Remote Sensing Symposium (IGARSS'93),* held in Tokyo, 18-21 August, vol. 2, 511-3, IEEE Press, Piscataway NJ

Key J., Maslanik J.A., Schweiger A.J. 1989. Classification of merged AVHRR and SMMR arctic data with neural networks, *Photogrammetric Engineering and Remote Sensing,* **55**(9):1331-8.

McClellan,G. E., DeWitt R. N., Hemmer T. H., Matheson L. N., Moe G. O. 1989. Multispectral image-processing with a three-layer backpropagation network, *Proceedings of the 1989 International Joint Conference on Neural Networks,* Washington, D.C.,151-3.

Wilkinson G.G., Fierens F., Kanellopoulos I. 1995. Integration of neural and statistical approaches in spatial data classification, *Geographical Systems*, **2**:1-20.
Zadeh L. 1965. Fuzzy sets, *Information and Control*, **8**:338-53.

Appendix A
In-Sample and Out-of-Sample Classification Error Matrices of the Classifiers

An error matrix is a square array of numbers set out in rows and columns which expresses the number of pixels assigned to a particular category relative to the actual category as verified by some reference (ground truth) data. The rows represent the reference data, the columns indicate the categorization generated. It is important to note that differences between the map classification and reference data might be not only due to classification errors. Other possible sources of errors include errors in interpretation and delineation of the reference data, changes in land use between the data of the remotely sensed data and the data of the reference data (temporal error), variation in classification of the reference data due to inconsistencies in human interpretation etc.

A.1. In-sample performance: Classification error matrices

(a) Fuzzy ARTMAP

Ground Truth Categories	C1	C2	C3	C4	C5	C6	C7	C8	Total
C1	167	0	0	0	0	0	0	0	167
C2	0	285	0	0	0	0	0	0	285
C3	0	0	128	0	0	0	0	0	128
C4	0	0	0	402	0	0	0	0	402
C5	0	0	0	0	102	0	0	0	102
C6	0	0	0	0	0	293	3	0	296
C7	0	0	0	0	0	5	148	0	153
C8	0	0	0	0	0	0	0	107	107
Total	167	285	128	402	102	298	251	107	1,640

(b) Multi-Layer Perceptron

Ground Truth Categories	C1	C2	C3	C4	C5	C6	C7	C8	Total
C1	157	10	0	0	0	0	0	0	167
C2	1	282	0	0	2	0	0	0	285
C3	0	0	128	0	0	0	0	0	128
C4	0	0	0	389	9	0	0	0	402
C5	0	0	2	2	98	0	0	0	102
C6	0	0	1	0	0	260	25	10	296
C7	0	0	0	0	0	60	93	0	153
C8	0	0	0	0	0	3	0	104	107
Total	162	292	131	391	109	323	118	114	1,640

(c) Gaussian Maximum Likelihood

Ground Truth Categories	C1	C2	C3	C4	C5	C6	C7	C8	Total
C1	161	5	0	1	0	0	0	0	167
C2	0	284	0	0	1	0	0	0	285
C3	0	0	124	0	4	0	0	0	128
C4	0	0	0	385	13	0	0	0	402
C5	0	0	0	0	102	0	0	0	102
C6	0	0	3	0	0	214	62	17	296
C7	0	0	0	0	0	37	116	0	153
C8	0	0	0	0	0	3	0	104	107
Total	161	293	127	386	120	254	178	121	1,640

A.2. Out-of-sample performance: Classification error matrices

(a) Fuzzy ARTMAP

Ground Truth	Classifier's Categories								
Categories	C1	C2	C3	C4	C5	C6	C7	C8	Total
C1	83	0	0	0	0	0	0	0	83
C2	0	142	0	0	0	0	0	0	142
C3	0	0	64	0	0	0	0	0	64
C4	0	0	0	200	0	0	0	0	200
C5	0	0	0	0	52	0	0	0	52
C6	0	0	0	0	0	146	2	0	148
C7	0	0	0	0	0	2	75	0	77
C8	0	0	0	0	0	0	0	54	54
Total	83	142	64	200	52	148	77	54	820

(b) Multi-Layer Perceptron

Ground Truth	Classifier's Categories								
Categories	C1	C2	C3	C4	C5	C6	C7	C8	Total
C1	79	4	0	0	0	0	0	0	83
C2	1	134	6	0	1	0	0	0	142
C3	0	0	64	0	0	0	0	0	64
C4	3	2	0	194	1	0	0	0	200
C5	0	3	0	0	49	0	0	0	52
C6	0	0	1	0	0	115	30	3	148
C7	0	0	0	0	0	29	48	0	77
C8	0	0	0	0	0	1	0	53	54
Total	83	142	70	194	51	145	78	56	820

(c) Gaussian Maximum Likelihood

Ground Truth	Classifier's Categories								
Categories	C1	C2	C3	C4	C5	C6	C7	C8	Total
C1	80	3	0	0	0	0	0	0	83
C2	0	141	0	0	1	0	0	0	142
C3	0	0	62	0	1	1	0	0	64
C4	1	3	0	191	5	0	0	0	200
C5	0	5	0	0	47	0	0	0	52
C6	0	0	1	0	2	73	64	8	148
C7	0	0	0	0	0	24	53	0	77
C8	0	0	0	0	0	2	0	52	54
Total	161	293	127	386	120	254	178	121	820

Appendix B
In-Sample and Out-of-Sample Map User's and Map Producers Accuracies of the Classifiers

B.1. In-sample map user's and map producer's accuracies

Category	Map User's Accuracy			Map Producer's Accuracy		
	Fuzzy ARTMAP	Multi-Layer Perceptron	Gaussian ML	Fuzzy ARTMAP	Multi-Layer Perceptron	Gaussian ML
C1	100.0	94.0	96.4	100.0	96.9	95.1
C2	100.0	98.9	99.6	100.0	96.9	96.9
C3	100.0	100.0	96.9	100.0	97.7	97.7
C4	100.0	96.8	95.8	100.0	99.5	97.7
C5	100.0	96.1	100.0	100.0	89.9	87.3
C6	99.0	87.8	72.3	98.3	80.5	79.8
C7	96.7	60.8	75.8	98.0	78.8	78.8
C8	100.0	97.2	97.2	100.0	91.2	85.8

Note: Map user's accuracies for land use categories are calculated by dividing the number of correctly classified pixels in each category [i.e. the main diagonal elements of the classification error matrix] by the row totals. Map producer's accuracies for land use categories are calculated by dividing the numbers of correctly classified pixels in each category [i.e. the main diagonal elements of the classification error matrix] by the columns totals.

B.2. Out-of-sample map user's and map producer's accuracies

Category	Map User's Accuracy			Map Producer's Accuracy		
	Fuzzy ARTMAP	Multi-Layer Perceptron	Gaussian ML	Fuzzy ARTMAP	Multi-Layer Perceptron	Gaussian ML
C1	100.0	95.2	96.4	100.0	95.2	98.8
C2	100.0	94.4	99.3	100.0	93.7	92.8
C3	100.0	100.0	96.9	100.0	91.4	98.4
C4	100.0	97.0	95.5	100.0	100.0	100.0
C5	100.0	94.2	90.4	100.0	96.1	83.9
C6	98.6	77.7	49.3	98.6	79.3	73.0
C7	97.4	62.3	68.8	97.4	61.5	45.3
C8	100.0	98.1	96.3	100.0	86.9	86.7

Note: Map user's accuracies for land use categories are calculated by dividing the number of correctly classified pixels in each category [i.e. the main diagonal elements of the classification error matrix] by the row totals. Map producer's accuracies for land use categories are calculated by dividing the numbers of correctly classified pixels in each category [i.e. the main diagonal elements of the classification error matrix] by the columns totals.

17 Feedforward Neural Network Models for Spatial Data Classification and Rule Learning

Yee Leung

Department of Geography and Center for Environmental Studies, The Chinese University of Hong Kong, Shatin, N.T., Hong Kong, FAX: (852) 26035006

17.1 Introduction

Spatial data classification has long been a major field of research in geographical analysis. Regardless of whether we are classifying statistical data into socio-economic patterns or remotely sensed data into land covers, our classification task is to group high dimensional data into separate clusters which represent distinguishable spatial features or patterns.

Over the years, we have developed or applied a large variety of techniques to spatial data classification. Statistical methods are perhaps the most dominant methodology we have in store for such a task. Discriminant analysis, cluster analysis, principal component analysis, factor analysis and other spatial statistical techniques, though with different assumptions and theoretical constructs, have been employed to classify spatial data. If used correctly, they are powerful and effective.

In recent years, a number of researchers has been advocating the neural network approach to spatial data classification or analysis with encouraging claimed or demonstrated results (see for example White, 1989; Halmari and Lundberg, 1991; Openshaw, 1992; Fischer and Gopal, 1994; Leung, 1994). Amidst the enthusiasm, one may wonder why neural networks are necessary while existing statistical methods appear to be handling rather adequately the task of high dimensional data classification. The answer to the question is essentially threefold.

First, classification problems are in general highly nonlinear. Mathematically speaking, data classification is basically a partitioning problem in high dimensional space. Our mission is to find hypersurfaces to separate data into clusters. For linearly separable problems, statistical methods are usually effective and efficient. The methods, however, fail when nonlinearity exists in spatial classification. Spatial data are ordinarily not linearly separable. While linear separability, for example, implies spherical separability and spherical separability implies quadratic separability, the reverse is not true. To be versatile, classification methods should also be able to handle nonlinearity. Neural networks, especially multilayer feedforward neural networks, are appropriate models for such a task. Learning algorithms in most neural networks can be viewed as a problem of hypersurface reconstruction which tries to approximate hypersurfaces partitioning nonlinearly separable clusters from training examples.

Second, most statistical methods assume certain types of probability distributions, e.g. Gaussian distribution in the maximum-likelihood Gaussian classifier. Nevertheless, a lot of the spatial data are non-Gaussian. In the classification of remotely sensed images, for example, data are multisource. Gaussian distribution is generally a wrong assumption when spectral (color, tone) and spatial (shape, size, shadow, pattern, texture, and temporal association) data are simultaneously employed as a basis of classification (see for example Benediktsson *et al.*, 1990; Hepner *et al.*, 1990). Neural network models, on the other hand, are exempt from such a restriction.

Third, on-line and real-time computing is a desirable feature of a classifier. Most statistical methods are not on-line and real-time. Neural network algorithms, on the other hand, usually strive for on-line or real-time computation. This is especially important when we are classifying a large volume of data and we do not wish to re-classify the data with, for example, a new datum, but still would like to train the classifier with that additional piece of information through a single computation. The adaptive mechanism of some neural-network learning algorithms can perform on-line or real-time computation.

Therefore, neural network models are not substitutes but complements or alternatives to statistical models for spatial data classification. For linearly separable and well-behaved, e.g. those with Gaussian distribution, data, statistical methods will be a good choice. When the previously discussed situations arise, however, neural networks will be a more appropriate alternative. Nevertheless, not all neural network models are suitable for spatial classification. The purpose of this paper is to review a basic class of neural network models, the feedforward (unidirectional) neural networks, suitable for such a task. Our discussion is by no means exhaustive but pedagogic.

I first review the multilayer feedforward neural networks in section 17.2. The application of genetic algorithms in revolving such type of networks is also discussed. Linear associative memories are then discussed in section 17.3. In section 17.4,

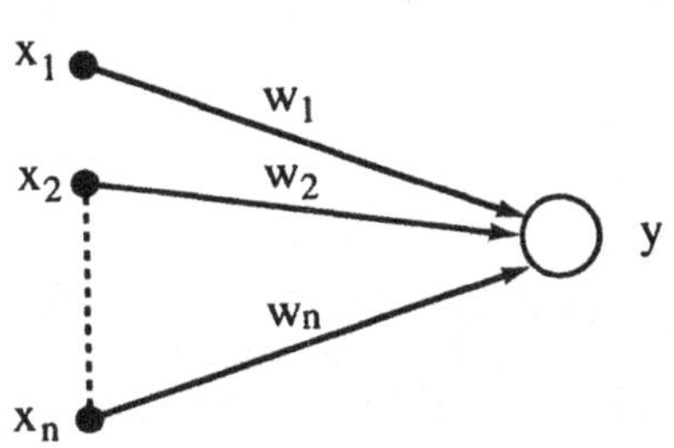

Fig.17.1. A Single-Layer Perceptron

radial basis function neural networks for rule learning are examined. Here, rule learning is similar to spatial data classification in high dimensional space. The paper is then concluded with a summary and out-look of the neural network approaches to spatial data classification in section 17.5.

17.2 Spatial Data Classification by Multilayer Feedforward Neural Networks

17.2.1 The Single-layer Perceptron

Among various neural network models, multilayer feedforward neural networks are perhaps one of the most powerful models for spatial data classification, especially when nonlinearity is encountered in the data set. A special case of multilayer networks is the single-layer perceptron which is capable of handling linear separability.

A single-layer perceptron comprises of an input and an output layer (Fig. 17.1). Let $X = \{x_1, x_2, ..., x_m\}$ and $Y = \{y_1, y_2, ..., y_n\}$ be the input and output layers respectively. In the limiting case, there can be only one node in the output layer. Then, given a set of input-output patterns, an initial weight matrix $W = \{w_{ij}\}$ and a threshold vector θ, the output y_j of the jth neuron can be obtained as

$$y_j = f\left(\sum_{i=1}^{m} w_{ij}x_i - \theta_j\right), \quad j = 1,2,...,n, \tag{17.1}$$

where, $f(\cdot)$ is the activation function which can be discrete, e.g. a step-function, or continuous, e.g. a sigmoid function; θ_j is a specified threshold value attached to node j; and w_{ij} is the connection weight from input i to output j.

A simple encoding scheme is a sign function

$$y_i = f\left(\sum_{i=1}^{m} w_{ij}x_i - \theta_j\right) = \begin{cases} 1, \text{ if } \sum_{i=1}^{m} w_{ij}x_i - \theta_j \geq 0, \\ 0, \text{ otherwise.} \end{cases} \tag{17.2}$$

To put it in the perspective of spatial data classification, the input layer X encodes relevant data, e.g. spectral reflectance in various bands, and the output layer Y encodes spatial patterns, e.g. land types. A learning algorithm is used to learn the internal representation of the data, i.e. associating appropriate input and output patterns, by adaptively modifying the connection weights of the network.

For the single-layer perceptron, given the encoding scheme in (17.2), the initial weights can be recursively adjusted by

$$w_{ij}(t+1) = w_{ij}(t) + \Delta w_{ij}, \tag{17.3}$$

where $w_{ij}(t)$ is the connection weight from input i to neuron j at time t; and Δw_{ij} is the change in weight which may be obtained by the delta rule:

$$\Delta w_{ij} = \eta \delta_j x_i, \tag{17.4}$$

where η, $0 < \eta < 1$, is a learning rate (step size), and δ_j is the error at neuron j obtained as

$$\delta_j = y_j^e - y_j, \tag{17.5}$$

where y_j^e and y_j are respectively the desirable value and the output value of neuron j.

The iterative adjustment terminates when the process converges with respect to some stopping criteria such as a specified admissible error.

Compared with the maximum-likelihood Gaussian classifier, the advantages of the single-layer perceptron are that:

(a) The learning algorithm is distribution-free. No assumption is made on the underlying distribution of the spatial data. It is thus more robust in handling non-Gaussian or asymmetrically distributed data.

(b) The learning algorithm works directly on the errors without having to know in what form they appear.

(c) The learning algorithm is adaptive. Though we can make a maximum-likelihood Gaussian classifier adaptive, it will take much effort and more complex implementation and computation.

17.2.2 The Multilayer Perceptrons

A drawback of the single-layer perceptron is that it cannot handle nonlinearity. The famous XOR problem is a typical example of where the perceptron fails. For highly nonlinear data, we need to embed hidden layers within the perceptron framework and form the multilayer feedforward perceptron (Fig. 17.2). For example, nodes in the first layer help to create a hyperplane to form separation regions. Nodes in the second layer combine hyperplanes to form convex separation regions. More complex separation regions can be created by adding more hidden layers to the network.

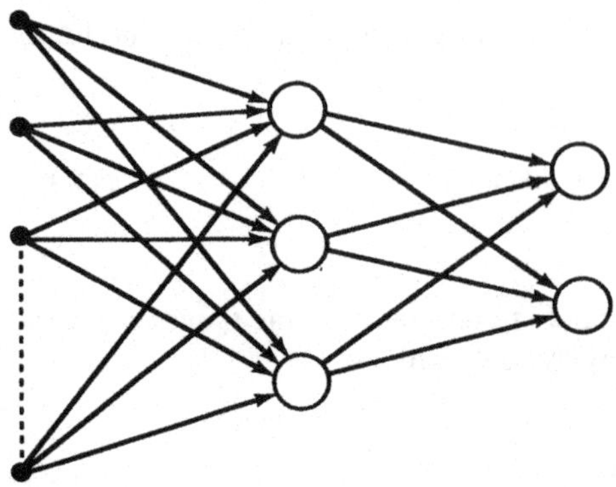

Fig. 17.2. A Multilayer Feedforward Neural Network

One of the advantages of using neural networks for spatial data classification is that the system can learn about its environment by dynamically adjusting its connection weights, thresholds, and topology. Mathematically speaking, learning is a problem of hypersurface reconstruction. It, for example, determines whether or not the internal representation to be learnt is linearly separable, spherically separable, or quadratically separable. Learning through the iterative adjustment of weights constitutes a good part of the research on learning algorithms.

For multilayer feedforward neural networks, a more powerful learning rule, called backpropagation (Rumelhart *et al.*, 1986), can be employed to recursively adjust the connection weights. The method is essentially a gradient descent procedure which searches for the solution of an error-minimization problem along the steepest descent, i.e. negative gradient, of the error surface with respect to the connection weights. The activation function is a sigmoid function.

In backpropagation, given a set of input-output patterns, an initial weight matrix W and a threshold vector θ, the value y_j of an output or hidden node j is obtained as

$$y_j = f\Big(\sum_{i=1}^{m} w_{ij}x_i - \theta_j\Big), \tag{17.6}$$

where

$$f\left(\sum_{i=1}^{m} w_{ij}x_i - \theta_j\right) = \frac{1}{1 + e^{-(\sum_{i=1}^{m} w_{ij}x_i - \theta_j)}}. \tag{17.7}$$

Adjustment of weight is to work recursively backward from the output node to the hidden layers by

$$w_{ij}(t+1) = w_{ij}(t) + \Delta w_{ij}, \tag{17.8}$$

where

$$\Delta w_{ij} = \eta \delta_j x_i, \tag{17.9}$$

and η is a learning rate. The error gradient δ_j at the output node j is obtained as

$$\delta_j = y_j(1 - y_j)(y_j^e - y_j), \tag{17.10}$$

and at the hidden node j is obtained as

$$\delta_j = y_j(1 - y_j) \sum_{k=1}^{\ell} \delta_k w_{jk}, \tag{17.11}$$

where δ_k is the error gradient at node k with a connection coming from hidden node j.

The iterative procedure stops when convergence is achieved with respect to a stopping criterion.

Feedforward neural networks have been constructed to process spatial information involving the recognition and classification of remotely sensed images. For example, multilayer feedforward neural networks with backpropagation have been applied to classify satellite images (Short 1991; Bischof *et al.*, 1992; Heermann and Khazenie, 1992; Tzeng *et al.*, 1994), to map land covers (Campbell *et al.*, 1989; Yoshida and Omatu, 1994; Wang, 1994), and to classify cloud types (Lee *et al.*, 1990).

Though there are slight variations in topologies, these network models generally consist of one or more hidden layers. The input layer comprises input units through which spectral information of selected channels and/or non-spectral information is supplied via the pixel-based and/or region-based (e.g. n × n window around a pixel) schemes. The output layer consists of land types or imagery types of a classification scheme. The hidden layer contains hidden nodes whose number can be dynamically or manually optimized. Classification is then accomplished by the backpropagation algorithm (Fig. 17.3).

Compared to statistical approaches, neural network models have been shown to be more powerful in the classification of multisource data consisting of spectral properties such as tone or color, and spatial properties such as shape, size, shadow, pattern, texture, and their associations across a temporal profile (Argialas and Harlow,

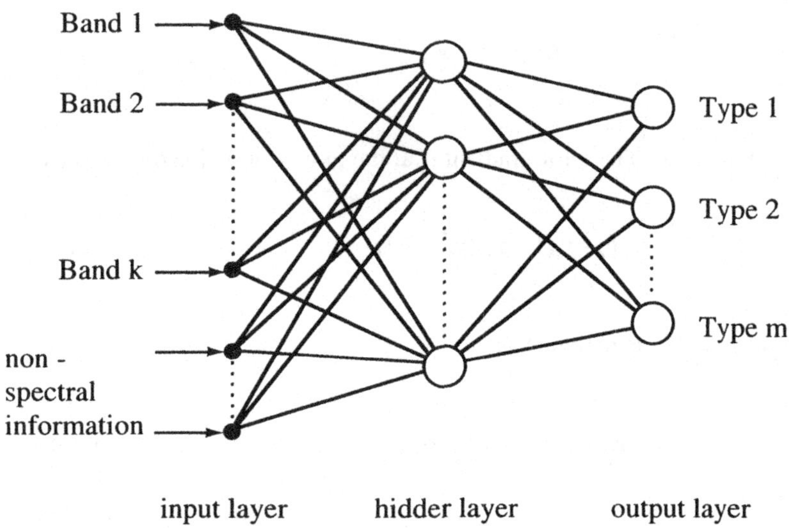

Fig. 17.3. Multilayer Feedforward Neural Network for Land Type Classification

1990; Benediktsson *et al.*, 1990; Hepner *et al.*, 1990; Civco, 1993). They, in general, can handle what statistical methods, e.g. maximum-likelihood Gaussian classifier, fail to manage, i.e. multitype data, unequal reliabilities of data sources, and non-Gaussian information distributions. So far, feedforward neural networks with backpropagation algorithm have been a common model applied to spatial information processing and to model telecommunication flows (Fischer and Gopal, 1994).

The advantage of backpropagation in learning multilayer neural networks is that it can handle nonlinearity. The problem about backpropagation is that it is prone to local minima and the determination of the right topology may take a long trial-and-error manual adjusting process. To alleviate the problem, methods such as genetic algorithms (Holland, 1975; Goldberg, 1989; Davis, 1991) can be combined to construct network topology and fine tune the weights. The following is an example of such an approach.

17.2.3 Evolving Multilayer Neural Networks by Genetic Algorithms

A difficulty in designing a multilayer feedforward neural network is the determination of its topology (architecture). The number of hidden layers and the number of hidden nodes in each layer of a feedforward neural network are usually determined by a manual trial-and-error process. Assume that domain knowledge about the topology is not available, genetic algorithms may provide a more effective way to evolve the topology of the neural network. We can first represent a multilayer feedforward neural network as a chromosome comprising the number of layers, number of nodes, connectivity, and other information about the network. A genetic algorithm can then be employed to evolve the initial topology into other topologies until the best topology (in terms of, for example, network complexity and learning speed) is obtained. One way to accomplish the task is to develop a genetic algorithm system such as GENNET depicted in Fig. 17.4 (Leung *et al.*, 1996a).

The system consists of two major components: a genetic algorithm engine and a neural network engine. Basically, the genetic algorithm engine encodes neural network topologies as chromosomes and evolve them through genetic operators. The evolved chromosome is then decoded into a neural network by the network decoder (generator) and is then fed to the neural network engine for training. Based on the given training patterns, the engine will train the given neural networks by *backpropagation* and the resulting networks will then be tested with the given testing patterns. Various statistics such as the size of the network, learning speed, and classification error will be recorded and passed back to the genetic algorithm Engine for fitness evaluation. Networks with high fitness will be selected and further processed by various genetic operators. The whole process will be repeated until a network with fitness value higher than the specified requirement is found.

Specifically, in the evolution of the spatial interaction neural network model (Leung *et al.*, 1996a), a multilayer feedforward neural network is encoded by the genetic algorithm engine as a 68-bit chromosome. The chromosome of 68 bits is encoded in Table 17.1.

For example, the following chromosome (string):

```
                 1         2         3         4         5         6
        12345678901234567890123456789012345678901234567890123456789012345678
string: 10111100000000000000000100100111010101010001000000000001000000000001
```

represents a network with the following structure:

Total no of hidden layer = 1
Within layer 0: No of node = 30
Connectivity = 1

The genetic algorithm engine then applies the genetic operators: selection, crossover, and mutation to evolve the current population of chromosomes in search of the best chromosome. First, the selection operator selects the chromosomes by their fitness values. Chromosomes with high fitness values are having a better chance to be selected for reproduction.

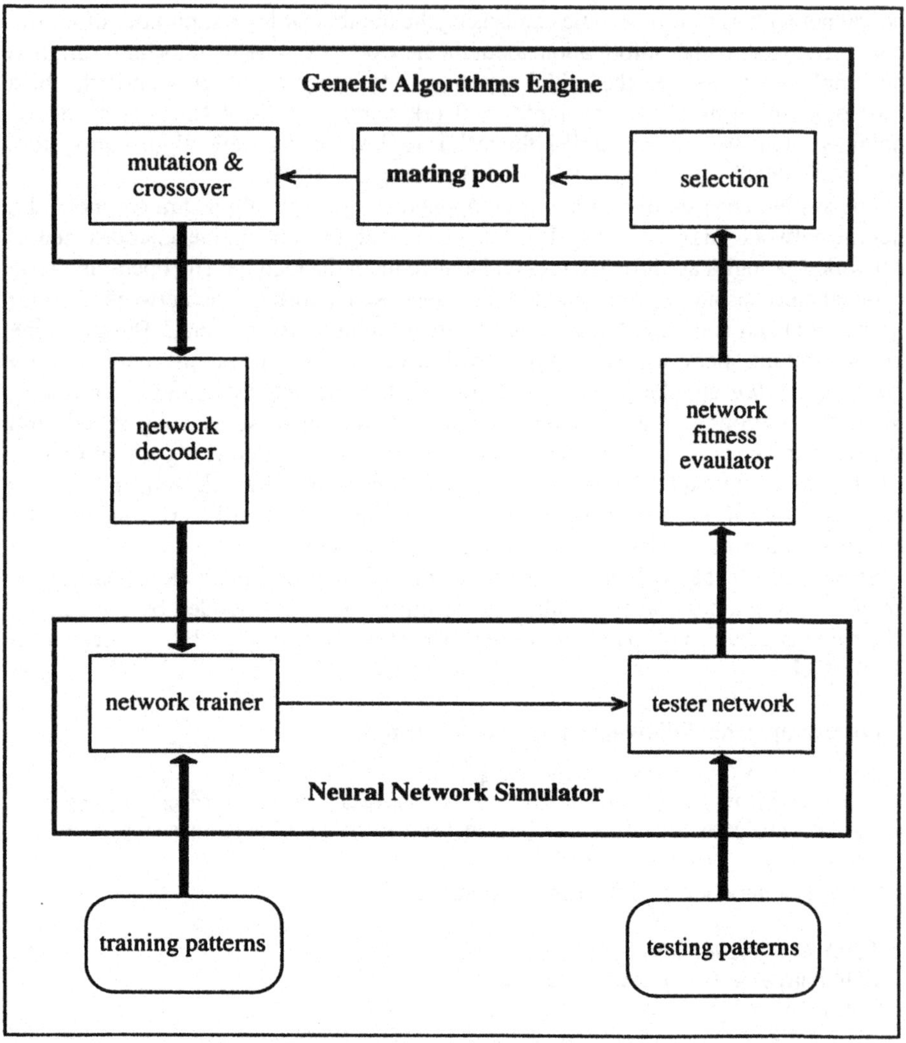

Fig. 17.4. GENNET: A Topology Evolving GA System

Table 17.1. Genetic-algorithm encoding of a multilayer feedforward neural network

Bits	Meaning
1	Present flag of layer 0
1-7	no of nodes in layer 0
8	no of nodes in layer 1
9-14	Present flag of layer 1
15	no of nodes in layer 2
16-21	Present flag of layer 2
22-24	determine upper weight limit (has 8 discrete levels)
25-27	determine lower weight limit (has 8 discrete levels)
28-30	momentum limit (has 8 discrete levels)
31-40	probability of connection of each layer
41-68	sigmoid variable(a) of sigmoid function

Once the fittest chromosomes are selected for reproduction, the crossover operator is then applied to generate offsprings. For example, let the following be two chromosomes before the one-point crossover:

```
                 1         2         3         4         5         6
         1234567 8901234567890123456789012345678901234567890123456789012345678
string1: 1001010||1000101000000000100100111010101010001000000000000100000000001
```

(the underlying network structure:
Total no of hidden layer = 2
Within layer 0: No of node = 10
Within layer 1: No of node = 5
Connectivity = 1)

```
                 1         2         3         4         5         6
         1234567 8901234567890123456789012345678901234567890123456789012345678
string2: 1001001||0000000000000000100100111010101010001000000000000100000000001
```

(the underlying network structure:
Total no of hidden layer = 1
Within layer 0: No of node = 9
Connectivity = 1)

After crossover, we obtain two new chromosomes (it can be one depending on the definition of the crossover operator) as offsprings in the next generation:

```
                  1         2         3         4         5         6
          1234567 8901234567890123456789012345678901234567890123456789012345678
string1': 1001010||0000000000000000100100111010101010001000000000000100000000001
```

(the underlying network structure:
Total no of hidden layer = 1

Within layer 0: No of node = 10
Connectivity = 1)

```
                 1         2         3         4         5         6
         1234567 890123456789012345678901234567890123456789012345678901234567 8
string2': 1001001||1000101000000001001001110101010100010000000000010000000000001
```

(the underlying network structure:
Total no of hidden layer = 2
Within layer 0: No of node = 9
Within layer 1: No of node = 5
Connectivity = 1)

If mutation is to take place, a mutation probability can be specified for evolution. For example, having mutation to take place in the first position of the following chromosome:

```
                 1         2         3         4         5         6
        12345678901234567890123456789012345678901234567890123456789012345678
string: 10111100000000000000000100100101010101010001000000000000100000000001
```

(the underlying network structure:
Total no of hidden layer = 1
Within layer 0: No of node = 30)

it gives rise to a new chromosome:

```
                 1         2         3         4         5         6
        12345678901234567890123456789012345678901234567890123456789012345678
string: 00111100000000000000000100100101010101010001000000000000100000000001
```

(the underlying network structure:
Total no of hidden layer = 0)

All chromosomes generated by the genetic algorithm engine are decoded as neural networks by the network decoder (generator). At the initialization of the system, the network decoder is initialized with the following information: maximum number of hidden layers, minimum and maximum number of nodes in each layer, as well as lower and upper bounds of the weights. Based on the information, the network decoder converts the chromosome into a multilayer feedforward neural network. Each connection weight is initialized with a random value between the lower and upper bounds of the weight. The decoded network is fed to the network trainer for training. The network trainer is responsible for the training of the neural network by backpropagation using the training set provided. The network tester is responsible for the testing of the trained network using the testing set provided. When the testing is finished, the Testing Fitness is obtained. The tested network is then evaluated by the Network Fitness Evaluator which is responsible for the evaluation of the overall fitness of the neural network.

Remark 1. Though multilayer feedforward neural networks are instrumental in handling non-linearity, they are not built to handle input noise in data classification. Linear associative memories, a single-layer feedforward neural network, is specifically constructed for such a purpose. The model should be useful in classifying, for example, remote sensing data in which noise exists. In what follows, I give a brief discussion of the model.

17.3 Linear Associative Memories for Spatial Data Classification

A linear associative memory (LAM) is a single-layer feedforward neural network which stores a given set of pattern pairs in the form of weighted connections between neuron-like elements (Kohonen, 1989). This type of networks is not only known to be the simplest model of associative memories, but has also found important applications, for instance, in pattern recognition (Wechsler and Zimmerman, 1988) and signal deconvolution (Eichmann and Stojancic, 1987).

In its applications, a LAM (Fig. 17.5) is expected to store a set of r pattern pairs $\{X^{(i)}, Y^{(i)}\}$, $X^{(i)} \in \mathbf{R}^n$, $Y^{(i)} \in \mathbf{R}^m$, i = 1, 2, ..., r, in such a way that whenever one of the stored vectors, $X^{(i)}$, is presented with noise, the LAM responds by producing the vector that is closest to the related vector, $Y^{(i)}$, as output. If $\mathbf{W} = (w_{ij})_{n\times m}$ denotes the matrix of connection weights (also referred to as encoding matrix or connection matrix) between the input and output layers of the LAM, where w_{ij} represents the connection strength from neuron i in the input layer to neuron j in the output layer, it is then required that whenever a vector of the form $\tilde{X}_i = X^{(i)} + \epsilon_i$ is presented as an input, where ϵ_i is an error or perturbation vector, the output of the LAM

$$\tilde{Y}_i = W\tilde{X}_i \tag{17.12}$$

should be as close to $Y^{(i)}$ as possible.

The closeness of $\tilde{Y}_i$ to $Y^{(i)}$, as commonly adopted, can be measured in terms of the Euclidean norm $| \tilde{Y}_i - Y^{(i)} |$. Therefore, the performance of a LAM can be measured by the mean-square-error (MSE):

$$\begin{aligned} MSE(W) &= E\left(\sum_{i=1}^{r} \| Y^{(i)} - \tilde{Y}_i \|^2\right) = \sum_{i=1}^{r} \left(E\left(tr((Y^{(i)} - \tilde{Y}_i)(Y^{(i)} - \tilde{Y}_i)^T)\right)\right) \\ &= E\left(\sum_{i=1}^{r} \| Y^{(i)} - W\tilde{X}_i \|^2\right) = E\left(\sum_{i=1}^{r} \| Y^{(i)} - W(X^{(i)} + \epsilon_i) \|^2\right) \end{aligned} \tag{17.13}$$

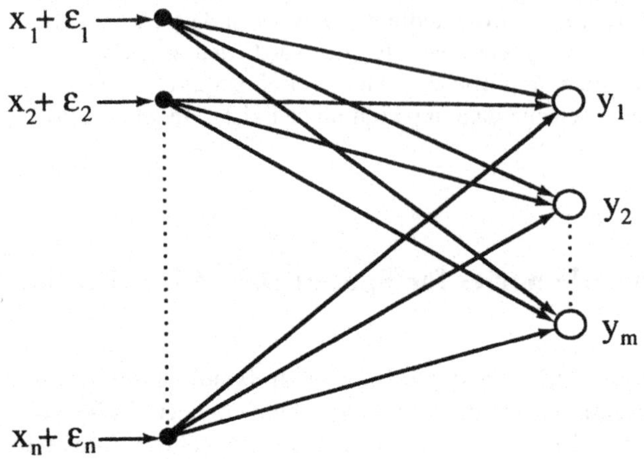

Fig. 17.5. A Linear Associative Memory

where E(.) denotes the statistical expectation of (.), and tr(A) is the trace of matrix A.

Along this line of reasoning, the basic problem of LAM can be stated as follows: Given a set of pattern pairs $\{X^{(i)}, Y^{(i)}\}$, and a distribution of errors $\{\epsilon_1, \epsilon_2, ..., \epsilon_r\}$ under a noise circumstance, find the encoding **W** by

$$\textit{Minimize } MSE(W) := E\left(\sum_{i=1}^{r} \| Y^{(i)} - W(X^{(i)} + \epsilon_i) \|^2 \right). \tag{17.14}$$

Let $\mathbf{X} = [X^{(1)}, X^{(2)}, ..., X^{(r)}]$ and $\mathbf{Y} = [Y^{(1)}, Y^{(2)}, ..., Y^{(r)}]$ be the set of input and output vectors respectively. We say that the encoding **W** is unbiased if it satisfies

$$\boldsymbol{Y} = \boldsymbol{WX}. \tag{17.15}$$

If **W** does not satisfy the equality in (17.15), it is then referred to as a biased encoding. It should be noted that a LAM with unbiased encoding always admits an error-free recall of all the stored $Y^{(i)}$'s if the input are all error-free.

LAM was first proposed by Anderson (1968), and was later studied by Anderson (1972) and Kohonen (1977). They all applied the encoding $\mathbf{W} = \mathbf{YX}^T$, i.e., the Hebbian rule (Hebb, 1949) to construct the LAM. Though the encoding is simple and natural, its validity is severely restricted to the situation where the stored patterns $\{X^{(1)}, X^{(2)}, ..., X^{(r)}\}$ must be normalized and orthogonal.

Wee (1968) and Kohonen and Ruohonen (1973) suggested the use of pseudo-inverse encoding:

$$\boldsymbol{W} =: \boldsymbol{W}_1 = \boldsymbol{YX}^+. \tag{17.16}$$

They showed that if **X** is of column-full rank (i.e., $\{X^{(1)}, X^{(2)}, ..., X^{(r)}\}$ is linearly independent), then the encoding is unbiased and optimal when the noise is white (i.e., when the distribution of the error vector ϵ_i, i = 1, 2, ..., r, satisfies $E(\epsilon_i) = 0$ and $Cov(\epsilon_i = \sigma^2 I)$. In this setting, Murakami and Aibara (1987) showed that the MSE value of the encoding is given by

$$MSE(W_1) = \sum_{i=t+1}^{r} \| Yq_i \|^2 + \sum_{i=1}^{t} \frac{\sigma^2 r}{\lambda_i} \| Yq_i \|^2 \tag{17.17}$$

where t is the rank of **X**; $\lambda_1 \geq \lambda_2 \geq ... \geq \lambda_t$ $(t \leq r)$ are the nonzero singular values of **X**; $\mathbf{P} = [p_1, p_2, ..., p_r]$, $\Lambda = \text{diag}\{\lambda_1, \lambda_2, ..., \lambda_r\}$, and $\mathbf{Q} = [q_1, q_2, ..., q_r]$ are the corresponding matrices of the singular value decomposition of **X** such that

$$X = P \Lambda Q^T. \tag{17.18}$$

The unbiased encodings have disadvantages not only in its sensitivity to input noise, but also in the LAM's loss of error-correction capability as the number of stored pattern pairs, r, approaches the dimension of the input vectors, n. To overcome these disadvantages of the unbiased encodings, Murakami and Aibara (1987) proposed an encoding $\mathbf{W}_2$ to perform biased association. In terms of the singular value decomposition of **X** introduced above, $\mathbf{W}_2$ is defined as

$$W_2 = Y\bar{X}^+ \tag{17.19}$$

where

$$\bar{X}^+ = \sum_{i=1}^{s} \lambda_i^{-1/2} q_i p_i^T, \tag{17.20}$$

and $s \in \{1, 2, ..., r\}$ is any fixed integer such that

$$\frac{\sigma^2 r}{\lambda_s} \leq 1 \leq \frac{\sigma^2 r}{\lambda_{s+1}}. \tag{17.21}$$

Under the white-noise circumstance, the corresponding MSE value is obtained as

$$MSE(W_2) = \sum_{i=s+1}^{r} \| Yq_i \|^2 + \sum_{i=1}^{s} \frac{\sigma^2 r}{\lambda_i} \| Yq_i \|^2. \tag{17.22}$$

The encoding is an improvement over the Wee-Kohonen encoding in the sense that $MSE(\mathbf{W}_2) \leq MSE(\mathbf{W}_1)$. Subsequent to the encoding in (17.19), there are some other improvements whose discussions are omitted here.

In terms of the MSE criterion, existing LAMs with biased encodings out-perform that with unbiased encodings under both the white-noise and colored-noise circumstances. Unlike the unbiased encodings, however, the problem of optimality has not been addressed in LAM with biased encoding. Two outstanding questions are then:

(a) Is there an optimal encoding for the LAM under the (i) white-noise and (ii) colored-noise circumstances?

(b) If the answer is positive, what then is the optimal encoding?

Leung *et al.* (1996b) give a theoretical analysis of these problems. The answers to both questions turn out to be affirmative and are summarized in the following theorems (Proofs are omitted here):

Theorem 1: Under white-noise input, the optimal biased encoding of the LAM exists and is given by

$$W_{opt} = YX^T(XX^T + \sigma^2 rI)^{-1}. \tag{17.23}$$

In this case, the minimum MSE value is given by

$$MSE(W_{opt}) = \sum_{i=1}^{r} \frac{r\sigma^2}{\lambda_i + r\sigma^2} \| Yq_i \|^2. \tag{17.24}$$

Theorem 2: Under colored-noise input, the optimal biased encoding, $\mathbf{W}^c_{opt}$, of the LAM exists and is given by

$$W^c_{opt} = YX^T(XX^T + rV)^{-1}. \tag{17.25}$$

In this case, the minimum MSE value is given by

$$MSE(W^c_{opt}) = \sum_{i=1}^{r} \frac{r}{\lambda_i^* + r} \| Yq_i^* \|^2. \tag{17.26}$$

In the sense of MSE, Leung *et al.* (1966b) have shown theoretically that their encodings are the best among existing LAM models. The analytical results have further been substantiated by computer simulations that show the optimality and superiority of the proposed models. Together with Wee-Kohonen's optimal unbiased encoding, the problem of optimality of the LAM under the white-noise and colored-noise situations have all been settled. Since associative memories are often embedded with noise, the LAM encodings proposed here are expected to be instrumental in real-life spatial pattern classification problems. Nevertheless, LAM may not perform well under non-linearity.

Remark 2. In addition to the above two classes of neural networks, there is another class of neural network called radial basis function (RBF) neural networks which are powerful, but yet not utilized, models for spatial data classification. Not only it is useful for classifying data into clusters, it is also instrumental in learning rules by examples. Mathematically, rule learning can be treated as high dimensional data cluster analysis, especially under nonlinearity. Each cluster can be interpreted as a rule specifying a particular relationship between input (premises) and output (conclusions). To make our discussion more diversified, I examine, in the following, RBF neural networks in the context of rule learning by training examples.

17.4 Radial Basis Function Neural Networks for Rule Learning

Feedforward neural networks with gradient-descent procedure, back-propagation algorithm, or least-mean-square algorithm, among other rule-learning techniques (see for example Kosko, 1992), have been constructed to acquire fuzzy rules from examples (Lin and George, 1991; Ishibuchi *et al.*, 1993; Nauck and Kruse, 1993; Sulzberger *et al.*, 1993). Though these models appear to render good learning procedures, they may not converge or may tend to have a very slow rate of convergence when the volume of learning examples is large, such as in remote sensing, and the variables involved are numerous.

To have a fast extraction of fuzzy and non-fuzzy IF-THEN rules, a RBF neural network has been developed for acquisition of knowledge from a large number of learning examples (Leung and Lin, 1996). A RBF neural network is essentially a feedforward neural network with a single hidden layer. The transformation (activation) function from the input space to the hidden-unit space is nonlinear, while that from the hidden-unit space to the output space is linear. Its advantage is that the hidden units provide a set of radial basis functions constituting an arbitrary basis for the expansion of the input patterns into the hidden-unit space. The RBF network differs from the multilayer perceptron in that it only has one hidden layer and the transformation from the hidden layer to the output layer is linear. It in a way is related to the single-layer perceptron but can implement nonlinear transformation of the input space.

While radial basis functions are first proposed for functional approximation (see the review by Powell, 1990), they have for example been exploited for the design of neural networks (Broomhead and Lowe, 1988) and fuzzy logic controllers (Steele *et al.*, 1995). Here, I introduce in brief how a competitive RBF network can be constructed for the fast extraction of fuzzy and non-fuzzy rules for spatial inference (Leung and Lin, 1966).

Without loss of generality, let the following be the ℓ-th rule of a set of IF-THEN rules to be learned:

$$\begin{aligned}&\textbf{If } x_1 \textbf{ is } \mu_{1\ell},\, x_2 \textbf{ is } \mu_{2\ell},\, \ldots,\, x_N \textbf{ is } \mu_{N\ell}\\ &\textbf{then } y_1 \textbf{ is } \nu_{\ell 1},\, y_2 \textbf{ is } \nu_{\ell 2},\, \ldots,\, y_p \textbf{ is } \nu_{\ell p},\end{aligned} \tag{17.27}$$

where $\nu_{\ell k}$, k = 1, 2, ..., p, can be a real number, a fuzzy subset (e.g. a fuzzy number), or a binary number. We first assume that $\nu_{\ell k}$ is a fuzzy subset.

Let $\{(X^{(i)}, Y^{(i)}), i = 1, 2, ..., M\}$, where $X^{(i)} = (x_1^{(i)}, x_2^{(i)}, ..., x_N^{(i)})$ and $Y^{(i)} = (y_1^{(i)}, y_2^{(i)}, ..., y_P^{(i)})$, be a set of M input-output patterns. Let s be the number of fuzzy subspaces partitioning each input and output space (for simplicity, we make the number of partitions in the input and output spaces be the same; the method to be introduced however holds for unequal number of partitions).

Let $[-\ell_j, \ell_j]$, j = 1, 2, ..., N, be the domain of input space j. Let the centers of the fuzzy input subspaces be

$$C_j(1), C_j(2), ..., C_j(s),\ j = 1, 2, ..., N. \tag{17.28}$$

Let $[-L_k, L_k]$, k = 1, 2, ..., p, be the domain of output space k. Let the centers of the fuzzy output subspaces be

$$\nu_k(1), \nu_k(2), ..., \nu_k(s),\ k = 1, 2, ..., p. \tag{17.29}$$

The RBF network first determines the centers of the input and output subspaces through an unsupervised competitive learning as follows:

(a) For each x_j, input $x_j^{(1)}$.
(b) Compute the degree of matching

$$d_{jr}^{(1)} = |x_j^{(1)} - C_j(r)|,\ r = 1, 2, ..., s. \tag{17.30}$$

(c) Select the best-match (winner) neuron and adjust its weight so that the node which is closest to $x_j^{(1)}$ has a greater chance to win (and thus achieve the clustering effect), i.e., Let

$$|x_j^{(1)} - C_j(r_1)| = \min_r d_{jr}^{(1)} = \min_r |x_j^{(1)} - C_j(r)|, \tag{17.31}$$

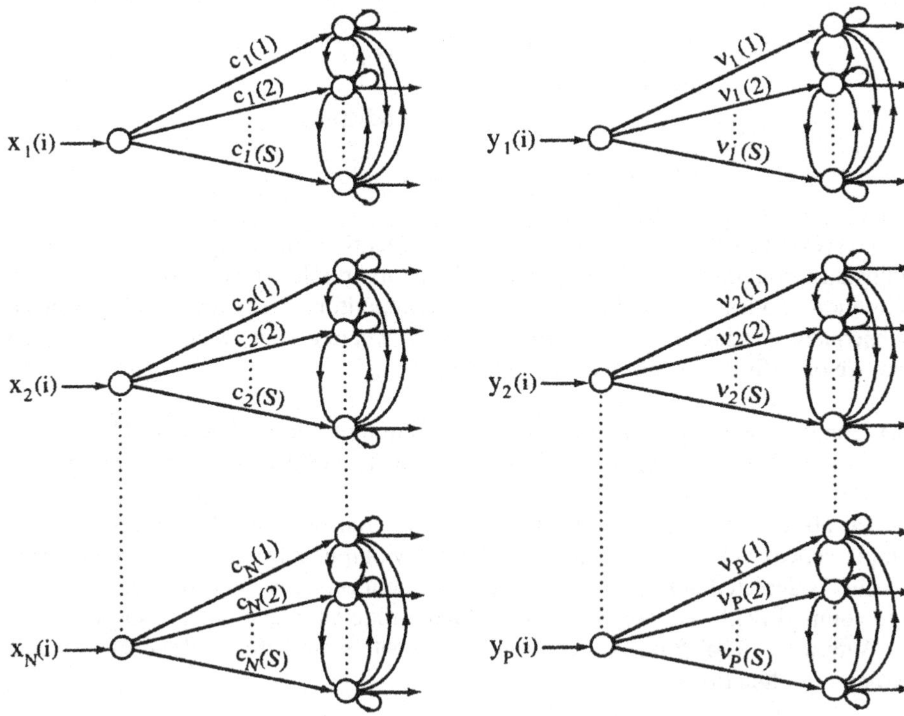

(a) Competitive learning of centers in the input space

(b) Competitive learning of centers in the output space

Fig. 17.6. Unsupervised Competitive Learning Process

then

$$\Delta C_j(r) = \begin{cases} \eta_1[x_j^{(1)} - C_j(r)], & \text{if } r = r_1, \\ 0, & \text{otherwise,} \end{cases} \tag{17.32}$$

where $\eta_1 \in (0, 1)$ is a coefficient of learning.

To prevent too large a fluctuation in the competitive clustering process, η_1 can be adjusted throughout. For example, we take a larger value of η_1 at the beginning so that more nodes can be involved in the competition. The value of η_1 can be subsequently reduced (e.g. we can take $\eta_1/\sqrt{i}$, where i is the rounds of training) so that we only fine tune the relevant center.

(d) Input $x_j^{(2)}$ and repeat steps (a)-(c) until all m inputs are exhausted. The derived $C_j(1)$, $C_j(2)$, ..., $C_j(s)$ are then the winner centers of the fuzzy subspaces of x_j.

The above unsupervised competitive learning procedure can also be applied to the output spaces to derive the centers of the fuzzy output subspaces. The competitive process is depicted in Fig. 17.6a, b. Here, the connection weights between the input and competitive layers are in fact the centers of the input and output subspaces.

Within the RBF framework, the derived centers are actually the centers of the radial basis functions:

$$G(|x - c|) = \exp[-(x - c)^2/2\sigma^2] \tag{17.33}$$

which are actually the membership functions of the linguistic terms of the fuzzy IF-THEN rules in (17.27).

The spread of G is σ which needs to be adaptively adjusted. For too large a σ, we would have too much of an overlapping of the fuzzy subspaces resulting in unclear classification of rules. For too small a σ, we would have too condense a radial basis function affecting the precision of computation. In place of the gradient descent procedure which tends to over tune σ, we use the following procedure:

$$\Delta\sigma_{jr} = \alpha\left[\frac{h_{jr}\cdot\sqrt{\rho}}{s} - \sigma_{jr}\right], \tag{17.34}$$

where

$$h_{jr} = \begin{cases} C_j(r+1) - C_j(r), & \text{if } r = 1, \\ C_j(r) - C_j(r-1), & \text{if } r = s, \\ [C_j(r+1) - C_j(r-1)]/2, & \text{if } 1 < r < s, \end{cases} \tag{17.35}$$

and α is an coefficient of adjustment. Here, by usual practice, the initial value of σ is selected as:

$$\sigma_r^0 = \frac{2\ell}{(s-1)s} \cdot \sqrt{\rho}, \tag{17.36}$$

where 2ℓ is the length of $[-\ell, \ell]$ and ρ is an experimental constant which by our experience can be selected as follows:

$$\rho = \begin{cases} 3, & \text{if } s = 5, \\ 6, & \text{if } s = 7, \\ 21, & \text{if } s = 13. \end{cases} \tag{17.37}$$

For s = 7, Fig. 17.7 depicts the classification of the subspaces of x_j after the unsupervised competitive learning and the adaptive adjustment of σ_{jr}.

Since the centers of all subspaces can be determined as above, then the centers of the clusters in $\mathbb{R}^N$ have coordinates

$$(C_1(i_1), C_2(i_2), ..., C_N(i_N)),\ i_1, i_2, ..., i_N \in \{1, 2, ..., S\}. \tag{17.38}$$

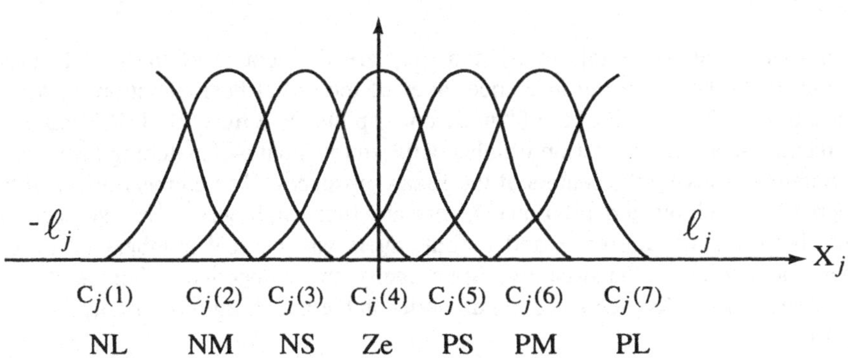

Fig. 17.7. Fuzzy Partitions of Input Space X_j

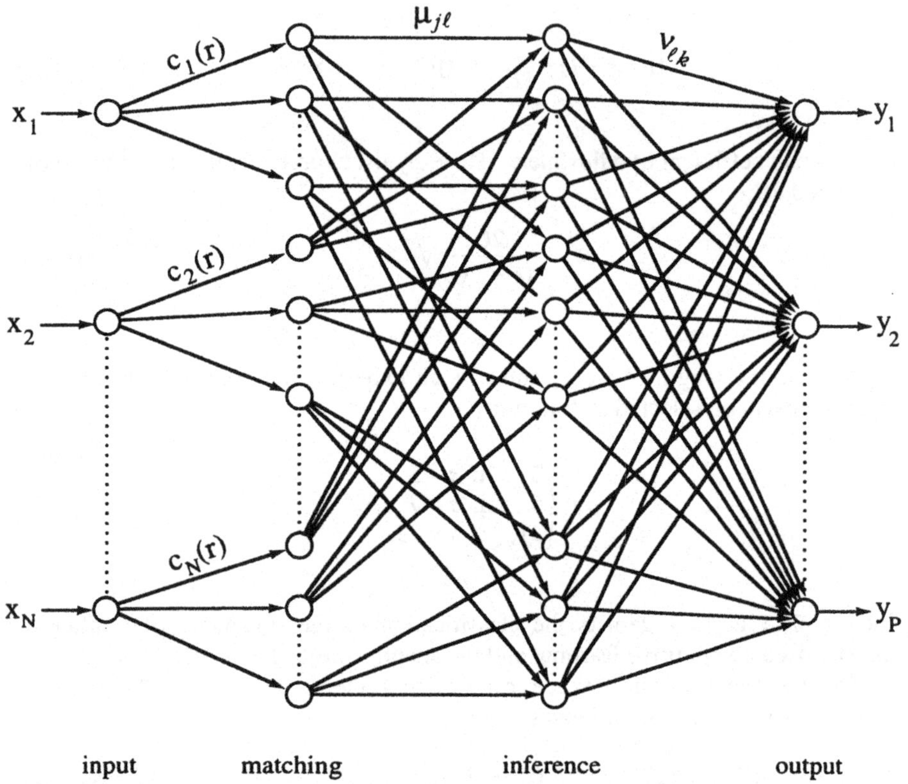

Fig. 17.8. Architecture of the RBF Network for Rule-Learning

The number of centers totals to S^N and they are the centers of the radial basis functions of the nodes in the inference layer of the radial basis function network depicted in Fig.17.8. Each of the S^N node corresponds to a fuzzy IF-THEN rule.

To recapitulate, the connection weights between the input and matching layers are real numbers specifying the centers of the fuzzy subspaces. The connection weights between the matching and inference layers are fuzzy numbers. The connection weights between the inference and output layers can be real numbers or fuzzy numbers, depending on the situation. Specifically, the connection weights between the ℓ-th node of the inference layer and all nodes in the matching layer are $\mu_{1\ell}, \mu_{2\ell}, \ldots, \mu_{N\ell}$, with

$$\mu_{j\ell}(x) = \exp\left[-\left(x - C_j(r_{j\ell})\right)^2 / 2\sigma_j^2 r_{j\ell}\right],\ j = 1, 2, \ldots, N,\ r_{j\ell} \in \{1, 2, \ldots, S\}. \quad (17.39)$$

Let $C_\ell = (C_1(r_{1\ell}), C_2(r_{2\ell}), \ldots, C_N(r_{N\ell}))$ be the center of the radial basis function corresponding to the ℓ-th node.
Let $X_\ell^{(i)} = (x_1(i_\ell), x_2(i_\ell), \ldots., x_N(i_\ell))$ be the input, among M inputs, closest to C_ℓ, i.e.

$$\|x_\ell^{(i)} - C_\ell\| = \min_i \|x^{(i)} - C_\ell\|. \tag{17.40}$$

Then, the connection weight between the ℓ-th node of the inference layer and the output layer can be taken as the output of the i_ℓ-th input, i.e.

$$v_{\ell 1} = y_1(i_\ell),\ v_{\ell 2} = y_2(i_\ell), \ldots,\ v_{\ell p} = y_{\ell p}(i_\ell). \tag{17.41}$$

Thus, the ℓ-th rule in (17.27) is extracted.

Without further elaboration, $v_{\ell k}$ (k = 1, 2, ..., p) can be a fuzzy number. If $v_{\ell k}$ is a real number, then the corresponding output is also a real number. By thresholding $v_{\ell k}$, we can also make a binary-valued output. Therefore, the RBF network can easily extract fuzzy rules and non-fuzzy rules from a large number of examples. The performance of the network has been supported by simulation results. Furthermore, it can also perform on-line computation, rule deletion, and rule tidying which are essential features in rule-based systems (Leung and Lin, 1996).

17.5 Conclusion

I have reviewed in this paper some basic feedforward neural network models suitable for spatial data classification and rule learning. Compared to statistical methods, they are more powerful and adaptive methods for handling nonlinearity and arbitrary distributions of data in spatial pattern classification. The models are instrumental in data and rule-set classifications through learning. I have also demonstrated how evolutionary computation, specifically genetic algorithms, can be employed to evolve feedforward neural networks.

Among all feedforward neural networks, the multilayer perceptron is the one which has been applied to spatial data classification. It is hoped that other models discussed in the present paper will receive more attention in the near future. Furthermore, recurrent neural networks, not discussed in this paper, which are useful for content addressable memories should also be explored for spatial data classification and learning, especially under noise. While there is a proliferation of neural network models, geographers should be prudent about their appropriateness in spatial analysis in general and spatial data classification and learning in particular.

Acknowledgement
This research is supported by the earmarked grant CUHK 8/93H of the Hong Kong Research Grants Council.

References

Anderson, J. 1968. A Memory Storage Model Utilizing Spatial Correlation Function, *Kybernetic*, **5**:113-119.

Anderson, J. 1972. Two Models for Memory Organization Using Interacting Traces, *Mathematical Biosciences*, **80**:137-160.

Argialas, D.P. and C.A. Harlow. 1990. Computational Image Processing Models: an Overview and Perspective, *Photogrammetric Engineering and Remote Sensing*, **56**:871-886.

Benediktsson, J.A., P.H. Swain and O.K. Esroy. 1990. Neural Network Approaches versus Statistical Methods in Classification of Multisource Remote Sensing Data, *IEEE Transactions on Geoscience and Remote Sensing*, **28**:540-552.

Bischof, H., W. Schneider and A.J. Pinz. 1992. Multispectral Classification of Landsat-Images using Neural Networks, *IEEE Transactions on Geoscience and Remote Sensing*, **30**:482-489.

Broomhead, D.S. and D. Lowe. 1988. Multivariable Functional Interpolation and Adaptive Networks, *Complex Systems*, **2**:321-355.

Campbell, W.J., S.E. Hill and R.F. Cromp. 1989. Automatic Labelling and Characterization of Objects using Artificial Neural Networks, *Telematics and Informatics*, **6**:259-271.

Civco, D.L. 1993. Artificial Neural Networks for Land-cover Classification and Mapping, *International Journal of Geographical Information Systems*, **7**:173-186.

Davis, L. (ed.). 1991. *Handbook of Genetic Algorithms*. N.Y.: Van Nostrand Reinbold

Eichmann, G.E. and M. Stojancic. 1987. Superresolving Signal and Image Restoration Using a Linear Associative Memory, *Applied Optics*, 1911-1918.

Fischer, M.M. and S. Gopal. 1994. Artificial Neural Networks: A New Approach to Modeling Interregional Telecommunication Flows, *Journal of Regional Science*, **34**:503-527.

Goldberg, D.E. 1989. *Genetic Algorithms in Search, Optimization and Machine Learning*. Reading: Addison-Wesley

Halmari, P.M. and C.G. Lundberg. 1991. *Bridging Inter- and Intra-corporate Information Flows with Neural Networks*, Paper presented at the Annual Meeting of the Association of American Geographers, Miami, April 13-17

Hebb, D.O. 1949. *The Organization of Behavior*. New York: John Wiley

Heermann, P.D. and N. Khazenie. 1992. Classification of Multispectral Remote Sensing Data using a Back-propagation Neural Network, *IEEE Transactions on Geoscience and Remote Sensing*, **30**:81-88.

Hepner, G.F., T. Logan, N. Rittner and N. Bryant. 1990. Artificial Neural Network Classification using a Minimal Training Set: Comparison to Conventional Supervised Classification, *Photogrammetric Engineering and Remote Sensing*, **56**:469-473.

Holland, J.H. 1975. *Adaptation in Natural and Artificial Systems*. Ann Arbor: University of Michigan

Ishibuchi, H., H. Tanaka and H. Okada. 1993. Fuzzy Neural Networks with Fuzzy Weights and Fuzzy Biases, *IEEE Internat. Conf. on Neural Networks*, 1650-1655.

Kohonen, T. 1989. *Self-organization and Associative Memory*. New York: Springer-Verlag

Kohonen, T. 1977. *Associative Memory – A System Theoretical Approach*. New York: Springer-Verlag

Kohonen, T. and M. Ruohonen. 1973. Representation of Associated Data by Matrix Operators, *IEEE Transaction on Computers*, C-**22**:701-702.

Kosko, B. 1992. *Neural Networks and Fuzzy Systems*. Englewood Cliffs: Prentice Hall

Lee, J., R.C. Weyer, S.K. Sengupta and R.M. Welch. 1990. A Neural Network Approach to Cloud Classification, *IEEE Transaction on Geoscience and Remote Sensing*, **28**:846-855.

Leung, Y. 1994. Inference with Spatial Knowledge: An Artificial Neural Network Approach, *Geographical Systems*, **1**:103-121.

Leung, Y., K.S. Leung, M.M. Fischer, W. Ng and M.K. Lau. 1996a. *The Evolution of Multilayer Feedforward Neural Networks for Spatial Interaction using Genetic Algorithms*. (unpublished paper)

Leung, Y., T.X. Dong and Z.B. Xu. 1996b. *The Optimal Encodings for Biased Association in Linear Associative Memories*. (unpublished paper)

Leung, Y. and X. Lin. 1996. *Fast Extraction of Fuzzy and Non-fuzzy IF-THEN Rules by a Radial Basis Function Network with Unsupervised Competitive Learning*. (unpublished paper)

Lin, C.T. and C.S. George. 1991. Neural-Network Based Fuzzy Logic Control and Decision Systems, *IEEE Transactions on Computers*, **40**(12):1320-1336.

Murakami, K. and T. Aibara. 1987. An Improvement on the Moore-Penrose Generalized Inverse Associative Memory, *IEEE Transactions on Systems, Man, and Cybernetics*, **17**:699-707.

Nauck, D. and R. Kruse. 1993. A Fuzzy Neural Network Learning Fuzzy Control Rules and Membership Function by Fuzzy Error Backpropagation, *IEEE Internat. Conf. on Neural Networks*, 1022-1027.

Openshaw, S. 1992. Modelling Spatial Interaction using a Neural Net, in M.M. Fischer and P. Nijkamp (eds.), *Geographical Information Systems, Spatial Modelling and Policy Evaluation*. Berlin et al.: Springer, pp. 147-164.

Powell, M.J.D. 1992. Radial Basis Functions in 1990, *Adv. Num. Anal.*, **2**:105-210.

Rumelhart, D.E., G.E. Hinton and R.J. Williams. 1986. Learning Internal Representation by Error Propagation, in D.E. Rumelhart, J.L. McClelland, and the PDP Research Group (eds.), *Parallel Distributed Processing: Exploration in the Microstructure of Cognition*, Vol. 1, Cambridge: MIT

Short, N. 1991. A Real-Time Expert System and Neural Network for the Classification of Remotely Sensed Data, *Proceedings of the Annual Convention on the American Society for Photogrammetry and Remote Sensing*, **3**:406-418.

Steele, N.C., C.R. Reeves, M. Nicholas and P.J. King. 1995. Radial Basis Function Artificial Neural Networks for the Inference Process in Fuzzy Logic Based Control, *Computing*, **54**:99-117.

Sulzberger, S.M., N.N. Tschichold-Gürman and S.J. Vestli.1993. FUN: Optimization of Fuzzy Rule Based Systems Using Neural Networks, *IEEE Internat. Conf. on Neural Networks*, 312-316.

Tzeng, Y.C., K.S. Chen, W.L. Kao and A.K. Fung. 1994. A Dynamic Learning Neural Network for Remote Sensing Applications, *IEEE Transactions on Geoscience and Remote Sensing*, **32**:1096-1102.

Wang, F. 1994. The Use of Artificial Neural Networks in a Geographical Information System for Agricultural Land-Suitability Assessment, *Environment and Planning A*, **26**:265-284.

Wechsler, H. and G.L. Zimmerman. 1988. 2-D Invariant Object Recognition Using Distributed Associative Memory, *IEEE Transactions on Pattern Analysis and Machine Intelligence*, **20**(6): 811-821.

Wee, W. 1968. Generalized Inverse Approach to Adaptive Multiclass Pattern Classification, *IEEE Transaction on Computers*, C-**17**:1157-1164.

White, R.W. 1989. The Artificial Intelligence of Urban Dynamics: Neural Net Modelling of Urban Structure, *Papers of the Regional Science Association* **67**:43-53.

Yoshida, T. and S. Omatu. 1994. Neural Network Approach to Land Cover Mapping, *IEEE Transaction on Geoscience and Remote Sensing*, **32**:1103-1109.

18 Building Fuzzy Spatial Interaction Models

Stan Openshaw

School of Geography, University of Leeds, Leeds LS2 9JT

18.1 Introduction

Since the work of Wilson (1970) there has been surprisingly little innovation in the design of spatial interaction models. The principal exceptions include the competing destinations version of Fotheringham (1983), the use of genetic algorithms to try and breed new forms of spatial interaction model, either directly (Openshaw, 1988) or by genetic programming (Turton et al., 1997) and the application of supervised artificial neural networks to model spatial interaction data (Openshaw, 1993; Fischer and Gopal, 1994). Clearly these latter methods will be developed much further over the next few years. The purpose of this Chapter is to initiate the development of a new class of spatial interaction models utilising the principles of fuzzy sets and fuzzy logic.

At first sight there may not seem to be much, if any, scope for fuzziness in spatial interaction models. The entropy maximising specification of the conventional model is a classic instance of an extremely crisp model that is free of fuzziness. However, the motives for contemplating the construction of a fuzzy model are: (1) to try and improve the model's empirical performance, (2) to investigate whether the spatial interaction model can itself be made more intelligent by improving its ability to handle qualitative knowledge about spatial interaction; (3) to create a model structure which can become sensitised and adaptive to localised trip patterning, and (4) to develop a flexible framework within which both macro and micro variables can be used in hybrid models of flow predictions.

From an AI perspective the conventional mathematical and statistical models of spatial interaction phenomena are essentially dumb because they are unable to incorporate much of the knowledge that is possessed about spatial interaction behaviour patterns other than that which can be encapsulated in the model equations. Indeed the input of human knowledge into such models is usually restricted to a burst of abstract mathematical manipulation when the model was first derived but this is usually non-geographical and fairly limited in what it can represent. One question is therefore what sort of knowledge do we have that is poorly handled by current models? The answer depends on how the available knowledge (or hypotheses) about spatial interaction is represented in the model. What would happen if an attempt is made to build a model that represents the basic distance decay hypothesis expressed qualitatively in words rather than quantitatively in algebraic equations? For instance, the distance decay model can be expressed linguistically as the rule of thumb that few trips go long distances. The gravity model is similarly described as indicating that few long distance trips occur unless the destination is highly attractive? Neither of these linguistic and qualitative models specify a precisely formulated log-linear distance decay or a negative exponential deterrence function or a Poisson trip distribution assumption as are found in conventional statistical and mathematical interaction models. Instead the linguistic models consists of vague concepts such as few trips are long distance and its corollary that most trips are short distance. Note the use of the vague or fuzzy terms *few trips*, *most trips*, *short* and *long distance* that are not defined in any precise way. However, the lack of a precise numeric definition does not stop the reader from understanding the structure of the model, even if each reader may well have quite different numeric values in mind. For instance, the UK concept of a *short distance* is probably much shorter from that of their US or Australian colleagues.

The conversion of this simple linguistic model into a conventional mathematical or statistical format involves removing any fuzziness. Otherwise the mathematical or statistical modelling methods would be unable to cope with it. However, this forces us to make various arbitrary assumptions that were not present in the original linguistic model; for instance, the use of a negative exponential deterrence function to represent the attenuation of flows with increasing distance. Fortunately this level of precision is no longer necessary as a fuzzy logic based model can now be built that retains the ambiguity and imprecision that characterised the linguistic version of the model. Fuzzy logic extends the domain of computers into areas that were previously considered to be exempt from computation. Zadeh (1995) writes "If I were asked to describe as succinctly as possible what fuzzy logic offers in the realms of systems analysis and design, I would answer: a methodology for computing with words". (page ix) The ability to develop soft computing applications that permit computer models to be specified and built from linguistic statements, using common-sense rules of thumb based on a mix of gut-feeling, intuition and qualitative theory, is potentially extremely important. It also offers a refreshingly new perspective on how to go about building better models of geographical systems by handling rather than ignoring or artificially removing the fuzziness within them.

Fuzzy logic is more than simply an AI tool but it is increasingly being viewed as an alternative paradigm for science and for building more intelligent modelling systems (Kosko, 1994; McNeil and Freiberger 1993). What started as a very simple idea of apparent limited interest is now being regarded as a major advance relevant to an increasing number of applications (Openshaw, 1996). Fuzzy logic modelling offers unique opportunities for the incorporation of knowledge and other intangible human skills and intuition into computer systems to help them behave in more human and in thus more intelligent ways. It attempts to emulate how the brain thinks rather than the processes by which it works. It offers a methodology for coming to terms with the pervasive imprecision of the real world. Somehow it is necessary to learn that being imprecise does not invalidate the science. It is important also to conquer the long standing and deeply rooted traditions that emphasise preciseness even in areas which are naturally imprecise. Such attitudes and beliefs are very profoundly held and are slow to shift since it is not immediately obvious that becoming less precise and less quantitative might actually be an advantage. Fuzzy sets and fuzzy logic make it possible to handle rather than ignore ambiguity.

The spatial interaction model is potentially a good candidate for a fuzzy model because: (1) the underlying concepts are fuzzy although the early adoption by Carey

(1858) of a particular mathematical specification (Newtonian gravity function) tended to disguise this; (2) only a small number of variables are involved which makes a fuzzy implementation straightforward; (3) there is a high degree of non-linearity in interaction phenomenon; (4) the availability of benchmark results which can be used in comparative evaluation; (5) a reasonably good understanding of what is going on that may be used to inform the fuzzy modelling process; and (6) continuing use of and research interest in a broad range of spatial interaction phenomena. How to build a fuzzy spatial interaction model is the subject of the remainder of the paper.

18.2 What is Fuzzy Logic?

18.2.1 We Live in a Fuzzy World

The term 'fuzzy' was first proposed by Zadeh (1962) and the first paper published in 1965; see Zadeh (1965). Much of the early work was dominated by attempts to create a theoretical foundation in a rigorous mathematical manner (Zadeh, 1971, 1974). As a result many of the more mathematically orientated papers are difficult for non-mathematicians to grasp. Additionally, the early applications were in control engineering which is an area unfamiliar to most geographers. Indeed, one of the first practical applications that demonstrated the potential of fuzzy logic was Mamdani and Assilian (1975) who showed how a small number of fuzzy IF-THEN rules could be used to control a laboratory steam engine. The basic idea was to incorporate the experience of a skilled human into a computer control algorithm based on a set of linguistic rules that represented various rules of thumb. There were four benefits: its use of knowledge, there was no need for a mathematical or statistical model of the process being controlled, it worked better than conventional control theory, and it was considerably simpler. Perhaps not surprisingly, the first commercial application in 1975 was a fuzzy controller developed for a cement kiln (Holmblad and Ostergaard, 1981, 1982). Again fuzzy logic was critical because it provided an easy way to make a computer model of a skilled kiln operator based on a small number of rules when the process itself was extremely complex. The principles and attractions of the fuzzy logic controller spread rapidly via widely read books on the subject (Gupta and Sanchez, 1982; Sugeno, 1985; Terano et al, 1989) Other much publicised applications of fuzzy logic control occurred in Japan, in a water treatment plant and particularly the highly publicised Sendai subway (Yasunobu and Miyamoto, 1985).

However, it is really only since the mid 1980s that fuzzy logic linked to computational methods has succeeded in transforming the basic mathematical concepts into a practically useful and widely applicable technology. This transformation was very rapid with over 4,000 papers being written on fuzzy logic applications by the late 1980s. The principal attraction being the realisation that due to incomplete knowledge and information, precise mathematics are not always sufficient to model the behaviour of complex systems. It suddenly became possible in the late 1980s to build micro-controllers that incorporated expert knowledge and human intuition that were also cheaper to implement, more flexible and more intelligent in their behaviour than more traditional mathematical control theory based approaches, and the new technology was simple enough to be incorporated into the available microprocessor hardware. As a result, intelligent control systems incorporating fuzzy logic are being embedded in an increasing number of domestic and industrial products. McNeil and Thro (1994) suggests that in 1994 Japan exported over $30 billion worth of products that had some form of fuzzy logic in them; ranging from washing machines to autofocusing systems on cameras to car braking systems. However, the potential applications go far beyond the design of intelligent controllers as the same technology is applicable to many data analysis, decision-making, and modelling problems relevant to geography and many other social sciences (Openshaw, 1996). It is important to recognise that a fuzzy controller is also a fuzzy systems model (Cox, 1994; p380). Fuzzy system modelling is therefore, directly applicable as an alternative modelling paradigm which is valid for urban and regional modelling as well as for many other data modelling applications in the social sciences. It would appear to be most useful whenever a powerful equation free universal approximator is needed that is capable of a plain English description. In theory this should produce easy to understand models that can incorporate whatever knowledge is possessed but also has an ability to learn or discover the rest from data. This would appear to offer the best of all possible approaches.

18.2.2 So What is Fuzzy Set Theory?

Fuzzy in this context has nothing to do with being confused, it is not that kind of fuzziness. It is all about moving beyond and away from crisp logic. Computers are based on a crisp binary or 0-1 logic. Normal science and logical positivism is also a reflection of a crisp black and white view of the world. With fuzzy logic the whole idea is to escape from the artificialness of two valued crisp logic and to think in terms of outcomes that can range anywhere between 0 (No) to 1 (Yes). A key feature is the use of words rather than numbers to describe or label the intermediate states. A fuzzy logic statement allows the quantification of concepts that have no clearly defined boundaries. Linguistic statements that express ideas, intuition, knowledge, feelings, rules of thumb and vague theories are often capable of being interpreted differently by different people because they involve fuzzy propositions. Somehow humans can cope and communicate despite the vagueness and imprecision of natural language. Fuzzy logic provides a computer technology that can cope with the imprecision in a manner akin to that of human beings.

Before Zadeh (1965), a number of other thinkers had toyed with the idea of a logic of vagueness. Lukasiewicz in the 1920s developed multi-valued logic with an infinity of values between 0 and 1 (Lukasiewicz, 1970). The next development occurred in 1937 with Black's general theory of vagueness. (Black, 1937). However it was left to Zadeh (1965) to assemble all the parts of the puzzle and put it all onto a sound mathematical basis. Zadeh successfully combined the concepts of crisp logic and the Lukasiewicz notion of multivalued sets by defining fuzzy sets with grades of membership. The change may appear trivial and obvious. It is surprising that the crisp approach ever came to be so dominant but crisp logic is deeply ingrained in people's belief systems and cultures. Nevertheless, fuzzy set theory provides a means of handling vague concepts defined by language or words in a mathematical manner, and as such it provides a much better basis for building models of complex systems where the level of understanding is at a linguistic level. It is this feature that makes them of such potential value to modelling geographical systems.

Fuzzy sets can be used to handle problems that have no sharp boundaries or situations in which events are fuzzily defined. For example, the statement "most trips are short distance" is not helpful in a crisp context because the definitions of 'most' and 'short' are undefined. It is no longer possible to represent the concept of a short distance trip by a simple probability, since there is no crisp boundary defining short distance that the probability can represent. The fuzzy concept of short distance can be broken down and expressed as sub-sets labelled short, medium, or long. Any spatial flow can belong to one or more of these fuzzy sets with grades of membership (i.e. probabilities of belonging) that range from 0.0 to 1.0.

Consider an example. What is meant by a long distance trip? Is 15 minutes a long distance trip? In crisp set theory it is necessary to select an arbitrary cut off point, at say 15 minutes. So every trip longer than 15 minutes is considered long distance. The membership graph for this set would display a sudden jump at the 15 minute boundary. So a trip of 14.9999 minutes is not long but one that is 15.0000 is, despite being 1/1000*th* of a minute greater. Is this sensible? Clearly this crisp representation only really works for non-continuous phenomenon that are lumpy in a meaningful way. The fuzzy set approach is to regard the membership function of the long distance set as being a gradation between definitely not long distance to definitely long distance. In Figure 18.1 the boundary of the long distance set is represented by an upwardly sloping line that stretches from 10 to 20 minutes. Any distance less than 10 is definitely not long distance, above 20 minutes it is long distance, and between 10 and 20 minutes the membership probabilities vary from 0 to 1.0. So for 14.9999 minutes, it is almost 0.5 whereas in the crisp set representation it would be zero. This illustrates another nice property of fuzzy sets. They incorporate crisp logic as special cases where there is no imprecision. As a result it provides an all embracing

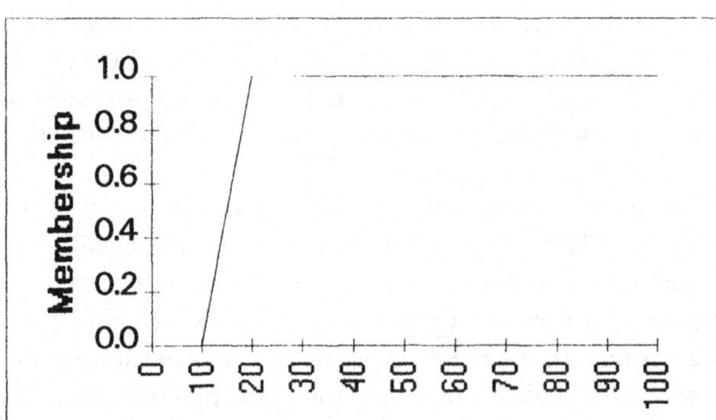

Fig. 18.1 A Fuzzy Representation of Long Distance

framework that keeps what we already have but significantly extends and adds flexibility to it. In a realistic application length of trip would be split into several fuzzy subsets; e.g. small, medium, long, and very long, the membership of which would overlap, see Figure 18.2 for one way of representing these fuzzy concepts.

18.2.3 Fuzziness Versus Probability

In essence fuzziness is an alternative to randomness as a basis for describing uncertainty, but is uncertainty the same as randomness? Kosko (1992) puts it "If we are not sure about something, is it only up to chance? Do the notions of likelihood and probability exhaust our notions of uncertainty?" (p264). Many statisticians believe so, particularly the Bayesianists. Others argue that fuzzy logic is really based on very fuzzy thinking and the problems it claims to handle are better solved using more traditional methods of logic and probability. Others are concerned that a focus on fuzziness represents a retreat from a scientific to an essentially non-scientific approach. McNeill and Freiberger (1994) write "Probability is today the great rival of fuzzy logic, and its champions claim it surpasses fuzzy logic in any task one can devise" (p58). The existence of such fairly inflexible and entrenched views partly explains why there has been so much opposition to fuzzy logic, particularly in the USA. However maybe this can be explained as von Altrock (1995) notes: "If you have a hammer in your hand, and it's your only tool, then suddenly everything looks like a nail!" (p4).

The key point to note is that randomness (from a probability perspective) and fuzziness are different things both conceptually and theoretically. Fuzzy set theory is not just another form of probability but representing a fundamentally different type of uncertainty. Kosko (1992) explains it like this: "Fuzziness describes event ambiguity. It measures the degree to which an event occurs, not whether an event occurs. Randomness describes the uncertainty of event occurrence. An event occurs or not, and you can bet on it. The issue concerns the occurring event: Is it uncertain in any way.... Whether an event occurs is "random". To what degree it occurs is fuzzy" (page 265). Fuzziness describes therefore the *degree to which an event occurs* not just *whether it occurs*. Probability deals with yes/no occurrences, it requires ignorance, and is statistical in nature. Fuzziness deals with degrees of occurrence, it does not require ignorance, and is non-statistical. For example, there may be a 20 per cent chance of light rain tomorrow, this is an example of a probability. Whether light rain occurs has a probability of 0.2, but the fuzziness here concerns the ambiguous nature of the expected event, viz. light rain. Neither more data nor a better model can make this ambiguity disappear. So whether the event occurs is the domain of probability but the degree to which it occurs is the domain of fuzzy logic. Of course the two can be combined so that there is a 0.2 probability of a fuzzy event labelled light rain occurring. Maybe some statisticians just do not like or see any need for their probability based world being embedded in a broader framework. However, unlike fuzziness, probability dissipates with increasing information. So when tomorrow comes it is possible to remove all uncertainty as to whether it rained or not, but the ambiguity about the nature of the event (viz. was it light rain or heavy rain) remains. The fuzziness is still there and has not been removed or diminished by complete information. Fuzziness is a genuine type of uncertainty.

18.3 Building a Fuzzy Spatial Interaction Model

18.3.1 A Basic Fuzzy Model

Fuzzy models usually start with some linguistic representation of the system of interest. The simplest fuzzy model is little more than a series of IF-THEN rules that when processed as fuzzy sets connect a set of inputs to a set of outputs in a non-linear and non-parametric way. It provides a way of allowing computers to reason with fuzzy numbers in the form of rules of thumb of the sort: viz. few trips are long distance. The knowledge or intelligence comes from associating the two fuzzy events, in this example few trips and distance travelled as these rules reflect knowledge about the system being modelled. For example, a simple distance decay relationship between a distance variable X and trip intensity Y can be represented as follows:

rule 1: if X is a *short* distance then Y is *massive*,
rule 2: if X is an *average* distance then Y is *some*,
rule 3: if X is a *big* distance then Y is *none*.

Note the use of subjective labels short, average, big, massive, some, and none for the three fuzzy sets representing distance and interaction. These three rules constitute a fuzzy model that relates interaction intensity to distance but without having to be explicit about the precise statistical or mathematical relationship being used. Change the rules and you can dramatically change the rate of distance decay the model represents; for example, a very rapid distance decay would have the rules

rule 1: if X is a *short* distance then Y is *massive,*
rule 2: if X is an *average* distance then Y is *zero,*
rule 3: if X is a *big* distance then Y is *zero,*

or if you would prefer a 'U' shaped model

rule 1: if X is a *short* distance then Y is *massive,*
rule 2: if X is an *average* distance then Y is *zero,*
rule 3: if X is a *big* distance then Y is *massive.*

Note that there is no statistical model or parameter estimation involved in these fuzzy models. The performance of the model depends on the appropriateness of rules that are used and on the fuzzy set definitions of the linguistic terms. Building a fuzzy spatial interaction model is in principle straightforward and can be divided into a number of simple steps.

18.3.2 *Step 1:* Specify the Nature of the System to be Modelled

This is the hardest part. You need to express in linguistic terms a model specification. This linguistic description of the system of interest encompasses your knowledge, intuition, theory, etc. about how the system of interest operates and the variables that are involved.

Consider an example involving a simple one origin distance decay model. Geographers have for a long time known that in general, as a rule of thumb, few trips (whether for work or pleasure) travel long distances and most only travel short distances. In other words, the number of trips diminishes rapidly with distance. This is a linguistic description of a mathematical distance decay model that might otherwise be expressed as a statistical model of the form

$$T_{1j} = b_1 + b_2 \log C_{1j} \quad j = 1, ..., n \tag{18.1}$$

where T_{1j} is the predicted number of trips going to a destination j from a particular origin 1, C_{1j} is the distance travelled b_1, b_2 are parameters to be estimated such that the predicted T_{1j} values match as best as possible the observed data, and n is the number of destinations.

A more sophisticated version of this simple distance decay model seeks to operationalise the notion that the intensity of the distance decay effect can also be moderated by the size of the destination zone so that whilst most trips only travel short distances some might travel long distances if the destination zone is sufficiently large. A statistical modeller could express this notion as a linear or loglinear equation that includes a destination zone attractiveness term

$$T_{1j} = b_1 + b_2 \log C_{1j} + b_3 D_j \quad j = 1, ..., n \tag{18.2}$$

where D_j is the attractiveness of destination zone j, b_1, b_2 b_3 are parameters estimated to minimise the errors between the observed and predicted values for T_{1j}.

A mathematical modeller might use an entropy-maximisation technique to derive an arguably more theoretically rigorous basis for this model to yield in this simple unconstrained case

$$T_{1j} = K O_i D_j \exp(-bC_{1j}) \quad j = 1, ..., n \tag{18.3}$$

where O_i is the total number of trips to be allocated by the model and K is a constant.

By contrast the fuzzy modeller would return to the basic soft ideas that underlies both the distance decay and gravity models. There is no longer any need to be precise about the nature of the mathematical specification and this is a tremendous advantage in nonlinear modelling. Even when a mathematical model exists it may be a poor representation of the underlying concepts that are more clearly expressed linguistically. A fuzzy model may provide a better basis for modelling despite the apparent imprecision because it may more faithfully represent the underlying conceptual model. For example, it is one thing to state that trip intensity decays with distance but it is quite another to claim that it follows a negative exponential function with a parameter b value of -0.256701 as could happen easily with equation (18.3).

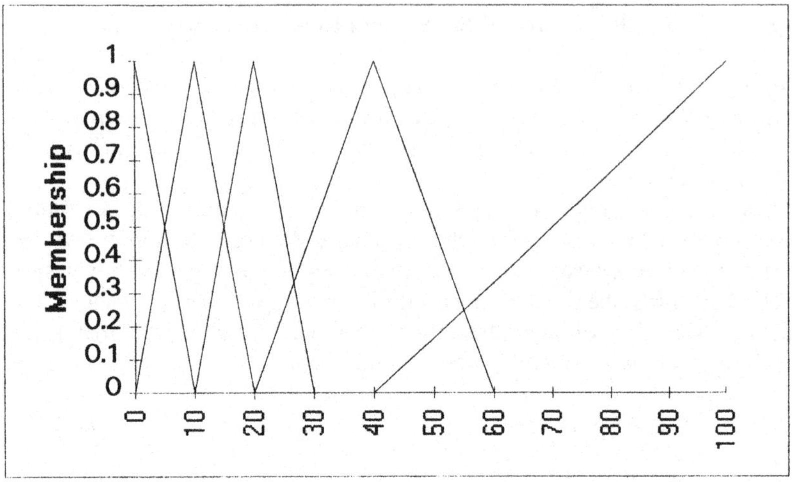

Fig. 18.2 A Fuzzy Set Representation of Travel Costs

18.3.3 *Step 2:* Identify and Label the Set Membership Functions for each Input Variable

The inputs now need to be given a fuzzy representation by specifying appropriate membership functions for each predictor variable. The only rules here are that the fuzzy membership set functions for the same variable should overlap to a large degree.

In the distance decay model there is only one input variable (C_{ij}). This is represented as shown in Figure 18.2 by five subsets with the following names, values, and shapes: for distances 0-10 the set label is *zero* (Z) and it is represented as it is a downward sloping line, for distances 0-20 it is labelled *short* distance (S) and is given a triangular shape, for distances 10-30 it is labelled *average* (A) and it is also triangular in shape for distances 20-60 likewise but the label is *big* (B), and finally for distances 40-100 the label is *long* (L) and the membership function is an upward sloping line. These shapes of membership functions are typical of those often used in fuzzy control theory (see for example, Driankov et al, 1996). It is useful that they overlap so as to capture the imprecision in the definitions of the meaning of the words zero, short, average, big, and long. Engineering experience also seems to suggest that for triangular shaped sets the overlap should be between 25 and 50% although this may depend on the intrinsic degree of imprecision associated with the two neighbouring states. If all this seems highly arbitrary then do not worry it is meant to be! You are supposed to know on the basis of knowledge, experience, intuition, skills as a geographer, etc. what are the most suitable values to use; however, this can always be changed later. Note also that the labels used does not affect the performance of the model, merely your attempts to understand and subsequently describe it.

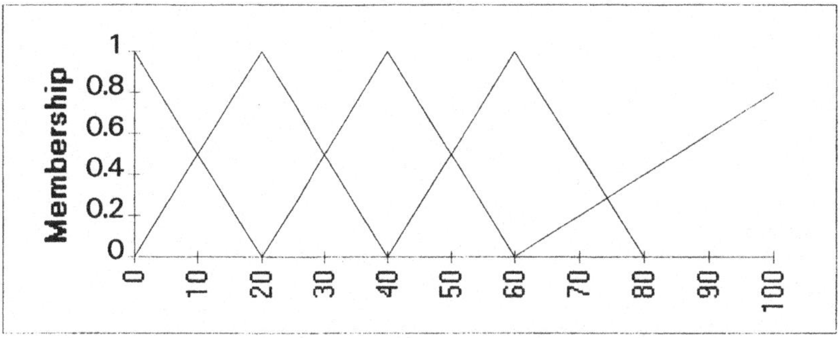

Fig 18.3 A Fuzzy Set Representation of Destination Zone Attractiveness

For the gravity model it is also necessary to produce membership functions for the D_j variable. Again five membership functions are used with the same shapes as for the C_{1j} variable but different values and labels. They are: 0-20 is *small* (S), 0-40 is *little* (L), 20-60 is *big* (R), 40-80 is *large* (L), 60-100 is *huge* (H); see Figure 18.3. Note that in both cases the numeric domain of the variables have been adjusted to be between 0 and 100.

18.3.4 *Step 3:* Identify and Label Membership Functions for the Fuzzy Outputs

The output from the fuzzy model is in the form of fuzzy sets which also require membership functions. These need to be realistically scaled in relation to the data being modelled.

Here the model is attempting to predict the magnitude of the trips. Again for simplicity five memberships functions are used with the same shapes as before. They are shown in Figure 18.4. The labels and ranges are as follows: 0-5 trips is labelled as *none* (N), 0-10 as *some* (S), 5-35 as *big* (B), 20-60 as *lots* (L), and 35-100 as *massive* (M).

There could be more or fewer fuzzy subsets. In an engineering context the numbers 3, 5, 7, 9, and 11 are often used. It depends on the level of accuracy required. This task is not so easy because the number ranges associated with the output fuzzy membership set functions determine the numerical outputs generated by the model. Careful thought is needed here and some experimentation may help. Alternatively you could re-visit these decisions later when evaluating model performance and wondering what to do to improve it.

18.3.5 *Step 4:* Creating the Rules

The heart of the fuzzy model is a series of IF-THEN rules that connect one or more fuzzy input membership sets to a fuzzy output membership set.

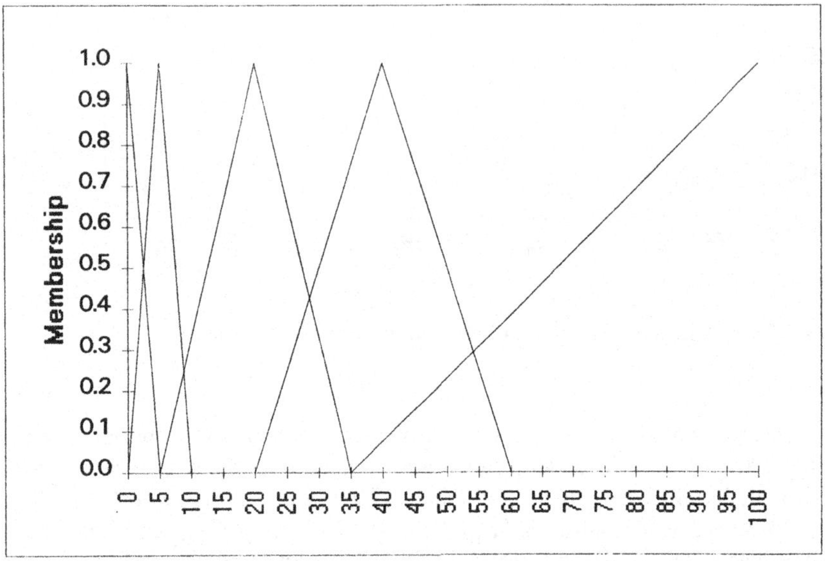

Fig. 18.4 A Fuzzy Set Representation of Spatial Interaction Intensity

If a purely distance decay model is desired then the following five rules may suffice:

rule 1:	if distance is *ZERO* then the interaction intensity is *MASSIVE*,
rule 2:	if distance is *SHORT* then the interaction intensity is *LOTS*,
rule 3:	if distance is *AVERAGE* then the interaction intensity is *BIG*,
rule 4:	if distance is *BIG* then the interaction intensity in *SOME*,
rule 5:	if distance is *LONG* then the interaction intensity is *NONE*.

This can be expressed in a table form as a Fuzzy Associative Memory (FAM); see Table 18.1. The term is borrowed from neural networks where an associative memory maps a set of inputs onto outputs. So a FAM is a transformation that maps fuzzy input sets onto fuzzy output sets. The values in the table are the fuzzy output set labels. The distance decay concept can be seen in the reduction in trip intensity with increasing distance. This table captures both the nonlinearity that is assumed to exist as well as the linguistic imprecision of the underlying concepts.

Whether the model works depends on how well these rules and membership functions represent the data. Note that most flows belong to more than one fuzzy input distance set, albeit to differing degrees. All the rules are evaluated (in parallel if need be) and more than one may be triggered to contribute to the output. This ensures that the fuzzy model's outputs responds to gradual changes of the inputs in a smooth way.

Table 18.1 A FAM for a Pure Distance Decay Model

Input fuzzy distance membership set	Output fuzzy trip intensity membership set
zero (Z)	*massive* (M)
short (S)	*lots* (L)
average (A)	*big* (B)
big (B)	*some* (S)
long (L)	*none* (N)

In this distance decay model there is only one input variable and this makes it very simple to see what is happening. But why is this useful? Why not simply use crisp, numerically precise, IF-THEN conditions? The answer is obvious in that there is insufficient knowledge to know precisely what numbers to use in the IF-THEN statements and even if we did then it would merely make the model extremely cumbersome. Additionally, the underlying concepts are linguistic and vague so that exact numerical definitions for the terms zero, short, average, etc. do not exist. If models of this simple linguistic type can be made to work then this would significantly extend what can be modelled at the same time it greatly eases the modellers task because no great degree of mathematical or statistical knowledge is involved in the modelling process. This may help 'open-up' computer modelling to non-quantitative geographers.

A more ambitious gravity model set of IF-THEN rules also needs to take into account the attractiveness of the destination zone which in a spatial interaction model context may moderate the pure distance decay effects. The fuzzy model rules now have to relate destination zone attractions and distance to fuzzy trip outputs. The full FAM is shown in Table 18.2 and the only second column is reproduced below:

rule 2: if D_j is *SMALL* and C_{ij} is *SHORT* then T_{ij} is *SOME,*
rule 7: if D_j is *LITTLE* and C_{ij} is *SHORT* then T_{ij} is *SOME,*
rule 12: if D_j is *BIG* and C_{ij} is *SHORT* then T_{ij} is *BIG,*
rule 17: if D_j is *LARGE* and C_{ij} is *SHORT* then T_{ij} is *LOTS,*
rule 22: if D_j is *HUGE* and C_{ij} is *SHORT* then T_{ij} is *MASSIVE.*

The general structure of the gravity model can be readily seen in Table 18.2. As distances increase from left to right interaction diminishes but at a different rate for different sizes of destination zone. These fuzzy rules determine how the model operates. It may not be necessary to fill all the boxes in the FAM but it usually helps to do so, if only for sake of completeness. The rules are fuzzy because the input variables may each belong to more than one fuzzy membership set and have differing degrees of membership associated with them.

Clearly the performance of this model now depends on both the nature and shape of the membership functions and the content of the FAM. It also requires that the data are suitably scaled given the definition of the membership functions. One very appealing feature of this fuzzy spatial interaction model is that different theories of spatial interaction can be included by changing the FAM composition. Furthermore, hybrid models are specified by merely adding in multiple sets of rules representing

Table 18.2 A FAM for a Gravity Model

		Input fuzzy distance membership sets				
		zero	*short*	*average*	*big*	*long*
Input	*small*	some	some	none	none	none
Destination	*little*	big	some	some	none	none
Size	*big*	lots	big	some	none	none
Membership	*large*	massive	lots	big	some	some
Set	*huge*	massive	massive	big	big	some

the different concepts. Likewise, it is not difficult to develop FAMs based on more than two input variables, although manually populating them with rules can rapidly become very tedious. Nevertheless, it is quite clear that FAMs provide a new and potentially very useful now approach to model representation that can be based on soft knowledge rather than rely solely on mathematical theory and statistical technology. However, if you lack this knowledge then there are tools that can help develop it for you, see Step 9. The rules can also be deterministic; for example, rule 1 might state that few trips go long distances and other conditional rules can then subsequently qualify this rule. Boolean rules can be mixed with fuzzy ones although care needs to be taken to ensure that the Boolean rules do not preempt the fuzzy ones.

18.3.6 *Step 5:* Apply the Model

The fuzzy rules or FAM define how the input(s) are converted into fuzzy output set membership probabilities. This is achieved by applying fuzzy inference to the fuzzy rules.

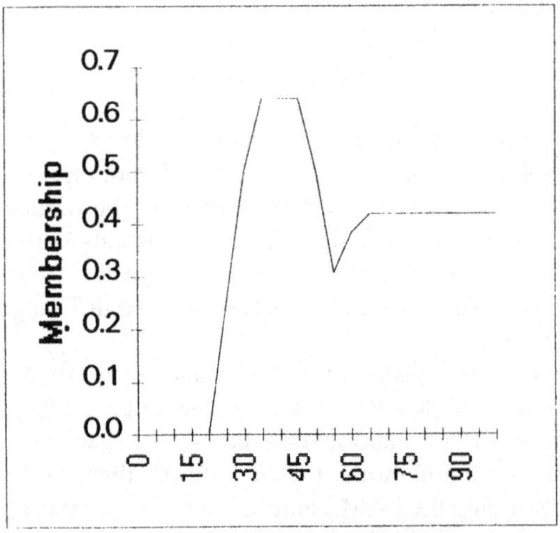

Fig. 18.5 Fuzzy Output for Distance Decay Model

In the case of the distance decay model it is very straightforward as a simple numerical example shows. For an input distance value of 8 this value lies in the input set *ZERO* distance set to the degree of 0.42 and also in the *SHORT* distance category to the degree of 0.64. So two rules in the FAM are applicable to it . This output partly belongs to the *MASSIVE* trip output set and partly to the *LOTS* trip set. The corresponding membership functions for these two output sets are truncated at values of 0.42 and 0.64 respectively; see Figure 18.5. This fuzzy output membership set would then be de-fuzzified in Step 6 to yield a fuzzy model crisp value of 62; or in other words for a distance of 8 units the model predicts 45 trips.

The gravity model is a little more complicated as there are two input variables and instead of five rules there are now twenty five rules to evaluate in Table 18.2. Additionally, the fuzzy logic reasoning process needs to take into account the AND part of the rules and infer the output contributed by each rule. The AND part is typically handled by a conjunction operator which in practice is a MIN-MAX method of fuzzy inference. If a rule fires then the output membership grade is the minimum of the two input variable's membership values. For example, consider the data for the values of distance of 8 destination size of 70. According to the FAM in Table 18.2 the following rules apply:

rule 6: if D_j is *LARGE* and C_{ij} is *ZERO* then T_{ij} are *MASSIVE* with membership grade of 0.42

rule 7:if D_j is *LARGE* and C_{ij} is *SHORT* then T_{ij} are *LOTS* with membership grade of 0.45

rule 11:if D_j is *HUGE* and C_{ij} is *ZERO* then T_{ij} are *MASSIVE* with membership grade of 0.25

rule 12:if D_j size is *HUGE* and C_{ij} is *SHORT* then T_{ij} trips are *MASSIVE* with membership grade of 0.25

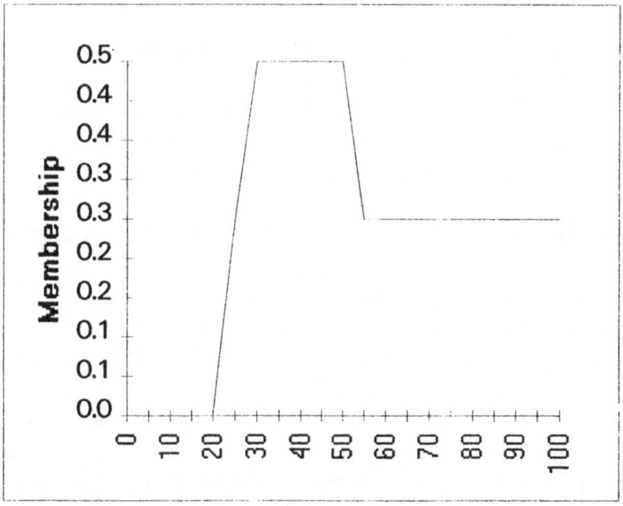

Fig. 18.6 Fuzzy Output for Gravity Model

The resulting fuzzy set membership output function is shown in Figure 18.6. This would then be de-fuzzed to yield a single crisp model prediction of 50. (viz. the Centre of gravity of the fuzzy output set).

18.3.7 *Step 6:* Defuzzify the Outputs to Obtain a Crisp Number

The result of applying fuzzy inference is an output fuzzy set. This output fuzzy set is an aggregation of one or more output membership sets that were selected by the FAM rules but then truncated at values determined by the fuzzy inferencing procedure that was used. To be useful this fuzzy output set has to be converted into a crisp number via a defuzzification process so that the model output is a single numeric value that best represents the information contained in the output fuzzy set.

There are various methods of defuzzification. The method used here is the centroid or centre of gravity technique. This is widely used and behaves in a manner akin to Bayesian estimates in that it defines a value that is supported by evidence that has been accumulated from over all the rules. In Figures 18.6 and 18.7 the crisp centroid estimates are respectively 45 and 50.

18.3.8 *Step 7:* Impose Accounting Constraints on the Model's Predictions

If the model is to be constrained then the appropriate balancing factors need to be applied to the defuzzified trips. How this is done is discussed in more detail later.

18.3.9 *Step 8:* Performance Evaluation

Fuzzy models like more conventional models need to be evaluated and their behaviour tested against validation data sets. It is a good idea to copy the neural network practice of keeping some data purely for validation purposes. Moreover, if required, a compute intensive statistical procedure could also be used to estimate model confidence intervals; for example, via a bootstrapping process.

Note though that there is no lengthy or complex model parameter calibration process, whether the model works well or poorly is mainly a function of the fuzzy set definitions and the membership functions that are used and the composition of the FAM. Whether a fuzzy model works well can be determined by comparing its performance with conventional alternatives. However, it may be worth accepting some loss of performance for a simpler non-mathematical interpretation. Also fuzzy systems modelling is still in its infancy, it may be too much to expect that the very first model you build should immediately outperform much more mature models on which many thousands of years of person effort and much research money have been expended. However, despite this caveat, maybe you will be pleasantly surprised at the levels of performance that can actually be achieved with a little effort.

18.3.10 *Step 9:* Manually Tuning the Model's Membership Functions and FAM

An extensive programme of manually fine tuning of either, or both the membership functions and FAM may be needed to obtain a good level of performance. The number and the definitions of the membership functions are more critical than their shape. This is inevitably a matter of informed guesswork or else it involves an extensive process of trial and error. The input data's fuzzy set and membership functions critically affect how the data are represented to the model whilst the nature of the output membership sets determine the numerical values generated by the model. Both tend to be application specific.

18.3.11 *Step 10:* Optimising Model Performance: Adaptive Fuzzy Systems Modelling

The simplest adaptive fuzzy modeller is to associate weights with each entry in the FAM. The weights determine how much each rule affects the final outcome of the model (von Altrock, 1995). These weights could be estimated using a neural network learning method to create a neurofuzzy systems modeller or else via a conventional parameter estimation method. Another approach to a fuzzy adaptive system involves changing the membership functions as a means of either improving performance or responding to changes in the data. The simplest method is to broaden or narrow the fuzzy sets according to feedback on model prediction error. For example, if the model overpredicts then for each fuzzy output set that was accessed narrow them slightly, that is move the left edge of the domain to the right and the right edge to the left; else broaden them. The problem now is that of overlap. The truth membership of a vertical line constructed through a region of overlap should not exceed 1.0 although some control systems violate this rule. It is not difficult to devise heuristic's to do this fuzzy set adjustment by an automated trial and error process.

The ultimate fuzzy modelling system is one that attempts to learn or discover the membership functions and the rules so as to optimise its predictive performance by itself. The integration of fuzzy logic with neural networks and genetic algorithms provides the basis for this type of extremely powerful approach to what is termed the model-free modelling of complex data. Ross (1995) argues that "The integration of fuzzy logic with neural networks and genetic algorithms is now making automated cognitive systems a reality in many disciplines. In fact, the reasoning power of fuzzy systems, when integrated with the learning capabilities of artificial neural networks and genetic algorithms, is responsible for new commercial products and processes that are reasonably effective cognitive systems (i.e. systems that can learn and reason)" page xv. In essence it is possible to build fuzzy logic equivalents to artificial neural networks in terms of their representational abilities but without many of the deficiencies.

A genetic algorithm (GA) provides an obvious way of estimating optimal FAMs and, or, membership functions to yield best performing models; see Karr (1991,1994). Quite simply the FAMs and the membership functions can be coded as bit strings and measures of fitness devised based on how well the resulting fuzzy model fits the data. Consider first a FAM only optimisation. The distance decay FAM in Table 18.1 can be optimised using a GA. The FAM entries correspond to integers in the range 1 to 5. The GA uses a binary representation so 4 bits would allow numbers in the range 1 to 15 to be used. This can easily be converted into integers between 1 and 5. Another possibility is to optimise simultaneously both the FAM and the input-output membership functions. This requires that the membership function can be parameterised and there are several ways of doing this that allows varying levels of flexibility for the GA to optimise. The main problem with hybrid GA fuzzy models is their need for extended computer run times so that optimal settings can be obtained. On the other hand, the results often have an intuitively obvious interpretation in both the shape of the membership functions, the nature of the FAM, and possibly also the strength associated with the various rules. Clearly this is an extremely powerful modelling technology that has much to commend it in appropriate circumstances. There is also considerable scope for further development.

18.4 A Family of Fuzzy Spatial Interaction Models

18.4.1 Model Description

A family of fuzzy spatial interaction models can now be formulated. The unconstrained model is straightforward

$$T_{ij} = K\,Fij \quad i,j \in n \tag{18.4}$$

where K is a parameter to ensure total trips equal a fixed value, and F_{ij} is the defuzzied intensity of interaction from origin i to destination j.

An origin constrained version is readily formulated

$$T_{ij} = Fij\, O_i / \Sigma_j F_{ij} \quad i,j \in n \tag{18.5}$$

Note that O_i variable is only used as part of the accounting constraint. The conventional model can also be reformulated in a similar manner; for example,

$$T_{ij} = O_i\, D_j\, A_i \exp(-bC_{ij}) \tag{18.6}$$

can be split into two parts; a trip prediction equation (T_{ij}*)

$$T_{ij}^* = D_j \exp(-bC_{ij}) \tag{18.7}$$

and a constraints mechanism which scales the T_{ij}* values to match the origin end totals

$$T_{ij} = T_{ij}^*\, O_i / \Sigma_j T_{ij}^* \tag{18.8}$$

The fuzzy component only replaces the trip prediction equation. The conventional entropy model conflated these two components. It is actually quite useful to view them as separate tasks so that the trip prediction function is independent of the constraints mechanism. It also has the virtue also of scaling the fuzzy model T_{ij} predictions in a realistic way.

There is a fuzzy spatial interaction model equivalent to the existing family of spatial interaction model, if the appropriate membership functions and rules are specified. However, it can also do much more because of its ability to offer extremely flexible and broad ranging diversity of representations. Competing destination and other hierarchical effects can be readily included. Various fuzzy and non-fuzzy hybrids can also be formulated. The limitations are those of the human modeller's ability to either specify meaningful rules of thumb or to devise methods so that the most applicable rules can be suggested by the data.

18.4.2 Empirical Results

A 73 by 73 journey to work data set is used to provide a preliminary analysis of the performance of a selection of fuzzy models. These data have been frequently used to benchmark different spatial interaction models, see Openshaw (1976). Attention is restricted to an origin constrained model. Table 18.3 offers the results for several conventional models with different deterrence functions. Performance seems to reach a limit when the residual standard deviation reaches about 15.

Table 18.3 Benchmark Conventional Singly Constrained Model Results

Type of deterrence function	Residual standard deviation
negative power	20.7
negative exponential	16.3
Tanner's	15.3
March function	15.2
Weibull function	15.2

The fuzzy model rules described in Table 18.2 has a residual standard deviation of 25 which is surprisingly good given the lack of thought applied to the construction of the rules. This task is made difficult by the need to carefully trade off input fuzzy set sizes (they interact) for D_j and C_{ij} with the output fuzzy sets for T_{ij}. Research is needed to provide useful design tools to assist this process. It could be improved by trial and error modification but here this was left to the GA. The results for a range of two and three variable input models is given in Table 18.4. Basically as the number of rules increases so does model performance but interpretability declines. Nevertheless, it would appear that in the limit levels of performance equivalent to that produced by a neural net might be achieved. There is considerable scope here for further experimentation particularly with the design of qualitative fuzzy models with the genetic optimisation being used to fine tune the model rules.

Table 18.4 Adaptive Fuzzy Model Results

Nature of model	Nature of optimisation	Total fuzzy rules	Residual standard deviation
$T_{ij}(4)\ D_j(4)\ C_{ij}(4)$	Rules only	16	42.2
$T_{ij}(8)\ D_j(4)\ C_{ij}(4)$	Rules only	16	37.2
$T_{ij}(8)\ D_j(8)\ C_{ij}(8)$	Rules only	64	25.5
$T_{ij}(4)\ D_j(4)\ C_{ij}(4)$	everything	16	17.2
$T_{ij}(5)\ O_i(5)\ D_j(5)\ C_{ij}(5)$	everything	5	23.1
$T_{ij}(5)\ O_i(5)\ D_j(5)\ C_{ij}(5)$	everything	7	22.3
$T_{ij}(5)\ O_i(5)\ D_j(5)\ C_{ij}(5)$	everything	16	20.4
$T_{ij}(8)\ D_j(4)\ C_{ij}(4)$	everything	16	19.4
$T_{ij}(8)\ D_j(8)\ C_{ij}(8)$	everything	64	16.8
$T_{ij}(16)\ D_j(4)\ C_{ij}(4)$	everything	16	17.2
$T_{ij}(8)\ O_i(3)\ D_j(3)\ C_{ij}(3)$	everything	27	19.8
$T_{ij}(8)\ O_i(4)\ D_j(4)\ C_{ij}(4)$	everything	64	13.1
$T_{ij}(8)\ O_i(4)\ D_j(4)\ C_{ij}(4)$	everything	14	21.9

Note: Figures in parenthesis are the number of fuzzy sets used, for example D_j (4) is the use of four fuzzy sets to represent D_j.

18.5 Some Other Types of Fuzzy Spatial Interaction Models

18.5.1 Different Types of Fuzzy Systems

Wang (1994) identifies three types of fuzzy logic systems: a pure fuzzy logic, a partly fuzzy system based on Takagi and Sugeno (1985), and a fuzzy logic system with fuzzifier and defuzzifier. Wang's pure fuzzy logic system produces fuzzy outputs which is not of much practical value. The third category of fuzzy system uses a fuzzifier and defuzzifier as interfaces to the crisp world and this is essentially the model previously described. However, the Takagi and Sugeno (1985) partly fuzzy model is also of potential interest in a spatial interaction modelling context. In this model only the IF part of the model rules is fuzzy, the THEN part is crisp. The advantages of this Takagi and Sugeno fuzzy modeller are that: (1) it has been successfully applied to a wide range of practical problems; (2) it incorporates conventional parameter estimation, (3) it readily permits hybridisation of spatial interaction modelling, (4) it produces a very compact model, (5) there is no need to de-fuzzify the outputs, and (6) it is a halfway house between a pure fuzzy approach and a crisp one. The principal disadvantage is that the THEN part is not fuzzy and this limits both the use of different fuzzy logic principles and the incorporation of fuzzy rules representing knowledge.

18.5.2 A Takagi - Sugeno (T-S) Type of Fuzzy Spatial Interaction Model

This approach can be used to build a hybrid spatial interaction model that combines knowledge in the form of fuzzy rules with a conventional set of one or more spatial interaction models. Consider the simplest case of an origin constrained model. It can be useful to have a different model for short distance trips than is used for long distance trips.

An obvious development is to use the fuzzy IF rules to suggest other types of model structure, not merely different model parameter values. This provides a means of building a hybrid fuzzy model based on a mix of different types of spatial interaction models. This can be built by using fuzzy D_j and C_{ij} sets to determine whether the model would be best predicted as:

1) a distance decay model $T_{ij}^* = \exp(-b_1 C_{ij})$
2) a gravity model $T_{ij}^* = O_i D_j \exp(-b_2 C_{ij})$
3) a competing destination gravity model $T_{ij}^* = O_i D_j Q_j^{b3} \exp(-b_4 C_{ij})$, and
4) an intervening opportunity model $T_{ij}^* = O_i (\exp(-b_5 X_{ij}) - \exp(-b_5 Z_{ij}))$

Note that the parameters b_1, .. b_5 are estimated using a non-linear least squares procedure, Q_j^{b3} is a competing destination terms, whilst the X_{ij} and Z_{ij} are intervening opportunity terms.

For example, if C_{ij} is small and D_j is big then use model 1. The rules can be constructed which specify the circumstances under which specific models might be expected to be most appropriate. However, it is important to remember that the rules once again overlap so that the predictions would be based on the weighted contributions from two (or more) different models. This is a very useful way of building a hybrid model. It also makes the spatial interaction model more sensitive (and thus maybe more intelligent) to the context that relates to particular trip pairs. The rules can also be optimised using a GA and this raises the utterly fascinating prospect that, if all four models are initially given identical IF rules and hence weightings for determining which conditions are best for each model, there is the possibility of one or more models either being deleted or included more than once.

18.5.3 Empirical Results

Table 18.5 shows the performance of the T-S type of partly fuzzy model. The results are quite remarkable as it is possible to obtain good levels of performance fairly readily. This would suggest that this type of fuzzy model may well offer a potentially very useful approach and yet be simple enough to run on a PC.

Table 18.5 Partly Fuzzy Model Results

Nature of model	Nature of optimisation	Total fuzzy rules	Residual standard deviation
$C_{ij}(8)\ D_j(8)$	everything	64	13.24
$C_{ij}(7)\ D_j(7)$	everything	49	12.61
$C_{ij}(4)\ D_j(4)$	everything	16	11.64
$C_{ij}(2)\ D_j(2)$	everything	4	12.65
$C_{ij}(8)\ D_j(8)$	everything	64	11.68
$C_{ij}(4)\ D_j(4)$	rules only	16	13.15
$C_{ij}(8)\ D_j(8)$	rules only	64	13.20

18.6 Conclusions

18.6.1 Fuzzy Modelling is in principle simple

Fuzzy systems modelling is not particularly difficult. The nicest aspect is that the technology is widely applicable to may different modelling situations, both numeric and non-numeric and that the models can be given an easy to understand interpretation because of the linguistic labels attached to the fuzzy sets.

18.6.2 Some Advantages

Fuzzy logic based modelling offers a number of potential benefits. They can be listed as follows:

a. provides a linguistic non-numerical, non-mathematics, and non-statistical based approach to modelling complex non-linear systems that are difficult or maybe even impossible to model by any other means,
b. a fairly simple approach with only a few variables or rules often being sufficient to handle systems modelling of considerable complexity,
c. inherently a non-linear modelling tool,
d. relates inputs to outputs without having to understand or have a complete and well specified model of the system with the prospect of improved levels of performance over more conventional approaches,
e. permits rapid prototyping, since a researcher does not have to know everything about the system before starting work nor is there any need to specify precise mathematical or statistical relationships,
f. offers a transparent box approach to modelling that is inherently more understandable than an artificial neural network (because you can readily look inside the box to find out what is going on),
g. ease of explanation to others since the model can be expressed in words,
h. robustness because of its ability to handle imprecision,
i. a means of incorporating existing intuition and knowledge, both soft and hard, in a direct manner in the fuzzy rules so that there is no need to pretend that we know

nothing about the system of interest even when we do,
j. a high degree of inherent parallelism,
k. can be used to create various hybrid modelling systems; and
l. possible to build models of systems for which there is little or no data.

These are, however, largely theoretical advantages and need to be proven in geographical applications.

18.6.3 Some Disadvantages

Fuzzy Logic is no magic panacea, it will not always work or work well. Simple linear systems are not much improved by adding fuzzy logic to them. While it is much less than a black box approach than neural networks, this may still cause problems; for example, if it is difficult to comprehend the fuzzy rules that produce good results. Other possible problems include:

a. the psychological and cultural prejudice that many people hold in favour of crisp systems or mathematically precise models may precondition the user to expect failure or to look for insurmountable problems even if there are none,
b. rule-based modelling is essentially simple and whilst useful is limited to those situations where this approach is possible and sensible,
c. there is an underlying curse of dimensionality whereby the number of rules is a power function of the number of input variables, which inevitably restricts fuzzy models to systems that can be characterised by few variables (although there are ways of addressing this problem),
d. fine tuning or optimisation of performance is often necessary (because of inadequacies in the knowledge that specified the system) but this is not always easy, although a genetic algorithmic approach is a useful aid,
e. models can be built of systems for which there is no good understanding of the processes, and
f. optimisation using GA may well require high performance computing or highly parallel hardware if there is a large number of input variables or a large number of cases in the training data set.

Perhaps the greatest risk with fuzzy logic modelling is that whilst this technology provides a powerful modelling tool able to represent the behaviour of complex systems it does not require any knowledge about why they operate in the way they do. Fuzzy rules of thumb based on generalising many years worth of experience of how a system works may be quite adequate for many purposes but it may also provide little or no real understanding of the mechanisms driving the systems of interest. However this criticism need not be as fundamental as it appears because modelling may well lead to enhanced understanding and explanation, particularly if fuzzy logic modelling is viewed as a means of hypothesis testing.

18.6.4 Fuzzy Modelling in Geography

Fuzzy models have a number of advantages over most other modelling tools. They are easy to set-up, highly flexible, computationally efficient once established, and robust. Fuzzy logic modelling has had a long gestation period and maybe its time has come. There is a great potential here but equally there are many avenues that still need to be explored, researched, and solutions developed. Fuzzy technology is proclaimed in Japan as the "keyword" for the 1990s. Although the technology involved is relatively new, it is of such a fundamental nature that it is highly unlikely that it will not be a widely used core research and modelling tool by the turn of the century.

References

Black, M., 1937, Vagueness: an exercise in logical analysis, *Philosophy of Science* **4**:427-455

Carey, H.C., 1858, *Principles of Social Science,* Lippincott, Philadelphia

Cox, E., 1994, *The fuzzy systems handbook.* A P Professional, Boston

Driankov, D., Hellendoorn, H., Reinfrank, M., 1996, *An introduction to Fuzzy Control,* Springer Verlag, Berlin

Gupta, M M., Sanchez, E (eds.), 1982 *Fuzzy Information and Decision Processes,* North-Holland, Amsterdam

Fischer, MM., Gopal, S., 1994, Artificial neural networks: a new approach to modelling interregional telecommunication flows, *Journal of Regional Science* **34**:503-527

Fotheringham, A.S., 1983, A new set of spatial interaction models: the theory of competing destinations, *Environment and Planning A,* **15**:15-36

Holmblad, L P., Ostergaard, J J, 1982, Control of a cement kiln by fuzzy logic, in M M Gupta and E Sanchez (eds.) *Fuzzy Information and Decision Processes.* North-Holland, Amsterdam 398-399

Holmblad, L P., Ostergaard, J J., 1981, Fuzzy logic control: operator experience applied in automatic process control, *ZKG International* **34**:127-133

Karr, C.L., 1991, Genetic algorithms for fuzzy logic controllers, *AI Expert* **6**:26-33

Karr, C. L., 1994, 'Adaptive control with fuzzy logic and genetic algorithms', in R.R. Yager, L.A. Zadeh *Fuzzy Sets, Neural Networks, and soft computing,* van Nostrand Reinhold, New York p.345-367

Kosko, B., 1990, Fuzziness vs. probability, *International Journal of General Systems* **17**:211-240.

Kosko, B., 1992, *Neural networks and fuzzy systems.* Prentice-Hall International, Englewood Cliffs

Kosko, B., 1993, Fuzzy systems as universal approximators, *Proceedings of First IEEE International Conference on Fuzzy Systems,* San Diego, 1153-1162

Kosko, B., 1994, *Fuzzy thinking.* Harper-Collins, London

Lukasiewicz, J., 1970, In defence of logic, Works in L Borkowski (ed) *Selected Works,* North-Holland, London

Mamdani, E., Assilian, S., 1975, An experiment in linguistic synthesis with a fuzzy logic controller, *International Journal of Man-Machine Studies* **7**:1-13

McNeill, D., Freiberger, P., 1993, *Fuzzy Logic.* Simon and Schuster, New York

McNeill, F.M., Thro, E., 1994 *Fuzzy Logic: a practical approach.* A P Professional, Boston

Openshaw, S., 1976, An empirical study of some spatial interaction models, *Environment and Planning A* **8**:23-41

Openshaw, S., 1988, Building an automated modelling system to explore a universe of spatial interaction models. *Geographical Analysis* **20**:31-46

Openshaw, S., 1993, Modelling spatial interaction using a neural net, in M M Fischer and P Nijkamp (eds) *GIS, Spatial Modelling and Policy.* Springer Berlin, 147-164

Openshaw, S., 1996. Fuzzy Logic as a new scientific paradigm for doing geography, *Environment and Planning A* **28**:761-768

Ross, T.J., 1995, *Fuzzy logic with engineering applications.* McGraw Hill, New York

Sugeno, M., (ed) 1985. *Industrial applications of fuzzy control.* North-Holland, Amsterdam

Takagi, T., Sugeno, M., 1985, Fuzzy identification of systems and its application to modelling and control; *IEEE Transactions on Systems, Man, and Cybernetics* SMC-**15**:116-132

Terano, T., Asai, K., Sugeno, M., 1989, *Applied Fuzzy Systems.* A P Professional, New York (English translation in 1994)

Turton, I., Openshaw, S., Diplock, GJ., 1997, A genetic programming approach to building new spatial models relevant to GIS, in Z Kemp (ed) *Innovations in GIS 4,* Taylor and Francis, London (forthcoming)

Von Altrock, C., 1995, *Fuzzy Logic and neurofuzzy applications explained,* Prentice Hall, Englewood Cliffs

Wang, L.X., 1994 *Adaptive fuzzy systems and control.* Prentice Hall, Englewood Cliffs, New Jersey

Webstead, S.T., *Neural network and fuzzy logic applications in C/C++,* Wiley, Chichester

Wilson, A.G., 1970, *Entropy in Urban and Regional Modelling,* Pion, London.

Yasunobu, S., Miyamoto, S., 1985, Automatic train operation system by predictive fuzzy control, in M Sugeno (ed.) *Industrial applications of fuzzy control,* North-Holland, Amsterdam 1-18

Zadeh, L.A., 1962, From circuit theory to system theory, *IRE Proceedings* **50**:856-865

Zadeh, L.A., 1965, Fuzzy sets, *Information and Control* **8**:338-353

Zadeh, L A., 1971, Towards a theory of fuzzy systems, In R E Kalman and N DeClaris (eds.) *Aspects of Network and Systems Theory.* Holt Rinehart, Winston

Zadeh, L A., 1974, A rationale for fuzzy control, *Journal of Dynamic Systems Measurement and Control* **3-4**

Zadeh, L.A., 1992, Foreword in *Neural networks and fuzzy systems,* Prentice-Hall International Englewood Cliffs, New Jersey, p xvii-xviii

Zadeh, L.A., 1995 Foreword in C von Altrock *Fuzzy Logic and Neuro fuzzy applications explained,* Prentice-Hall, Englewood Cliffs, New Jersey pxi-xii

19 Epilogue

Manfred M. Fischer[1] and Arthur Getis[2]

[1]Department of Economic & Social Geography, Vienna University of Economics and Business Administration, A-1090 Vienna, Augasse 2-6, Austria
[2]Department of Geography, San Diego State University, San Diego, CA 92182-4493

Having read 18 chapters that represent the frontier in spatial analysis, we ask ourselves the following questions: What can be said about the state of knowledge in this area? What directions appear to be the most promising? In this conclusion, the editors attempt to answer these questions.

As evidenced in this volume, a number of research trends strike us as representative of new interests among spatial scientists. The first trend to capture our attention is the conscious effort by experienced researchers to adjust their thinking to take advantage of rapidly advancing technology. As one might expect, this trend is more in evidence in Part A, on spatial data analysis, and Part C, on computation intelligence, than in Part B, on behavioural modelling. Often, spatial data are provided in large data sets where each datum represents a small portion of the study area. There has been a strong response to the need to come to grips with the problems inherent in large data sets, such as heterogeneity and spatial dependence. By devising statistics that are particularly well suited to fine scale data analysis, but require the storage and manipulative attributes of a geographic information system, researchers are attacking the problems in a number of different types of data environments. Indeed, many of the models being developed and reported on in Part A lend themselves to computationally intensive estimation. Part C makes it clear that neural networks owe their success to the use of high speed, high capacity computer routines. To project that this trend will continue with vigor into the next decade is only to say that we can expect even more value to be extracted from the more flexible, high-capacity, computer technology currently under development.

A second trend is the juxtaposition of spatial modelling and exploratory analysis. At one time, spatial modelling was divorced from exploration in the sense that the model represented a deterministic, monolithic representation of spatial phenomena *à la* Christaller and Lösch. Exploration was frowned upon as something not done in a public forum, although it could be relegated to the privacy of one's office. Thirty years ago, exploration of data was considered the pastime of rudderless empiricists who had little or no appreciation of theory. Now, the trend is clear. Computer technology makes it possible to consider data from many different representational and visual points of view. The research reported on in this volume takes advantage of this technology to develop models that fit the data rather than the other way around (data-driven as opposed to model-driven). No wonder Bayesian and maximum likelihood approaches are becoming increasingly popular.

A number of analysts have found that combining time and space series in the same model adds insight and predictive capability. This third trend, time-space modelling, is a good example of several points of view being brought to bear on problems that previously were studied actively in either a time or space milieu. In many respects, time-space analysis, although touted as the wave of the future for a half century, is today really an infant industry. The models, valuable as they appear to be, are still in the formative stages of development. From the perspective of the spatial scientist, good time-space theory, including spatial dynamics, has yet to be developed. This shortcoming is another reason why exploratory analysis of patterns and trends is so important to the model-building enterprise. Note that this emphasis is represented in all three substantive sections of the volume. Whether it be time-space models of the incidence of diseases, consumer behaviour, or spatial interaction, the dynamic element is becoming much more a part of the spatial analyst's thinking.

The fourth trend, most noticeable in Part A, is the interest in health related examples of spatial modelling. Perhaps spatial scientists are moved by the opportunity to make humanitarian contributions by developing and applying models of the spatial form, structure, and pattern of serious diseases. Earlier work in the patterns of disease occurrence have been either ecological in theoretical structure or time series in methodology. Time-space modelling lends itself both to traditional interests, such as the prediction of the expected size and the time of an outbreak, and to newer interests in the spatial pattern of disease transmission and the location of the infected individual relative to other variables. In this age of rapid transportation and hence rapid disease transmission to all parts of the world, the fear of devastating diseases has intensified. Knowledge of the spatial transmission of disease is a high societal priority.

The fifth trend, evident in the behavioural modelling field, is the development of innovative extensions to traditional models of human movement and decision making. New approaches to questions of continuing interest appear uppermost in importance. The issues are mainly the same: migration, travel and destination choice, and retail location strategies, but the techniques of analysis now include computational process models, travel-time constraint models, and personal history evaluation. There is now more concern with public policy (store shopping hours), agricultural decision-making, and strategic planning. Perhaps of greatest interest is the professed need to study individual, disaggregate behaviour. Data sets in this area are difficult to obtain, and in this respect, there appears to be a limit on the progress researchers can make in testing theory. Much data is proprietary, in the hands of marketing research firms who have only a passing interest in theory development. There is no shortage, however, of the development of new ideas and high quality research, much of which provides deeper insights into individual marketplace behaviour. In the chapters presented in this volume, the relatively few, good data sets are exploited for this purpose.

Finally, the sixth trend is specifically associated with the movement toward the entire field of computational intelligence. Such a large portion of a volume on new advances in spatial modelling would not have been possible ten years ago. Included in this volume are among the first substantive articles in the spatial field. Neural modelling is gaining acceptance as its usefulness becomes apparent. Two roles for neural modelling are: as a classification device to identify the underlying structural elements in large and complex spatial data sets, and as a way to come to grips with unclear or fuzzy data. In the first instance, neural methods are gaining acceptance as classifiers of remotely sensed pixel data; and in the second instance, data that in past years may have been disregarded because of their inconclusive nature are being evaluated using neural modelling approaches. This trend is likely to continue not only because of the need to make sense of difficult data sets but also because the methods are tied directly to the new technology that allows for computationally intensive analysis.

In summary, the previous 18 chapters in this volume represent the leading edge of new ideas, new computational routines, and new approaches that will eventually become known as the new spatial science. Not only are new journals springing up to represent these fields of research, but older journals are changing their emphases as the combination of technology, large data sets, and innovative models provides a powerful basis for meritorious contributions by spatial analysts into the 21st century.

Bibliography

Abrahamson E. 1991. Managerial fads and fashions: the diffusion and rejection of innovations. *Academy of Management Review*, **16**:586-612.

Abuelgasim A. and Gopal S. 1994. Classification of multiangle and multispectral ASAS data using a hybrid neural network model, *Proceedings International Geoscience and Remote Sensing Symposium (IGARSS 94)*, held at Pasadena, 8-12 August, IEEE Press, Piscataway, NJ, Vol. 3, 1670-1672.

Aitken M. 1996. A General Maximum Likelihood Analysis of Overdispersion in Generalized Linear Models, *Statistics and Computing* (forthcoming).

Altman D. 1988. Legitimation through disaster: AIDS and the gay movement in *AIDS: the burden of history*. Eds E Fee & D M Fox (Berkeley: University of California Press) pp 301-15.

Ambergey T.L., Kelly D., Barnett W.P. 1993. Resetting the clock: the dynamics of organizational change and failure. *Administrative Science Quarterly*, **38**:51-73.

Ambergey T.L., and Miner A.S. 1992. Strategic momentum: the effects of repetitive, positional, and contextual momentum on merger activity. *Strategic Management Journal*, **13**:335-48.

Anas A. 1990. General Economic Principles for Building Comprehensive Urban Models, in C.S. Bertuglia, G. Leonardi and A.G. Wilson eds. *Urban Dynamics: Designing an Integrated Model*, Routledge, London, 7-44.

Anderberg M.R. 1973. *Cluster analysis for applications*. Academic Press, New York.

Anderson R.M. 1988. The epidemiology of HIV infection: variable incubation plus infectious periods and heterogeneity in sexual activity. *Journal of the Royal Statistical Society A* **151**: 66-93.

Anderson R.M. and May R.M. 1988. The epidemiological parameters of HIV transmission. *Nature* **333**:314-9.

Anderson R.M. and May R.M. 1991. *Infectious Diseases of Humans: Dynamics and Control*, Oxford University Press.

Anderson W. P. and Papageorgiou Y.Y. 1994. An analysis of migration streams for the Canadian regional system 1952-1983 1. Migration Probabilities. *Geographical Analysis* **26**:15-36.

Anderson J. 1968. A Memory Storage Model Utilizing Spatial Correlation Function, *Kybernetic*, **5**:113-119.

Anderson J. 1972. Two Models for Memory Organization Using Interacting Traces, *Mathematical Biosciences*, **80**:137-160.

Anderson J.R. 1990. *The adaptive character of thought*. Hillsdale, NJ: Erlbaum.

Andrews K. R. 1971. *The concept of corporate strategy*. Homewood, IL: Dow-Jones Irwin.

Anselin L. 1988. *Spatial Econometrics: Methods and Models*, Dordrecht: Kluwer Academic.

Anselin L. 1990. What is Special About Spatial Data? Alternative Perspectives on Spatial Data Analysis in: D.A. Griffith, ed., *Spatial Statistics, Past, Present and Future*, Ann Arbor, MI:Institute of Mathematical Geography, pp. 63-77

Anselin L.1992 Spatial Dependence and Spatial Heterogeneity: Model Specification Issues in the Spatial Expansion Paradigm, pp.334- 354 in *Applications of the Expansion Method*, J.P. Jones III and E. Casetti eds., Routledge, London and New York.

Anselin L. 1992. *SpaceStat: A Program for the Analysis of Spatial Data*, National Center for Geographic Information and Analysis, University of California, Santa Barbara, CA.

Anselin L. 1994. Exploratory Spatial Data Analysis and Geographic Information Systems, in: M. Painho ed., *New Tools for Spatial Analysis*, Luxembourg: Eurostat, pp. 45-54.

Anselin L. 1995. Local Indicators of Spatial Association: LISA, *Geographical Analysis* **27**:93-115.

Anselin L. 1995. *SpaceStat Version 1.80 User's Guide*, Regional Research Institute, West Virginia University, Morgantown, WV.

Anselin L. 1996. The Moran Scatterplot as an ESDA Tool To Assess Local Instability in

Spatial Association, in: M. Fischer, H. Scholten and D. Unwin, eds., *Spatial Analytical Perspectives on GIS in Environmental and Socio-Economic Sciences*, London: Taylor & Francis.

Anselin L. 1997. GIS Research Infrastructure for Spatial Analysis of Real Estate Markets, *Journal of Housing Research* **8**.

Anselin L. 1997. Interactive Techniques and Exploratory Spatial Data Analysis, in: P. Longley, M. Goodchild, D. Maguire and D. Rhind eds., *Geographical Information Systems: Principles, Techniques, Management and Applications*, Cambridge: Geoinformation International.

Anselin L. and Bao S. 1996. *SpaceStat User's Guide*, Research Paper 9628, Regional Research Institute, West Virginia University, Morgantown, WV.

Anselin L., Dodson R. and Hudak S. 1993. Linking GIS and Spatial Data Analysis in Practice, *Geographical Systems* **1**:3-23.

Anselin L. and Getis A. 1992. Spatial Statistical Analysis and Geographic Information Systems, *The Annals of Regional Science* **26**:19-33.

Arentze T.A., Borgers A.W.J. , and Timmermans H.J.P. 1996. A Generic Spatial Decision Support System for Planning Retail Facilities, in M. Craglia and H. Couclelis eds. *Geographic Information Research: Bridging the Atlantic*, Forthcoming.

Arentze T.A., Borgers A.W.J. , and Timmermans H.J.P. 1996. A Knowledge-Based Model for Developing Location Strategies in a DSS for Retail Planning, *Proceedings of the 3rd Design & Decision Support Systems in Architecture & Urban Planning Conference*, Spa, Belgium, 17-38.

Arentze T.A., Borgers A.W.J. , and Timmermans H.J.P. 1996. Design of a View-based DSS for Location Planning, *International Journal of Geographical Information Systems*, **10**:219-236.

Arentze T.A., Borgers A.W.J. , and Timmermans H.J.P. 1996d. The Integration of Expert Knowledge in Decision Support Systems for Facility Location Planning, *Computers, Environment and Urban Systems*, **19**:227-247.

Argialas D.P. and Harlow C.A. 1990. Computational Image Processing Models: an Overview and Perspective, *Photogrammetric Engineering and Remote Sensing*, **56**:871-886.

Argyris C. 1977. Double loop learning in organizations. *Harvard Business Review*, **55**:115-25.

Armstrong M.P. and Densham P.J. 1990. Database Organization Strategies for Spatial Decision Support Systems, *International Journal of Geographical Information Systems* **4**:3-20.

Armstrong M.P., Rushton G., Hoeney R., Dalziel B.T., Lolonis P., De S., Densham P.J. 1991. Decision Support for Regionalization: A Spatial Decision Support System for Regionalizing Service Delivery Systems, *Computers, Environment and Urban Systems, 15*:37-53.

Axhausen K.W., and Gärling T. 1992. Activity-based approaches to travel analysis: Conceptual frameworks, models, and research problems. *Transport Reviews*, **12**(4): 323-341.

Bacon R. 1971. An approach to the theory of consumer shopping behaviour, *Urban Studies*, **5**:55-64.

Bailey A.J. 1993. Migration history migration behavior, and electivity. *The Annals of Regional Science* **27**:315-326.

Bailey A.J. 1993. Migration and unemployment duration among young adults. *Papers in Regional Science* **73**:289-307.

Bailey N.T.J. 1957. *The Mathematical Theory of Epidemics*, Griffin: London.

Bailey N.T.J. 1975. *The Mathematical Theory of Infectious Diseases*, Griffin: London.

Bailey T.C. 1994. A Review of Statistical Spatial Analysis in Geographical Information

Systems, in: S. Fotheringham and P. Rogerson eds., *Spatial Analysis and GIS*, London: Taylor & Francis, pp. 13-44.

Bailey T. C. and Gatrell A.C. 1995. *Interactive Spatial Data Analysis*, Harlow: Longman Scientific and Technical.

Baker R.G.V. 1985. A dynamic model of spatial behaviour to a planned suburban shopping centre, *Geographical Analysis*, **17**:331-338.

Baker R.G.V. 1993. The regionalising of consumer behaviour: Sydney and beyond. Paper presented at the *Australian and New Zealand Regional Science Association Conference*, Armidale, December, 1993.

Baker R.G.V. 1994. An assessment of the space-time differential model for aggregate trip behaviour to planned suburban shopping centres, *Geographical Analysis*, **26**:341-362.

Baker R.G.V. 1994. The impact of trading hour deregulation on the retail sector and the Australian community, *Urban Policy and Research*, **12**:104-117.

Baker R.G.V. 1996. Multi-purpose shopping behaviour at planned suburban shopping centres: a space-time analysis, *Environment and Planning A*, **28**:611-630.

Baker R.G.V. 1996. *On the development of the SASTV model in estimating retail trading hours and its application to market area analysis*, Unpublished paper, available from the author.

Banerjee K. 1989. Rising prevalence of antibodies against human immunodeficiency virus (HIV-1) in Western Maharashtra, India. *Abstracts of the Fifth International Conference on AIDS*, Montreal, Abstract TG022.

Bao S. and Henry M. 1996. Heterogeneity Issues in Local Measurements of Spatial Association, *Geographical Systems* **3**:1-13.

Bao S., Henry M., Barkley D. and Brooks K. 1995. RAS: A Regional Analysis System Integrated with ARC/INFO, *Computers, Environment and Urban Systems* **18**: 37-56.

Barkley D., Henry M., Bao S. and Brooks K. 1995. How Functional are Economic Areas? Tests for Intra-Regional Spatial Association using Spatial Data Analysis, *Papers in Regional Science* **74**:297-316.

Barnett T. and Blaikie P. 1992. *AIDS in Africa: its present and future impact*. London: Belhaven Press.

Barnett W.P. and Carroll G.R. 1995. Modeling internal organizational change. *Annual Review of Sociology*, **21**:217-36.

Barnsley M. 1993. Monitoring urban areas in the EC using satellite remote sensing, *GIS Europe*, **2**(8):42-4.

Barr P. 1991. *Organization stress and mental models*. Unpublished doctoral dissertation, University of Illinois.

Barré-Sinoussi F., Chermann J.C., Rey F., Nugeyre M.T., Charnaret S., Gruest J., Danduet C., Axler-Blin C., Vézinet-Brun F., Rouzioux C., Rozenbaum W. and Montagnier L. 1983. Isolating of a T-lymphotropic retro-virus from a patient at risk from acquired immune deficiency syndrome. *Science* **220**:865-71.

Bartlett M.S. 1957. Measles periodicity and community size. *Journal of the Royal Statistical Society A*, **120**:48-70.

Bartlett M.S. 1960. *Stochastic population models in ecology and epidemiology*. London: Methuen.

Bartlett M.S. 1960. The critical community size for measles in the United States. *Journal of the Royal Statistical Society A*, **123**:37-44.

Bassett M.T. and Mhloyi M. 1991. Women and AIDS in Zimbabwe: the making of an epidemic. *International Journal of Health Services* **21**:143-56.

Batty M. and XieY. 1994. Modelling Inside GIS: Part I. Model Structures, Exploratory Spatial Data Analysis and Aggregation. *International Journal of Geographical Information Systems* **8**:291 307

Beaumont J.R. 1981. The Dynamics of Urban Retail Structure: Some Exploratory Results Using Difference Equations and Bifurcation Theory, *Environment and Planning A*, **13**:1473-1483.

Beaumont J.R. 1987. Location-Allocation Models and Central Place Theory, in Ghosh A., Rushton G. eds. *Spatial Analysis and Location-Allocation Models*, Von Nostrand Reinhold Company, New York, 21-53.

Beavon K.S.O. 1977. *Central Place Theory: A Reinterpretation.* Longman, London.

Becker R. and Cleveland W.S. 1987. Brushing scatterplots, *Technometrics* **29**:127-142.

Belsley D.A., Kuh E., and Welsch R.E. 1980. *Regression Diagnostics: Identifying Influential Data and Sources of Collinearity*, Wiley.

Ben-Akiva M. and Lerman S.R. 1978. Disaggregate Travel and Mobility-Choice Models and measures of Accessibility, in A. Hensher and P.R. Stopher eds. *Behavioural Travel Modelling*, Croom Helm, London, 654-679.

Ben-Akiva M. and Boccara. B. 1995. Discrete Choice Models with Latent Choice Sets. *International Journal of Research in Marketing*, **12**:9-24.

Ben-Akiva M. and Bowman J.L. 1995. Activity based disaggregate travel demand model system with daily activity schedules. Paper presented at the *International Conference on Activity based Approaches: Activity Scheduling and the Analysis of Activity Patterns*, May 25-28. Eindhoven University of Technology, The Netherlands.

Ben-Akiva M. and Lerman S. 1985. *Discrete choice analysis.* Cambridge, MA: MIT Press.

Benediktsson J.A., Swain P.H., Ersoy O.K. 1990. Neural network approaches versus statistical methods in classification of multisource remote sensing data, *IEEE Transactions on Geoscience and Remote Sensing*, **28**:540-52.

Benediktsson J. A., Swain P. H. and Ersoy O. K. 1993. Conjugate gradient neural networks in classification of multisource and very-high dimensional remote sensing data. *International Journal of Remote Sensing*, **14**(15):2883-2903.

Bernard A. C. and Wilkinson G. G. 1995. *Neural network classification of mixed pixels*, Proc. International Workshop on "Soft Computing in Remote Sensing Data Analysis", Milano, Italy, 4-5 December 1995, Organized by Consiglio Nazionale Delle Ricerche, Italy.

Bertuglia C.S. and Leonardi G. 1980. Heuristic Algorithms for the Normative Location of Retail Activities Systems, *Papers of the Regional Science Foundation*, **44**:149-159.

Bertuglia C.S., Clarke G.P. and Wilson A.G. 1994. Models and Performance Indicators in Urban Planning: The Changing Policy Context, in C.S. Bertuglia, G.P.Clarke and A.G. Wilson eds. *Modelling the City: Performance, Policy and Planning*, Routledge, London, 20-36.

Bezdak J.C.and Pal S.K. eds. 1992. *Fuzzy Models for Pattern Recognition*, IEEE, New York.

Birkin M., Clarke G., Clarke M. and Wilson A. 1994. Applications of Performance Indicators in Urban Modelling: Subsystems Framework, in C.S. Bertuglia, G.P. Clarke and A.G. Wilson eds.. *Modelling the City: Performance, Policy and Planning*, Routledge, London, 121-150.

Birkin M., Clarke G., Clarke M., Wilson A. 1996. *Intelligent GIS: Location Decisions and Strategic Planning*, Geoinformation International, Cambridge.

Bischof H., Schneider W., Pinz, A.J. 1992. Multispectral classification of Landsat-images using neural networks, *IEEE Transactions on Geoscience and Remote Sensing*, **30**(3): 482-90.

Bithell J.F. 1990. An application of density estimation to geographical epidemiology. *Statistics in Medicine* **9**:691-701.

Black F.L. 1966. Measles endemicity in insular populations; critical community size and its evolutionary implication. *Journal of Theoretical Biology*, **11**:207-11.

Black M. 1937. Vagueness: an exercise in logical analysis, Philosophy of Science **4**:427-455.

Black W.C. 1984. Choice-Set Definition in Patronage Modeling. *Journal of Retailing*. **60**:63-85.

Bogner W.C., Pandian J.R., Thomas H. 1994. Modeling Strategic Group Movements. In H. Daems and H. Thomas eds., *Strategic groups, strategic moves and performance*. Tarrytown, New York: Elsevier Science.

Borgers A.W.J., and Timmermans H.J.P . 1987. Choice Model Specification, Substitution and

Spatial Structure Effects: A Simulation Experiment. *Regional Science and Urban Economics,* **17**:29-47.

Borgers A.W.J. and Timmermans H.J.P. 1991. A Decision Support and Expert System for Retail Planning. *Computers, Environment and Urban Systems*, **15**:179-188.

Bowen J.L. 1994. *Investigating the Relationship Between Foreign Aid and Economic Growth in Recipient Countries*, Ph.D. Dissertation, UMI: Ann Arbor, Michigan

Bowen J.L. 1995. Foreign Aid and Economic Growth: An Empirical Analysis, *Geographical Analysis*, **27**:249-261.

Bowman A.W. 1984. An Alternative Method of Cross-Validation for the Smoothing of Density Estimates, *Biometrika* **71**: 353-360.

Box G.E.P. and Jenkins G.M. 1970. *Time series analysis, forecasting and control.* San Francisco: Holden-Day.

Box G.E.P. and Tiao G.C. 1975. Intervention analysis with applications to economic and environmental problems. *Journal of the American Statistical Assosciation*, **70**:70-79.

Bradley R. and Haslett J. 1992. High Interaction Diagnostics for Geostatistical Models of Spatially Referenced Data. *The Statistician* **41**:371-380.

Brookmeyer R. and Gail M.H. 1994. *AIDS Epidemiology: A Quantitative Approach*, Oxford University Press: New York Oxford.

Broomhead D.S. and Lowe D. 1988. Multivariable Functional Interpolation and Adaptive Networks, *Complex Systems*, **2**:321-355.

Brown D. and Rothery P.1993. *Models in Biology: Mathematics, Statistics and Computing*, Wiley.

Brown M 1995. Ironies of distance: an ongoing critique of the geographies of AIDS *Environment and Planning D* **13**:159-83.

Brunsdon C.F. 1991. Estimating Probability Surfaces in GIS: an Adaptive Technique, in *Proceedings of the Second European Conference on Geographical Information Systems*, eds Harts, J., Ottens H.F. and Scholten, H.J., 155-163, Utrecht : EGIS Foundation.

Brunsdon C.F. 1995. Estimating Probability Surfaces for Geographical Point Data: An Adaptive Kernel Algorithm, *Computers and Geosciences*, **21**: 877-894.

Buja A., Cook D. and Swayne D.F. 1996. Interactive High-Dimensional Data Visualization, *Journal of Computational and Graphical Statistics* **5**:78-99.

Buja A., McDonald J.A. , Michalak J. and Stuetzle W. 1991. Interactive Data Visualization using Focusing and Linking, in: G.M. Nielson and L. Rosenblum, eds., *Proceedings of Visualization '91*, Los Alamitos, CA: IEEE Computer Society Press, pp. 155-162.

Burnet M. and White D.O. 1972. *Natural History of Infectious Diseases (Fourth Edition)*. Cambridge: Cambridge University Press.

Burnett K.P. 1980. Spatial Constraints-Oriented Modelling as an Alternative Approach to Movement. Microeconomic Theory and Urban Policy. *Urban Geography*, **1**:151-66.

Burnett K.P. and Hanson S. 1979. Rationale for an Alternative Mathematical Approach to Movement as Complex Human Behavior. *Transportation Research Record*, **723**:11-24.

Burnett, K.P., and S. Hanson. 1982. The Analysis of Travel as an Example of Complex Human Behavior in Spatially Constrained Situations: Definitions and Measurement Issues. *Transportation Research A*, **16**:87-102.

CEO. 1995. *Centre for Earth Observation (CEO) Concept Document,* European Commission. Joint Research Centre. Document Reference CEO/160/1995.

Campbell W.J., Hill S.E. and Cromp R.F. 1989. Automatic Labelling and Characterization of Objects using Artificial Neural Networks, *Telematics and Informatics*, **6**:259-271.

Can A. 1996. Weight Matrices and Spatial Autocorrelation Statistics Using a Topological Vector Data Model, *International Journal of Geographical Information Systems* **10**: 1009-1017.

Carey H.C. 1858 *Principles of Social Science*. Lippincott, Philadelphia.

Carlstein T., Parkes D., and Thrift N. 1978. *Timing space and spacing time*, 3 Volumes. London: Edward Arnold.

Carpenter G.A. 1989. Neural network models for pattern recognition and associative memory, *Neural Networks*, **2**:243-57.

Carpenter G.A. and Grossberg S. 1985. Category learning and adaptive pattern recognition, a neural network model. *Proceedings of the Third Army Conference on Applied Mathematics and Computing*, ARO-Report 86-1, 37-56.

Carpenter G.A. and Grossberg S. 1987. A massively parallel architecture for a self-organizing neural pattern recognition machine, *Computer Vision, Graphics, and Image Processing*, **37**:54-115.

Carpenter G.A. and Grossberg S. 1987. ART 2 Stable self-organizing of pattern recognition codes for analog input patterns, *Applied Optics*, **26**:4919-30.

Carpenter G.A. and Grossberg S. 1995. Fuzzy ART, in Kosko B. ed. *Fuzzy Engineering*, Prentice Hall., Carmel.

Carpenter G.A. and Grossberg S. eds. 1991. *Pattern Recognition by Self-organizing Neural Networks*, MIT Press, Cambridge MA.

Carpenter G.A., Grossberg S., Markuzon N., Reynolds J.H. and Rosen D. B. 1992. Fuzzy ARTMAP A neural network architecture for incremental supervised learning of analog multidimensional maps, *IEEE Transactions on Neural Networks*, **3**:698-713.

Carpenter G.A., Grossberg S. and Reynolds J.H. 1991. ARTMAP Supervised real-time learning and classification of nonstationary data by a self-organizing neural network, *Neural Networks*, **4**:565-88.

Carpenter G.A., Grossberg S. and Rosen, D.B. 1991. Fuzzy ART Fast stable learning and categorization of analog patterns by an adaptive resonance system, *Neural Networks*, **4**:759-71.

Carpenter G.A., Grossberg S. and Ross W.D. 1993. ART-EMAP: A neural network architecture for object recognition by evidence accumulation, *Proceedings of the World Congress on Neural Networks (WCNN-93)*, Lawrence Ertbaum Associates, III, Hillsdak N.J., 643-56.

Casetti E. 1972. Generating Models by the Expansion Method: Applications to Geographic Research, *Geographical Analysis*, **4**:81-91.

Casetti E. 1991. The Investigation of Parameter Drift by Expanded Regressions: Generalities and a 'Family Planning' Example, *Environment and Planning A*, **23**:1045-1061.

Casetti E. 1993. Spatial Analysis: Perspectives and Prospects, Urban Geography, 14:526-537.

Casetti E.1995. Spatial Mathematical Modeling and Regional Science, *Papers In Regional Science*, **74**:3-11.

Casetti E.1996. *The Expansion Method, Mathematical Modeling and Spatial Econometrics*, Manuscript.

Casetti E. and Jones J.P. III. 1992. *Applications of the Expansion Method* London: Routledge.

Cease K.B. and Berzofsky J.A. 1988. Antigenic structures recognised by T-cells: towards the rational design of an AIDS vaccine. *AIDS* **2**(1):95-101.

Chao C.-H. and Dhawan A.P. 1994. Edge detection using a Hopfield neural network. *Optical Engineering* **33**(11):3739-3747.

Chirimuuta R.C. and Chirimuuta R.J. 1987. *AIDS, Africa and racism*. Bretby, Derbyshire: Chirimuuta.

Chown E., Kaplan S., and Cortenkamp D. 1995. Prototypes, location, and associative networks (PLAN): Towards a unified theory of cognitive mapping. *Cognitive Science*, **19**:1-51.

Christaller W. 1966. *Central Places in Southern Germany*. Translated by C.W. Baskin Prentice Hall, Englewood Cliffs, NJ.

Civco D. L. 1993. Artificial neural networks for land-cover classification and mapping, *International Journal of Geographical Information Systems*, **7**(2):173-186.

Clark W.A.V. 1992. Comparing cross-sectional and longitudinal analyses of residential mobility and migration.*Environment and Planning A* **45**:1291-1302.

Clark W.A.V. Deurloo M.C. and Dieleman F.M. 1994. Tenure change in the context of micro-level family and macro-level economic shifts. *Urban Studies* **31** :137-154.

Clarke C. and Clarke M. 1995. The Development and Benefits of Customized Spatial Decision Support Systems, in P. Longley and G. Clarke eds. *GIS for Business and Service Planning*. Geoinformation International, Cambridge, 227-254.

Clarke M. and Wilson A.G. 1983. The Dynamics of Urban Spatial Structure: Progress and Problems. *Journal of Regional Science*, **23**:1-18.

Clayton D. and Kaldor J. 1987. Empirical Bayes estimates of age-standardized relative risks for use in disease mapping. *Biometrics* **43**:671-681.

Cleveland W.S. 1979. Robust Locally Weighted Regression and Smoothing Scatterplots, *Journal of the American Statistical Association* **74**: 829-836.

Cleveland W.S. 1993. *Visualizing Data*, Summit, NJ: Hobart Press.

Cleveland W.S. and McGill M.E. 1988. *Dynamic Graphics for Statistics*, Pacific Grove, CA: Wadsworth.

Cleveland W.S. and Devlin S.J. 1988. Locally Weighted Regression: An Approach to Regression Analysis by Local Fitting, *Journal of the American Statistical Association* **83**: 596-610.

Cliff A.D. and Haggett P. 1988. *Atlas of Disease Distributions*. Oxford: Blackwell Reference Books.

Cliff A.D. and Haggett P. 1989. Spatial aspects of epidemic control. *Progress in Human Geography* **13**:315-47.

Cliff A.D. and Ord J.K. 1973. *Spatial Autocorrelation*. Pion, London.

Cliff A.D. and Ord J.K. 1981. *Spatial Processes: Models and Applications*, London: Pion.

Cliff A.D., Haggett P. and Smallman-Raynor M.R. 1993. *Measles: an historical geography of a major human viral disease from global expansion to local retreat, 1840-1990*. Oxford: Blackwell Reference Books.

Cliff A.D., Haggett P., Ord J.K. and Versey G.R. 1981. *Spatial diffusion: an historical geography of measles epidemics in an island community*. Cambridge: Cambridge University Press.

Cliff A.D., Haggett P., and Ord J. K.1986. *Spatial aspects of influenza epidemics*. London: Pion.

Cliff A.D., Haggett P., Ord J.K., Bassett and Davies. 1975. *Elements of spatial structure*. Cambridge University Press.

Coelho J.D., and Wilson A.G. 1976. The Optimum Location and Size of Shopping Centres, *Regional Studies*, **10**:413-421.

Cook D., Majure J., Symanzik J. and Cressie N. 1996. Dynamic Graphics in a GIS: Exploring and Analyzing Multivariate Spatial Data Using Linked Software, *Computational Statistics* **11**:467-480.

Cool K.O. and Dierickx I. 1993. Rivalry, strategic groups and firm profitability. *Strategic Management Journal*, **14**:47-59.

Courgeau D. 1990. Migration, family and careet: A life course approach, in: *Life Span Development and Behavior*, eds. P.B. Baltes, D.L. Featherman and R. Lerner Hillsdale, N.J.: Lawrence Erlbaum Associates. pp. 219-255.

Courgeau D. and Lelièvre E. 1992. *Event History Analysis in Demography* Oxford: Clarendon Press.

Courgeau D. and Lelièvre E. 1992. Interrelations between first home-ownership, constitution of the family, and professional occupation in France, in: *Demographic Applications of Event History Analysis*, eds. J. Trussell, R. Hankinson, and J. Tilton Oxford: Clarendon Press. pp. 120-140.

Cox D.R. 1972. Regression models and life tables. *Journal of the Royal Statistical Society Series B* **34**:187-220.

Cox E. 1994. *The fuzzy systems handbook*. A P Professional, Boston.

Cressie N. 1984. Towards resistant geostatistics. In G.Verly, et al eds. *Geo-statistics for natural resources characterization*. 21-44. Reidel, Dordrecht.

Cressie N. 1991. *Statistics for spatial analysis*. John Wiley and Sons, New York.

Cressie N. 1993. *Statistics for Spatial Data*, New York: Wiley.

Cromley R.G. 1982. The von Thünen model and environmental uncertainty. *Annals of the Association of American Geographers*, **72**:404-410.

Cromley R.G. and Hanink D.M. 1989. A financial-economic von Thünen model. *Environment and Planning A*, **21**:951-960.

Crompton J.L. 1992. Structure of Vacation Destination Choice Sets. *Journal of Tourism Research*, **19**:420-434.

Crompton J.L., and Ankomah P.K. 1993. Choice Set Propositions in Destination Decisions. *Annals of Tourism Research*, **20**:461-476.

Curry L. 1962. The geography of service centres within towns: the elements of an operational approach, *Lund Studies in Geography*, Series B, **24**:31-53.

DIBE. 1995. *Portability of neural classifiers for large area land cover mapping by remote sensing*, final report on Contract 10420-94-08 ED ISP I, to Joint Research Centre, European Commission, by Department of Biophysical and Electronic Engineering (DIBE), University of Genova, Italy.

Damm D., and Lerman S. 1981. A theory of activity scheduling behavior. *Environment and Planning A*, **13**:703-718.

Davenport W. 1960. Jamaican fishing: a game theory analysis. *Yale University Publications in Anthropology*, **59**:3-11.

Davies R.B. and Flowerdew R. 1992. Modelling migration careers using data from a British survey. *Geographical Analysis* **24**:35-57.

Davies R.B. and Pickles A.R. 1985. Longitudinal versus cross-sectional methods for behavioral research: A first round knock out. *Environment and Planning A* **17**: 1315-1329.

Davis L. ed. 1991. *Handbook of Genetic Algorithms*. N.Y.: Van Nostrand Reinbold.

Dawson M.S., Fung A.K., Manry M.T. 1993. Surface parameter retrieval using fast learning neural networks, *Remote Sensing Reviews*, **7**:1-18.

Densham P.J. 1991. Spatial Decision Support Systems, in D.J. Maguire, M.F.Goodchild and D.W. Rhind eds., *Geographical Information Systems: Principles*, John Wiley & Sons, New York, 403-412.

Densham P.J. 1994. Integrating GIS and spatial modelling: Visual interactive modelling and location selection, *Geographical Systems*, **1**:203-221.

Densham P.J., and Rushton G. 1988. Decision Support Systems for Locational Planning, in R. Golledge and H. Timmermans eds. *Behavioural Modelling in Geography and Planning*, Croom-helm, London, 65-90.

Department of Transportation. 1990 *Highway Network and TAZ Documentation*. Minnesota Department of Transportation, Office of Transportation Data Analysis, St. Paul, MN.

Desbarats J. 1983. Spatial Choice and Constraints on Behavior. *Annals of the Association of American Geographers*, **73**:340-357.

Deurloo M.C., Clark W.A.V. and Dieleman F.M. 1994. The move to housing ownership in temporal and regional contexts. *Environment and Planning A* **26**:1659-1670.

Dial R.B. 1996. Multicriterion equilibrium traffic assignment: Basic theory and elementary algorithms. Part I - T2: The bicriterion model. *Transportation Science* accepted.

Dieleman F.M. 1992. Struggling with longitudinal data and modelling in the analysis of residential mobility. *Environment and Planning A* **24**: 1527-1530.

Dieleman F.M. 1995. Using panel data: Much effort, little reward? *Environment and Planning A* **27**:676-682.

Dieleman F.M., Clark W.A.V., and Deurloo M.C. 1994. Tenure choice: Cross-sectional and longitudinal analyses. *Netherlands Journal of Housing and the Built Environment* **9**:229-246.

Ding Y. and Fotheringham A.S. 1992. The Integration of Spatial Analysis and GIS, *Computers, Environment and Urban Systems* **16**:3-19.

Dreyer P. 1993. Classification of land cover using optimized neural nets on SPOT data, *Photogrammetric Engineering and Remote Sensing*, **59**(5):617-621.

Drezner Z. ed. 1995. *Facility Location: A Survey of Applications and Methods*, Springer, New York.

Driankov D., Hellendoorn H., Reinfrank M., 1996. *An introduction to Fuzzy Control*, Springer Verlag, Berlin.

Dunn Jr., E.S. 1955. The equilibrium of land-use patterns in agriculture. *Southern Economic Journal*, **21**:173-187.

Durbin J.1953. A Note on Regression when there is Extraneous Information about One of the Coefficients, *Journal of the American Statistical Association*, **48**:799-808.

Dyck I. 1990. Context, culture and client: geography and the health for all strategy. *Canadian Geographer* **34**:338-41.

ESRI. 1995. *ArcView 2.1 The Geographic Information System for Everyone* Redlands CA: Environmental Systems Research Institute.

ESRI. 1995. *ArcView Version 2 Shapefile Technical Description.* White Paper, Redlands, CA: Environmental Systems Research Institute.

ESRI. 1994. *Avenue, Customization and Application Development for ArcView*, Redlands, CA: Environmental Systems Research Institute.

ESRI. 1995. *Understanding GIS, The ARC/INFO Method*, Redlands, CA: Environmental Systems Research Institute.

Earickson R.J. 1990. International behavioural responses to a health hazard: AIDS. *Social Science and Medicine* **31**:951-62.

Eaton B. and Lipsey R. 1982. An economic theory of central places, *Economic Journal*, **92**:56-72.

Eichmann G.E. and Stojancic M. 1987. Superresolving Signal and Image Restoration Using a Linear Associative Memory, *Applied Optics*, 1911-1918.

Elder G.H. 1985. *Life Course Dynamics*. Ithaca, N.Y. Cornell University Press.

Eldridge J.D. and Jones J.P. III. 1991. Warped Space: a Geography of Distance Decay, *Professional Geographer* **43**:500-511.

Elliott P., Cuzick J., English D. and Stern R. 1992. *Geographical and environmental epidemiology : methods for small area studies*. Oxford University Press.

Ellis J., Warner M. and White R. G. 1994. Smoothing SAR images with neural networks, *Proc. International Geoscience and Remote Sensing Symposium (IGARSS 94)*, held at Pasadena, 8-12 August, IEEE Press, Piscataway, NJ, 4, 1883-1885.

Epstein S.L. 1996. *Spatial representation for pragmatic navigation*. Personal Communication, Department of Computer Science, Hunter College and the Graduate School of The City University of New York.

Ettema D. 1995. *SMASH: A model of activity scheduling and travel behavior*. Unpublished manuscript, Eindhoven University of Technology, The Netherlands.

Ettema D., Borgers A. and Timmermans H.J.P. . 1993. A Simulation Model of Activity Scheduling Behavior. *Transportation Research Record,* **1413**:1-11.

Ettema D., Borgers A. and Timmermans H.P.J. 1995. *SMASH (Simulation Model of Activity Scheduling Heuristics): Empirical test and simulation issues*. Paper presented at the International Conference on Activity based Approaches: Activity Scheduling and the Analysis of Activity Patterns, May 25-28. Eindhoven University of Technology, The Netherlands.

Ettema D., and Timmermans H.P.J. 1993. Using interactive experiments for investigating activity scheduling behavior. *Proceedings of the PTRC 21st Summer Annual Meeting*, Manchester, England, Vol. P366, 267-281.

Evans P.E., Rutherford G.W., Amory J.W., Ilessol N.A., Bolan G.A., Herring M. and Werdegar D. 1988. Does health education work? Publically funded health education in San Francisco, 1982-1986. *Abstracts of the Fourth International Conference on AIDS*. Stockholm, Abstract 6044.

Farley J.A., Limp W.F. and Lockhart J. 1990. The Archeologist's Workbench: Integrating GIS,

Remote Sensing, EDA and Database Management, in: K. Allen, F. Green and E. Zubrow eds., *Interpreting Space: GIS and Archaeology*, London: Taylor & Francis, pp. 141-164.

Fenner F., Henderson D.A., Arita I., Jezek Z. and Ladnyi I.D. 1988. *Smallpox and its eradication*. Geneva: World Health Organization.

Fiegenbaum A. 1987. *Dynamic aspects of strategic groups and competitive strategy: Concepts and empirical examination in the insurance industry*. Unpublished doctoral dissertation, University of Illinois, Urbana-Champaign.

Fierens F., Wilkinson G. G. and Kanellopoulos I. 1994. Studying the behaviour of neural and statistical classifiers by interaction in feature space, *SPIE Proceedings*, **2315**:483-493.

Fischer M.M. 1993. Travel demand. in Polak J and Heertie J. eds. *European Transport Economics*, Blackwell, Oxford (UK), 6-32.

Fischer M.M. 1997. Spatial analysis: Retrospect and prospect,.in Longley P., Goodchild M., Maguire D., and Rhind D., eds., *Geographical Information Systems: Principles, Techniques, Management and Applications*. GeoInformation International, Cambridge.

Fischer M.M., Gopal S., Staufer P., Steinnocher K. 1994. *Evaluation of neural pattern classifiers for a remote sensing application*, Paper presented at the 34th European Congress of the Regional Science Association, Groningen, August 1994.

Fischer M.M. and Nijkamp P. 1993. *Geographic Information Systems, Spatial Modelling and Policy Evaluation*, Berlin: Springer-Verlag.

Fischer M.M., Nijkamp P., and Papageorgiou Y.Y. 1990. Current trends in behavioural modelling. in Fischer M.M., Nijkamp P., Papageorgiou Y.Y. ,eds. *Spatial Choices and Processes*, North-Holland, Amsterdam, 1-14.

Fischer M.M., Scholten H. and Unwin D. 1996. *Spatial Analytical Perspectives on GIS in Environmental and Socio-Economic Sciences*, London: Taylor & Francis.

Fischer M.M., Scholten H.J. and Unwin D. 1997. Geographic information systems, spatial data analysis and spatial modelling: Problems and possibilities,.in Fischer M.M., Scholten H.J. and Unwin D., eds. *Spatial Analytical Perspectives on GIS*, Taylor & Francis, London, (in press).

Fischer M. M. and Gopal S. 1993. Neurocomputing and spatial information processing -from general considerations to a low dimensional real world application, *Proc. Eurostat/DOSES Workshop on New Tools for Spatial Data Analysis*, Lisbon, November. EUROSTAT, Luxembourg, Statistical Document in Theme 3, Series D, 55-68.

Fischer, M.M. and Gopal S. 1994. Artificial Neural Networks: A New Approach to Modeling Interregional Telecommunication Flows, *Journal of Regional Science*, **34**:503-527.

Fitch J. P. et al. 1991. Ship wake detection procedure using conjugate grained trained artificial neural networks, *IEEE Transactions on Geoscience and Remote Sensing*, GE-**29**(5):718-726.

Flowerdew R. and Green M. 1991. Data Integration: Statistical Methods for Transferring Data Between Zonal Systems, in: I. Masser and M. Blakemore eds., *Handling Geographical Information*, London: Longman, pp. 38-54.

Fomby T.B., Hill R.C. and Johnson S.R.1984. *Advanced Econometrics Methods*, Springer-Verlag.

Foody G. M. 1995. Using prior knowledge in artificial neural network classification with a minimal training set, *International Journal of Remote Sensing*, **16**(2):301-312.

Foody G. M. 1996. Approaches for the production and evaluation of fuzzy land cover classifications from remotely-sensed data, *International Journal of Remote Sensing*, **17**(7):1317-1340.

Foody G. M. 1996. Relating the land-cover composition of mixed pixels to artificial neural network classification output, *Photogrammetric Engineering and Remote Sensing*, **62**(5):491-499.

Foody G. M., McCulloch M. B. and Yates W. B. 1994. Crop classification from C-band polarimetric radar data, *International Journal of Remote Sensing*, **15**(14):2871-2885.

Foster S.A. and Gorr W.L. 1986. An Adaptive Filter for Estimating Spatially Varying

Parameters: Application to Modeling Police Hours Spent in Response to Calls for Service, *Management Science* **32**: 878-889.

Fotheringham A.S. and Charlton M. 1994. GIS and Exploratory Spatial Data Analysis: An Overview of some Research Issues, *Geographical Systems* **1**:315-327.

Fotheringham A.S. and Rogerson P. 1994. *Spatial Analysis and GIS*, London: Taylor & Francis.

Fotheringham A.S. and Wong D.W.S. 1991. The modifiable areal unit problem in multivariate statistical analysis. *Environment and Planning A*. **23**:1025-1044.

Fotheringham A.S. 1982. Distance decay parameters: a reply, *Annals Association of American Geographers*, **72**:552-554.

Fotheringham A.S. 1983. A New Set of Spatial Interaction Models: The Theory of Competing Destinations. *Environment and Planning A*, **15**:15-36.

Fotheringham A.S. 1985. Spatial competition and agglomeration in urban modelling, *Environment and Planning A*, **17**:213-230.

Fotheringham A.S. 1986. Modelling hierarchical destination choice. *Environment and Planning A* **18**:401-418.

Fotheringham A.S. 1988. Consumer Store Choice and Choice Set Definition. *Marketing Science*, **7**:299-310.

Fotheringham A.S. 1992. Exploratory Spatial Data Analysis and GIS, *Environment and Planning A* **25**:156-158.

Fotheringham A.S. 1994. On the Future of Spatial Analysis: The Role of GIS, *Environment and Planning A* Anniversary Issue:30-34.

Fotheringham A.S. and Rogerson P.A. 1993. GIS and Spatial Analytical Problems, *International Journal of Geographic Information Systems* **7**: 3-19.

Fotheringham A.S. and Pitts T.C. 1995. Directional Variation in Distance-Decay, *Environment and Planning A* **27**: 715-729.

Fotheringham A.S., Charlton M. and Brunsdon C.F. 1996. The Geography of Parameter Space: An Investigation into Spatial Non-Stationarity, *International Journal of Geographical Information Systems* **10**: 605-627.

Fotheringham A.S., Charlton M and Brunsdon C.F. 1996. Two Techniques for Exploring Non-Stationarity in Geographical Data, *Geographical Systems* (forthcoming).

Foust J.B. and de Souza A.R. 1978. *The Economic Landscape*. Charles Merrill, Columbus, Ohio.

Frank J.A., Orenstein W.A., Bart K.J., Bart S.W., El-Tantawy N., David R.M. and Hinman A.R. 1985. Major impediments to measles elimination. *American Journal of Diseases of Children*, **139**:881-8.

Furby S. 1996. *A comparison of neural network and maximum likelihood classification*, Technical Note No. I.96.11, Joint Research Centre, European Commission, Ispra, Italy.

Gagnon J.H. and Simon W. 1974. *Sexual contacts: the social sources of human sexuality* . Chicago: Aldine.

Gallo R.C., Salahuddin S.Z., Popvic M., Shearer G.M., Kaplan M., Haynes B.F., Palker T.J., Redfield R., Oleske J., Safai B., White G., Foster P. and Markham P.D. 1984. Frequent detection and isolation of cytopathic retroviruses (HTLV III) from patients with AIDS and at risk of AIDS. *Science* **224**:500-3.

Gatrell A.C. 1989. On the spatial representation and accuracy of address-based data in the United Kingdom. *International Journal of Geographical Information Systems* **3**:335-48.

Gaudry M.J.I., and Dagenais M.G. 1979. The Dogit Model. *Transportation Research B*, **13**:105-111.

Gersick C.J.G. 1991. Revolutionary change theories: a multilevel exploration of the punctuated equilibrium paradigm. *Academy of Management Review*, **16**:10-36.

Getis A. and Ord J.K. 1992. The analysis of spatial association by use of distance statistics. *Geographical Analysis* **24**:75-95.

Ghosh A. 1984. Parameter Nonstability in Retail Choice Models. *Journal of Business*

Research, **12**:425-436.

Ghosh A. and Rushton G. eds. 1987. *Spatial Analysis and Location-Allocation Models*, Von Nostrand Reinhold Company, New York.

Ghosh A. and McLafferty S.L. 1987. Optimal Location and Allocation with Multipurpose shopping, in Ghosh A. and Rushton G. eds. *Spatial Analysis and Location-Allocation Models*, Von Nostrand Reinhold Company, New York, 55-75.

Ghosh A. and S.L. McLafferty 1987. *Location Strategies for Retail and Service Firms*, Lexington Books, Lexington.

Ghosh A., McLafferty S. and Craig C.S. 1995. Multifacility Retail Networks, in Z. Drezner ed. *Facility Location: A Survey of Applications and Methods*, Springer, New York, 301-330.

Gioia D. and Sims. H., Jr. eds. 1986. *The thinking organization.* San Francisco: Jossey-Bass.

Goldberg. D.E. 1989. *Genetic Algorithms in Search, Optimization and Machine Learning.* Reading: Addison-Wesley.

Goldberger A.S. 1968. *Topics in Regression Analysis*, Macmillan: London.

Goldstein H. 1987. *Multilevel Models in Educational and Social Research*, London: Oxford University Press.

Golledge R.G. 1995. *Defining the criteria used in path selection.* Paper presented at the International Conference on Activity based Approaches: Activity Scheduling and the Analysis of Activity Patterns, May 25-28. Eindhoven University of Technology, The Netherlands.

Golledge R.G. 1995. Path selection and route preference in human navigation: A progress report. In A.U. Frank and W. Kuhn Eds.., *Spatial information theory: A theoretical basis for GIS.* Proceedings, International Conference COSIT 95. Semmering, Austria, September. Berlin: Springer-Verlag, pp. 207-222.

Golledge R.G., Kwan M-P.. and Gärling T. 1994. Computational process modeling of household travel decisions using a geographic information system. *Papers in Regional Science*, **73**:2:99-117.

Golledge R.G., Ruggles A.J., Pellegrino J.W., and Gale N.D. 1993. Integrating route knowledge in an unfamiliar neighborhood: Along and across route experiments. *Journal of Environmental Psychology*, **13**:4: 293-307.

Golledge R.G., Smith T.R., Pellegrino J.W., Doherty S., and Marshall S.P. 1985. A conceptual model and empirical analysis of children's acquisition of spatial knowledge. *Journal of Environmental Psychology*, **5**:125-152.

Golob T.F., and Meurs H. 1988. Development of structural equations models of the dynamics of passenger travel demand. *Environment and Planning A* **20**:1197-1218.

Good I.J. 1983. The Philosophy of Exploratory Data Analysis, *Philosophy of Science* **50**:283-295

Goodchild M.F. 1987. A Spatial Analytical Perspective on Geographical Information Systems, *International Journal of Geographical Information Systems* **1**:327-334.

Goodchild M.F. 1992. Geographical Information Science, *International Journal of Geographical Information Systems* **6**:31-45.

Goodchild M.F., Haining R.P., Wise S., et al. 1992. Integrating GIS and Spatial Analysis - Problems and Possibilities, *International Journal of Geographical Information Systems* **6**:407-423.

Goodwin P.B. 1981. The Usefulness of Travel Budgets. *Transportation Research A*, **15**:97-106.

Gopal S., Klatzky R.L. and Smith T.R. 1989. NAVIGATOR: A psychologically based model of environmental learning through navigation. *Journal of Environmental Psychology*, **9**:4: 309-332.

Gopal S., Sklarew D.M., and Lambin E. 1993. Fuzzy neural networks in multitemporal classification of landcover change in the Sahel. *Proc. Eurostat/DOSES Workshop on New Tools for Spatial Data Analysis*, Lisbon, November. EUROSTAT, Luxembourg, Statistical

Document in Theme 3, Series D, 69-81.

Gopal S., and Smith T.R. 1990. NAVIGATOR: An AI-based model of human way-finding in an urban environment. In M.M. Fischer and Y.Y. Papageorgiou Eds., *Spatial choices and processes*. North-Holland: Elsevier Science Publishers, pp. 168-200.

Gorr W.L. and Olligschlaeger A.M. 1994. Weighted Spatial Adaptive Filtering: Monte Carlo Studies and Application to Illicit Drug Market Modeling. *Geographical Analysis* **26**: 67-87.

Gotlieb M.S., Schroff R., Schanker H.M., Weisman J.D., Fan P.T., Wolf R.A. and Saxon A.S. 1981. Pneumocystis carinii pneumonia and mucosal candidiasis in previously healthy homosexual men: evidence of a new acquired cellular immunodeficiency, *New England Journal of Medicine*, **305**:1425-31.

Gould P.R. 1963. Man against his environment: a game theoretic framework. *Annals of the Association of American Geographers*, **53**:290-297.

Grant R.M., Wiley J.A. and Winkelstein W. 1987. Infectivity of the Human Immunodeficiency Virus of homosexual men. *Journal of Infectious Diseases* **156**:189-93.

Greig D.M., 1980. *Optimisation*, Longman:London.

Griffith D.A. 1973. The effect of measles vaccination on the incidence of measles in the community. *Journal of the Royal Statistical Society A*, **136**:441-49.

Griffith D.A. 1993. Which Spatial Statistics Techniques Should Be Converted to GIS Functions? in: M.M. Fischer and P. Nijkamp eds., *Geographic Information Systems, Spatial Modelling and Policy Evaluation*, Berlin: Springer-Verlag, pp. 101-114.

Grossberg S. 1969. Some networks that can learn, remember, and reproduce any number of complicated space-time patterns, *Journal of Mathematics and Mechanics*, **19**:53-91.

Grossberg S. 1976. Adaptive pattern classification and universal recoding, I: Parallel development and coding of neural feature detectors, *Biological Cybernetics*, **23**:121-34.

Grossberg S. 1976. Adaptive pattern classification and universal recoding, II: Feedback, expectation, olfaction and illusion, *Biological Cybernetics*, **23**:187-202.

Grossberg S. 1988. Nonlinear neural networks Principles, mechanisms, and architectures, *Neural Networks*, **1**:17-61.

Grover J. Z., 1988. AIDS, keywords and cultural work in *Cultural studies*. Eds L. Grossberq, C. Nelson, and P. Treichler. Cambridge: MIT Press, pp 295-337.

Gupta MM., Sanchez E. eds., 1982 *Fuzzy Information and Decision Processes*. North-Holland, Amsterdam.

Gärling T., Brännäs K., Garvill J., Golledge R.G., Gopal S., Holm E., and Lindberg E. 1989. Household activity scheduling. In: *Transport policy, management and technology towards 2001*: Selected proceedings of the fifth world conference on transport research, Volume IV. Ventura, CA: Western Periodicals, pp. 235-248.

Gärling T., Kwan M-P., and Golledge R.G. 1994. Computational-process modelling of household activity scheduling. *Transportation Research B*, 28B, **5**:355-364.

Gärling T., and Gärling E. 1988. Distance minimization in downtown pedestrian shopping behavior. *Environment and Planning A*, **20:**547-554.

Hachen D.S. 1988. The competing risks model: A method for analyzing processes with multiple types of events. *Sociological Methods and Research* **17**:21-54.

Haggett P. 1990. *The geographer's art*. Oxford: Blackwell.

Haggett P. 1992. Sauer's 'Origins and dispersals': its implications for the geography of disease. *Transactions of the Institute of British Geographers* **17**:387-98.

Haggett P. 1994. Prediction and predictability in geographical systems. *Transactions of the Institute of British Geographers* **19**:6-20.

Haining R. 1990. *Spatial Data Analysis in the Social and Environmental Sciences*. Cambridge: Cambridge University Press.

Haining R. 1994. Designing Spatial Data Analysis Modules for Geographical Information Systems. in: S. Fotheringham and P. Rogerson eds., *Spatial Analysis and GIS*, London: Taylor and Francis, pp. 45-63.

Haining R.P, Wise S.M. and Ma J. 1996. The design of a software system for interactive spatial statistical analysis linked to a GIS. *Computational Statistics* (in press).

Haining R.P. 1996. *Spatial statistics and the analysis of health data.* Paper presented to the GISDATA workshop on GIS and health. Helsinki, June 1996.

Haining R.P., Wise S.M. and Blake M. 1994. Constructing regions for small area analysis: material deprivation and colorectal cancer. *Journal of Public Health Medicine* **16**:429-438.

Halmari P.M. and Lundberg C.G. 1991. *Bridging Inter- and Intra-corporate Information Flows with Neural Networks*, Paper presented at the Annual Meeting of the Association of American Geographers, Miami, April 13-17.

Hannon M.T. and Freeman J. 1984. Structural inertia and organizational change. *American Sociological Review*, **49**:149-64.

Hanson S. 1980. Spatial diversification and multi-purpose travel, *Geographical Analysis*, **12**:245-257.

Hara Y., Atkins R.G., Yueh S.H., Shin R.T., Kong J. A. 1994. Application of neural networks to radar image classification, *IEEE Transactions on Geoscience and Remote Sensing*, **32**:100-9.

Harris B., and Wilson A.G. 1978. Equilibrium Values and Dynamics of Attractiveness Terms in Production-Constrained Spatial-Interaction Models, *Environment and Planning A*, **10**:371-388.

Hartigan J. 1975. *Clustering Algorithms*, Wiley, New York.

Harvey A.C. 1989. *Forecasting, structural time series models and the Kalman filter.* Cambridge: Cambridge University Press.

Harvey A.C. and Durbin J. 1986. The effects of seat belt legislation on British road casualties: a case study in structural time series modelling. *Journal of the Royal Statistical Society A*, **149**:187-227.

Harvey A.C. and Koopman S.J. 1996. Structural time series models in medicine. *Statistical Methods in Medical Research*, **5**:23-49.

Haslett J. 1992. Spatial Data Analyis Challenges, *The Statistician* **41**:271-284.

Haslett J. and Power G.M. 1995. Interactive Computer Graphics for a more Open Exploration of Stream Sediment Geochemical Data, *Computers and Geosciences* **21**:77-87.

Haslett J., Bradley R., Craig P., Unwin A. and Wills C. 1991. Dynamic Graphics for Exploring Spatial Data with Applications to Locating Global and Local Anomalies, *The American Statistician* **45**:234-242.

Haslett J., Bradley R., Craig P.S., Will G. and Unwin A.R. 1991. Dynamic graphics for exploring spatial data with application to locating global and local anomalies. *American Statistician*, **45**:234-42.

Haslett J., Wills G. and Unwin A. 1990. SPIDER - An Interactive Statistical Tool for the Analysis of Spatially Distributed Data, *International Journal of Geographical Information Systems* **4**:285-296.

Hastie T. and Tibshirani, R. 1990. *Generalized Additive Models*, Chapman and Hall, London.

Hauser J.R., and Westerfelt B. 1990. An Evaluation Cost Model of Consideration Sets. *Journal of Consumer Research*, **16**:393-408.

Hayes-Roth B. and Hayes-Roth F. 1979. A cognitive model for planning. *Cognitive Science*, **3**:275-310.

Hebb D.O. 1949. *The Organization of Behavior*. New York: John Wiley.

Heermann P.D. and Khazenie N. 1992. Classification of Multispectral Remote Sensing Data using a Back-propagation Neural Network, *IEEE Transactions on Geoscience and Remote Sensing*, **30**:81-88.

Hellwich O. 1994. Detection of linear objects in ERS-1 SAR images using neural network technology, *Proc. International Geoscience and Remote Sensing Symposium (IGARSS 94)*, held at Pasadena, California, 8-12 August, IEEE Press, Piscataway, NJ, **4**: 1886-1888.

Hendriksson B. and Ytterberg H. 1992. *Sweden: the power of the moral(istic) left in AIDS in the industrialised democracies: passions, politics and policies*, Eds D.L. Kirp and R. Bayer, New Brunswick, New Jersey: Rutgers University Press, pp 317-38.

Hepner G.F., Logan T., Ritter, N., Bryant N. 1990. Artificial neural network classification using a minimal training set comparison of conventional supervised classification, *Photogrammetric Engineering and Remote Sensing*, **56**:469-73.

Hertz J., Krogh A., Palmer, R.G. 1991. *Introduction to the Theory of Neural Computation*, Addison-Wesley, Redwood City CA.

Herzog H.W., Schlottmann A.M. and Boehm T.P. 1993. Migration as spatial job-search: A survey of empirical findings, *Regional Studies* **37**:327-340.

Hirtle S.C. and Gärling T. 1992. Heuristic rules for sequential spatial decisions. *Geoforum*, **23**:2: 227-238.

Hodgson M.J. 1978. Toward More Realistic Allocation in Location-Allocation Models: An Interaction Approach, *Environment and Planning A*, **10**:1273-1285.

Holland J.H. 1975. *Adaptation in Natural and Artificial Systems*. Ann Arbor: University of Michigan.

Holmblad L.P., Ostergaard J.J, 1982. Control of a cement kiln by fuzzy logic, in M.M. Gupta and E. Sanchez eds. *Fuzzy Information and Decision Processes*. North-Holland, Amsterdam 398-399.

Holmblad L.P., Ostergaard J.J., 1981. Fuzzy logic control: operator experience applied in automatic process control, *ZKG International* **34**:127-133.

Horn M.E.T. 1995. Solution techniques for large regional partitioning problems. *Geographical Analysis* **27**:230-248.

Horowitz J.L. 1988. Specification Tests for Probabilistic Discrete Choice Models of Consumer Behaviour. In Golledge, R.G., and Timmermans H.J.P. , Ed., *Behavioural Modelling in Geography and Planning*, London: Croom Helm, pp. 124-137.

Horowitz J.L. 1991. Modeling the Choice of Choice Set in Discrete-Choice Random-Utility Models. *Environment and Planning A*, **23**:1237-1246.

Horowitz J.L. and Louviere J. 1995. What is the Role of Consideration Sets in Choice Modeling? *International Journal of Research in Marketing*, **12**:39-54.

Howard J.A., and Seth J.N. 1969. *The Theory of Buying Behavior*. New York: John Wiley.

Howard J.A.1961. *Marketing Management Analysis and Planning*. Homewood, IL: Irwin.

Huber P. J. 1985. Projection pursuit, *Annals of Statistics*, **13**(2):435-475.

Huff A. S. 1982. Industry influences on strategy reformulation. *Strategic Management Journal*, **3**:119-131.

Huff J. O. and Clark, W. A. V. 1978. Cumulative stress and cumulative inertia: A behavioral model of the decision to move. *Environment and Planning A*, **10**:1101-19.

Huff J. O. , Huff A. S. and Thomas H. 1994. In H. Daems and H. Thomas eds., *Strategic groups, strategic moves and performance*. Tarrytown, New York: Elsevier Science.

Huff J.O. and Huff A.S. 1995. Stress, inertia, opportunity, and competitive position: A SIOP model of strategic change in the pharmaceuticals industry. *Best Papers Proceedings,*. Vancouver: Academy of Management.

Hägerstrand T. 1970. What about People in Regional Science? *Papers of the Regional Science Association*, **24**:7-21.

Isham V. 1988. Mathematical modelling of the transmission dynamics of HIV infection and AIDS. *Journal of the Royal Statistical Society A* **151**:5-30.

Isham V. and Medley G. eds. 1996. *Models for Infectious Human Deseases: Their Structure and Relation to Data*. Cambridge University Press: Cambridge.

Ishibuchi H., H. Tanaka and H. Okada. 1993. Fuzzy Neural Networks with Fuzzy Weights and Fuzzy Biases, *IEEE Internat. Conf. on Neural Networks*, 1650-1655.

Jacquez J.A., Simon C.P., Koopman J., Sattenspiel L. and Perry T. 1988. Modelling and analyzing HIV transmission: the effect of contact patterns. *Mathematical Biosciences* **92**: 119-99.

Jewell N.P., Dietz K. and Farewell V.T. eds. 1992. *AIDS Epidemiology: Methodological*

Issues, Birkhauser: Boston Basel Berlin

Jimenez L. and Landgrebe D. 1994. High dimensional feature reduction via projection pursuit, *Proc. International Geoscience and Remote Sensing Symposium (IGARSS '94)*, held at Pasadena, California, 2:1145-1147, IEEE Press, Piscataway, NJ.

Johnson G. 1988. Rethinking incrementalism. *Strategic Management Journal*, **9**:75-91.

Jones A., McGuire W. and Witte A. 1978. A reexamination of some aspects of von Thünen's model of spatial location. *Journal of Regional Science*, **18**:1-15.

Jones D.W. 1983. Location, agricultural risk, and farm income diversification. *Geographical Analysis*, **15**:231-246.

Jones D.W. 1991. An introduction to the Thünen location and land use model. *Research in Marketing*, **5**:35-70.

Jones K 1991. Specifying and Estimating Multilevel Models for Geographical Research, *Transactions of The Institute of British Geographers* **16**:148-159.

Jones P., Koppelman F., and Orfeuil J.-P. 1990. Activity analysis: State-of-the-art and future directions. In P. Jones ed., *Developments in dynamic and activity-based approaches to travel analysis*. Aldershot: Avebury, pp. 34-55.

Jones P.M., Dix M.C., Clarke M.I., and Heggie I.G. 1983. *Understanding travel behavior*. Aldershot: Gower.

Judge G.G., Griffith W.E., Hill R.C., Lutkepohl H., and Lee T.C. 1985. *The Theory and Practice of Econometrics*, New York: Wiley.

Kalbfleisch, J.D. and Prentice, J.L. 1980. *The Statistical Analysis of Failure Time Data*. New York: John Wiley.

Kalman R.E. 1960. A new approach to linear filtering and prediction problems. *Transactions of the ASME Journal of Basic Engineering*, **82**:34-45.

Kanaya F. and Miyake S. 1991. Bayes statistical classifier and valid generalization of pattern classifying neural networks, *IEEE Transactions on Neural Networks*, **2**(4):471-475.

Kanellopoulos I., Wilkinson G.G., Mégier J. 1993. Integration of neural network and statistical image classification for land cover mapping, *Proceedings of the International Geoscience and Remote Sensing Symposium (IGARSS'93)*, held in Tokyo, 18-21 August, vol. 2, 511-3, IEEE Press, Piscataway NJ.

Kanellopoulos I., Varfis A., Wilkinson G. G. and Mégier J. 1991. Neural network classification of multi-date satellite imagery, *Proc. International Geoscience and Remote Sensing Symposium (IGARSS'91)*, 3-6 June, Espoo, Finland, IEEE, Piscataway, NJ., Vol IV, 2215-2218.

Kanellopoulos I., Varfis A., Wilkinson G. G., and Mégier J. 1992. Land cover discrimination in SPOT imagery by artificial neural network -a twenty class experiment, *International Journal of Remote Sensing*, **13**(5):917-924.

Kanellopoulos I., Wilkinson G. G. and Chiuderi A. 1994. Land cover mapping using combined Landsat TM imagery and textural features from ERS-1 Synthetic Aperture Radar imagery, *Proc. European Symposium on Satellite Remote Sensing*, held in Rome, 26-30 September 1994, European Optical Society / International Society for Optical Engineering.

Kaplan E.H. 1989. Can bad models suggest good policies? Sexual mixing and the AIDS epidemic. *The Journal of Sex Research* **26**:301-14.

Kaplan E.H. and Lee Y.S. 1989. How bad can it get? Bounding worst case endemic heterogeneous mixing models of HIV/AIDS. *Mathematical Biosciences* **99**:157-80.

Karr C. L., 1994. Adaptive control with fuzzy logic and genetic algorithms', in R.R. Yager, L.A. Zadeh *Fuzzy Sets, Neural Networks, and soft computing*, van Nostrand Reinhold, New York p.345-367.

Karr C.L., 1991. Genetic algorithms for fuzzy logic controllers, *AI Expert* **6**:26-33

Kearns R.A. 1996. AIDS and medical geography: embracing the Other. *Progress in Human Geography* **20**:123-31.

Keil G. and Haberkern G.P. 1994. *A review of the methodology and implication of the work of Dr R.G.V. Baker on retail trading hours*, Report Commissioned by Coles-Myer,

Marketshare, Brisbane.

Key J., Maslanik J.A., Schweiger A.J. 1989. Classification of merged AVHRR and SMMR arctic data with neural networks, *Photogrammetric Engineering and Remote Sensing*, **55**(9):1331-8.

Khosaka H. 1989. A Spatial Search-Location Model of Retail Centers, *Geographical Analysis*, **21**:338-349.

Khosaka H. 1993. A Monitoring and Locational Decision Support System for Retail Activity, *Environment and Planning A*, **25**:197-211.

Kirp D.L. and Bayer R. 1992. *AIDS in the industrialized democracies: passions, politics and policies*. New Brunswick, New Jersey: Rutgers University Press.

Kitamura R. 1984. A model of daily time allocation to discretionary out-of-home activities and trips. *Transportation Research B*, **18**:255-266.

Kitamura R. 1988. An evaluation of activity-based travel analysis. *Transportation*, **15**: 9-34.

Kitamura R., Kazuo N., and Goulias K. 1990. Trip chaining behavior by central city commuters: A causal analysis of time-space constraints. In P. Jones Ed.., *Developments in dynamic and activity-based approaches to travel analysis*. Aldershot: Avebury, pp. 145-170.

Kitamura R., and Goulias K.G. 1989. *MIDAS: A travel demand forecasting tool based on dynamic model system of household car ownership and mobility*. Unpublished manuscript.

Kitamura R., and Kermanshah M. 1984. Sequential Model of Interdependent Activity and Destination Choices. *Transportation Research Record*, **987**:81-89.

Kitamura R. and Lam T.N. 1984. A Model of Constrained Binary Choice. In Volmuller, J., and R. Hamerslag (Editors) *Proceedings of the 9th international Symposium on Transportation and Traffic Theory*, Utrecht: VNU Science Press, pp. 493-512

Knox E.G. 1986. A transmission model for AIDS. *European Journal of Epidemiology* **2**:165-77.

Knox E.G., MacArthur C. and Simons K.J. 1993. *Sexual behaviour and AIDS in Great Britain*. London: HMSO.

Kochanowski P. 1990. The Expansion Method as a Tool of Regional Analysis, *Regional Science Perspectives*, **20**:52-65.

Kohonen T. 1977. *Associative Memory - A System Theoretical Approach*. New York: Springer-Verlag.

Kohonen T. 1984. *Self Organization and Associative Memory*, Springer Series in Information Sciences Vol. 8, Springer-Verlag, Berlin.

Kohonen T. and Ruohonen M. 1973. Representation of Associated Data by Matrix Operators, *IEEE Transaction on Computers*, C-**22**:701-702.

Kosko B. 1992. *Neural Networks and Fuzzy Systems*. Englewood Cliffs: Prentice Hall.

Kosko B. 1990. Fuzziness vs. probability, *International Journal of General Systems* **17**:211-240.

Kosko B., 1993. Fuzzy systems as universal approximators, *Proceedings of First IEEE International Conference on Fuzzy Systems*, San Diego, 1153-1162.

Kosko B., 1994. *Fuzzy thinking*. Harper-Collins, London.

Koutsoyiannis A. 1977. *Theory of Econometrics*, 2nd ed, Macmillan.

Krieger N. and Appleman R. 1994. The politics of AIDS. in *AIDS: the politics of survival*, Eds N. Krieger and G. Margo, New York: Baywood, pp 3-54.

Krieger N. and Margo G. 1994. *AIDS: the politics of survival*, New York: Baywood.

Kristensen G. and Tkocz Z. 1994. The Determinants of Distance to Shopping Centers in an Urban Model Context. *Journal of Regional Science*, **34**:425-443.

Kuhn T.S. 1970. *The structure of scientific revolutions*. Chicago: The University of Chicago Press.

Kuipers B.J. 1978. Modelling spatial knowledge. *Cognitive Science*, **2**:129-153.

Kung S. Y. 1993. *Digital Neural Networks*, Prentice Hall, Englewood Cliffs, NJ.

Kwan M-P. 1995. *GISICAS: An activity-based travel decision support system using a*

GIS-interfaced computational-process model. Paper presented at the International Conference on Activity based Approaches: Activity Scheduling and the Analysis of Activity Patterns, May 25-28. Eindhoven University of Technology, The Netherlands.

Kwan M.-P. 1994. *GISICAS: A GIS-interfaced Computational-Process Model for Activity Scheduling in Advanced Traveler Information Systems.* Unpublished Ph.D. dissertation, University of California, Santa Barbara.

Kwok R. et al. 1991. Application of neural networks to sea ice classification using polarimetric SAR images, *Proc. International Geoscience and Remote Sensing Symposium (IGARSS 91)*, held at Espoo, Finland, 3-6 June, IEEE Press, Piscataway, NJ, **1**: 85-88.

Lam N., Fan M. and Liu K-B. 1996. Spatial-Temporal Spread of the AIDS Epidemic, 1982 1990: A Correlogram Analysis of Four Regions of the United States, *Geographical Analysis* **28**:93-107.

Lancaster T. 1990. *The Econometric Analysis of Duration Data.* Cambridge: Cambridge University Press.

Landau U., Prashker J.N. and Hirsh M. 1981. The Effect of Temporal Constraints on Household Travel Behavior. *Environment and Planning A*, **13**:435-448.

Lange S. 1978. The role of consumer behaviour in the distribution of shopping centres, in Funck, R. and Parr, J.B. (Editors), The Analysis of Regional Structure: Essays in Honour of August Losch, *Karlsruhw Papers in Regional Science*, No.2, Pion, London., 62-73.

Lankford P.M. 1969. Regionalisation: Theory and Alternative Algorithms. *Geographical Analysis* **1**:196-212.

Latif A.S., Bassett M.T. and Mhloyi M. 1989. Genital ulcers and transmission of HIV among couples in Zimbabwe. *AIDS* **3**:519-23.

Lave C. A. and March J. G. 1975. *An introduction to models in the social sciences.* New York: Harper and Row.

Lawless J.E. 1982. *Statistical Models and Methods for Lifetime Data.* New York: John Wiley.

Leamer E.E. 1983. Let's Take the Con Out of Econometrics, *American Economic Review,* **73**:31-43.

Leamer E.E. and Leonard H.1983. Reporting the Fragility of Regression Estimates, *Review of Economic and Statistics*, **65**:306-317.

Leamer E.E., 1985. Sensitivity Analyses would Help, *American Economic Review,* **75**:308-313.

Learned E. P., Christensen C. R., Andrews K. R. and Guth W. D. 1965. *Business Policy.* Homewood, IL: Irwin.

Lee J., Weyer R.C., Sengupta S.K. and Welch R.M. 1990. A Neural Network Approach to Cloud Classification, *IEEE Transaction on Geoscience and Remote Sensing,* **28**:846-855.

Leiser D., and Zilberschatz A. 1989. The TRAVELLER: A computational model of spatial network learning. *Environment and Behavior*, **21**(4):435-463.

Lelièvre E., and Bonvalet C. 1994. A compared cohort history of residential mobility, social change and home-ownership in Paris and the est of France. *Urban Studies* **31**:1647-1665.

Lenntorp B. 1976. Paths in Space-Time Environments: A Time Geographic Study of Movement Possibilities of Individuals. *Lund Studies in Geography*, Series B, **44**.

Lenntorp B. 1978. A time-geographic simulation model of individual activity programmes. In T. Carlstein, D. Parkes and N. Thrift eds., *Human activity and time geography*. London: Edward Arnold, pp. 162-180.

Leonardi G. 1978. Optimum Facility Location by Accessibility Maximizing, *Environment and Planning A*, **10**:1287-1305.

Leonardi G. 1981. A Unifying Framework for Public Facility Location Problems-Part 1: A Critical Overview and Some Unsolved Problems, *Environment and Planning A*, **13**:1001-1028.

Leonardi G. 1981. A Unifying Framework for Public Facility Location Problems-Part 2: Some New Models and Extensions, *Environment and Planning A*, **13**:1085-1108.

Leung Y. 1994. Inference with Spatial Knowledge: An Artificial Neural Network Approach, *Geographical Systems*, **1**:103-121.

Leung Y. and Lin X. 1996. *Fast Extraction of Fuzzy and Non-fuzzy IF-THEN Rules by a Radial Basis Function Network with Unsupervised Competitive Learning.* (unpublished paper).

Leung Y., Leung K.S., Fischer M.M., Ng W. and Lau M.K. 1996. *The Evolution of Multilayer Feedforward Neural Networks for Spatial Interaction using Genetic Algorithms.* (unpublished paper).

Leung Y., Dong T.X. and Xu Z.B. 1996. *The Optimal Encodings for Biased Association in Linear Associative Memories.* (unpublished paper).

Levine R. and Renelt D. 1992. A Sensitivity Analysis of Cross-Country Growth Regressions, *The American Economic Review*, **82**:942-963.

Li W.H., Tanimura M. and Sharp P.M. 1988. Rates and dates of divergence between AIDS virus nucleotide sequences. *Molecular Biology and Evolution* **54**:313-30.

Liaw K-L, and Ledent J. 1987. Nested logit model and maximum quasi-likelihood methods: A flexible methodology for analyzing interregional migration patterns. *Regional Science and Urban Economics* **17**: 67-88.

Liaw K-L. 1990. Joint effects of personal factors and ecological variables in the interprovincial migration pattern of young adults in Canada: A nested logit analysis. *Geographical Analysis* **22**:189-208.

Lichter D.T. 1980. Household migration and the labor market position of married women. *Social Science Research* **9**:83-97.

Lin C.T. and George C.S. 1991. Neural-Network Based Fuzzy Logic Control and Decision Systems, *IEEE Transactions on Computers*, **40**(12):1320-1336.

Liu Z. K. and Wilkinson G. G. 1992. *A neural network approach to geometrical rectification of remotely-sensed satellite imagery*, Technical Note No. 1.92.118, Joint Research Centre, Commission of the European Communities, Ispra, Italy.

Lombardo S.T. and Rabino G.A. 1989. Urban Structures, Dynamic Modelling and Clustering, in J. Hauer, H. Timmermans and N. Wrigley eds. *Urban Dynamics and Spatial Choice Behaviour*, Kluwer Academic Publishers, Dordrecht, 203-217.

Losch A. 1954. *The Economics of Location.* Yale University Press, New Haven.

Loung G. and Tan Z. 1992. Stereo matching using artificial neural networks, *International Archives of Photogrammetry and Remote Sensing*, **29**:B3, 417-421.

Lukasiewicz J. 1970. In defence of logic, Works in L Borkowski ed. *Selected Works*, North-Holland, London.

Löytönen M. 1991. The spatial diffusion of human immunodeficiency virus type 1 in Finland, 1982-1997. *Annals of the Association of American Geographers* **81**:127-51.

MacDougall E.B. 1991. A Prototype Interface for Exploratory Analysis of Geographic Data, *Proceedings of the Eleventh Annual ESRI User Conference*, vol. 2,. Redlands, CA: Environmental Systems Research Institute, Inc., pp. 547-553.

Majure J. and Cressie N. 1997. Dynamic Graphics for Exploring Spatial Dependence in Multivariate Spatial Data, *Geographical Systems* (forthcoming).

Majure J., Cook D., Cressie N., Kaiser M., Lahiri S. and Symanzik J. 1996. Spatial CDF Estimation and Visualization with Applications to Forest Health Monitoring, *Computing Science and Statistics* **27**:93-101.

Majure J., Cressie N., Cook D. and Symanzik J. 1996. GIS, Spatial Statistical Graphics, and Forest Health, in: *Proceedings, Third International Conference/Workshop on Integrating GIS and Environmental Modeling*, Santa Fe, NM, January 21-26, 1996, Santa Barbara, CA, National Center for Geographic Information and Analysis (CD ROM).

Malczewski J. and Ogryczak W. 1990. An Interactive Approach to the Central Facility Location Problem: Locating Pediatric Hospitals in Warsaw, *Geographical Analysis*, **22**:244-258.

Malczewski J. and Ogryczak W. 1995. The Multiple Criteria Location Problem: 1. A

Generalized Network Model and the Set of Efficient Solutions, *Environment and Planning A*, **28**:1931-1960.

Malczewski J. and Ogryczak W. 1996. The Multiple Criteria Location Problem: 2. Preference-Based Techniques and Interactive Decision Support, *Environment and Planning A*, **28**:69-98.

Mamdani E. and Assilian S.. 1975. An experiment in linguistic synthesis with a fuzzy logic controller, *International Journal of Man-Machine Studies* **7**:1-13.

Manski C.F. 1977. The Structure of Random Utility Models. *Theory and Decision*, **8**:229-254.

Maslanik J., Key J., and Schweiger A. 1990. Neural network identification of sea-ice seasons in passive microwave data, *Proc. International Geoscience and Remote Sensing Symposium (IGARSS `90)*, held in Maryland, USA, IEEE Press, Piscataway, NJ, 1281-1284.

MathSoft. 1996. *S+Gislink*, Seattle: MathSoft, Inc.

MathSoft. 1996. *S+Spatialstats User's Manual*, Version 1.0, Seattle: MathSoft, Inc.

May R.M., Anderson R.M. and Blower S.M. 1989. The epidemiology and transmission dynamics of HIV/AIDS. *Daedalus* **118**:163-201.

McCalla G.I., Reid L. and Schneider P.K. 1982. Plan creation, plan execution, and knowledge execution in a dynamic micro-world. *International Journal of Man-Machine Studies*, **16**:89-112.

McCarty H.H. and Lindberg J.B. 1966. *A Preface to Economic Geography*. Prentice-Hall, Englewood Cliffs, N.J.

McClellan G.E., DeWitt R.N., Hemmer T.H., Matheson L.N. and Moe G.O. 1989. Multispectral image-processing with a three-layer backpropagation network, *Proceedings of the 1989 International Joint Conference on Neural Networks*, Washington, D.C.,151-3.

McFadden D. 1978. Modelling the Choice of Residential Location. In Karlquist, A., Lundqvist, L., Snickars, F., and J.W. Weibull (Editors) *Spatial Interaction Theory and Planning Models*, Amsterdam: North Holland, pp. 75-96.

McFadden D. 1979. Quantitative methods for analyzing travel behavior of individuals: Some recent developments. In D. Hensher and P. Stopher eds., *Behavioural travel modelling*. London: Croom Helm, pp. 279-319.

McNeill D., Freiberger, P., 1993. *Fuzzy Logic*. Simon and Schuster, New York.

McNeill F.M. and Thro E., 1994 *Fuzzy Logic: a practical approach*. A P Professional, Boston

McQueen J. 1967. Some Methods for Classification and Analysis of Multivariate Observations. *Proceedings of the 5th Berkeley Symposium on Mathematical Statistics and Probability*, 1:281-297.

Metropolitan Council.1992. *Home Interview Survey. Methodology and Results*. Publication No. 550-92-061, Metropolitan Council, St. Paul, MN.

Meyer A. D., Brooks, G. R. and Goes, J. B. 1989. *Environmental jolts and industry revolutions: Organizational responses to discontinuous change*. Working Paper, Graduate School of Management, University of Oregon.

Meyer R. 1980. Theory of Destination Choice-Set Formation under Informational Constraints. *Transportation Research Record*, **750**:6-12.

Meyer R.J. and Eagle T.C. 1982. Context-Induced Parameter Instability in a Disaggregate-Stochastic Model of Store Choice. *Journal of Marketing Research*, **19**:62-71.

Miller E.J. and O'Kelly M.E. 1983. Estimating Shopping Destination Models from Travel Diary Data. *Professional Geographer*, **35**:440-449.

Minsky M. 1977. Frame-system theory. In P. N. Johnson-Laird and P. C. Wason, *Thinking*. Cambridge: Cambridge University Press.

Minsky M. L. and Papert S. A. 1969. *Perceptrons*. MIT Press, Cambridge, Mass.

Molho I. 1986. Theories of migration: A review. *Scottish Journal of Political Economy* **33**:396-419.

Mollison D. 1995. *Epidemic models: their structure and relation to data.* Publications of the Isaac Newton Institute, 5. Cambridge: Cambridge University Press.

Monmonier M. 1989. Geographic Brushing: Enhancing Exploratory Analysis of the Scatterplot Matrix, *Geographical Analysis* **21**:81-84.

Moody A., Gopal S., Strahler A. H., Borak J. and Fisher P. 1994. A combination of temporal thresholding and neural network methods for classifying multiscale remotely-sensed image data, *Proc. International Geoscience and Remote Sensing Symposium (IGARSS 94)*, held at Pasadena, California, 8-12 August, IEEE Press, Piscataway, NJ, Vol. 4, 1877-1879.

Moss W.G. 1979. A note on individual choice models of migration. *Regional Science and Urban Economics* **9**: 333-343.

Mulder C.H., and Wagner M. 1993. Migration and marriage in the life course: A method for studying synchronized events, *European Journal of Population* **9**: 55-76.

Murakami K. and Aibara T. 1987. An Improvement on the Moore-Penrose Generalized Inverse Associative Memory, *IEEE Transactions on Systems, Man, and Cybernetics*, **17**:699-707.

Murray G.D. and Cliff A.D. 1977. A stochastic model for measles epidemics in a multi-region setting, *Transactions of the Institute of British Geographers*, **2**:158-74.

Nagar A.L. and Kakwani N.C. 1964. The Bias and Moment Matrix of a Mixed Regression Estimator, *Econometrica*, **32**:389-402.

Nauck D. and Kruse R. 1993. A Fuzzy Neural Network Learning Fuzzy Control Rules and Membership Function by Fuzzy Error Backpropagation, *IEEE Internat. Conf. on Neural Networks*, 1022-1027.

Nelson R. R. and Winter S. G. 1982. *An evolutionary theory of economic change.* Cambridge: Cambridge University Press.

Nerlove M.L. and Sadka E. 1991. Von Thünen's model of the dual economy. *Journal of Economics*, 54:97-123.

Newell A. 1992. *Unified theories of cognition.* Cambridge, MA: Harvard University Press.

Newell A. and Simon H.A. 1972. *Human problem solving.* Englewood Cliffs, NJ: Prentice-Hall.

Nowak M., Anderson R.M., McLean A.R., Wolfs T., Goudsmit J. and May R.M. 1991. Antigenic diversity thresholds and the development of AIDS. *Science* **254**:963-9.

Nystuen J. 1967. A theory and simulation of intraurban travel, in Garrison, W., and Marble, D. (Editors), *Quantitative Geography Pt.1: Economic and Cultural Topics*, Northwestern University Press, Evanston.

O'Loughlin J. and Anselin L. 1996. Geo-Economic Competition and Bloc Formation: U.S., German and Japanese Trade Development, 1968-1992, *Economic Geography* **72**:131-160.

Oden N.L. 1984., Assessing the Significance of a Spatial Correlogram, *Geographical Analysis* **16**:1-16.

Odland J. 1996. Longitudinal analysis of migration and mobility: Spatial behavior in explicitly temporal contexts. forthcoming in: *Spatial and Temporal Reasoning* eds. M.J. Egenhofer and R.G. Golledge.

Odland J. and Bailey A.J. 1990. Regional out-migration rates and migration histories: A longitudinal analysis. *Geographical Analysis* **22**:158-170.

Odland J. and Shumway J.M. 1993. Interdependencies in the timing of migration and mobility events. *Papers in Regional Science* **72**:221-237.

Oliver M.A. and Webster R. 1989. A geostatistical basis for spatial weighting in multivariate classification. *Mathematical Geology* **21**:15-35.

Openshaw S. 1976. An empirical study of some spatial interaction models. *Environment and Planning A* **8**:23-41

Openshaw S. 1978. An optimal zoning approach to the study of spatially aggregated data. in Masser I. and Brown P.J. eds. *Spatial Representation and Spatial Interaction.* Martinus Nijhoff.Leiden. 95-113.

Openshaw S. 1984. The modifiable areal unit problem. *Concepts and Techniques in Modern*

Geography **38**.GeoAbstracts, Norwich.

Openshaw S. 1988. Building an automated modelling system to explore a universe of spatial interaction models, *Geographical Analysis* **20**:31-46.

Openshaw S. 1991. Developing Appropriate Spatial Analysis Methods for GIS, in: D. Maguire, M.F. Goodchild and D. Rhind eds., *Geographical Information Systems: Principles and Applications*, Vol 1, London: Longman, pp. 389-402.

Openshaw S. 1992. Modelling Spatial Interaction using a Neural Net, in M.M. Fischer and P. Nijkamp eds., *Geographical Information Systems, Spatial Modelling and Policy Evaluation*. Berlin et al.: Springer, pp. 147-164.

Openshaw S. 1993. Exploratory Space-Time-Attribute Pattern Analysers, in *Spatial Analysis and GIS*, eds. Fotheringham A.S. and Rogerson, P.A, 147-163. London: Taylor and Francis.

Openshaw S. 1994. Two exploratory space-time attribute pattern analysers relevant to GIS. in Fotheringham S. and Rogerson P., eds., *Spatial Analysis and GIS*, Taylor and Francis, London, 83-104.

Openshaw S. 1996. Fuzzy Logic as a new scientific paradigm for doing geography, *Environment and Planning A* **28**:761-768.

Openshaw S. and Fischer M.M. 1995. A Framework for Research on Spatial Analysis Relevant to Geo-Statistical Information Systems in Europe, *Geographical Systems* **2**:325-337.

Openshaw S. and Rao L. 1994. *Re-engineering 1991 census geography: serial and parallel algorithms for unconstrained zone design.* Manuscript.

Oppewal H. 1995. *Conjoint Experiments and Retail Planning*, Ph.D-Dissertation, Eindhoven University of Technology, Eindhoven.

Ord J. K. and Getis A. 1995. Local Spatial Autocorrelation Statistics: Distributional Issues and Applications, *Geographical Analysis* **27**:286-306.

Painho M. 1994. *New Tools for Spatial Analysis*, Luxembourg: Eurostat.

Paola J. D. and Schowengerdt R. A. 1995. A review and analysis of backpropagation neural networks for classification of remotely-sensed multi-spectral imagery, *International Journal of Remote Sensing*, **16**(16):3033-3058.

Papageorgiou Y.Y. 1982. Some thoughts about theory in the social sciences. *Geographical Analysis*, **14**:340-6.

Papageorgiou Y.Y. and Brummell A.C. 1975. Crude inferences on spatial behaviour, *Annals Association of American Geographers*, **65**:1-12.

Parzen E. 1962. On the Estimation of a Probability Density Function and the Mode, *Annals of Mathematical Statistics* **33**:1065-1076.

Pederson C., Nielsen C.M., Vestergaard B.F., Gerstoft N., Krogsgaard K. and Nielsen J.O. 1987. Temporal relation of antigenaemia and loss of antibodies to core antigens to development of clinical disease in HIV infection. *British Medical Journal*, **295**:567-569.

Peet R. 1969. The spatial expansion of commercial agriculture in the nineteenth century: a von Thünen interpretation. *Economic Geography*, **45**:283-301.

Pellegrini P.A., Fotheringham A.S. and Lin G. 1997. Parameter Sensitivity to Choice Set Definition in Shopping Destination Choice Models. *Papers in Regional Science*, forthcoming.

Peterman T.A., Stoneburner R.L., Allen J.R., Jaffe H.W. and Curran J.W. 1988. Risk of HIV transmission from heterosexual adults with transfusion-associated infections. *Journal of the American Medical Association* **259**:53-63.

Petersen T. 1995. Analysis of event histories, in: *Handbook of Statistical Modeling for the Social and Behavioral Sciences* eds. G.Arminger, C. C. Clogg, and M.E. Sobel New York, Plenum Press. pp. 453-517.

Pierce L.E., Sarabandi K., and Ulaby F.T. 1994. Application of an artificial neural network in canopy scattering inversion, *International Journal of Remote Sensing*, **15**:3263-3270.

Pitt M. and Johnson G. 1987. Managing strategic change. In G. Johnson ed., *Business strategy and retailing*. Chichester: Wiley.

Porter M. E. 1980. *Competitive strategy*. New York: Free Press.

Powell M.J.D. 1992. Radial Basis Functions in 1990, *Adv. Num. Anal.*, **2**:105-210.

Pyle G. 1986. *The diffusion of influenza: patterns and paradigms* Totowa: Rowan and Littlefield.

Ramacher U., Raab W., Anlauf J., Hachmann U. and Wesseling M. 1994. *SYNAPSE-1 -a general purpose neurocomputer*, Technical Report, Corporate Research and Development Division, Siemens-Nixdorf.

Rao C.R. and Toutenburg H. 1995. *Linear Models: Least Squares and Alternatives*, Springer Verlag: New York

Raper J., Rhind D.W and Shepherd J.W. 1990. *Postcodes: the new geography*. Longman, Harlow.

Recker W.W., McNally M., and Root G.S. 1986. A model of complex travel behavior: Part I: Theoretical development. *Transportation Research A*, **20**:4: 307-318.

Recker W.W., McNally M., and Root G.S. 1986. A model of complex travel behavior: Part II: An operational model. *Transportation Research A*, **20**:4: 319-330.

Rees P. 1995. Putting the Census on the Researcher's Desk, Chapter 2 in *The Census Users' Handbook*, ed. S. Openshaw, 27-81, GeoInformation International: Cambridge.

Reger R. K. 1988. *Competitive positioning in the Chicago banking market: Mapping the mind of the strategist*. Unpublished doctoral dissertation, University of Illinois, Urbana-Champaign.

Richardson A. 1982. Search Models and Choice Set Generation. *Transportation Research A*, **16**:403-419.

Rijk F.J.A. and Vorst A.C.F. 1983. On the Uniqueness and Existence of Equilibrium Points in an Urban Retail Model, *Environment and Planning A*, **15**:475-482.

Roberts J. and P. Nedungadi. 1995. Studying Consideration in the Consumer Decision Process: Progress and Challenges. *International Journal of Research in Marketing*, **12**:3-7.

Rogers E. M. 1962. *Diffusion of innovations*. New York: Free Press.

Root G.S. and Recker W.W. 1983. Towards a dynamic model of individual activity pattern formation. In S. Carpenter and P. Jones eds., *Recent advances in travel demand analysis*. Aldershot: Gower.

Rosenberg P.S., Gail M.H. and Carroll R.J. 1992. Estimating HIV prevalence and projecting AIDS incidence in the United States: a model that accounts for therapy and changes in the surveillance definitions of AIDS. *Statistics in Medicine* **11**:1633-55.

Rosenberg B. 1973. A Survey of Stochastic Parameter Regression, *Annals of Economic and Social Measurement* **1**:381-397.

Rosenfeld A. and Kak A. 1982. *Digital picture processing*. Academic Press, London.

Rosing K.E., E.L. Hillsman, H. Rosing-Vogelaar 1979. The Robustness of Two Common Heuristics for the p-Median Problem, *Environment and Planning A*, **11**:373-380.

Ross T.J., 1995. *Fuzzy logic with engineering applications*. McGraw Hill, New York.

Roy J.R. 1995. The Use of Spatial Interaction Theory in Spatial Economic Modeling, in R. Wyatt and H. Hossain eds. *Proceedings off the 4th International Conference on Computers in Urban Planning and Urban Management*, Melbourne, Australia, 139-150.

Roy J.R. and Johansson B. 1984. On Planning and Forecasting the Location of Retail and Service Activity, *Regional Science and Urban Economics*, **14**:433-452.

Roy J.R., and Anderson M. 1988. Assessing Impacts of Retail Development and Redevelopment, in P.W. Newton, M.A.P. Taylor and R. Sharpe eds. *Desktop Planning: Microcomputer Applications for Infrastructure and Services Planning and Management*, Hargreen, Melbourne, 172-179.

Rumelhart D.E., Hinton G.E. and WilliamsR.J. 1986. Learning Internal Representation by Error Propagation, in D.E. Rumelhart, J.L. McClelland, and the PDP Research Group eds., *Parallel Distributed Processing: Exploration in the Microstructure of Cognition*, Vol. 1, Cambridge: MIT

Rumelt R. P. 1984. Towards a strategic theory of the firm. In R. Lamb. ed., *Competitive strategic management.* Englewood Cliffs, NJ: Prentice-Hall.

Ryan T. W., Sementelli P., Yuen P. and Hunt B. R. 1991. Extraction of shoreline features by neural nets and image processing, *Photogrammetric Engineering and Remote Sensing*, **57**(7):947-955.

Sammons R. 1978. A simplistic approach to the redistricting problem. in Masser I. and Brown P.J. eds. *Spatial Reprsentation and Spatial Interaction.* Martinus Nijhoff, Leiden, 71-94.

Samuelson P.A. 1983. Thünen at two hundred. *Journal of Economic Literature*, **21**:1468-1488.

Sandefur G.S. and Scott W. 1981. A dynamic analysis of migration: An assessment of the effects of age, family and career variables. *Demography* **18**:355-368.

Santana S., Faas L. and Wald K. 1991. Human immunodeficiency virus in Cuba: the public response of a third world country. *International Journal of Health Services* **21**:511-37.

Schenzle D. and Dietz K. 1987. Critical population sizes for endemic virus transmission. In Fricke W. and Hinz E. eds. *Räumliche Persistenz und Diffusion von Krankheiten*, 83. Heidelberg: Heidelberg Geographical Studies, 31-42.

Schwenk C. and Tang M. 1989. Persistence in questionable strategies: explanations from the economic and psychological perspectives. *OMEGA: International Journal of Management Science*, **17**:559-570.

Semple R.K. and Green M.B. 1984. Classification in Human Geography. in G.L.Gaile and C.J.Wilmott eds. *Spatial statistics and models*, Reidel, Dordrecht, 55-79.

Senser D.J., Dull H.B. and Langmuir A.D. 1967. Epidemiological basis for the eradication of measles. *Public Health Reports*, **82**:253-6.

Shang C. and Brown K. 1994. Principal features-based texture classification with neural networks, *Pattern Recognition*, **27**(5):675-687.

Sheldon A. 1980. Organizational paradigms. *Organization Dynamics*, **8**:61-71.

Sheppard E.S. 1980. The Ideology of Spatial Choice. *Papers of the Regional Science Association*, **45**:197-213.

Sheppard E.S. 1984. The Distance-Decay Gravity Model Debate. In G.L. Gaile, and C.J. Wilmott (Editors), *Spatial Statistics and Models*, Dordrecht: Reidel, pp. 367-388.

Shocker A.D., Ben-Akiva M., Boccara B. and Nedungadi P. 1991. Consideration Set Influences on Consumer Decision-Making and Choice: Issues, Models, and Suggestions. *Marketing Letters*, **2**(3):181-197.

Short N. 1991. A Real-Time Expert System and Neural Network for the Classification of Remotely Sensed Data. *Proceedings of the Annual Convention on the American Society for Photogrammetry and Remote Sensing*, **3**:406-418.

Shumway J.M. 1993. Factors influencing unemployment duration with a special emphasis on migration: An investigation using SIPP data and event history methods. *Papers in Regional Science* **72**:159-176.

Silverman B.W. 1986. *Density Estimation for Statistics and Data Analysis*, London:Chapman and Hall.

Simon H.A. 1945. *Administrative behavior.* New York: MacMillan.

Simon H.A. 1955. A behavioral model of rational choice. *Quarterly Journal of Economics*, **69**:99-118.

Simon H.A. 1990. Invariants of human behavior. *Annual Review of Psychology*, **41**:1-19.

Simpson P. K. 1990. *Artificial Neural Systems: Foundations, Paradigms, Applications and Implementations*, Pergamon Press, New York.

Sjaastad L. 1962. The costs and returns of human migration, *Journal of Political Economy* **70**:80-93.

Smallman-Raynor M.R. and Cliff A.D. 1990. Acquired Immunodeficiency Syndrome (AIDS): literature, geographical origins and global patterns. *Progress in Human Geography* **14**: 157-213.

Smith T.R. 1977. Uncertainty, diversification, and mental maps in spatial choice problems. *Geographical Analysis*, **10**:120-140.

Smyth F.M. 1995. *Social and epidemiological constructions of HIV/AIDS in Ireland*, Unpublished PhD thesis, University of Manchester.

Smyth F.M. and Thomas R.W. 1996. Preventative action and the diffusion of HIV/AIDS. *Progress in Human Geography* **20**:1-22.

Smyth F.M. and Thomas R.W. 1996. Controlling HIV/AIDS in Ireland: the implications for health policy of some epidemic forecasts. *Environment and Planning A* **27**:99-118.

Spath H. 1980. *Cluster Analysis Algorithms*. John Wiley and Sons, New York.

Spina M.S., Schwartz, M.J., Staelin, D.H., and Gasiewski, A.J. 1994. Application of multilayer feedforward neural networks to precipitation cell-top altitude estimation. *Proc. International Geoscience and Remote Sensing Symposium (IGARSS '94)*, held at Pasadena, California, 8-12 August, IEEE Press, Piscataway. NJ., Vol. 4, 1870-1872.

Spjotvoll E. 1977. Random Coefficients Regression Models, A Review, *Mathematische Operationsforschung und Statistik* **8**: 69-93.

Steele N.C., Reeves C.R. , Nicholas M. and King P.J. 1995. Radial Basis Function Artificial Neural Networks for the Inference Process in Fuzzy Logic Based Control, *Computing*, **54**:99-117.

Stein J. ed. 1966. *The Random House dictionary of the English language*. New York: Random House.

Stetzer F.C. and Phipps A.G. 1977. Spatial Choice Theory and Spatial Indifference: A Comment. *Geographical Analysis*, **9**:400-403.

Stuetzle W. 1987. Plot windows, *Journal of the American Statistical Association* **82**:466-475.

Sugeno M. ed. 1985. *Industrial applications of fuzzy control*. North-Holland, Amsterdam.

Sulzberger S.M., Tschichold-Gürman N.N. and Vestli S.J. 1993. FUN: Optimization of Fuzzy Rule Based Systems Using Neural Networks, *IEEE International Conference on Neural Networks*, 312-316.

Supernak J. 1992. Temporal utility profiles of activities and travel: Uncertainty and decision making. *Transportation Research B*, **26**:61-76.

Sutter R.W., Markowitz S.E., Bennetch J.M., Morris W., Zell E.R. and Preblud S.R. 1991. Measles among the Amish: a comparative study in primary and secondary cases in households. *Journal of Infectious Diseases*, **163**:12-16.

Swait J.D., and Ben-Akiva M.E. 1987.Incorporating Random Constraints in Discrete Models of Choice Set Generation. *Transportation Research B*, **21**:91-102.

Swait J.D., and Ben-Akiva M.E. 1987. Empirical Test of a Constrained Choice Discrete Model: Mode Choice in Sao Paulo, Brazil. *Transportation Research B*, **21**:103-115.

Symanzik J., Majure J. and Cook D. 1996. Dynamic Graphics in a GIS; A Bidirectional Link between ArcView 2.0 and XGobi, *Computing Science and Statistics* **27**:299-303.

Symanzik J., Majure J., Cook D. and Cressie N. 1994. Dynamic Graphics in a GIS: A Link between Arc/Info and XGobi, *Computing Science and Statistics* **26**:431-435.

Takagi T. and Sugeno M.. 1985. Fuzzy identification of systems and its application to modelling and control; *IEEE Transactions on Systems, Man, and Cybernetics* SMC-**15**:116-132.

Talen E. 1997. Visualizing Fairness: Equity Maps for Planners, *Journal of the American Planning Association* (forthcoming).

Talen E. and Anselin L. 1997. Assessing Spatial Equity: The Role of Access Measures, *Environment and Planning A* (forthcoming).

Taylor P.J. 1969. The location variable in taxonomy. *Geographical Analysis* **1**:181-195.

Teitz M.B. and Bart P. 1968. Heuristic Methods for Estimating the Generalised Vertex Median of a Weighted Graph, *Operations Research*, **16**:955-961.

Terano T., Asai K., Sugeno M., 1989. *Applied Fuzzy Systems*. A P Professional, New York (English translation in 1994)

Theil H. 1971. *Principles of Econometrics*, Wiley.

Theil H. and Goldberger A.S. 1961. On Pure and Mixed Statistical Estimation in Economics, *International Economic Review*, **2**:65-78.

Theil H.1963. On the Use of Incomplete Prior Information in Regression Analysis, *Journal of the American Statistical Association*, **58**:401-414.

Thill J.-C. and Thomas I. 1987. Towards conceptualizing trip-chaining behavior: A review. *Geographical Analysis*, **19**:1-17.

Thill J.-C. 1992. Choice Set Formation for Destination Choice Modelling. *Progress in Human Geography*, **16**:361-382.

Thill J.-C. 1995. Modeling Store Choices with Cross-Sectional and Pooled Cross-Sectional Data: A Comparison. *Environment and Planning A*, **27**:130-1315.

Thill J.-C., and Horowitz J.L. 1991. Estimating a Destination-Choice Model from a Choice-Based Sample with Limited Information. *Geographical Analysis*, **23**:298-315.

Thill J.-C., and Horowitz J.L.. 1997. Travel-Time Constraints on Destination Choice Sets. *Geographical Analysis*, forthcoming.

Thomas R.W. 1988. Stochastic carrier models for the simulation of Hodgkin's disease in a system of regions. *Environment and Planning A* **20**:1575-601.

Thomas R.W. 1992. Geomedical systems: intervention and control. London: Routledge.

Thomas R.W. 1993. Source region effects in epidemic disease modelling: comparisons between influenza and HIV. *Papers in Regional Science* **72**:257-82.

Thomas R.W. 1994. Forecasting global HIV/AIDS dynamics: modelling strategies and preliminary simulations. *Environment and Planning A* **26**:1147-66

Thomas R.W. 1996. Alternative population dynamics in selected HIV/AIDS modelling systems: some cross-national comparisons. *Geographical Analysis* **28** :108-25

Thomas R.W. 1996. Modelling space-time HIV/AIDS dynamics: applications to disease control. *Social Science and Medicine* **41**(in press).

Timmermans H.J.P., and Golledge R.G. 1990. Applications of behavioral research on spatial problems II: Preference and choice. *Progress in Human Geography*, **14**:311-354.

Toutenburg H. 1982. *Prior Information in Linear Models*, Wiley.

Tukey J.W. 1977. *Exploratory Data Analysis*, Reading MA: Addison-Wesley.

Turton I., Openshaw S. and Diplock GJ., 1997. A genetic programming approach to building new spatial models relevant to GIS, in Z Kemp ed. *Innovations in GIS 4*, Taylor and Francis, London (forthcoming).

Tushman M. L. and Romanelli E. 1985. Organizational evolution. *Research in Organization Behavior*, **7**:171-222.

Tversky A., and Kahneman D. 1991. Loss aversion in riskless choice: A reference-dependent model. *The Quarterly Journal of Economics*, **106**:1039-1061.

Tversky A., and Kahneman D. 1992. Advances in prospect theory: Cumulative representation of uncertainty. *Journal of Risk and Uncertainty*, **5**:4: 297-323.

Tzeng Y.C., Chen K.S., Kao W.L. and Fung A.K. 1994. A Dynamic Learning Neural Network for Remote Sensing Applications. *IEEE Transactions on Geoscience and Remote Sensing*, **32**:1096-1102.

U.S. Department of Commerce.1994. *USA Counties 1994 CD-ROM*, Washington, D.C.: Bureau of the Census.

Unwin A. 1994. REGARDing Geographic Data, in; P. Dirschedl and R. Osterman eds., *Computational Statistics*, Heidelberg: Physica Verlag, pp. 345-354.

Unwin A. 1996. Exploratory Spatial Analysis and Local Statistics, *Computational Statistics* **11**:387-400.

Upton G. J. and Fingleton B. 1985. *Spatial Data Analysis by Example*, New York: Wiley.

van der Heijden R.E.C.M. 1986. *A Decision Support System for the Planning of Retail Facilities: Theory, Methodology and Application*, Ph.D-Dissertation, Eindhoven University of Technology, Eindhoven.

van der Hoorn, T. 1983. Experiments with an Activity-Based Travel Model. *Transportation*, **12**:61-77.

van der Maas H. L. J., Vershure P. F. M. J., and Molenaar P. C. M. 1990. A note on chaotic behaviour in simple neural networks, *Neural Networks*, **3**:119-122.

Van Druten J.A.M., Reintjes A.G.M., Jager J.C., Heisterkamp S.H., Poos M.J.J.C., Coutinho R.A., Dijkgraaf M.G.W. and Ruitenberg E.J. 1990. Infection dynamics and intervention experiments in linked risk groups. *Statistics in Medicine* **9**:721-36.

Venables W. N and Ripley B.D. 1994. *Modern Applied Statistics with S-Plus*, New York: Springer-Verlag.

Vinod H.D. and Ullah A.1981. *Recent Advances in Regression Methods*, Dekker: New York

Von Altrock C., 1995. *Fuzzy Logic and neurofuzzy applications explained*, Prentice Hall, Englewood Cliffs.

Wald A. 1950. *Statistical Decision Functions*. John Wiley, New York.

Waldorf B.S. 1994. Assimilation and attachment in the context of international migration: The case of guest workers in Germany, *Papers in Regional Science* **73**:241-266.

Waldorf B.S., and Esparza A. 1991. A parametric failure time model of international return migration. *Papers in Regional Science* **70**:419-438.

Wan E. A. 1990. Neural network classification: a Bayesian interpretation, *IEEE Transactions on Neural Networks*, **1**(4):303-305.

Wan W. and Fraser D. 1994. A self-organizing map model for spatial and temporal contextual classification, *Proc. International Geoscience and Remote Sensing Symposium (IGARSS 94)*, held at Pasadena, California, 8-12 August, IEEE Press, Piscataway, NJ, Vol. 4, 1867-1869.

Wand M.P. and Jones M.C. 1995. *Kernel Smoothing*, London:Chapman and Hall.

Wang F. 1994. The Use of Artificial Neural Networks in a Geographical Information System for Agricultural Land-Suitability Assessment, *Environment and Planning A*, **26**:265-284.

Wang L.X., 1994 Adaptive fuzzy systems and control. Prentice Hall, Englewood Cliffs, New Jersey.

Warnes A.M. and Daniels P.W. 1979. Spatial aspects of an intrametropolitan central place hierarchy, *Progress in Human Geography*, **3**:384-406.

Watzlawick P., Weakland J. H. and Fisch R. 1974. *Change*. New York: Norton.

Webstead S.T. 1987. *Neural network and fuzzy logic applications in C/C++*, Wiley, Chichester.

Wechsler H. and Zimmerman G.L. 1988. 2-D Invariant Object Recognition Using Distributed Associative Memory, *IEEE Transactions on Pattern Analysis and Machine Intelligence*, **20**(6): 811-821.

Wee W. 1968. Generalized Inverse Approach to Adaptive Multiclass Pattern Classification, *IEEE Transaction on Computers*, C-**17**:1157-1164.

Werbos P. 1974. *Beyond regression: new tools for prediction and analysis in the behavioural sciences*, Unpublished PhD thesis, Harvard University, Cambridge, Mass.

White R.W. 1989. The Artificial Intelligence of Urban Dynamics: Neural Net Modelling of Urban Structure, *Papers of the Regional Science Association* **67**:43-53.

Wigan M.R. and Morris J.M. 1981. The Transport Implications of Activity and Time Budget Constraints. *Transportation Research A*, **15**:63-86.

Wilkinson G.G., Fierens F. and Kanellopoulos I. 1995. Integration of neural and statistical approaches in spatial data classification, *Geographical Systems*, **2**:1-20.

Wilkinson G.G., Folving S., Kanellopoulos I., McCormick N., Fullerton K. and Mégier J. 1995. Forest mapping from multi-source satellite data using neural network classifiers -an experiment in Portugal, *Remote Sensing Reviews*, **12**:83-106.

Wilkinson G.G., Kontoes C. and Murray C. N. 1993. Recognition and inventory of oceanic clouds from satellite data using an artificial neural network technique, *Proc. International Symposium on "Dimethylsulphide, Oceans, Atmosphere and Climate"*, held at Belgirate.

Italy, 13-15 October 1992. G. Restelli and G. Angeletti (editors), Kluwer Academic Publishers, Dordrecht, 393-399.

Willekens F. J. 1991. Understanding the interdependence between parallel careers, in: *Female Labor Market Behavior and Fertility* eds. J.J. Siegers, J. de Jong-Gierveld, and E. Van Imhoff Berlin: Springer-Verlag. pp. 2-31.

Williams I., Limp W., Briuer F. 1990. Using Geographic Information Systems and Exploratory Data Analysis for Archeological Site Classification and Analysis, in: K. Allen, F. Green and E. Zubrow eds., *Interpreting Space: GIS and Archaeology*, London: Taylor and Francis, pp. 239-273.

Williams H.C.W.L., and Ortuzar J.D. 1982. Behavioural Theories of Dispersion and the Misspecification of Travel Demand Models. *Transportation Research B*, **16**:167-219.

Williamson O.E. 1979. Transaction-cost economics: The governance of contractual relations. *Journal of Law and Economics*, **22**:233-260.

Wilson A.G. and Bennett R.J. 1985. *Mathematical models in human geography and planning*. Chichester: Wiley.

Wilson A.G. 1990. Services 1: A Spatial Interaction Approach, in C.S. Bertuglia, G. Leonardi and A.G. Wilson eds. *Urban Dynamics: Designing an Integrated Model*, Routledge, London, 251-273.

Wilson A.G., 1970. *Entropy in Urban and Regional Modelling*, Pion, London.

Winkelstein W. Jr., Samuel M., Padain N.S., Wiley J.A., Lane W., Anderson R.B. and Levy J.A. 1987. The San Francisco men's health study III: reduction in human immunodeficiency virus transmission among homosexual/bisexual men, 1982-1986. *American Journal of Public Health* **76**:685-9.

Winston G.C., 1987. Activity choice: A new approach to economic behavior. *Journal of Economic Behavior and Organization*, **8**:567-585.

Wolpert J. 1964. The decision process in a spatial context. *Annals of the Association of American Geographers*, **54**:537-558.

Wrigley N. 1994. After the store wars. Towards a new era of competition in UK food retailing, *Retailing and Consumer Services*, **1**:1-17.

Yamaguchi K. 1991. *Event History Analysis* Newbury Park, CA: Sage Publications..

Yancey T.A., Judge G.G. and Bock M.E. 1974. A Mean Square Error Test When Stochastic Restrictions Are Used in Regression, *Communications in Statistics*, **3**:755-768.

Yasunobu S. and Miyamoto S., 1985. Automatic train operation system by predictive fuzzy control, in M Sugeno ed. *Industrial applications of fuzzy control*, North-Holland, Amsterdam 1-18

Yoon B. 1995. *An estimation of the returns to migration of male youth in the United States: A longitudinal analysis.* Ph.D. dissertation, Indiana University.

Yorke J.A., Hethcote H.W. and Nold A. 1978. Dynamics and control of the transmission of gonorrhoea. *Sexually Transmitted Diseases* **5**:51-156.

Yoshida T. and Omatu S. 1994. Neural Network Approach to Land Cover Mapping, *IEEE Transaction on Geoscience and Remote Sensing*, **32**:1103-1109.

Zadeh L.A. 1965. Fuzzy sets, *Information and Control*, **8**:338-53.

Zadeh L A. 1971. Towards a theory of fuzzy systems, In R E Kalman and N DeClaris eds. *Aspects of Network and Systems Theory*. Holt Rinehart, Winston.

Zadeh L A. 1974. A rationale for fuzzy control, *Journal of Dynamic Systems Measurement and Control* **3-4**.

Zadeh L.A. 1962. From circuit theory to system theory, *IRE Proceedings* **50**:856-865.

Zadeh L.A. 1992. Foreword in *Neural networks and fuzzy systems*, Prentice-Hall International Englewood Cliffs. New Jersey, p xvii-xviii.

Zadeh L.A. 1995 Foreword in C von Altrock *Fuzzy Logic and Neuro fuzzy applications explained*, Prentice-Hall, Englewood Cliffs. New Jersey pxi-xii.

Zajac E.J. and Kraatz M.S. 1993. A diametric forces model of strategic change: assessing the antecedents and consequences of restructuring in the higher education industry. *Strategic*

Management Journal, **14**:83-102.

Zhang A., Yu H. and Huang S. 1994. Bringing Spatial Analysis Techniques Closer to GIS Users: A User-Friendly Integrated Environment for Statistical Analysis of Spatial Data, in: T. C. Waugh and R. G. Healy eds., *Advances in GIS Research*, London: Taylor and Francis, pp. 297-313.

Zhang M. and Scofield R.A. 1994. Artificial neural network technique for estimating heavy convective rainfall and recognizing cloud mergers from satellite data, *International Journal of Remote Sensing* **15**(16):3241-3261.

Author Index

Subject Index

Contributors

LUC ANSELIN, Regional Research Institute and Department of Economics, West Virginia University, Morgantown WV 26506-6825, USA

THEO A. ARENTZE, Eindhoven University of Technology, Urban Planning Group, P.O. Box 513, 5600 MB Eindhoven, The Netherlands

ROBERT G. V. BAKER, Department of Geography and Planning, University of New England, Armidale, 2351, Australia.

SHUMING BAO, Data Analysis Products Division, MathSoft Incorporated, Seattle WA 98109, USA

ALOYS W. J. BORGERS, Eindhoven University of Technology, Urban Planning Group, P.O. Box 513, 5600 MB Eindhoven, The Netherlands

CHRIS BRUNSDON, Department of Town and Country Planning, University of Newcastle, Newcastle-upon-Tyne, NE1 7RU, UK

EMILIO CASETTI, Department of Economics, Odense University, Odense M, Denmark

MARTIN CHARLTON, Department of Geography, University of Newcastle, Newcastle-upon-Tyne, NE1 7RU, UK

A.D. CLIFF, Department of Geography, University of Cambridge, Downing Place, Cambridge, CB2 3EN, UK

MANFRED M. FISCHER, Department of Economic & Social Geography, Vienna University of Economics and Business Administration, A-1090 Vienna, Augasse 2-6, Austria and Institute for Urban and Regional Research, Austrian Academy of Sciences, A-1010 Vienna, Postgasse 4, Austria

A. STEWART FOTHERINGHAM, Department of Geography, University of Newcastle, Newcastle-upon-Tyne, NE1 7RU, UK

ARTHUR GETIS, Department of Geography, San Diego State University, San Diego, CA, 92182-4493, USA

REGINALD G. GOLLEDGE, Department of Geography & Research Unit on Spatial Cognition and Choice, University of California at Santa Barbara, Santa Barbara, CA 93106-4060, USA

SUCHARITA GOPAL, Department of Geography, Boston University, Boston, MA 02215, USA

ROBERT HAINING, Sheffield Centre for Geographic Information and Spatial Analysis, Department of Geography, University of Sheffield, Sheffield S10 2TN, UK.

JOEL L. HOROWITZ, Department of Economics, University of Iowa, Iowa City IA 52242, USA

ANNE S. HUFF, College of Business Administration, University of Colorado, Boulder, CO 80309, USA

JAMES O. HUFF, Department of Geography, University of Colorado, Boulder, CO 80309-0260, USA

MEI-PO KWAN, Department of Geography, The Ohio State University, Columbus, OH 43210-1361, USA

YEE LEUNG, Department of Geography and Center for Environmental Studies, The Chinese University of Hong Kong, Shatin, N.T., Hong Kong

J.D. LOGAN, Statistical Laboratory, University of Cambridge, Mill Lane, Cambridge, CB2 1SB, UK

JINGSHENG MA, Sheffield Centre for Geographic Information and Spatial Analysis, Department of Geography, University of Sheffield, Sheffield S10 2TN, UK.

GORDON F. MULLIGAN, Department of Geography and Regional Development, University of Arizona, Tucson, AZ 85721, USA

JOHN ODLAND, Department of Geography, Indiana University, Bloomington IN 47405, USA

STAN OPENSHAW, School of Geography, University of Leeds, Leeds, LS29JT, UK

FIONA SMYTH, School of Geography, University of Manchester, Manchester, M13 9PL, UK

JEAN-CLAUDE THILL, Department of Geography and National Center for Geographic Information and Analysis, State University of New York, Amherst NY 14261, USA

RICHARD THOMAS, School of Geography, University of Manchester, Manchester, M13 9PL, UK

HARRY J. P. TIMMERMANS, Eindhoven University of Technology, Urban Planning Group, P.O. Box 513, 5600 MB Eindhoven, The Netherlands

GRAEME G. WILKINSON, Joint Research Centre, European Commission, 21020 Ispra, Varese, Italy

STEVE WISE, Sheffield Centre for Geographic Information and Spatial Analysis, Department of Geography, University of Sheffield, Sheffield S10 2TN, UK

Made in the USA
Las Vegas, NV
08 November 2024

11359474R00249